T0299207

A MODERN COURSE IN TRANSPORT PHENOMENA

This advanced text presents a unique approach to studying transport phenomena. Bringing together concepts from both chemical engineering and physics, it makes extensive use of nonequilibrium thermodynamics, discusses kinetic theory, and sets out the tools needed to describe the physics of interfaces and boundaries. More traditional topics such as diffusive and convective transport of momentum, energy, and mass are also covered. This is an ideal text for advanced courses in transport phenomena, and for researchers looking to expand their knowledge of the subject.

Also included:

- Novel applications such as complex fluids, transport at interfaces, and biological systems
- Approximately 250 exercises with solutions (included separately) designed to enhance understanding and reinforce key concepts
- End-of-chapter summaries

DAVID C. VENERUS is a Professor of Chemical Engineering in the Department of Chemical and Biological Engineering at Illinois Institute of Technology in Chicago. His research interests are in the areas of transport phenomena in soft matter, polymer science, and the rheology of complex fluids. Professor Venerus has received numerous teaching awards both within the Department and at the College of Engineering at Illinois Institute of Technology. He is a member of the American Institute of Chemical Engineers and of the Society of Rheology.

HANS CHRISTIAN ÖTTINGER is Professor of Polymer Physics at the ETH Zürich. His main research interest is in developing a general framework of nonequilibrium thermodynamics as a tool for describing dissipative classical and quantum systems. He is the author of *Stochastic Processes in Polymeric Fluids* (Springer, 1996), *Beyond Equilibrium Thermodynamics* (Wiley, 2005), and *A Philosophical Approach to Quantum Field Theory* (Cambridge, 2017).

A MODERN COURSE IN TRANSPORT PHENOMENA

DAVID C. VENERUS

Illinois Institute of Technology, USA

HANS CHRISTIAN ÖTTINGER

ETH Zürich, Switzerland

CAMBRIDGE
UNIVERSITY PRESS

University Printing House, Cambridge CB2 8BS, United Kingdom

One Liberty Plaza, 20th Floor, New York, NY 10006, USA

477 Williamstown Road, Port Melbourne, VIC 3207, Australia

314–321, 3rd Floor, Plot 3, Splendor Forum, Jasola District Centre, New Delhi - 110025, India

79 Anson Road, #06-04/06, Singapore 079906

Cambridge University Press is part of the University of Cambridge.

It furthers the University's mission by disseminating knowledge in the pursuit of education, learning and research at the highest international levels of excellence.

www.cambridge.org
Information on this title: www.cambridge.org/9781107129207
DOI: 10.1017/9781316416174

First published 2018

A catalogue record for this publication is available from the British Library

Library of Congress Cataloging-in-Publication data
Names: Venerus, David, author. | Ottinger, Hans Christian, 1958– author.
Title: A modern course in transport phenomena / David Venerus, Illinois Institute of Technology, Hans Christian Ottinger, ETH Zentrum, Switzerland.
Description: New York, NY, USA : Cambridge University Press, [2017] | Includes bibliographical references and indexes.
Identifiers: LCCN 2017041983 | ISBN 9781107129207 (hardback : alk. paper)
Subjects: LCSH: Transport theory – Textbooks.
Classification: LCC QC175.25 .V46 2017 | DDC 530.4/75 – dc23 LC record available at https://lccn.loc.gov/2017041983

ISBN 978-1-107-12920-7 Hardback

To our families

Contents

Preface

The subject of 'Transport Phenomena' is almost synonymous with the names Bird, Stewart, and Lightfoot (BSL). When these authors published their pioneering textbook in 1960, their goal was to establish this subject as one of the key engineering sciences:[1] "Knowledge of the basic laws of mass, momentum, and energy transport has certainly become important, if not indispensable, in engineering analysis. In addition, material in this text may be of interest to some who are working in physical chemistry, soil physics, meteorology, and biology." BSL certainly reached their declared goal, and *Transport Phenomena* became a scientific bestseller known to every chemical engineer. Based on their enormous success, with the second edition, BSL proceeded to expand into new fields of technological importance:[2] "While momentum, heat, and mass transfer developed independently as branches of classical physics long ago, their unified study has found its place as one of the fundamental engineering sciences. This development, in turn, less than half a century old, continues to grow and to find applications in new fields such as biotechnology, microelectronics, nanotechnology, and polymer science."

In view of the outstanding book by BSL (and a number of additional good books on transport phenomena), why would we write another textbook on the subject of transport phenomena? Most of the existing textbooks have been written by chemical engineers and focus on developing problem-solving skills for engineering purposes. As a team of a chemical engineer (DCV) and a theoretical physicist (HCO), we would like to develop transport phenomena into a subject that appeals to both engineers and natural scientists. Ultimately, the unified approach to "momentum, heat, and mass transfer developed independently as branches of classical physics" should return as a rich, coherent, and attractive core subject to physics, offering many highly

---

[1] Preface of first edition of Bird, Stewart & Lightfoot, *Transport Phenomena* (Wiley, 1960).
[2] Preface of second edition of Bird, Stewart & Lightfoot, *Transport Phenomena* (Wiley, 2001).

relevant applications in science and engineering, ranging from developing new key technologies to understanding the origin of life. Hence, we focus on basic ideas and principles, conceptual clarity, illuminating interconnections, modern applications, and deep understanding.

There are a number of distinguishing features of *A Modern Course in Transport Phenomena* compared with previous books on this subject.

- We begin with a discussion of the transport of probability and the general mathematical form of diffusion equations. This approach offers a deeper understanding of the ubiquitous diffusion equations as well as versatile stochastic simulation techniques.
- We make extensive use of nonequilibrium thermodynamics, which we consider to be an integral part of transport phenomena. Analysis of the entropy production in terms of thermodynamic forces and fluxes guides the formulation of constitutive equations for the fluxes to be used in balance equations.
- We develop a new, more intuitive approach to interfacial balance and transport equations. By focusing on the motion of Gibbs dividing surfaces and interpreting the ambiguity behind their precise location as a gauge degree of freedom, we gain conceptual clarity and insight. In particular, we develop the tools for describing the physics of interfaces and boundaries, going far beyond the usual formulation of boundary conditions.
- We develop a number of novel applications and material characterization techniques that include field-flow fractionation, complex fluids or soft matter, relativistic hydrodynamics (which turns out to be intimately related to complex fluids), fiber spinning, Czochralski crystal growth, molecular motors, ion pumps, microbead rheology, and dynamic light scattering.

The structure, contents, and style of *A Modern Course in Transport Phenomena* arose as the product of blending the very different perspectives of an engineer and a physicist. Not surprisingly, this process was dissipative at times. However, more often, it forced the authors to broaden their perspectives, leading to a deeper understanding of the subject. Several chapters are based on lectures given over the past 25 years in advanced undergraduate and graduate courses on transport phenomena for chemical engineering students at the IIT Chicago. This material has been revised and significantly expanded in connection with a new two-semester course for graduate students in the materials program at the ETH Zürich, offered for the first time in 2012–13. There are, of course, several important topics we have chosen to exclude from this book. Our decision to exclude, or touch only briefly, topics such as turbulence, boundary layer theory, stability analysis, etc., was based on a combination of factors that include space limitations and the preferences and expertise of the authors.

Our approach to the subject of transport phenomena has been inspired by Ravel's *Boléro*.[3] The basic theme appears throughout the book, but more instruments, maybe more adequately referred to as tools in the present context, join in from chapter to chapter. The faint beginning is with one-dimensional transport of probability, for which the basic theme is already there. The dimensionality is then increased and the transport of mass, momentum, and energy join in harmonically. Once the constitutive equations have been found by means of thermodynamics, one can really start to enjoy the piece by solving a number of problems with a variety of methods. Complex fluids or soft matter add a modern sound. Nonlinearity requires powerful new instruments. Interfaces come in as trumpets – our possibilities open up enormously. The orchestration becomes complete when kinetic theories join in with the theme of transport phenomena. Fully solved exercises, études of a kind, invite the audience to not just listen but join in with the theme. A striking difference between Ravel's *Boléro* and our transport phenomena is that we need more than 17 minutes from our audience.

How to Teach from This Book

The most traditional audience for *A Modern Course in Transport Phenomena* is chemical engineers, but interesting courses for several other areas of study are easy to envision. These areas include physics, biology, physical chemistry, and applied mathematics, as well as biomedical, environmental, materials, mechanical, molecular, and nuclear engineering. The material in this book is more than enough to cover two, one-semester, graduate-level courses on transport phenomena. Prior exposure to the subject of transport phenomena is useful, but not essential; the book assumes that the reader has a basic knowledge of classical physics, physical chemistry, vector and tensor analysis, and applied mathematics.

By intention, the book has a non-rigid, non-traditional structure. Each chapter can be covered in roughly 2–3 hours of lecture time. The introductory Chapters 1, 2, and 3 provide motivation and a healthy warm-up. The chapters that follow form the main structure of the book, which is based on blocks of related chapters. Two fundamental blocks contain indispensable core chapters on thermodynamics, balance equations, and constitutive

---

[3] *Boléro* is a one-movement orchestral piece composed by Maurice Ravel (1875–1937), performed for the first time in 1928. It is based on a single theme repeated over and over again; instead of developing the theme, Ravel gradually increases the orchestra to make the piece interesting in a captivatingly dramatic way.

equations: Chapters 4–6 for bulk phases; Chapters 13–15 for interfaces (chapters inside [...] may be omitted or skimmed by readers with prior knowledge of these topics).

| [4] Equilibrium thermodynamics |
| 5 Balance equations |
| 6 Forces and fluxes |

| 13 Thermodynamics of interfaces |
| 14 Interfacial balance equations |
| 15 Interfacial force–flux relations |

Each of the two fundamental blocks is supplemented by a block of applications:

| 7 Measuring transport coefficients |
| 8 Pressure-driven flow |
| 9 Heat exchangers |
| 10 Gas absorption |
| 11 Driven separations |
| 12 Complex fluids |

| 16 Polymer processing |
| 17 Transport around a sphere |
| 18 Bubble growth and dissolution |
| 19 Semi-conductor processing |

The application blocks contain chapters involving fluid mechanics, heat transfer, mass diffusion, and chemical reaction. A number of applications to choose from are offered in these non-monolithic blocks. Chapters 7–10 focus on classical problems in transport phenomena, and Chapter 11 covers more modern applications. The modern subject of complex fluids is presented in Chapter 12 and provides essential background for later chapters. The block on the right containing Chapters 16–19 covers important applications where interfacial transport is emphasized.

Taken together the four blocks above reflect a natural structure of the *phenomenological approach* to the subject. Horizontal pairs of blocks consist of foundation and application chapters; vertical pairs of blocks contain chapters that emphasize either bulk or interfacial transport phenomena.

The last two blocks of chapters are once more grouped in foundations and applications, where the unifying theme now is the *molecular approach*, that is, the ambitious goal of bridging scales. The first block begins with Chapter 20 and provides background (if necessary) for kinetic theories, which are developed for gases and polymeric liquids in Chapters 21 and 22. The second block containing Chapters 23–26 is offered as a menu to choose from; it includes both theoretical developments and experimental methods.

Approximately 250 *exercises* can be found throughout this book. Exercises found within the text of a given section serve the purpose of reinforcing concepts, while exercises at the end of a section either allow for the streamlined development of concepts that involve tedious intermediate calculations, or give applications of new concepts. This book has two *appendices*. Appendix A contains several compact and systematically organized tables of thermodynamic relations useful for defining thermodynamic properties and the relationships between them. Appendix B contains tables of the most commonly used differential operators expressed in rectangular, cylindrical, and spherical coordinate systems, which can be used to express balance and evolution equations for particular applications.

Acknowledgments

We have greatly benefited from discussions about this book with a number of colleagues and their feedback has guided us throughout the writing process. In particular, we would like to thank Patrick Anderson, Dick Bedeaux, Manuel Laso, and Jay Schieber for their careful reading of the manuscript at various stages of its development – their feedback had a significant impact on the final version of this book.

The material in this book has been used for several years in graduate courses given at the IIT Chicago and the ETH Zürich. The teaching process has led to important improvements resulting from interactions both with the teaching assistants involved and with the students taking our courses. In particular, we would like to acknowledge the feedback from our teaching assistants Patrick Ilg, David Nieto Simavilla, Maksym Osmanov, Meisam Pourali, Marco Schweizer, Pavlos Stephanou, Laura Stricker, David Taj, and Carl Zinner. Of the many students who have taken our courses, we would like to acknowledge Nora Zimerli and Whitney Fowler for their careful reading of the manuscript, which helped eliminate many typographical errors and improve our grammar. DCV would like to thank Ernest Venerus for his constant support while writing this book, and much more.

Symbols and Notation

$\boldsymbol{B}$	Magnetic field vector
b	Scattering length
b_1, b_2, b_{11}	Second-order fluid model parameters
c	Molar density
c	Speed of light (2.99792458×10^8 m/s)
$c_{\hat{s}}, c_T$	Speed of sound at constant entropy, temperature
$\hat{c}_{\hat{v}}, \hat{c}_p$	Heat capacity per unit mass at constant volume, pressure
c_α	Molar density of species α
$\boldsymbol{c}$	Relative position vector of non-inertial coordinate system
$\boldsymbol{c}$	Conformation tensor
$D, \boldsymbol{D}$	Diffusion coefficient, tensor
D	Diameter
$D_{\alpha\beta}$	Diffusivity of species α relative to species β
$D_{q\alpha}$	Thermal diffusion coefficient of species α
D_T	Thermal diffusion coefficient normalized by temperature
$D_{\text{eff}}, \boldsymbol{D}_{\text{eff}}$	Effective diffusivity, tensor in porous media
$\mathfrak{D}_{\alpha\beta}$	Maxwell–Stefan diffusivity of species α relative to species β
d	Particle diameter
d	Number of dimensions
E	Energy
E_{tot}	Total energy
E_j	Energy of microstate j
E^2	Operator in axisymmetric spherical coordinates
$\tilde{E}_{\text{act}}$	Activation energy
$\boldsymbol{E}$	Electric field vector
$\boldsymbol{E}$	Finite strain tensor
e	Mathematical constant (2.71828)
e	Energy density
e	Elementary charge [$1.6021766208(98) \times 10^{-19}$ C]
$\boldsymbol{e}_i$	Eigenvectors ($i = 1, 2, 3$) for evolution of fluctuating fields in dynamic light scattering
F	Helmholtz free energy
F	Force exerted by molecular motor on filament
$\tilde{F}$	Faraday constant [$\tilde{F} = \tilde{N}_A z_{\text{el}} = 9.648533289(59) \times 10^4$ C/mol]
F_L	Take-up force in fiber spinning
$\boldsymbol{\mathcal{F}}_{\text{s}}$	Force exerted by fluid on solid
f	Helmholtz free energy density
f	Dimensionless force exerted by molecular motor on filament
f_{s}	Friction factor

$f(\boldsymbol{r}, \boldsymbol{p})$	Probability density in Boltzmann's equation
$\boldsymbol{f}_\alpha$	External force on species α
f	Number of independent intensive variables in Gibbs phase rule
$\boldsymbol{f}_\mathrm{B}$	Brownian force
$\mathfrak{f}_\alpha$	Fugacity of species α
G	Gibbs free energy
G	Relaxation modulus
G_0	Relaxation modulus at $t = 0$
G', G''	Storage, loss modulus
G^*	Complex modulus
$\tilde{G}(\gamma)$	Reaction coordinate-dependent molar Gibbs free energy
$\mathcal{G}_\alpha, \tilde{\mathcal{G}}_\alpha$	Total mass, molar transfer rate of species α
g	Gibbs free energy density
$\boldsymbol{g}$	Gravitational acceleration
H	Enthalpy
H	Height
H	Hookean spring constant
H_α, H_ω	Positive constants
H_B	Functional in Boltzmann's H-theorem
$\mathcal{H}$	Heaviside step function
h	Enthalpy density
h	Film thickness
h	Heat transfer coefficient
h_g	Heat transfer coefficient in gas
h_m	Heat transfer coefficient in melt
I_A	Spectral density of autocorrelation function of A
$I_\mathrm{i}, I_\mathrm{s}$	Intensity of incident, scattered wave
i	$\sqrt{-1}$
i	Integer indicating state $(0, 1)$ of molecule conformation
$\boldsymbol{i}$	Electric flux (current)
$J, \boldsymbol{J}$	Probability flux, vector
$\boldsymbol{J}_\alpha$	Diffusive molar flux of species α relative to $\boldsymbol{v}$
$\boldsymbol{J}_\alpha^*$	Diffusive molar flux of species α relative to $\boldsymbol{v}^*$
$\boldsymbol{j}_a$	Diffusive flux of a
$\boldsymbol{j}_\alpha$	Diffusive mass flux of species α relative to $\boldsymbol{v}$
$\boldsymbol{j}_\alpha^\dagger$	Diffusive mass flux of species α relative to $\boldsymbol{v}^\dagger$
$\boldsymbol{j}_\mathrm{el}$	Mass flux of electric charge relative to $\boldsymbol{v}$
$\boldsymbol{j}_q, \boldsymbol{j}_q'$	Diffusive energy flux, modified
$\boldsymbol{j}_s$	Diffusive entropy flux

K	Equilibrium constant for chemical reaction
K	Power-law model viscosity parameter
K_d	Dispersion coefficient
K_i	Dispersion theory functions $(i = 1, 2, ...)$
K_{tot}	Total kinetic energy
k	Spatial Fourier transform variable
k	Number of components in mixture
k, k'	Rate constants
$k_m, \tilde{k}_m$	Mass transfer coefficients
k_B	Boltzmann's constant $[1.3806505(24) \times 10^{-23} \text{ J/K}]$
k_H	Henry's law constant
k_w	Scaled inverse of Henry's law constant
k_{eff}	Effective reaction rate constant in porous media
$\boldsymbol{k}_i, \boldsymbol{k}_s$	Wave vector of incident, scattered wave
L	Length (characteristic)
L_b	Length of fixed bed reactor
L_{ext}	Length of extruder channel
L_i	Phenomenological (transport) coefficient $(i = q, \tau, \Gamma)$
L_g	Length of cubic volume element
L_p	Injection pulse length
L_v	Entrance length for velocity
L_T	Entrance length for temperature
L_{ij}	Matrix of phenomenological (transport) coefficients $(i, j = \alpha, \beta, q, \mathcal{A})$
l_{cap}	Capillary length
l_f	Characteristic pore size
l_{md}	Mean distance between particles
l_{mfp}	Mean free path of a particle
l_{ij}	Matrix of phenomenological (transport) coefficients $(i, j = \alpha, \beta, q, \mathcal{A})$
ℓ	Displacement of interface surface
M	Mass
M	Memory function
M_{tot}	Total mass
$M_{\alpha,tot}$	Total mass of species α
$\tilde{M}$	Molecular weight (of mixture)
$\tilde{M}_\alpha$	Molecular weight of species α
$\tilde{M}_q, \tilde{M}_{q_j}$	Molecular weight of reactants/products for reaction, reaction j
$\tilde{M}_w$	Molecular weight (weight-average) of polymer

$M_{\text{cycle}}^{(i)}$	ith moment of molecular motor displacement
$\boldsymbol{M}$	Momentum
$\boldsymbol{M}_{\text{tot}}$	Total momentum
$\boldsymbol{\mathcal{M}}_{\text{s}}$	Torque exerted by fluid on solid
m	Mass of a particle, bead
m	Number of fundamental dimensions in n physical quantities
m_{eff}	Effective mass of bead
m_{el}	Mass of electron [$9.10938215(45) \times 10^{-31}$ kg]
$\boldsymbol{m}$	Momentum density
N	Number of moles
N	Number of cycles
N_α	Number of moles of species α
$\tilde{N}_{\text{A}}$	Avagadro's number [$6.022140857(74) \times 10^{23}$ mol^{-1}]
N_1, N_2	First, second normal stress differences
$(N_1)_R$	First normal stress difference at capillary wall
N_w	Dimensionless driving force parameter for bubble growth
$N_{\Delta p}$	Dimensionless pressure difference for bubble growth
N_γ	Dimensionless interfacial tension for bubble growth
N_η	Dimensionless viscosity for bubble growth
$\mathcal{N}$	Prefactor integral of $\Phi(\gamma)$
$\boldsymbol{N}_\alpha$	Molar flux of species α relative to stationary frame
n	Number of independent physical quantities
n	Power-law model index parameter
n	Number of particles in system
n, n'	Ratios of phenomenological coefficients
n_{p}	Number density of bead–spring elements
$\boldsymbol{n}$	Unit normal vector
n_α	Elementary charge of species α
$\boldsymbol{n}_\alpha$	Mass flux of species α relative to stationary frame
P_{el}	Electrical power per unit length
P_z	Transition probability
$\mathcal{P}$	Modified pressure
p	Probability density
p	Pressure
p^{L}	Pseudo-pressure
p_n	Eigenfunction of probability density
$\boldsymbol{p}$	Momentum of particle
Q	Heat
$\dot{Q}$	Total rate of heat transfer

Q_{jk}	Orthogonal rotation matrix
$\boldsymbol{Q}$	Structural (connector) vector
q	Coupling coefficient for ion pump
$\boldsymbol{q}$	Momentum of particle
$\boldsymbol{q}$	Scattering vector
R	Radius
$\tilde{R}$	Ideal gas constant $[\tilde{R} = \tilde{N}_A k_B = 8.3144598(48)$ J/(mol K)]
R_b	Radius of fixed bed reactor
R_{el}	Electrical resistance per unit length
R_{hyd}	Hydraulic radius
R_K^{Is}, R_K^{IIs}	Kapitza resistance of phase I, II
R_Q	Autocorrelation function of Q
R_{ij}	Matrix of phenomenological (resistance) coefficients $(i, j = \alpha, \beta, q, \mathcal{A})$
$\boldsymbol{R}$	Position of detector relative to scattering volume
$\boldsymbol{R}_1, \boldsymbol{R}_2$	Bead position vectors
$\mathcal{R}_\alpha, \tilde{\mathcal{R}}_\alpha$	Total rate of species α mass, moles produced by chemical reaction
r	Radial coordinate in cylindrical coordinate system
r	Radial coordinate in spherical coordinate system
r	Randomness factor
$\boldsymbol{r}$	Position vector
$\boldsymbol{r}'$	Position vector in non-inertial coordinate system
$\boldsymbol{r}_b$	Bead position vector
S	Entropy
$S(\boldsymbol{q}, \omega)$	Dynamic structure factor
$\dot{S}$	Total rate of entropy transfer
S_{tot}	Total entropy
S_T	Soret coefficient
s	Laplace transform variable
s	Entropy density
s_1	Sedimentation coefficient
T	Temperature
T_{fl}	Temperature of fluid away from surface
T_g	Temperature of fluid in gas
T_m	Temperature of fluid in melt
$|\boldsymbol{\nabla}T|_\infty$	Temperature gradient (magnitude) around sphere
$T^{\mu\nu}$	Energy-momentum tensor $(\mu, \nu = 0, 1, 2, 3)$
$\boldsymbol{T}$	Maxwell electromagnetic stress tensor
t	Time

t'	Lorentz transformed time
$\boldsymbol{t}$	Unit tangent vector
U	Internal energy
u	Internal energy density
$\tilde{u}, \hat{u}$	Internal energy per mole, mass
u^μ	Velocity four-vector ($\mu = 0, 1, 2, 3$)
$\boldsymbol{u}$	Displacement vector
$\boldsymbol{u}$	Diffusion-induced velocity in porous media
V	Volume
V	Velocity (characteristic)
V_{con}	Volume of continuous phase
V_{dis}	Volume of dispersed phase
V_{eff}	Effective volume
V_{f}	Fluid phase volume within V_{eff}
V_{s}	Solid phase volume within V_{eff}
$\hat{v}$	Volume per unit mass
$\hat{v}_\alpha$	Partial specific volume of species α
v_n^{s}	Normal part of interface velocity $\boldsymbol{v}^{\text{s}}$
v_{na}^{s}	Normal part of interface velocity $\boldsymbol{v}^{\text{s}}$ for which $a^{\text{s}} = 0$
$\boldsymbol{v}$	Mass-average (barycentric) velocity
$\boldsymbol{v}^*$	Molar-average velocity
$\boldsymbol{v}^\dagger$	Volume-average velocity
$\boldsymbol{v}_\alpha$	Species α velocity
$\boldsymbol{v}_{\text{b}}$	Sphere or bead velocity
$\boldsymbol{v}^{\text{s}}$	Interface velocity
$\boldsymbol{v}_{\text{tr}}^{\text{s}}$	Translational part of interface velocity $\boldsymbol{v}^{\text{s}}$
$\boldsymbol{v}_{\text{def}}^{\text{s}}$	Deformational part of interface velocity $\boldsymbol{v}^{\text{s}}$
$\boldsymbol{v}_\parallel^{\text{s}}$	Tangential part of interface velocity $\boldsymbol{v}^{\text{s}}$
W	Work
$\dot{W}$	Total rate of work
$\mathcal{W}$	Total mass flow rate
$\boldsymbol{W}_j$	d'-dimensional column vector of independent Gaussian random variables
w	Transition rate for momentum exchange
w_α	Mass fraction of species α
$w_{\alpha\text{fl}}, x_{\alpha\text{fl}}$	Mass, mole fraction of species α of fluid away from surface
$\boldsymbol{w}$	Vorticity vector
$\boldsymbol{X}_j$	d-dimensional column vector of random variables
$x, \boldsymbol{x}$	Position, vector
$\boldsymbol{x}$	Vector of fluctuating fields in dynamic light scattering

x_i	Cartesian coordinates ($i = 1, 2, 3$)
x_i'	Cartesian coordinates ($i = 1, 2, 3$) in non-inertial coordinate system
x_i'	Lorentz transformed spatial coordinates ($i = 1, 2, 3$)
x_α	Mole fraction of species α
x^μ, x_μ	Time and space coordinates ($\mu = 0, 1, 2, 3$)
$\boldsymbol{y}$	Position vector within V_{eff}
Z	Thermoelectric figure of merit
Z	Canonical partition sum
Z	Entropy production parameter for ion pump
$\mathcal{Z}$	Prefactor integral of $\Phi(\gamma)$
z	Axial coordinate in cylindrical coordinate system
z	Integer indicating position along filament
z_α	Charge per unit mass of species α
z_{el}	Charge per unit mass of electron

Greek Symbols

α	Angle
$\alpha_t, \boldsymbol{\alpha}_t$	Mean of Gaussian distribution
$\alpha_{\hat{s}}, \alpha_p$	Thermal expansion coefficient at constant entropy, pressure
α_i	Eigenvalues ($i = 1, 2, 3$) for evolution of fluctuating fields in dynamic light scattering
$\alpha^{\mu\nu}$	Structure tensor ($\mu, \nu = 0, 1, 2, 3$)
$\overline{\alpha}_{\mu\nu}, \overset{\circ}{\alpha}_{\mu\nu}$	Trace and trace-free parts of $\alpha^{\mu\nu}$
$\boldsymbol{\alpha}$	Dyadic product of $\boldsymbol{Q}$
$\beta, \bar{\beta}$	Dimensionless parameter
$\Gamma, \tilde{\Gamma}$	Mass, molar rate of chemical reaction per volume
Γ	Attenuation coefficient
$\Gamma_{\text{A}}^{\pm}$	Transition rate between states z, i
Γ_{tot}	Total transition rate
$\boldsymbol{\Gamma}_t$	Phase-space trajectory
γ	Ratio of heat capacities ($= \hat{c}_p / \hat{c}_{\hat{v}}$)
γ	Shear strain
$\dot{\gamma}$	Shear strain rate
γ	Lorentz factor
γ	Interfacial tension
γ	Reaction coordinate
$\bar{\gamma}_\alpha$	Activity coefficient for species α
γ_T	Derivative of interfacial tension γ with respect to T

γ	Strain tensor
$\dot{\gamma}$	Rate of strain tensor
δ	Dirac's delta function
δ_T	Thermal boundary layer thickness
$\boldsymbol{\delta}, \delta_{ij}$	Identity (unit) tensor, Kronecker delta
$\boldsymbol{\delta}_i$	Base vector
$\boldsymbol{\delta}_i'$	Base vector in non-inertial coordinate system
$\boldsymbol{\delta}_\parallel$	Tangential identity tensor
ϵ	Internal energy density (Section 12.6 only)
ϵ	Extensional strain
ϵ	Ratio of spherical particle density to surrounding fluid density
ϵ	Porosity
$\dot{\epsilon}$	Extensional strain rate
$\boldsymbol{\epsilon}, \epsilon_{ijk}$	Alternating identity (unit) tensor, permutation symbol
ε	Dimensionless (perturbation) parameter
$\varepsilon_{\mathrm{el}}$	Seebeck coefficient
ζ	Parameter for streamline
ζ	Friction coefficient
η	Shear viscosity
η	Efficiency of ion pump
η	Effectiveness factor for porous catalyst
η_{d}	Dilatational viscosity
η_0, η_∞	Zero-, infinite-shear rate viscosity
η_j	Viscosity for relaxation time λ_j
$\eta_{\mathrm{B}}, \eta_{\mathrm{E}}$	Equibiaxial, uniaxial elongational viscosities
η_{eff}	Effective viscosity
η', η''	Real, imaginary part of complex viscosity
η^*	Complex viscosity
$\eta^{\mu\nu}, \eta_{\mu\nu}$	Minkowski tensors $(\mu, \nu = 0, 1, 2, 3)$
$\hat{\eta}^{\mu\nu}$	Spatial projection of $\eta^{\mu\nu}$ $(\mu, \nu = 0, 1, 2, 3)$
θ	Azimuthal angle in cylindrical coordinate system
θ	Polar angle in spherical coordinate system
θ	Contact angle
$\boldsymbol{\Theta}$	Second moment of connector vector $\boldsymbol{Q}$
$\Theta_t, \boldsymbol{\Theta}_t$	Variance of Gaussian distribution
$\kappa_{\hat{s}}, \kappa_p$	Compressibility at constant entropy, pressure
$\boldsymbol{\kappa}$	Transpose of gradient of velocity $[= (\boldsymbol{\nabla v})^{\mathrm{T}}]$
$\kappa, \boldsymbol{\kappa}$	Permeability, tensor

Λ	Dimensionless parameter characterizing deviation from equilibrium at interface
Λ_p	Dimensionless parameter for ultracentrifugation
Λ_T	Dimensionless parameter for thermal field flow fractionation
Λ_ξ	Dimensionless slip coefficient
Λ_ϕ	Dimensionless parameter for electric field flow fractionation
λ, λ'	Thermal conductivity, modified
$\boldsymbol{\lambda}$	Thermal conductivity tensor
λ_{eff}	Effective thermal conductivity
$\lambda_0, \lambda_1, \lambda_2$	Relaxation times
λ, λ_i	Relaxation time, spectrum $(i = 1, 2, ...)$
$\lambda_0, \boldsymbol{\lambda}_1, \lambda_2$	Parameters in Maxwellian distribution
λ_n	Eigenvalue of probability density
$\tilde{\mu}, \hat{\mu}$	Chemical potential per mole, mass
$\tilde{\mu}, \hat{\mu}$	Flat-average chemical potential per mole, unit mass
$\tilde{\mu}_\alpha, \hat{\mu}_\alpha$	Chemical potential of species α per mole, unit mass
$\hat{\mu}'_\alpha$	Electrochemical potential of species α per unit mass
ν	Kinematic viscosity
ν_{l}	Longitudinal kinematic viscosity
$\nu_\alpha, \nu_{\alpha_j}$	Coefficient for species α in chemical reaction, j
$\tilde{\nu}_\alpha, \tilde{\nu}_{\alpha_j}$	Stoichiometric coefficient for species α in chemical reaction, j
ξ, ξ_j	Extent of reaction, reaction j
ξ	Similarity transform variable
ξ_n	Mobility coefficient for normal velocity jump at interface
$\boldsymbol{\xi}_{\parallel}$	Mobility coefficient for tangential velocity jump at interface
π	Number of phases in Gibbs phase rule
π	Mathematical constant (3.14159)
π_{el}	Peltier coefficient
$\boldsymbol{\pi}$	Pressure (momentum flux) tensor
$\boldsymbol{\pi}_n$	Normal projection of pressure tensor (stress vector)
ρ	Mass density
ρ_α	Mass density of species α
$\dot{\Sigma}$	Total rate of entropy production
σ	Entropy production rate per volume
σ	Yield stress
σ_{el}	Electrical conductivity
τ_Γ	Relaxation time for chemical reaction
τ_{flow}	Characteristic time for flow
τ_R	Shear stress at capillary wall

$\boldsymbol{\tau}$	Extra pressure (stress) tensor
$\Upsilon_{\alpha k}, \tilde{\Upsilon}_{\alpha k}$	Relative mass, molar adsorption for species α
Φ	Velocity potential
Φ	Potential function for bead–spring element
Φ_{tot}	Total potential energy
$\Phi(\gamma)$	Reaction coordinate-dependent potential energy
ϕ	Azimuthal angle in spherical coordinate system
ϕ	Gravitational potential
ϕ	Rotation angle in Q_{jk} matrix
ϕ	Volume fraction
$\phi^{(e)}$	External force potential
ϕ_α	External force potential for species α
$\bar{\phi}_\alpha$	Fugacity coefficient of species α
φ^{s}	Function describing surface
χ	Thermal diffusivity
Ψ_t	Integrating factor for diffusion equation
Ψ_1, Ψ_2	First, second normal stress coefficients
ψ	Stream function
Ω	Angular velocity
Ω	Number of microstates
ω	Oscillation frequency
ω	Time Fourier transform variable
ω^μ	Structure vector ($\mu = 0, 1, 2, 3$)
$\boldsymbol{\omega}$	Angular velocity vector

Other Symbols

δa	Fluctuation of a
$\langle a \rangle$	Average of a
$\tilde{a}$	Spatial Fourier transform of a
$\bar{a}$	Laplace transform of a
(1)	Subscript indicating convected time derivative
a	Subscript indicating surface area average of quantity
eq	Subscript indicating quantity at equilibrium
i	Subscript indicating intrinsic average of quantity
f	Subscript indicating quantity in comoving local reference frame
s	Subscript indicating superficial average of quantity
0	Subscript/superscript indicating initial or reference value
s	Superscript indicating interfacial quantity

$*$	Superscript indicating complex conjugate
I, II	Superscript indicating quantity in phase I, II at interface
$(0), (1), \ldots$	Superscript indicating zero-, first-, ... order perturbation solution
$'$	Superscript indicating quantity in non-inertial coordinate system
T	Superscript indicating transpose of tensor
$\bar{x}$	Rescaled variable x
Δa	Difference in a
∇	Nabla operator
$\nabla_{\parallel}$	Nabla operator for surface
∇^2	Laplacian operator

Dimensionless Groups

$N_{\mathrm{Bi}} = hL/\lambda$	Biot number
$N_{\mathrm{Br}} = \eta V^2/(\lambda\,\Delta T)$	Brinkman number
$N_{\mathrm{Da}} = kL^2/D_{12}$	Damköhler number
$N_{\mathrm{De}} = \lambda/\tau_{\mathrm{flow}}$	Deborah number
$N_{\mathrm{Fr}} = V^2/(gL)$	Froude number
$N_{\mathrm{Gr}} = gL^3\alpha_p\,\Delta T/\nu^2$	Grashof number
$N_{\mathrm{Ka}} = R_{\mathrm{K}}\lambda/L$	Kapitza number
$N_{\mathrm{Kn}} = l_{\mathrm{mfp}}/L$	Knudsen number
$N_{\mathrm{Le}} = \chi/D_{12}$	Lewis number
$N_{\mathrm{Ma}} = V/c_T$	Mach number
$N_{\mathrm{Na}} = V^2|\partial\eta/\partial T|/\lambda$	Nahme–Griffith number
$N_{\mathrm{Nu}} = hL/\lambda_{\mathrm{fl}}$	Nusselt number
$N_{\mathrm{Pe}} = \rho V L/D_{12}$	Péclet number for mass transfer
$N'_{\mathrm{Pe}} = \rho V L/\chi$	Péclet number for energy transfer
$N_{\mathrm{Pr}} = \nu/\chi$	Prandtl number
$N_{\mathrm{Ra}} = gL^3\alpha_p\,\Delta T/(\nu\chi)$	Rayleigh number
$N_{\mathrm{Re}} = VL/\nu$	Reynolds number
$N_{\mathrm{Sc}} = \nu/D_{12}$	Schmidt number
$N_{\mathrm{Sh}} = k_{\mathrm{m}}L/(\rho D_{12})_{\mathrm{fl}}$	Sherwood number
$N_{\mathrm{St}} = \hat{c}_p\,\Delta T/\Delta\hat{h}$	Stefan number
$N_{\mathrm{Th}} = L\sqrt{a_{\mathrm{fs}}k^{\mathrm{s}}/D_{\mathrm{AB}}}$	Thiele number (modulus)
$N_{\mathrm{We}} = \rho V^2 L/\gamma$	Weber number
$N_{\mathrm{Wi}} = \lambda\dot{\gamma}$	Weissenberg number

1

Approach to Transport Phenomena

"A good grasp of transport phenomena is essential for understanding many processes in engineering, agriculture, meteorology, physiology, biology, analytical chemistry, materials science, pharmacy, and other areas. Transport phenomena is a well-developed and eminently useful branch of physics that pervades many areas of applied science. [...] The subject of transport phenomena includes three closely related topics: fluid dynamics, heat transfer, and mass transfer. Fluid dynamics involves the transport of *momentum*, heat transfer deals with the transport of *energy*, and mass transfer is concerned with the transport of *mass* of various chemical species." Let's approach the field of transport phenomena by trying to appreciate and elaborate this introductory quote on the importance and ubiquity of transport phenomena in science and engineering from the pioneering textbook of Bird, Stewart and Lightfoot.[1] Let's look at the topic of transport phenomena from various perspectives.

1.1 The First Three Minutes

We wake up in the morning, switch on the light, and take a shower. Our day usually begins with the flow of electric energy and water to our house. We immediately find ourselves right in the middle of transport phenomena. What a luxury.

Let's add a little drama by scaling things up. Consider (i) crude oil pipelines and (ii) high-voltage electric power transmission lines. (i) For example, the Druzhba[2] pipeline built in the early 1960s carries crude oil from the Russian heartland all the way to Germany (some 4 000 km). Some 200 000 m³ of crude oil per day are pumped through steel tubes with a

---

[1] See p. 1 of Bird, Stewart & Lightfoot, *Transport Phenomena* (Wiley, 2001).
[2] Druzhba is the Russian word for friendship.

diameter of up to one meter, which implies a characteristic speed of $3\,\mathrm{m/s}$. Every three days one could fill a supertanker. Pumping stations are typically required every $100\,\mathrm{km}$ (give or take a factor of two). (ii) Electric energy transmission in overhead power lines takes place above $100\,\mathrm{kV}$. For proper load balancing and to satisfy the demand of a flourishing electricity trading market, electric energy is transported over large distances. The largest power grid, with a total power generation of 667 gigawatts, is the Synchronous Grid of Continental Europe, which serves 400 million customers in 24 countries.[3]

The length scale associated with these impressive transport challenges is of the order of the size of countries or continents. The same length scale is also involved in long-term weather forecasts or the prediction of climate changes, where large-scale transport in the atmosphere and oceans matters.

Very clearly, our life depends on transport phenomena; not just in the sense of convenience considered so far, but also in the sense of biological functions. We depend on the never-ceasing flow of blood and oxygen through our body, and drugs need to be targeted to the appropriate places in our body. On the level of cells, ions need to be transported, often against concentration gradients, which requires energy from a chemical reaction. Also muscle contraction relies on transport phenomena. These biological functions are associated with transport phenomena on a molecular length scale.

Let's add even more drama to the story. Let's not think about the first three minutes of our day or the first three minutes of this course on transport phenomena, but about the first three minutes of the Universe. Even the origin of the Universe is largely dominated by the transport of mass and energy in the presence of violent reactions between particles.[4]

The ubiquity and importance of transport phenomena should be clear from these few obvious examples. Moreover, a thorough understanding of transport phenomena is of self-evident economic and environmental importance.

Exercise 1.1   Transport of Crude Oil and Money

The Trans-Alaska Pipeline (see Figure 1.1) built in the 1970s carries crude oil from Prudhoe Bay to Valdez, Alaska, a distance of roughly $1300\,\mathrm{km}$. Typically, $100\,000\,\mathrm{m}^3$ of crude oil per day are pumped through steel tubes with a diameter of more than one meter. According to the current market price of crude oil, usually given in dollars per barrel, what is the value of the crude oil passing through the Trans-Alaska Pipeline in one day?

---

[3] According to `en.wikipedia.org/wiki/Synchronous_grid_of_Continental_Europe`.
[4] Weinberg, *The First Three Minutes* (Basic Books, 1977).

Figure 1.1 The Trans-Alaska Pipeline starts at Pump Station 1 in Prudhoe Bay. (Figure from `www.clui.org/ludb/site/alaska-pipeline-origin`.)

1.2 Complex Systems

In spite of their striking importance, the transport phenomena discussed in the preceding section are of a rather basic type, at least, as they have been presented. In modern applications, we usually deal with far more complex systems. A few examples should suffice to illustrate this point and to indicate some important implications for this course.

What happens with crude oil after it has been transported? It enters a chemical plant where it is processed and refined into more useful products. In such an oil refinery, we encounter transport in a huge number of pipes carrying various streams of different fluids between chemical processing units, such as distillation towers and cracking units, where separation processes and chemical reactions are taking place during the flow. Among the components separated in an oil refinery are the feedstocks for the syntheses of various kinds of polymers, or plastics. Finally, very complex flow situations of highly viscoelastic fluids occur in polymer processing operations, such as injection molding or film blowing. There is a really long way to go for crude oil.

In meteorology, atmospheric flow phenomena on a wide range of length scales are affected by a number of other systems and phenomena, such as ocean circulations. Nontrivial boundary conditions play an important role and phase transitions (evaporation, condensation, melting) are ubiquitous phenomena, the occurrence of which may be of particular interest.

We have already seen that transport processes in biological systems occur on very different length scales ranging from the meter scale of our body to the 1–100 µm scale of cells. These transport phenomena are coupled to many further biophysical and biochemical processes over a large range of length scales.

The briefly sketched complexity of many transport phenomena of interest has some immediate implications. Once we have understood the basic structure of balance equations and the constitutive assumptions for the fluxes occurring in these equations, we are faced with a number of serious challenges. How do we obtain all the material properties, such as transport coefficients, occurring in these equations? How do we actually solve rather large sets of coupled equations? How do we formulate physically meaningful boundary conditions, and how do we get the additional material information possibly contained in them? In this course we build up increasing knowledge about these questions and consider increasingly complex applications.

The experimental measurement of material properties related to transport is a challenging topic. In Chapter 7, we describe the most basic ideas for measuring transport coefficients, such as viscosity, thermal conductivity, and diffusivity. Microbead rheology provides another alternative to investigate momentum transport, as discussed in Chapter 25. By developing the theory of dynamic light scattering in Chapter 26, we obtain an alternative tool for measuring transport coefficients. Thorough foundations for discussing boundary conditions and expressing the physics happening at boundaries and interfaces are laid in Chapters 13–15.

Statistical mechanics offers an alternative theoretical path to obtain material properties, which allows the development of basic ideas for the kinetic theories of gases (Chapter 21) and polymeric liquids (Chapter 22). Further applications of methods for bridging scales include porous media (Chapter 23) as well as molecular motors and ion pumps (Chapter 24).

Once we have the equations, the boundary conditions, and all the required material information, we need to solve the equations. We typically deal with nonlinear and often coupled partial differential equations. Finite element and finite difference methods are important tools to solve such equations, and can be implemented using several commercial software packages. In some cases, the lattice Boltzmann method offers an interesting alternative (inspired by the kinetic theory of gases, but going far beyond it). To solve some illustrative problems with very simple simulation programs, we introduce Brownian dynamics at an early stage (Chapter 3).

1.3 Classical Field Theories

Transport implies the flow of certain quantities in space, and it takes time. For example, mass can move from point to point in space. To describe this transport of mass, it is convenient to assign a mass density to every point in space so that we end up with mass density fields that change in space and time. Similarly, momentum and energy density fields arise naturally, and velocity and temperature fields may be convenient alternatives. In modeling transport phenomena, we are interested in the time-evolution equations for such hydrodynamic fields, that is, we actually deal with classical field theories. The most well-known field equations of transport theory are the Navier–Stokes–Fourier equations of hydrodynamics (to be developed in Chapters 5 and 6 and briefly summarized in Section 7.1), which provide the basic and quite universal description of the flow of mass, momentum, and energy. As a consequence of convection, these field equations are *nonlinear* in the hydrodynamic fields and hence exhibit a very rich behavior of solutions, including turbulence. Complex fluids, often referred to as soft matter, require even more complicated field theories, typically with additional fields describing the local structure of the complex fluid (Chapter 12).

All field theories are idealizations and must be expected to lead to difficulties when variations from point to point are taken too literally. Below a certain length scale, the atomic nature of matter must affect the validity of field theories. For a gas under normal conditions, the mean free path of the gas atoms or molecules between collisions is of the order of 0.1 μm so that nonlocality effects must be taken into account below this length scale. If we consider volume elements of the size of a few nanometers, even for liquids instead of gases, the number of atoms or molecules contained in such a volume becomes so small that fluctuation effects need to be taken into account. Modified hydrodynamic theories work down to amazingly small length scales, but eventually field theories for continua need to be given up and atoms or molecules and their interactions need to be considered explicitly. As even atoms consist of much smaller particles,[5] these interactions are not the fundamental ones, but typically van der Waals forces. The even more surprising success of the hydrodynamic approach in extracting information on the properties and dynamics of quark–gluon plasmas created in relativistic heavy-ion collisions calls for explanation.[6]

---

[5] The famous Rutherford experiment (published in 1911) revealed that an atom possesses a very small positively charged nucleus with a diameter of the order of 10^{-14} m and, a century later, the Large Hadron Collider in Geneva (officially inaugurated in 2008) allows us to probe the structure of matter down to some 10^{-20} m.

[6] Baier *et al.*, *Phys. Rev. C* **73** (2006) 064903.

The above discussion of continuum field theories and length scales has two important practical implications for our treatment of transport phenomena. (i) There are further challenges for investigating transport phenomena, namely those posed by *nonlocality and fluctuation effects*. To understand these effects, we need to stretch the theory to its limits. (ii) All our equations for transport phenomena result from coarse graining and hence must account for the *emergence of irreversibility and dissipation*. This observation has important consequences for the structure of "good" equations for modeling transport phenomena. Nonequilibrium thermodynamics,[7] ideally supported by statistical mechanics, provides the proper setting for formulating equations for coarse grained systems.

- Large-scale transport has a direct impact on our everyday life, e.g. transport in pipelines, electric power grids ...
- Transport is at the heart of many engineering tasks: chemical reactor design, materials processing, aerodynamical optimization (airplanes, vehicles) ...
- Transport is at the heart of many problems in the natural sciences: atmospheric sciences (climatology, meteorology), life sciences (blood circulation, muscle contraction, ecosystems), cosmology (expansion of the Universe, galaxy formation and evolution) ...
- Transport often occurs in complex coupled systems involving a wide range of length scales and a variety of dynamic material properties and boundary conditions.
- As transport takes place in space and time, we deal with partial differential equations for the evolution of coupled fields; the laws of nonequilibrium thermodynamics set the structure of the field equations and necessitate fluctuation effects.

---

[7] de Groot & Mazur, *Non-Equilibrium Thermodynamics* (Dover, 1984); Öttinger, *Beyond Equilibrium Thermodynamics* (Wiley, 2005).

2

The Diffusion Equation

One of the most famous equations in the field of transport phenomena is the diffusion equation. Its wide-ranging importance is underlined by the fact that, depending on the context, it is known by various names. In a probabilistic interpretation, it is usually referred to as a Fokker–Planck equation, which is a special type of Kolmogorov's forward equation for memoryless stochastic processes.[1] In the context of Brownian motion, the name Smoluchowski equation is most appropriate. The variety of names nicely indicates that this equation is not only useful for describing the transport phenomenon of mass diffusion; we will actually encounter it many times, in particular, also in the description of momentum and heat transport and in polymer kinetic theory. In the present chapter, we introduce it to describe the flow of probability. In doing so, we present the basic theme of transport phenomena, including some important concepts, tools, and results.

2.1 A Partial Differential Equation

By a first glance at the Fokker–Planck or diffusion equation in one space dimension,

$$\frac{\partial p(t,x)}{\partial t} = -\frac{\partial}{\partial x}\Big[A(t,x)p(t,x)\Big] + \frac{1}{2}\frac{\partial^2}{\partial x^2}\Big[D(t,x)\,p(t,x)\Big], \qquad (2.1)$$

one recognizes a second-order partial differential equation for the evolution of some function $p(t,x)$ of two real arguments involving coefficient functions $A(t,x)$ and $D(t,x)$. Here we simply assume $D(t,x) \geq 0$, but later will show that this follows from the second law of thermodynamics (see Exercise 2.4). In many cases, the given coefficient functions A and D are independent of their first argument, t. The goal of this chapter is to bring the reader

---

[1] See, for example, Chapter 3 of Gardiner, *Handbook of Stochastic Methods* (Springer, 1990).

from this superficial perspective to a deep understanding of the physical meaning and implications of the diffusion equation (2.1). The reader shall recognize the physical meaning of the coefficient functions A and D, and will develop a feeling for the significance of the occurrence of first- and second-order derivatives on the right-hand side of the evolution equation (2.1). It is important to develop a sound physical understanding of such second-order partial differential equations because they are at the heart of transport phenomena.

2.2 Probability Flux, Drift, and Diffusion

To assign intuitive names to the unknown function p and the given coefficient functions A and D in the diffusion equation (2.1), we consider diffusion from the perspective of Brownian motion, that is, the motion of a large particle in a surrounding fluid consisting of much smaller particles. The historical prototype of that kind of system is pollen in water, for which in 1828 the botanist Robert Brown (1773–1858) published his observation of a very irregular and unpredictable motion of particles in the absence of external forces (other researchers had observed this "Brownian motion" more than a century before Brown, but he was the first to establish Brownian motion as an important phenomenon and to investigate it in more detail[2]). The origin of this motion was much later discovered to lie in the enormously frequent collisions between the Brownian particle and the many surrounding fluid particles which are in incessant thermal motion.

A Brownian particle moves on a stochastic trajectory that is continuous but so irregular that its velocity cannot be defined (see Figure 2.1). The mass density is concentrated in the time-dependent position of the Brownian particle, and the mass flux is an even more singular object. On the other hand, the probability density for the location of the Brownian particle is a smooth function and there occurs a smooth flux of probability that leads to a smearing and broadening of the distribution. In this chapter, we focus on the evolution of the probability density, or on the flow of probability.

Marian Smoluchowski was the first to introduce a diffusion equation of the form (2.1) to describe Brownian motion in 1906. In this context, one interprets $p(t, x)$ as the probability density for finding the Brownian particle around the position x at time t. If we are interested in the probability of finding the Brownian particle between fixed positions x_1 and x_2 (where we

---

[2] An interesting survey of the history of Brownian motion can be found in §§2–4 of the monograph *Dynamical Theories of Brownian Motion* by Edward Nelson (Princeton, 1967). The early history of the stochastic description of Brownian motion has been reviewed by Subrahmanyan Chandrasekhar in *Rev. Mod. Phys.* **15** (1943) 1.

Figure 2.1 Trajectory of a Brownian particle.

assume $x_1 < x_2$), we have the evolution equation

$$\frac{d}{dt}\int_{x_1}^{x_2} p(t,x)dx = -\int_{x_1}^{x_2}\frac{\partial}{\partial x}J(t,x)dx = J(t,x_1) - J(t,x_2), \qquad (2.2)$$

with the quantity $J(t,x)$ read off from the diffusion equation (2.1),

$$J(t,x) = A(t,x)p(t,x) - \frac{1}{2}\frac{\partial}{\partial x}\Big[D(t,x)\,p(t,x)\Big]. \qquad (2.3)$$

We can thus interpret $J(t,x_1)$ as the influx and $J(t,x_2)$ as the outflux of probability, as illustrated in Figure 2.2. More generally, $J(t,x)$ is the probability flux at position x and time t. If the right-hand side of an evolution equation for the density of some quantity is written in the derivative form, the quantity behind the derivative is the corresponding flux. Proper sign conventions need to be chosen; we here use a positive sign for a flux in the direction of increasing x. It is natural to assume that the probability density and flux vanish at infinity. According to (2.2), the total probability $\int p(t,x)dx = 1$ is then conserved in time.

To learn more about the coefficient functions A and D, we define averages of suitable functions $f(x)$ performed with the probability density $p(t,x)$,

$$\langle f\rangle_t = \int_{-\infty}^{\infty} f(x)p(t,x)dx. \qquad (2.4)$$

Assuming that the function $f(x)$ is sufficiently smooth (for example, that it possesses piecewise continuous second-order derivatives), we obtain the following evolution equation for averages from the diffusion equation (2.1) after some integrations by parts and neglecting the probability density and

The Diffusion Equation

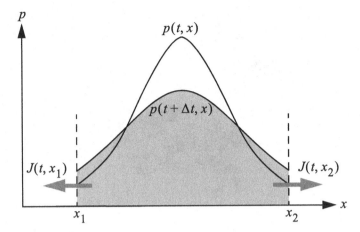

Figure 2.2 In the time interval from t to $t + \Delta t$, probability flows out of the interval $[x_1, x_2]$ (note that the flux $J(t, x_1)$ is negative).

flux at infinity,

$$\frac{d \langle f \rangle_t}{dt} = \left\langle A \frac{df}{dx} \right\rangle_t + \frac{1}{2} \left\langle D \frac{d^2 f}{dx^2} \right\rangle_t. \tag{2.5}$$

Of particular interest are the moments $\langle x^n \rangle_t$, where n is a positive integer. If we use $f(x) = x$ in (2.5), we find the following evolution equation for the first moment or average position,

$$\frac{d \langle x \rangle_t}{dt} = \langle A \rangle_t. \tag{2.6}$$

On average, the coefficient function A describes the velocity of the Brownian particle, say due to the presence of a gravitational or some other external field. This is a systematic, deterministic effect known as drift. The intuitive concept of velocity is given by the rate of change of the position resulting from the motion of a particle. Equation (2.6) expresses this idea on average. Equation (2.3) suggests a different concept of velocity in terms of the probability flux: velocity = (non-diffusive) probability flux/probability density. This alternative concept leads to a velocity field without referring to individual particle trajectories. If we consider $f(x) = x^2$ in (2.5), we further obtain

$$\frac{d \langle x^2 \rangle_t}{dt} = 2 \langle Ax \rangle_t + \langle D \rangle_t. \tag{2.7}$$

Equation (2.7) is the key to interpreting D. Even in the absence of systematic drift effects (that is, for $A = 0$), the second moment $\langle x^2 \rangle_t$ increases with the rate $\langle D \rangle_t$. We hence arrive at the interpretation of D as the diffusion coefficient which, by definition, governs the linear relationship between mean-square displacement and time.

In conclusion, the drift term characterized by A in (2.1) describes a deterministic motion with velocity A, whereas the diffusion term characterized by D describes stochastic smoothing or smearing effects. Systematic deterministic evolution is associated with first-order derivatives; stochastic smearing is associated with second-order derivatives. In the diffusion equation (2.1) and in the corresponding equation for averages (2.5), drift and diffusion are nicely separated.

2.3 Some Exact Solutions in Terms of Gaussians

For general drift and diffusion terms A and D, one cannot determine the first and second moments from the evolution equations (2.6) and (2.7) because the right-hand sides of these equations involve more complicated averages. However, for

$$A(t, x) = A_0(t) + A_1(t)x, \qquad D(t, x) = D_0(t) + D_2(t)[x - D_1(t)]^2, \quad (2.8)$$

the moment equations (2.6) and (2.7) for $\langle x \rangle_t$ and $\langle x^2 \rangle_t$ form a closed set of equations. For $D_0 \geq 0$ and $D_2 \geq 0$, the diffusion coefficient $D(t, x)$ is nonnegative for all x. In the following, we restrict ourselves to the case $D_2 = 0$, that is, the diffusion coefficient $D(t, x) = D_0(t)$ is independent of position but possibly dependent on time. We refer to this case as the linear problem (meaning linear in x; note that all diffusion equations (2.1) are linear in the probability density p). The evolution equations for the first and second moments are then given by

$$\frac{d \langle x \rangle_t}{dt} = A_1(t)\langle x \rangle_t + A_0(t), \tag{2.9}$$

and

$$\frac{d}{dt}\left(\langle x^2 \rangle_t - \langle x \rangle_t^2 \right) = 2A_1(t)\left(\langle x^2 \rangle_t - \langle x \rangle_t^2 \right) + D_0(t). \tag{2.10}$$

These equations can be solved according to the general theory of linear ordinary differential equations. We first consider the useful special case of A_1 being independent of time. The homogeneous equations obtained by assuming $A_0(t) = D_0(t) = 0$ are then solved by multiples of $e^{A_1 t}$ and $e^{2A_1 t}$, respectively. The solution of the full inhomogeneous equations is obtained by a method known as variation of constants, which gives

$$\alpha_t = \langle x \rangle_t = \langle x \rangle_0 e^{A_1 t} + \int_0^t e^{A_1(t-t')} A_0(t')dt', \tag{2.11}$$

and

$$\Theta_t = \langle x^2 \rangle_t - \langle x \rangle_t^2 = \left(\langle x^2 \rangle_0 - \langle x \rangle_0^2 \right) e^{2A_1 t} + \int_0^t e^{2A_1(t-t')} D_0(t') dt', \quad (2.12)$$

for the mean α_t and the variance Θ_t.

If $A_1(t)$ depends on time, we introduce the fundamental solution of the homogeneous equation (2.9) as

$$\Psi_t = \exp \left\{ \int_0^t A_1(t') dt' \right\}. \tag{2.13}$$

The explicit solutions of the moment equations (2.9) and (2.10) are then given by the following generalizations of (2.11) and (2.12),

$$\alpha_t = \Psi_t \left[\langle x \rangle_0 + \int_0^t \Psi_{t'}^{-1} A_0(t') dt' \right], \tag{2.14}$$

and

$$\Theta_t = \Psi_t^2 \left[\langle x^2 \rangle_0 - \langle x \rangle_0^2 + \int_0^t \Psi_{t'}^{-2} D_0(t') dt' \right]. \tag{2.15}$$

The explicit expressions (2.14) and (2.15) for the mean α_t and the variance Θ_t actually characterize the full probability density p which, for the linear problem with suitable initial conditions, is given by the Gaussian probability density

$$p_{\alpha_t \Theta_t}(x) = \frac{1}{\sqrt{2\pi \Theta_t}} \exp \left\{ -\frac{(x - \alpha_t)^2}{2\Theta_t} \right\}. \tag{2.16}$$

If $p_{\alpha_t \Theta_t}(x)$ is used as a trial solution in the diffusion equation (2.1) for the linear problem, an explicit calculation shows that the time dependence of the parameters α_t and Θ_t is governed by the equations (2.9) and (2.10) with the solutions (2.14) and (2.15) (see Exercise 2.1). This is plausible because the time derivative, the drift term, and the diffusion term evaluated for the Gaussian trial solutions are all quadratic polynomials times Gaussians, so that only the coefficients need to be matched. We have thus found the exact solution of the diffusion equation for the linear problem. Of course, if we wish to use Gaussian trial solutions, the initial condition must be of the Gaussian type.

The Gaussian solution of the linear problem with constant diffusion coefficient in the absence of a drift ($A = 0$, $D = 1$) is shown in Figure 2.3. Note that the variance $\Theta_t = t$ is proportional to time so that the width of the curve in Figure 2.3 scales with the square root of time. As an initial condition for the solution shown in Figure 2.3 we have assumed a Gaussian

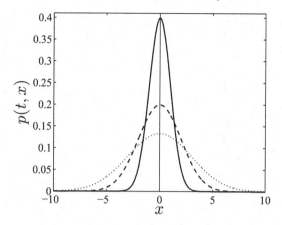

Figure 2.3 Gaussian solution of the diffusion equation (2.1) with drift $A = 0$ and diffusion coefficient $D = 1$ for $t = 1$ (continuous line), $t = 4$ (dashed line), and $t = 9$ (dotted line); the initial condition at $t = 0$ is assumed to be a δ function or a Gaussian of zero width located at the origin.

of infinitely narrow width, which may be considered as a regularization of Dirac's δ function or as a deterministic initialization at the origin.

Also the solution for non-Gaussian initial conditions can be found explicitly. In view of the linearity of the diffusion equation in p, its solution for arbitrary initial conditions can be found as a superposition of the Gaussian solutions obtained for deterministic initial conditions. These Gaussian solutions may be considered as transition probability densities from deterministic initial states and they must be superimposed according to the initial probability density. An example of such a superposition is shown in Figure 2.4. Again we consider the linear problem with constant diffusion coefficient and without drift ($A = 0$, $D = 1$). The initial probability density of the rectangular form is rounded off very rapidly. Whereas the transition probability densities are Gaussian, the time-dependent solution shown in Figure 2.4 clearly is not. It actually is the difference of two error functions. Only when the width of the curves becomes large compared with that of the initial distribution does the solution approach a Gaussian. The omnipresence of Gaussians is very natural because Brownian motion is the result of many independent collisions with small particles and the central limit theorem of probability theory hence assures Gaussian behavior.[3]

---

[3] A consequence of the central limit theorem is that the uncertainty of the average of n independent measurements scales as $1/\sqrt{n}$. The fluctuations around the average have a Gaussian distribution, which is reflected by the fact that the uncertainty associated with one standard deviation is 68.3%. For further discussion, see Section 2.8.2 of Gardiner, *Handbook of Stochastic Methods* (Springer, 1990).

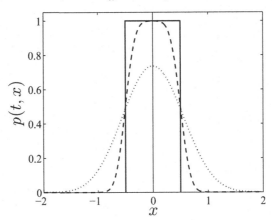

Figure 2.4 Solution of the diffusion equation (2.1) with drift $A = 0$ and diffusion coefficient $D = 1$ for $t = 0$ (continuous line), $t = 0.02$ (dashed line), and $t = 0.2$ (dotted line); the initial condition at $t = 0$ is chosen as a rectangular probability density.

Exercise 2.1   Verification of Gaussian Solutions
Calculate the time derivative, the drift term, and the diffusion term in the diffusion equation for the linear problem for the Gaussian trial solutions (2.16). Verify that the probability density $p_{\alpha_t \Theta_t}(x)$ defined in (2.16) solves the diffusion equation (2.1) for the linear problem, provided that the time dependence of the parameters α_t and Θ_t is governed by the equations (2.9) and (2.10).

Exercise 2.2   Rectangular Initial Condition
Show that the time-dependent curves in Figure 2.4 can be obtained as the difference of two error functions.

Exercise 2.3   Triangular Initial Condition
Produce the counterpart of Figure 2.4 for a triangular instead of rectangular initial probability density with corners at $(-1, 0)$, $(0, 1)$, and $(1, 0)$.

2.4 Equilibrium Probabilities

When the drift and diffusion coefficients are independent of time, we can look for steady-state solutions of the diffusion equation (2.1). For such equilibrium states, the probability flux (2.3) must vanish, that is,

$$J(x) = A(x)p(x) - \frac{1}{2}\frac{d}{dx}\Big[D(x)p(x)\Big] = 0. \qquad (2.17)$$

For constant D, we can rewrite this condition as

$$\frac{2A(x)}{D} = \frac{d}{dx}\ln p(x). \qquad (2.18)$$

If the drift term appearing on the left-hand side of (2.18) is expressed in the derivative form with a suitable dimensionless function $\Phi(x)$,

$$\frac{2A(x)}{D} = -\frac{d}{dx}\Phi(x), \qquad (2.19)$$

then one can integrate the equilibrium condition (2.18) to obtain the following simple exponential form of the probability density,

$$p(x) = p_{\text{eq}}(x) \propto e^{-\Phi(x)}. \qquad (2.20)$$

Equilibrium probabilities come in the form of Boltzmann factors, provided that the drift term is of the form (2.19). Normalizability of the probability density (2.20) requires that the Brownian particle is essentially restricted to a finite domain.

To generalize the above argument when D depends on x, instead of (2.17), one would prefer to write a probability flux of the form

$$J(x) = \bar{A}(x)p(x) - \frac{1}{2}D(x)\frac{dp(x)}{dx}. \qquad (2.21)$$

In the diffusion equation (2.1), $D(x)$ would then be positioned between the two space derivatives rather than behind them. Such a rearrangement is always possible if we are willing to modify the drift term as follows,

$$\bar{A}(x) = A(x) - \frac{1}{2}\frac{dD(x)}{dx}. \qquad (2.22)$$

If the representation (2.19) of the modified drift term is generalized to

$$\bar{A}(x) = -\frac{1}{2}D(x)\frac{d}{dx}\Phi(x), \qquad (2.23)$$

the diffusion equation can be rewritten in the alternative form

$$\frac{\partial p(t,x)}{\partial t} = \frac{1}{2}\frac{\partial}{\partial x}\left\{p(t,x)D(x)\frac{\partial}{\partial x}\left[\Phi(x) + \ln p(t,x)\right]\right\}, \qquad (2.24)$$

suggesting the general flux expression

$$\begin{aligned}
J(t,x) &= -\frac{1}{2}p(t,x)D(x)\frac{\partial}{\partial x}\left[\Phi(x) + \ln p(t,x)\right] \\
&= -\frac{1}{2}p(t,x)D(x)\frac{\partial}{\partial x}\ln\left(\frac{p(t,x)}{p_{\text{eq}}(x)}\right).
\end{aligned} \qquad (2.25)$$

Let us try to develop some deeper insight into the significance of the equations of this section. We hence conclude with three observations.

- As $\Phi(x)$ is a dimensionless potential, (2.23) states that the modified drift velocity is proportional to a force. With the idea "force = friction coefficient $\times$ velocity" in mind, we realize that the friction coefficient should be inversely proportional to the diffusion coefficient. As a diffusion coefficient describes fluctuations and a friction coefficient implies energy dissipation, this is our first encounter with a *fluctuation–dissipation relation*.
- Equation (2.25) can be regarded as a constitutive equation for the probability flux, in which $\ln(p/p_{eq})$ plays a special role. The same logarithm appears as a factor in the relative entropy used as a non-symmetric measure of the difference between the probability densities p and p_{eq} in information theory: $-\int p \ln(p/p_{eq})dx$.
- Different but equivalent ways of writing the diffusion equation can be natural and convenient for different purposes. The original formulation (2.1) with D behind the derivatives leads to simple relations between the coefficients A and D on the one hand and the first and second moments on the other hand. If D is positioned between the derivatives, the discussion of equilibrium states becomes simpler.

Exercise 2.4   *Entropy Production and Sign of Diffusion Coefficient*
Derive the following formula for the entropy production:

$$\frac{d}{dt}\left[-\int p\ln(p/p_{eq})dx\right] = \frac{1}{2}\int pD\left[\frac{\partial}{\partial x}\ln(p/p_{eq})\right]^2 dx. \qquad (2.26)$$

As the sign of the local contribution to the entropy production is given by the sign of D, we conclude that the condition $D \geq 0$ must be imposed.

2.5 Eigenfunction Methods

The diffusion equation (2.1) is a linear time-evolution equation for the probability density $p(t,x)$. If we consider (2.1) as a set of time-evolution equations for $p(t,x)$ with a label x, all the equations for different x are coupled through the derivative operators on the right-hand side of (2.1) so that the problem is nontrivial in spite of its linearity. In fact, the problem is as difficult as solving the linear Schrödinger equation of quantum mechanics, where one needs to find properly normalized eigenfunctions of the Hamiltonian for given boundary conditions. Actually, even many of the methods for solving transport problems are closely related to the methods known from quantum mechanics: in particular, we use eigenfunction methods, separation of variables, and perturbation theory.

To eliminate the coupling of linear equations by derivative operators, we introduce Fourier transforms because they are known to turn derivative operators into multiplications. We hence introduce the Fourier representation[4]

$$p(t, x) = \frac{1}{2\pi} \int_{-\infty}^{\infty} e^{ikx} \tilde{p}(t, k) dk. \tag{2.27}$$

If there is no further occurrence of x in the coefficient functions A and D, substitution of the Fourier representation (2.27) into the diffusion equation (2.1) leads to the nicely decoupled equations

$$\frac{\partial \tilde{p}(t, k)}{\partial t} = \left[-ikA(t) - \frac{1}{2}k^2 D(t) \right] \tilde{p}(t, k), \tag{2.28}$$

one separate equation for each value of k. The drift term $A(t)$ causes oscillations; the diffusion term $D(t)$ implies decay with the rate $k^2 D(t)/2$. For constant D, the solution of (2.28) contains a single exponential decay factor with rate $k^2 D/2$. Small-scale features characterized by large values of k decay rapidly; large-scale features characterized by small values of k decay slowly. The k^2 dependence of the decay rate is a characteristic of dissipative transport processes. A countable set of eigenfunctions and decay rates is usually singled out by boundary conditions.

When the coefficient functions A and D in the diffusion equation (2.1) depend on x, it becomes much more complicated to find the eigenfunctions and decay rates. The case where $A(x)$ and $D(x)$ are independent of t has been discussed by Gardiner.[5] One then looks for functions $p_n(x)$ solving the equation

$$-\frac{\partial}{\partial x} \left[A(x) p_n(x) \right] + \frac{1}{2} \frac{\partial^2}{\partial x^2} \left[D(x) p_n(x) \right] = -\lambda_n p_n(x) \tag{2.29}$$

so that $p_n(x)$ (for $n = 0, 1, 2, \ldots$) is an eigenfunction of the second-order differential operator occurring in the diffusion equation and $-\lambda_n$ is the corresponding eigenvalue. Note that $p_n(x) e^{-\lambda_n t}$ is a solution of the diffusion equation (2.1), where the variables x and t occur in a separated form as arguments of two different factors. The discrete nature of the spectrum is

---

[4] The Fourier transform of the function $f(t, x)$ and its derivative $\partial f / \partial x$ are given by

$$\int_{-\infty}^{\infty} f(t, x) e^{-ikx} dx = \tilde{f}(t, k), \qquad \int_{-\infty}^{\infty} \frac{\partial f(t, x)}{\partial x} e^{-ikx} dx = ik \tilde{f}(t, k),$$

where $i = \sqrt{-1}$. The second result is obtained using integration by parts with $f(t, \pm\infty) = 0$.

[5] See Section 5.2.5 of Gardiner, *Handbook of Stochastic Methods* (Springer, 1990).

a consequence of the boundary conditions, where one often considers those of the reflecting or absorbing type, to be satisfied by each $p_n(x)$. If the boundary conditions admit the equilibrium probability density $p_{eq}(x)$ of Section 2.4, it is an eigenfunction with eigenvalue 0. All other eigenvalues must be positive[6] so that the equilibrium state is the stable long-time solution. This search for eigenfunctions is completely analogous to the passage from the time-dependent Schrödinger equation to the time-independent eigenvalue problem for the Hamilton operator in quantum mechanics.

The most general solution of the diffusion equation for given boundary conditions is given by weighted sums $\sum_n c_n p_n(x) e^{-\lambda_n t}$ with coefficients c_n depending on the initial conditions. In particular, we can write the transition probability density $p(t, x|0, x_0)$ from the initial state x_0 at $t = 0$ to the state x at time t as

$$p(t, x|0, x_0) = \sum_n c_n(x_0) p_n(x) e^{-\lambda_n t}, \qquad (2.30)$$

where the coefficients $c_n(x_0)$ in the eigenfunction representation (2.30) still need to be determined. As a highlight of the development in Section 5.2.5 of Gardiner, a completeness relation for the eigenfunctions $p_n(x)$ is used to find the coefficient functions $c_n(x_0) = p_n(x_0)/p_{eq}(x_0)$. As a consequence, we realize that the joint equilibrium distribution,

$$p(t, x|0, x_0) p_{eq}(x_0) = \sum_n p_n(x_0) p_n(x) e^{-\lambda_n t}, \qquad (2.31)$$

is symmetric in x_0 and x, and we obtain a very elegant expression for the two-time correlation function of a quantity $Q(x)$ at equilibrium, which is also known as the autocorrelation function of Q,

$$R_Q(t) = \int_{-\infty}^{\infty} \int_{-\infty}^{\infty} Q(x) p(t, x|0, x_0) Q(x_0) p_{eq}(x_0) dx\, dx_0$$

$$= \sum_n \left[\int_{-\infty}^{\infty} Q(x) p_n(x) dx \right]^2 e^{-\lambda_n t}. \qquad (2.32)$$

All autocorrelation functions decay with the same spectrum of relaxation times, but the weights of the different contributions depend on the overlap integral of the particular quantity Q with the respective eigenfunctions.

We will repeatedly use eigenfunction methods throughout the book. For example, they are used to solve a master equation for molecular motors (see Section 24.1) and the equations of linearized hydrodynamics (see Section 26.3). Also the separation of time and space variables, which we will

---

[6] More generally, these eigenvalues must have a positive real part.

encounter in a number of specific transport problems (see, for example, Sections 7.2, 7.3, and 7.4), is intimately related to eigenfunction methods.

Exercise 2.5   *Toy Model of Drug Release*
Solve the diffusion equation (2.1) with drift $A = 0$ and diffusion coefficient $D = 1$ on the interval $[0, 1]$ for a uniform initial condition $p(0, x) = 1$ and the boundary conditions $p(t, 0) = p(t, 1) = 0$ by means of an eigenfunction method. The result describes the diffusive release of a substance from a confined space; the boundary conditions express the efficient removal of the substance when it reaches the boundary. Calculate the fraction of the substance released as a function of time.

2.6 Increasing Dimensionality

So far, we have considered diffusion in one dimension only. If we are interested in describing Brownian motion in three space dimensions then we need a multivariate generalization of the diffusion equation (2.1). The natural generalization of (2.1) to d dimensions is

$$\frac{\partial p(t, \boldsymbol{x})}{\partial t} = -\sum_{j=1}^{d} \frac{\partial}{\partial x_j} \Big[A_j(t, \boldsymbol{x}) p(t, \boldsymbol{x}) \Big] + \frac{1}{2} \sum_{j,k=1}^{d} \frac{\partial^2}{\partial x_j \, \partial x_k} \Big[D_{jk}(t, \boldsymbol{x}) p(t, \boldsymbol{x}) \Big],$$

(2.33)

where the position vector $\boldsymbol{x}$ has d components and the coefficient functions A_j and D_{jk} are the components of a d-dimensional drift vector and diffusion tensor, respectively.[7] Only the symmetric part of the diffusion tensor matters in (2.33), so we can assume the symmetry property $D_{jk} = D_{kj}$. The generalization of the condition $D \geq 0$ is to assume that D_{jk} represents a positive semidefinite tensor.[8] In a more elegant and compact vector–tensor notation, the multivariate diffusion equation (2.33) can be rewritten as

$$\frac{\partial p(t, \boldsymbol{x})}{\partial t} = -\frac{\partial}{\partial \boldsymbol{x}} \cdot \Big[\boldsymbol{A}(t, \boldsymbol{x}) p(t, \boldsymbol{x}) \Big] + \frac{1}{2} \frac{\partial}{\partial \boldsymbol{x}} \frac{\partial}{\partial \boldsymbol{x}} : \Big[\boldsymbol{D}(t, \boldsymbol{x}) p(t, \boldsymbol{x}) \Big]. \quad (2.34)$$

Comparing the last two equations, we see that a single dot ($\cdot$) indicates a single contraction in (2.34), which is expressed as a summation over the index j in (2.33); a double dot ($:$) indicates a double contraction in (2.34), which is expressed as a double summation over the indices j and k in (2.33).

As in the special case of one dimension, the diffusion equation (2.34)

---

[7] In general, $\boldsymbol{x}$ is a d-dimensional vector of independent variables that here is position. In later chapters, when referring to three-dimensional spatial position, we use $\boldsymbol{r}$ (cf. footnote on p. 58).

[8] An $d \times d$ symmetric tensor $\boldsymbol{D}$ is positive semidefinite if $\boldsymbol{a}^{\mathrm{T}} \cdot \boldsymbol{D} \cdot \boldsymbol{a} \geq 0$ for all d-component vectors $\boldsymbol{a}$. The conditions $D_{ii} \geq 0$ and $\det[D_{ij}] \geq 0$ are necessary but, in general, not sufficient for $\boldsymbol{D}$ to be a positive semidefinite tensor. Only for $d = 1, 2$ are they also sufficient.

describes the flux of probability resulting from drift and diffusion. It can be written in the divergence form

$$\frac{\partial p(t, \boldsymbol{x})}{\partial t} = -\frac{\partial}{\partial \boldsymbol{x}} \cdot \boldsymbol{J}(t, \boldsymbol{x}), \qquad (2.35)$$

with the probability flux vector

$$\boldsymbol{J}(t, \boldsymbol{x}) = \boldsymbol{A}(t, \boldsymbol{x}) p(t, \boldsymbol{x}) - \frac{1}{2} \frac{\partial}{\partial \boldsymbol{x}} \cdot \left[\boldsymbol{D}(t, \boldsymbol{x}) p(t, \boldsymbol{x}) \right]. \qquad (2.36)$$

In order to relate the rate of change of the probability in a finite volume to the flux through its surface one now needs to apply Gauss's divergence theorem to obtain the proper generalization of (2.2). It is natural that the drift contribution to the flux of probability is of the form "drift velocity vector times probability density." By means of the product rule for the drift term in the diffusion equation (2.34),

$$-\frac{\partial}{\partial \boldsymbol{x}} \cdot \left[\boldsymbol{A}(t, \boldsymbol{x}) p(t, \boldsymbol{x}) \right] = -\boldsymbol{A}(t, \boldsymbol{x}) \cdot \frac{\partial}{\partial \boldsymbol{x}} p(t, \boldsymbol{x}) - p(t, \boldsymbol{x}) \frac{\partial}{\partial \boldsymbol{x}} \cdot \boldsymbol{A}(t, \boldsymbol{x}), \quad (2.37)$$

one can distinguish the convection of probability with velocity vector $\boldsymbol{A}$ and the density effect due to the volume change associated with the divergence of $\boldsymbol{A}$. In describing the transport of a scalar rather than a scalar density, the latter term would be absent because a scalar is affected only by displacement but not by volume changes. Placing $\boldsymbol{A}$ before or behind the derivative makes the difference between convecting a scalar or a scalar density. The diffusive flux contribution in (2.36) counteracts the gradient of the probability density and hence smoothes it. Again, this thermodynamic effect becomes more transparent if the diffusion tensor $\boldsymbol{D}$ stands before the derivative in the flux expression (2.36), that is, between the derivatives in the diffusion equation (2.34).

The multivariate generalization of the evolution equation (2.5) for averages is given by

$$\frac{d \langle f \rangle_t}{dt} = \left\langle \boldsymbol{A} \cdot \frac{\partial f}{\partial \boldsymbol{x}} \right\rangle_t + \frac{1}{2} \left\langle \boldsymbol{D} : \frac{\partial^2 f}{\partial \boldsymbol{x} \, \partial \boldsymbol{x}} \right\rangle_t, \qquad (2.38)$$

and the generalizations of expressions (2.14) and (2.15) for the mean and variance of the Gaussian solution to the linear problem can be found in Section 3.3.2 of Öttinger (1996).[9] Instead of the factors Ψ_t^2 and $\Psi_{t'}^{-2}$ in (2.15), there occur two factors of Ψ_t and $\Psi_{t'}^{-1}$ from the left and from the right, where the solution Ψ_t of the homogeneous equation is given by a time-ordered

---

[9] Öttinger, *Stochastic Processes in Polymeric Fluids* (Springer, 1996).

matrix exponential[10] (see Exercise 2.6). The corresponding multivariate Gaussian probability density with mean vector $\boldsymbol{\alpha}$ and positive definite covariance matrix $\boldsymbol{\Theta}$ is given by

$$p_{\boldsymbol{\alpha\Theta}}(\boldsymbol{x}) = \frac{1}{\sqrt{(2\pi)^d \det(\boldsymbol{\Theta})}} \exp\left\{-\frac{1}{2}(\boldsymbol{x} - \boldsymbol{\alpha}) \cdot \boldsymbol{\Theta}^{-1} \cdot (\boldsymbol{x} - \boldsymbol{\alpha})\right\}. \qquad (2.39)$$

Whereas the representation (2.19) of the drift term is always possible in one dimension, the existence of a potential for the drift term in $d > 1$ dimensions is restrictive because it implies integrability conditions. Otherwise, the ideas and arguments of Section 2.4 can be generalized in a straightforward way to the multivariate case. We recommend that the reader spells out all the multivariate equations to become familiar with the notation and to repeat the basic notions for describing transport phenomena. The reader should also look back at the diffusion equation (2.1) and try to see much more in it than a second-order partial differential equation. Retell the story of drift, diffusion, probability flux, equilibrium solutions, transition probability densities, and the qualitative behavior of time-dependent solutions.

Exercise 2.6   *Time-Ordered Matrix Exponential*
Write down the third-order Taylor expansion of time-ordered exponential $\mathcal{T} - \exp\{\boldsymbol{M}(t_1) + \boldsymbol{M}(t_2)\}$ for two non-commuting matrices $\boldsymbol{M}(t_1)$ and $\boldsymbol{M}(t_2)$, where $t_2 > t_1$. What is the difference compared with $\exp\{\boldsymbol{M}(t_1) + \boldsymbol{M}(t_2)\}$?

Exercise 2.7   *Verification of Multivariate Gaussian Solutions*
Practice the vector/matrix and/or index notation by generalizing Exercise 2.1 to the multivariate case.

Exercise 2.8   *A Concrete Multivariate Gaussian*
Consider the multivariate Gaussian distribution $p_{\boldsymbol{\alpha\Theta}}(\boldsymbol{x})$ given in (2.39) for the special case of two dimensions, $\boldsymbol{x} = (x_1, x_2)$, with $\boldsymbol{\alpha} = \boldsymbol{0}$ and the 2×2 matrix $\boldsymbol{\Theta}$ given by $\Theta_{11} = 0.4$, $\Theta_{22} = 0.6$, and $\Theta_{12} = \Theta_{21} = 0.3$. Use Mathematica® or MATLAB® to produce a three-dimensional plot of $p_{\boldsymbol{\alpha\Theta}}(\boldsymbol{x})$. Use the "rotate" option to view the probability density from different angles.

Exercise 2.9   *Factorization of Multivariate Gaussians*
Explain why, in suitable coordinates, any Gaussian probability density in d dimensions is a product of d one-dimensional Gaussians.

---

[10] Exponentials of matrices can be introduced through Taylor expansions; in a time-ordered exponential, time-dependent factors are ordered such that the time arguments decrease from left to right.

- The diffusion equation for a probability density is a partial differential equation involving first and second derivatives, intuitively associated with drift and diffusion, respectively.
- As probability is conserved, it is natural to introduce a probability flux such that the rate of change of the probability density is given by the divergence of the flux; this observation is the essence of local conservation laws.
- Linear diffusion equations on unbounded domains possess Gaussian solutions, which are fully characterized in terms of first and second moments.
- Diffusion equations are linear in the probability density and can hence be analyzed with the eigenfunction methods known from quantum mechanics.

3

Brownian Dynamics

We have already learned that smooth probability densities and irregular trajectories of Brownian particles are two sides of the same coin. The evolution of the probability densities illustrated in Figures 2.3 and 2.4 is governed by a diffusion equation. Is there a corresponding equation governing the irregular trajectories of the type shown in Figure 2.1? Can a general equivalence between such evolution equations and diffusion equations be established? Can one use such equations to obtain simulation methods for solving diffusion equations? The answers to these questions are given by the theory of stochastic differential equations and by the idea of Brownian dynamics simulations, as we elaborate in this chapter. We hence encounter a first numerical method for solving transport equations on a computer. But we gain much more than this because stochastic differential equations offer a different viewpoint and additional insight into the physical content of diffusion equations.

3.1 Stochastic Difference Equations

With the goal of developing computer simulations in mind, it is most convenient to introduce stochastic differential equations by means of their time-discretized versions, that is, stochastic difference equations. We consider a time interval $[t_0, t]$ and a mesh of points t_j such that $t_0 < t_1 < t_2 < \ldots < t_M = t$. A series of random column vectors $\boldsymbol{X}_j$, each of which consists of d real random variables, is constructed by the following iterative scheme,

$$\boldsymbol{X}_{j+1} = \boldsymbol{X}_j + \boldsymbol{A}(t_j, \boldsymbol{X}_j)\Delta t_j + \boldsymbol{B}(t_j, \boldsymbol{X}_j) \cdot \boldsymbol{W}_j \sqrt{\Delta t_j}, \qquad (3.1)$$

where the functions $\boldsymbol{A}(t, \boldsymbol{x})$ and $\boldsymbol{B}(t, \boldsymbol{x})$ are d-component column vectors and $d \times d'$ matrices, respectively, $\Delta t_j = t_{j+1} - t_j$, and the column vectors $\boldsymbol{W}_j$ consist of d' independent Gaussian random variables with zero first and unit second moments,

$$\langle \boldsymbol{W}_j \rangle = \boldsymbol{0}, \qquad \langle \boldsymbol{W}_j \boldsymbol{W}_k \rangle = \delta_{jk} \boldsymbol{\delta}. \tag{3.2}$$

The averages indicated by angle brackets are multivariate generalizations of the definition (2.4), where the integrals are performed with the Gaussian distributions of $\boldsymbol{W}_j$ and the Gaussian joint distributions of $\boldsymbol{W}_j$ and $\boldsymbol{W}_k$. The Kronecker δ_{jk} implies that, for different time steps, the Gaussian random vectors are independent, and the unit matrix $\boldsymbol{\delta}$ implies that also the d' different components of each random vector $\boldsymbol{W}_j$ are independent. The random vectors $\boldsymbol{X}_j$ are meant to represent a stochastic process at the discrete times t_j. For a given initial condition $\boldsymbol{X}_0$, which is assumed to be stochastically independent from the Gaussian random vectors $\boldsymbol{W}_j$, the difference equation (3.1) can readily be used to construct discrete random trajectories on a computer.

For $\boldsymbol{B} = \boldsymbol{0}$, the iterative scheme (3.1) coincides with the explicit Euler integration scheme for the deterministic ordinary differential equation $d\boldsymbol{X}/dt = \boldsymbol{A}(t, \boldsymbol{X})$. For $\boldsymbol{B} \neq \boldsymbol{0}$, a stochastic noise term with amplitude $\boldsymbol{B}$ is added. To understand the occurrence of $\sqrt{\Delta t_j}$ in the noise term, which has some far-reaching consequences to be revealed in the remainder of this chapter, we consider the special case $d = d' = 1$, $A = 0$, $B = 1$, and $X_{t_0} = 0$ with the explicit Gaussian solution $X_t = W_0 \sqrt{\Delta t_0} + \cdots + W_{M-1} \sqrt{\Delta t_{M-1}}$. The average of X_t vanishes and the variance $\langle X_t^2 \rangle = \Delta t_0 + \cdots + \Delta t_{M-1} = t - t_0$ of this sum of independent random numbers is independent of the choice of the time steps; the occurrence of $\sqrt{\Delta t_j}$ is crucial for the latter property.

How does the average of an arbitrary function $f(\boldsymbol{X}_j)$ change with time? In view of the occurrence of $\sqrt{\Delta t_j}$ in (3.1), we need a second-order Taylor expansion for the change of the function during a time step,

$$f(\boldsymbol{X}_{j+1}) = f(\boldsymbol{X}_j) + \frac{\partial f(\boldsymbol{X}_j)}{\partial \boldsymbol{X}_j} \cdot \left[\boldsymbol{A}(t_j, \boldsymbol{X}_j) \Delta t_j + \boldsymbol{B}(t_j, \boldsymbol{X}_j) \cdot \boldsymbol{W}_j \sqrt{\Delta t_j} \right]$$

$$+ \frac{1}{2} \frac{\partial^2 f(\boldsymbol{X}_j)}{\partial \boldsymbol{X}_j \partial \boldsymbol{X}_j} : \left[\boldsymbol{B}(t_j, \boldsymbol{X}_j) \cdot \boldsymbol{W}_j \boldsymbol{W}_j \cdot \boldsymbol{B}^{\mathrm{T}}(t_j, \boldsymbol{X}_j) \Delta t_j \right]. \tag{3.3}$$

By averaging this equation, we obtain

$$\langle f(\boldsymbol{X}_{j+1}) \rangle = \langle f(\boldsymbol{X}_j) \rangle$$

$$+ \left\langle \frac{\partial f(\boldsymbol{X}_j)}{\partial \boldsymbol{X}_j} \cdot \left[\boldsymbol{A}(t_j, \boldsymbol{X}_j) \Delta t_j + \boldsymbol{B}(t_j, \boldsymbol{X}_j) \cdot \langle \boldsymbol{W}_j \rangle \sqrt{\Delta t_j} \right] \right\rangle$$

$$+ \frac{1}{2} \left\langle \frac{\partial^2 f(\boldsymbol{X}_j)}{\partial \boldsymbol{X}_j \partial \boldsymbol{X}_j} : \left[\boldsymbol{B}(t_j, \boldsymbol{X}_j) \cdot \langle \boldsymbol{W}_j \boldsymbol{W}_j \rangle \cdot \boldsymbol{B}^{\mathrm{T}}(t_j, \boldsymbol{X}_j) \Delta t_j \right] \right\rangle,$$

$$\tag{3.4}$$

where we have used the linearity of the averaging procedure (the average of the sum of random variables is the sum of averages, and the average of a deterministic multiple of a random variable is the multiple of the average) as well as the factorization of averages for products of independent random variables. By using the moments (3.2) and going to the limit of small time step Δt_j, we obtain

$$\frac{d \langle f \rangle}{dt} = \left\langle \boldsymbol{A} \cdot \frac{\partial f}{\partial \boldsymbol{x}} \right\rangle + \frac{1}{2} \left\langle \boldsymbol{D} : \frac{\partial^2 f}{\partial \boldsymbol{x} \, \partial \boldsymbol{x}} \right\rangle, \tag{3.5}$$

where we have introduced the $d \times d$ matrix

$$\boldsymbol{D}(t, \boldsymbol{x}) = \boldsymbol{B}(t, \boldsymbol{x}) \cdot \boldsymbol{B}^{\mathrm{T}}(t, \boldsymbol{x}), \tag{3.6}$$

and the superscript T indicates transposition. Note that (3.5) coincides with (2.38), which is wonderful news. The stochastic generalization (3.1) of the Euler integration scheme gives us the possibility to calculate the time-dependence of all averages performed with the solution of the multivariate diffusion equation (2.34) from the random variables $\boldsymbol{X}_j$. Therefore, we have established a popular general simulation technique known as Brownian dynamics. And the good news continues. There actually is a mathematically profound way of introducing stochastic differential equations such that (3.1) can be recovered as a numerical integration scheme, as discussed in the following section.

3.2 Stochastic Differential Equations

The stochastic differential equation associated with the difference equation (3.1) is often written in the symbolic form

$$d\boldsymbol{X}_t = \boldsymbol{A}(t, \boldsymbol{X}_t)dt + \boldsymbol{B}(t, \boldsymbol{X}_t) \cdot d\boldsymbol{W}_t. \tag{3.7}$$

How can the solution $\boldsymbol{X}_t$ of the stochastic differential equation (3.7) be defined in a mathematically rigorous way? As the axioms of probability theory are chosen such that at most a countably infinite number of random variables can be handled, the definition of a continuous stochastic process $\boldsymbol{X}_t$ solving (3.7) has to rely on limits of discretizations. Stochastic calculus is hence developed in a way that is very similar to the difference equation approach in (3.1).

The stochastic limits can be performed only on the level of random variables, not on the level of individual trajectories. From the difference equation (3.1), we see that the deterministic and stochastic increments are of order Δt_j and $\sqrt{\Delta t_j}$, respectively. For small Δt_j, both increments are small, so we expect continuous trajectories for the solutions of stochastic differential equations. However, if we divide by Δt_j to evaluate the time derivative, we encounter $1/\sqrt{\Delta t_j}$ in the contribution from the stochastic increments so that, in the continuous time limit, the trajectories develop infinite slope and cannot be differentiable. This is the origin of the "wildness" of the trajectories that we have already observed in Figure 2.1, and prevents us from having a trajectorywise definition of solutions. This is also the reason why we have not divided by dt in the notation (3.7) for stochastic differential equations. In view of this non-differentiability, we cannot define a particle velocity for the trajectory of a Brownian particle illustrated in Figure 2.1. Only the average particle velocity and the velocity field associated with the probability flux exist, as discussed after (2.6).

Not only is there a rigorous mathematical approach to stochastic calculus, but also there are powerful theorems for the existence and uniqueness of solutions to stochastic differential equations. As in the theory of deterministic differential equations, the Lipschitz continuity conditions

$$|\boldsymbol{A}(t, \boldsymbol{x}) - \boldsymbol{A}(t, \boldsymbol{y})| \le c|\boldsymbol{x} - \boldsymbol{y}|, \tag{3.8}$$

$$|\boldsymbol{B}(t, \boldsymbol{x}) - \boldsymbol{B}(t, \boldsymbol{y})| \le c|\boldsymbol{x} - \boldsymbol{y}|, \tag{3.9}$$

and the linear growth conditions

$$|\boldsymbol{A}(t, \boldsymbol{x})| \le c(1 + |\boldsymbol{x}|), \tag{3.10}$$

$$|\boldsymbol{B}(t, \boldsymbol{x})| \le c(1 + |\boldsymbol{x}|), \tag{3.11}$$

for some constant c are sufficient to guarantee the existence of a unique solution to (3.7).[1] Whereas these conditions may look familiar in the theory of ordinary differential equations, through the equivalence of stochastic differential equations and diffusion equations, they are also useful to guarantee the existence of unique solutions in the theory of partial differential equations of the Fokker–Planck type (2.34).

There is also a mathematical theory of numerical integration schemes for stochastic differential equations.[2] The simplest integration scheme, which is given by (3.1), is known as the Euler–Maruyama scheme. According to the central limit theorem, the Gaussian random numbers in (3.1) can be replaced by computationally less expensive random numbers with the same first and second moments. Implicit, semi-implicit, or higher-order schemes

---

[1] For a detailed discussion and useful references, see Section 3.3.1 of Öttinger, *Stochastic Processes in Polymeric Fluids* (Springer, 1996).

[2] Kloeden & Platen, *Numerical Solution of SDEs* (Springer, 1992); Öttinger, *Stochastic Processes in Polymeric Fluids* (Springer, 1996).

cannot be generalized from deterministic to stochastic differential equations as easily as the explicit Euler scheme, as discussed in the following section.

3.3 Itô versus Stratonovich

Assume that we would like to improve the stability of the Euler–Maruyama scheme (3.1) for stochastic differential equations. In order to do so, we might be tempted to borrow the implicit Euler scheme from the theory of deterministic differential equations,

$$X_{j+1} = X_j + \frac{1}{2}\Big[A(t_j, X_j) + A(t_{j+1}, X_{j+1})\Big]\Delta t_j$$
$$+ \frac{1}{2}\Big[B(t_j, X_j) + B(t_{j+1}, X_{j+1})\Big] \cdot W_j \sqrt{\Delta t_j}. \quad (3.12)$$

We would now like to repeat the steps that led us from (3.3) to (3.5). However, X_{j+1} is not independent of W_j. An additional term of order Δt_j hence arises from the Taylor expansion

$$B(t_{j+1}, X_{j+1}) = B(t_j, X_j) + W_j \sqrt{\Delta t_j} \cdot B^{\mathrm{T}}(t_j, X_j) \cdot \frac{\partial}{\partial x} B(t_j, X_j). \quad (3.13)$$

If we calculate the evolution of averages resulting from the implicit scheme (3.12) with the replacement (3.13) in exactly the same way as we did in (3.3) for the explicit scheme, we obtain the result

$$\frac{d\langle f\rangle}{dt} = \left\langle A \cdot \frac{\partial f}{\partial x}\right\rangle + \frac{1}{2}\left\langle \left(B^{\mathrm{T}} \cdot \frac{\partial}{\partial x}\right) \cdot \left(B^{\mathrm{T}} \cdot \frac{\partial}{\partial x}\right) f\right\rangle. \quad (3.14)$$

The evolution equations (3.5) and (3.14) coincide only if $B(t, x)$ is independent of x, a case known as *additive noise*. If $B(t, x)$ depends on x, that is, for *multiplicative noise*, the explicit and implicit Euler schemes lead to different results for the averages. This surprising difference compared with deterministic differential equations is a consequence of the anticipating character of $B(t_{j+1}, X_{j+1})$ in (3.12), where the random variable X_{j+1} in the prefactor of W_j depends already on W_j.

The nonanticipating integrator (3.1) corresponds to Itô calculus, whereas the anticipating integrator (3.12) corresponds to Stratonovich calculus. These two different versions of stochastic calculus appear frequently in the literature on stochastic differential equations. Stratonovich calculus is popular among physicists because it leads to differentiation rules that look like those of deterministic calculus, but the nonanticipating character of Itô calculus makes it more appropriate for rigorous proofs and mathematical developments.

Note that the difference between (3.5) and (3.14) is a term in which one derivative acts on $\boldsymbol{B}^{\mathrm{T}}$ and the other derivative acts on f. This term can be compensated for by modifying the drift term. Therefore, implicit integrators can be constructed even for stochastic differential equations with multiplicative noise, provided that proper care is taken in modifying the drift term.

3.4 Boundary Conditions

So far, we have not paid any attention to boundary conditions. What happens, for example, to a diffusing particle hitting a wall? On the level of stochastic trajectories, the two most natural possibilities are (i) that the diffusing particle sticks to the wall, or (ii) that it is simply reflected back from the wall. We will recognize in the subsequent section that "impenetrable" is a better description than "reflecting."

What are the corresponding boundary conditions on the level of probability densities? If diffusing particles stick to the wall, the neighborhood of the wall must be depleted of particles because wild fluctuations (see Figure 2.1) would otherwise have made nearby particles stick to the wall. The absorbing boundary condition (i) hence corresponds to zero probability at the wall. For a reflecting wall, the flux of particles to the wall must be equal to the flux of particles reflected back from the wall. The reflecting boundary condition (ii) hence corresponds to a vanishing normal component of the flux (2.36) at the wall. In short, the equivalence of diffusion equations and stochastic differential equations is thus supplemented by an equivalence of boundary conditions for absorbing and reflecting boundaries characterized by the following conditions to be imposed at the boundary,

$$p = 0 \quad \text{for absorbing boundaries,} \tag{3.15}$$

$$\boldsymbol{n} \cdot \boldsymbol{A}p = \frac{1}{2}\boldsymbol{n}\frac{\partial}{\partial \boldsymbol{x}} : (\boldsymbol{D}p) \quad \text{for reflecting boundaries,} \tag{3.16}$$

where the vector $\boldsymbol{n}$ is perpendicular to the boundary.

We have now established the full equivalence of diffusion equations and stochastic differential equations. The stochastic evolution equations may be considered as the simplified molecular or kinetic theory behind the phenomenologically introduced diffusion equations. However, we here consider the stochastic approach merely to obtain a simulation technique for solving diffusion equations. Kinetic theory will be considered only in later chapters.

3.5 Single-Particle Diffusion in One Dimension

To illustrate how Brownian dynamics actually works in all detail, we consider a single diffusing particle in one space dimension. We first revisit the problem with constant diffusion coefficient and without drift ($A = 0$, $D = 1$) for an initial probability density of the rectangular form, which we had previously studied in Section 2.3. More precisely, we try to reproduce the curve for $t = 0.02$ in Figure 2.4 by simulation. The corresponding special case of the stochastic difference equation (3.1) reads

$$X_{j+1} = X_j + W_j\sqrt{\Delta t},\qquad(3.17)$$

where we have assumed a constant time step Δt.

The core of our simulation program consists of the following steps. We initialize NTRA trajectories by choosing initial values according to a uniform distribution over the interval $[-0.5, 0.5]$. The trajectories are then evolved according to (3.1) for $A = 0$ and $B = 1$ in one dimension, where we choose the small constant time steps $\Delta t_j = 0.01$ (parameter DT). Of course, one might argue that this time step is not sufficiently small because the number of time steps NTIME required to reach the final time $t = 0.02$ is only 2. However, the arguments given in Section 3.1 to explain the occurrence of $\sqrt{\Delta t_j}$ in the noise term of (3.1) imply that, in our simple example, the result is independent of the time step. From the trajectories at $t = 0.02$, we construct a histogram in the range from XMIN to XMAX with steps DX; we cover the range from -1 to 1 with bins of width 0.05. We actually repeat the construction of NTRA trajectories NHIST ($= 100$) times and collect the properly normalized histograms as the rows of a matrix p.

A simple statistical analysis of the ensemble of normalized histograms allows us to calculate the average and error bars for the probability density (simulation results without error bars should be recognized as rather useless!). These simulation ideas are implemented in the following MATLAB® code:

```
% Simulation parameters
NTRA=1000; NTIME=2; NHIST=100; DT=0.01;
XMIN=-1.; DX=0.05; XMAX=1.;
edges=XMIN:DX:XMAX;
centers=XMIN+DX/2:DX:XMAX-DX/2;

for K=1:NHIST
    % Generation of NTRA trajectories x
    x=random('Uniform',-0.5,0.5,[1,NTRA]);
    for J=1:NTIME
        x=x+random('Normal',0,sqrt(DT),[1,NTRA]);
    end
```

```
    % Collection of NHIST histograms in matrix p
    p(K,:)=histc(x,edges)/(DX*NTRA);
end

% Plot of simulation results
errorbar([centers NaN],mean(p),std(p)/sqrt(NHIST),'LineStyle','none')
```

Note that the core part of this simulation code consisting of the two nested for...end loops comprises only three actual command lines. The outer loop is for the desired number of histograms (`NHIST`), each produced from a number of trajectories (`NTRA`); the inner loop performs `NTIME` time steps. The result of this short simulation code is shown by the error bar symbols in Figure 3.1, where also the rigorous result given by the difference of two error functions is shown. The agreement between simulation results and exact curve is perfect. Even on a desktop computer, much bigger ensembles could be simulated within seconds so that smaller error bars or better resolution could be achieved easily. The reader should play with this MATLAB® program, for example, to verify that the result for a correspondingly larger number of smaller time steps remains unchanged. In general, however, a detailed analysis of the effect of the time step is required. Sophisticated extrapolation procedures for that purpose have been described in Exercise 4.11 of Öttinger (1996).[3]

To make our example of a diffusing particle in one dimension more interesting, we switch on a drift term. We choose $A = -1$ so that the particle drifts to the left and the stochastic difference equation (3.17) becomes

$$X_{j+1} = X_j - \Delta t + W_j \sqrt{\Delta t}. \tag{3.18}$$

In order to keep the particle from moving to $-\infty$, we let it start at $x = 1$ and put a reflecting wall at $x = 0$.[4] We need only minor modifications of the previous MATLAB® code to simulate this case:

```
% Simulation parameters
NTRA=5000; NTIME=300; NHIST=100; DT=0.001;
XMIN=0; DX=0.05; XMAX=2;
edges=XMIN:DX:XMAX;
centers=XMIN+DX/2:DX:XMAX-DX/2;

for K=1:NHIST
    % Generation of NTRA trajectories x
    x=ones(1,NTRA);
    for J=1:NTIME
        x=x-DT+random('Normal',0,sqrt(DT),[1,NTRA]);
        x=(x+abs(x))/2;
    end
```

---

[3] Öttinger, *Stochastic Processes in Polymeric Fluids* (Springer, 1996).
[4] See Exercise 3.34 of Öttinger, *Stochastic Processes in Polymeric Fluids* (Springer, 1996).

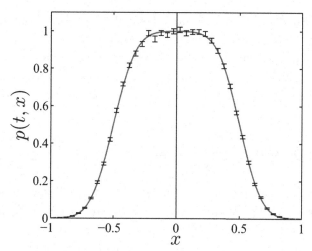

Figure 3.1 Solution of the diffusion equation (2.1) with drift $A = 0$ and diffusion coefficient $D = 1$ for $t = 0.02$ obtained from Brownian dynamics simulations (error bar symbols). For comparison, the exact solution is included (continuous line), and the result is a smeared version of the rectangular initial probability density.

```
% Collection of NHIST histograms in matrix p
  p(K,:)=histc(x,edges)/(DX*NTRA);
end

% Plot of simulation results
errorbar([centers NaN],mean(p),std(p)/sqrt(NHIST),'LineStyle','none')
```

In the inner for...end loop, we have added −DT as the deterministic increment of x resulting from $A = -1$. The crucial new line for the handling of reflecting boundary conditions is x=(x+abs(x))/2, which replaces negative values by zero and leaves positive values unchanged. Due to the presence of the drift we cannot perform the true reflection x=abs(x) because, during the reflection period, the drift term would have the wrong sign.[5] We only push back from negative values to zero. For nonzero drift, a term like "impenetrable," "impermeable," or "repelling" would actually be more appropriate than "reflecting."

For checking the results of our Brownian dynamics simulation, we would like to find the analytical solution of the equivalent diffusion equation with repelling boundary conditions. This is considerably more work than changing one line of code and adding another one in the simulation code and, in the end, we still need to perform a numerical integration.

---

[5] See comments on pp. 118f of Öttinger, *Stochastic Processes in Polymeric Fluids* (Springer, 1996).

In the absence of the boundary, typical solutions of the diffusion equation (2.1) with $A = -1$ and $D = 1$ would drift to the left with speed 1 and simultaneously widen according to diffusion. From the treatment of the boundary condition in the simulation we expect a solution in terms of additional Gaussians spreading around 0 to avoid a loss of probability at the boundary,

$$p(t, x) = p_{\mathrm{f}}(t, x) + \int_0^t \frac{a(t')}{\sqrt{2\pi(t - t')}} \exp\left\{ -\frac{1}{2} \frac{(x + t - t')^2}{t - t'} \right\} dt'. \quad (3.19)$$

Here, $p_{\mathrm{f}}(t, x)$ is the free solution obtained from the initial condition at $t = 0$ in the absence of the boundary at $x = 0$, and the function $a(t)$ is to be chosen such that the probability of the interval $[0, \infty[$ is conserved. The calculation of $p_{\mathrm{f}}(t, x)$ for any initial condition is as simple as our previous example without boundary conditions. By integrating over all x from 0 to ∞, differentiating with respect to t, and using the diffusion equation for the time derivative of the probability densities, we obtain the following condition for the function $a(t)$ (see Exercise 3.2):

$$a(t) - \int_0^t \frac{a(t')}{\sqrt{2\pi(t - t')}} e^{-(t-t')/2} dt' = \left. \frac{\partial p_{\mathrm{f}}(t, x)}{\partial x} \right|_{x=0} + 2 p_{\mathrm{f}}(t, 0). \quad (3.20)$$

By combining (3.19) and (3.20), we get a nicely simple relation between $a(t)$ and $p(t, 0)$,

$$a(t) = \left. \frac{\partial p_{\mathrm{f}}(t, x)}{\partial x} \right|_{x=0} + p_{\mathrm{f}}(t, 0) + p(t, 0). \quad (3.21)$$

For large times, $p_{\mathrm{f}}(t, x)$ and its derivative vanish.

As the condition (3.20) for $a(t)$ involves a convolution, it is most conveniently solved in terms of the Laplace transform[6]

$$\bar{a}(s) = \int_0^\infty a(t) e^{-st} dt, \quad (3.22)$$

---

[6] The Laplace transform of the function $f(t)$ is given by

$$\bar{f}(s) = \int_0^\infty f(t) e^{-st} dt.$$

If the Laplace transform of the function $f(t)$ can be written as $\bar{f}(s) = \bar{u}(s)\bar{v}(s)$, then the inverse Laplace transform can be written as a convolution,

$$f(t) = \int_0^t u(t') v(t - t') dt',$$

where $u(t)$ and $v(t)$ are the inverse Laplace transforms of $\bar{u}(s)$ and $\bar{v}(s)$, respectively.

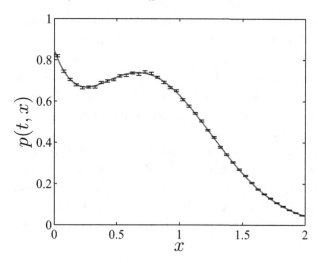

Figure 3.2 Solution of the diffusion equation (2.1) with drift $A = -1$ and diffusion coefficient $D = 1$ for $t = 0.3$ obtained from Brownian dynamics simulations (error bar symbols). For comparison, the solution (3.19) is included (continuous line) where the particles start at $x = 1$ for $t = 0$.

for which we obtain the explicit solution

$$\bar{a}(s) = \left(1 - \frac{1}{\sqrt{1+2s}}\right)^{-1} \int_0^\infty \left[\frac{\partial p_f(t,x)}{\partial x}\bigg|_{x=0} + 2p_f(t,0)\right] e^{-st}dt. \quad (3.23)$$

For our deterministic initial condition $p(0,x) = \delta(x-1)$, the integral on the right-hand side of (3.23) can be evaluated analytically. We thus obtain a closed-form expression for $\bar{a}(s)$,

$$\bar{a}(s) = \left(1 - \frac{1}{\sqrt{1+2s}}\right)^{-1} \left(1 + \frac{1}{\sqrt{1+2s}}\right) e^{1-\sqrt{1+2s}}. \quad (3.24)$$

Our result for $\bar{a}(s)$ is so complicated that we do not expect to find an explicit expression for $a(t)$. We hence rely on Zakian's approximate method for inverting Laplace transforms (see Exercise 3.3). The resulting $a(t)$ starts at 0 for $t = 0$, goes through a flat maximum around $t = 0.6$ (the maximum value of $a(t)$ is less than 2.3), and rapidly approaches the asymptotic value of 2, which follows from (3.21) and the steady-state solution $p_{eq}(x) = 2e^{-2x}$.[7] After performing the remaining numerical integration in (3.19) by means of Mathematica®, we finally obtain the desired solution $p(t,x)$.

In Figure 3.2, we show the results of the simulation and the solution of the diffusion equation for $t = 0.3$. Again, we find perfect agreement. At

[7] See Exercise 3.34 of Öttinger, *Stochastic Processes in Polymeric Fluids* (Springer, 1996).

$t = 0.3$, a maximum of the probability density is still visible and has drifted from 1 to around 0.7. With significant probability, a particle has reached the boundary and has been repelled. To compensate the deterministic drift to the left by a diffusive flux to the right, the slope at the wall has to be negative, so that the interesting nonmonotonic curve of Figure 3.2 arises. For larger times, the solution rapidly reaches the monotonically decreasing equilibrium solution $p_{eq}(x) = 2e^{-2x}$.

In contrast to our previous example, the simulation result now depends very significantly on the time step. Obtaining reliable results near the wall requires very small time steps. The error is actually of the order of $\sqrt{\Delta t}$. For the small time step $\Delta t = 10^{-3}$ displayed in the above MATLAB® code, the result for the first bin is significantly off; we have hence produced the simulation results in Figure 3.2 with the even smaller time step $\Delta t = 10^{-5}$.

Exercise 3.1   Simulation with Triangular Initial Condition
Develop a Brownian dynamics simulation to solve Exercise 2.3.

Exercise 3.2   Conservation of Probability at an Impenetrable Wall
Elaborate the derivation of (3.20).

Exercise 3.3   Deterministic Solution to a Diffusion Problem
Calculate the inverse Laplace transform of the function $\bar{a}(s)$ defined in (3.24) and plot the result for $a(t)$. Use this result in (3.19) to reproduce the continuous curve in Figure 3.2.

- Diffusion equations for smooth probability densities are equivalent to stochastic differential equations for everywhere-continuous and nowhere-differentiable trajectories; this equivalence can be extended to absorbing and impenetrable boundary conditions.
- The existence and uniqueness of solutions to stochastic differential equations is guaranteed by the continuity and linear growth conditions known from the theory of deterministic ordinary differentiable equations.
- The non-differentiability of the trajectories renders stochastic differential equations fundamentally different from deterministic ones, where the Itô–Stratonovich problem associated with nonanticipating versus anticipating integrands is the most well-known consequence.
- The theory of higher-order explicit and implicit numerical integration schemes for stochastic differential equations is well-developed.
- The numerical integration of stochastic differential equations leads to Brownian dynamics simulations, which can be implemented by a few lines of MATLAB® code, even in the presence of highly nontrivial boundary conditions.

4

Refreshing Topics in Equilibrium Thermodynamics

The reader may be asking why a book on the subject of transport phenomena has a chapter about thermodynamics. The answer is that the two subjects are intimately related, and one purpose of this book is for the reader to develop a deep understanding of this relationship. Most books on transport phenomena do not dedicate an entire chapter to thermodynamics, but rather use concepts and results from it when necessary.

Even though we give the subject of equilibrium thermodynamics its own chapter, we will have room only to cover some of its highlights. Readers interested in more complete discussions should consult textbooks devoted to the subject. We highly recommend the classic book by Herbert B. Callen,[1] an executive summary of which has been attempted in an article by Jongschaap and Öttinger,[2] and a more recent book by de Pablo and Schieber.[3]

It will be useful before we go much further to make a few general comments concerning thermodynamics. According to Callen, thermodynamics provides a universal framework for the macroscopic description of the properties of matter that follow from the fundamental laws of physics. Equilibrium thermodynamics deals with systems that do not change in time, even over infinitely long periods of time. As elaborated by Callen,[4] there is a close relationship between the variables of equilibrium thermodynamics and the symmetries of a system. The natural candidates for variables that do not change in time are conserved quantities. According to Noether's theorem,[5] such conserved quantities result directly from the continuous symmetries of a system. For example, symmetry under time translation implies the conservation of energy.

---

[1] Callen, *Thermodynamics* (Wiley, 1985).
[2] Jongschaap & Öttinger, *J. Non-Newtonian Fluid Mech.* **96** (2001) 5.
[3] de Pablo & Schieber, *Thermodynamics* (Cambridge, 2014).
[4] See Chapter 21 of Callen, *Thermodynamics* (Wiley, 1985).
[5] See Section 12-7 of Goldstein, *Classical Mechanics* (Addison-Wesley, 1980).

Figure 4.1 Emmy Noether, 1882–1935.

The framework of equilibrium thermodynamics can be constructed entirely within an axiomatic approach, which was pioneered by the Greek mathematician Constantin Carathéodory (1873–1950). An elegant and useful set of postulates extracted from the axiomatic approach can be found in Callen's book. More specifically, the framework of equilibrium thermodynamics places restrictions on the possible states that a system may assume, and on the processes by which a system may go from one state to another.

Already, we have used a number of terms whose meaning in the context of thermodynamics should be made clear. The system is simply the matter within a volume that we wish to describe in terms of a few macroscopic variables, and everything else is considered the surroundings. In equilibrium thermodynamics, the properties of the matter, mass density for example, are uniform throughout a given phase and are independent of time. The system and surroundings communicate with each other through walls, which may, or may not, allow the transfer of matter, and/or the transfer of energy. An isolated system is one where there is neither mass nor energy exchange with the surroundings, and a closed system is one where there is no exchange of matter.

In the remainder of this chapter, we present the structure of equilibrium thermodynamics and develop results that will be used in subsequent chapters of this book. This begins with a discussion about familiar concepts

like energy and entropy, and work and heat. We then introduce thermodynamic functions from which the properties of matter can be obtained.
The thermodynamic functions are put into forms that describe the behavior
of matter subjected to changes in temperature, pressure and composition.
These expressions, which are known as thermodynamic constitutive equations or equations of state, when combined with equations formulated in
subsequent chapters, provide the necessary instruments to play in the field
of transport phenomena. In the final section, we briefly review equilibrium
states of chemically reacting systems.

4.1 Energy and Entropy

Callen's first postulate of thermodynamics states that for a system of volume
V comprised of k chemically distinct components with mole numbers N_α
$(\alpha = 1, 2, \ldots, k)$, there exist *equilibrium states* that are completely characterized by the specification of an extensive macroscopic *state variable* called
the *internal energy* U.

The first postulate of thermodynamics implies that the internal energy is
subject to a conservation principle. This is the first law of thermodynamics,
which can be expressed as

$$dU = dQ + dW, \tag{4.1}$$

where dU expresses a differential change in the internal energy, and dQ and
dW express the transfer of energy to the system in the form of heat and work,
respectively. The symbol d used in dQ and dW indicates that these are not
differentials of a state variable like dU is, but rather are small amounts of
transferred heat and work.

To illustrate how work and heat are transferred to a system, we consider
N moles of a gas in a volume V. Work can be done on the gas by changing
the volume or the number of particles, where we assume that these changes
happen quasi-statically. If the volume of the gas is changed by dV, the
corresponding amount of work on the gas is given by $dW = -p\,dV$, where
p is the uniform pressure in the gas. If the number of moles in the gas is
changed by dN, the corresponding amount of work is given by $dW = \tilde{\mu}\,dN$,
where $\tilde{\mu}$ is the uniform chemical potential per unit mole. The total work
transferred during these two processes is given by

$$dW = -p\,dV + \tilde{\mu}\,dN. \tag{4.2}$$

We assume that an equilibrium state of the gas depends only on the internal energy U, volume V, and number of moles of gas N of the system.

The question we have not yet addressed is how the system knows what equilibrium state it should assume. For example, a system might prefer to exist in a single phase or in two coexisting phases. The answer is provided by the next postulate of thermodynamics, which says the unconstrained variables of a system arrange themselves to achieve an equilibrium state such that an extensive function called *entropy S is maximized*. This probably sounds like the second law of thermodynamics the reader has learned in a previous course. Energy transfer in the form of heat is associated with a change in entropy dS. The identification of entropy and absolute temperature T as another pair of conjugate variables represents a long and painful development in thermodynamics. If the heat transfer process takes place quasi-statically, we can compute the amount of heat transfer by the expression

$$\dJ Q = T\, dS. \tag{4.3}$$

If heat is not transferred to or from the system the process is adiabatic.

Substitution of (4.2) and (4.3) into (4.1) gives

$$dU = T\, dS - p\, dV + \tilde{\mu}\, dN, \tag{4.4}$$

which expresses the idea that, for quasi-static processes, changes in the internal energy U result from changes in entropy S, volume V, and number of moles N. The reader should appreciate the different types of energy transfer appearing on the right-hand side of (4.4). The last two terms associated with work are the result of processes that are mechanistically controlled.

In equilibrium thermodynamics, one usually considers systems at rest. If a thermodynamic system is allowed to move with velocity $\boldsymbol{v}$, the total energy, including the kinetic energy, is given by

$$E = \frac{1}{2} N \tilde{M} \boldsymbol{v}^2 + U, \tag{4.5}$$

where $\tilde{M}$ is the molar mass. We then have

$$dE = \frac{1}{2}\tilde{M}d\frac{(N\boldsymbol{v})^2}{N} + dU$$
$$= \boldsymbol{v} \cdot d\boldsymbol{M} + T\, dS - p\, dV + \left(\tilde{\mu} - \frac{1}{2}\tilde{M}\boldsymbol{v}^2\right) dN, \tag{4.6}$$

where $\boldsymbol{M} = N\tilde{M}\boldsymbol{v}$ is the total momentum of the thermodynamic system. We note that the velocity and the total momentum form a natural pair consisting of an intensive variable $(\boldsymbol{v})$ and an extensive variable $(\boldsymbol{M})$ that contributes to the change of energy in exactly the same way as the other pairs of intensive and extensive variables. Moreover, the chemical potential gets modified by a kinetic-energy contribution.

4.2 Fundamental Equations

Equation (4.4) is known as *Gibbs' fundamental form*. Its generalization to the case of k chemically distinct components with chemical potentials $\tilde{\mu}_1, \tilde{\mu}_2, \ldots, \tilde{\mu}_k$ reads

$$dU = T\,dS - p\,dV + \sum_{\alpha=1}^{k} \tilde{\mu}_\alpha\,dN_\alpha. \tag{4.7}$$

From the intuitive discussion of the previous section, it should be clear that Gibbs' fundamental form is a compact combination of the first and second laws of thermodynamics. We next establish its central importance and enormous usefulness in the framework of equilibrium thermodynamics.

A consequence of the postulates introduced in the previous section is that the thermodynamic behavior of a system at equilibrium is completely characterized by the specification of the entropy S for given values of internal energy U, volume V, and mole numbers of all the distinct chemical species N_α ($\alpha = 1, 2, \ldots, k$),

$$S = S(U, V, N_1, N_2, \ldots, N_k), \tag{4.8}$$

which is known as a *fundamental equation*. The next postulates of equilibrium thermodynamics describe the properties of entropy and its dependence on the arguments in (4.8). More specifically, the postulates tell us that S is extensive, and is a homogeneous function of U, V, and N_α ($\alpha = 1, 2, \ldots, k$). Furthermore, S is a continuous and monotonically increasing function of U. These properties mean that the fundamental equation (4.8) can be inverted to give

$$U = U(S, V, N_1, N_2, \ldots, N_k), \tag{4.9}$$

which is another fundamental equation. It is important to emphasize that the fundamental equation, expressed in terms of entropy (4.8), or in terms of energy (4.9), contains *complete* thermodynamic information about a system at equilibrium. In order to verify this statement, we consider the total differential of (4.9),

$$dU = \left(\frac{\partial U}{\partial S}\right)_{V,N_\alpha} dS + \left(\frac{\partial U}{\partial V}\right)_{S,N_\alpha} dV + \sum_{\alpha=1}^{k} \left(\frac{\partial U}{\partial N_\alpha}\right)_{S,V,N_{\beta(\beta\neq\alpha)}} dN_\alpha. \tag{4.10}$$

Now, if we compare (4.10) and (4.7), we can identify the temperature, pressure, and chemical potential per unit mole of species α as follows,

$$T = \left(\frac{\partial U}{\partial S}\right)_{V,N_\alpha}, \tag{4.11}$$

$$p = -\left(\frac{\partial U}{\partial V}\right)_{S,N_\alpha}, \tag{4.12}$$

$$\tilde{\mu}_\alpha = \left(\frac{\partial U}{\partial N_\alpha}\right)_{S,V,N_{\beta(\beta\neq\alpha)}}. \tag{4.13}$$

The expressions in (4.11), (4.12), and (4.13) are known as thermodynamic *equations of state* and, in contrast to the fundamental equation (4.9), each of them contains only partial information about the thermodynamic properties of a system.

Despite the fact that (4.7) and (4.10) look similar, it is important to understand the difference in the way the two equations were derived. The former, (4.7), was obtained by writing expressions for work and heat transfer terms on the right-hand side of (4.1) and contains deep physics. The latter, (4.10), was obtained by calculating the total differential of internal energy, a state variable, and is a result of elementary mathematics. The comparison of (4.7) and (4.10) then leads to the identification of thermodynamic equations of state.

The total differential of the entropy expression (4.8), when compared with the fundamental equation (4.7) rewritten in the form

$$dS = \frac{1}{T}dU + \frac{p}{T}dV - \sum_{\alpha=1}^{k} \frac{\tilde{\mu}_\alpha}{T}dN_\alpha, \tag{4.14}$$

leads to alternative expressions for T, p, and $\tilde{\mu}_\alpha$,

$$\frac{1}{T} = \left(\frac{\partial S}{\partial U}\right)_{V,N_\alpha}, \tag{4.15}$$

$$\frac{p}{T} = \left(\frac{\partial S}{\partial V}\right)_{U,N_\alpha}, \tag{4.16}$$

$$\frac{\tilde{\mu}_\alpha}{T} = -\left(\frac{\partial S}{\partial N_\alpha}\right)_{U,V,N_{\beta(\beta\neq\alpha)}}. \tag{4.17}$$

Take note of (4.15)–(4.17), which we will soon see play a central role as driving forces for the transport of energy and mass.

As discussed above, the variables $S, U, V, N_1, N_2, \ldots, N_k$ are proportional to the size of the system. Hence, we can write (4.9) as

$$\lambda U(S, V, N_1, N_2, \ldots, N_k) = U(\lambda S, \lambda V, \lambda N_1, \lambda N_2, \ldots, \lambda N_k), \tag{4.18}$$

where λ is an arbitrary constant. If we differentiate both sides of (4.18) with respect to λ holding $S, V,$ and N_α constant, using the chain rule and the expressions in (4.11)–(4.13), and then set λ equal to unity, we obtain

$$U = TS - pV + \sum_{\alpha=1}^{k} \tilde{\mu}_\alpha N_\alpha, \qquad (4.19)$$

which is known as the *Euler equation*. It is important to appreciate that (4.19) follows from the extensive nature of the variables $U, S, V, N_1, N_2, \ldots, N_k$. Taking the differential of each side of (4.19) and combining the result with (4.7), we obtain the *Gibbs–Duhem equation*,

$$S\, dT - V\, dp + \sum_{\alpha=1}^{k} N_\alpha\, d\tilde{\mu}_\alpha = 0, \qquad (4.20)$$

which is a constraint on the intensive variables $T, p, \tilde{\mu}_1, \tilde{\mu}_2, \ldots, \tilde{\mu}_k$ for a system at equilibrium.

So far, we have introduced the fundamental equation for U that is a function of $S, V, N_1, N_2, \ldots, N_k$, all of which are extensive thermodynamic variables. It is possible to swap the roles played by internal energy and entropy, that is, to make S the dependent variable and U an independent variable. We have also seen how variables on this list can be supplemented by the intensive variables $T, p, \tilde{\mu}_1, \tilde{\mu}_2, \ldots, \tilde{\mu}_k$, where the position of a variable in the list of these conjugate variables should be preserved. This list of variables and their structure now makes it possible to describe the matter within a volume, for example a gas, as a *thermodynamic system*. We encourage the reader to fully appreciate this structure as it will be used in the remainder of this chapter and, as promised, throughout this book.

Exercise 4.1   Ideal Gas Equations of State
The fundamental entropy equation for a one-component, ideal, monatomic gas can be written as

$$S(U, V, N) = N\tilde{s}_0 + N\tilde{R}\ln\left[\left(\frac{U}{N\tilde{u}_0}\right)^{3/2}\left(\frac{V}{N\tilde{v}_0}\right)\right], \qquad (4.21)$$

where $\tilde{s}_0$, $\tilde{u}_0$, and $\tilde{v}_0$ are constants.[6] Use (4.15)–(4.17) to obtain the well-known equations of state,

$$U = \frac{3}{2}N\tilde{R}T, \qquad V = N\tilde{R}T/p, \qquad \tilde{\mu} = \tilde{\mu}^0(T, p) + \tilde{R}T\ln p, \qquad (4.22)$$

---

[6] Actually only one combination of the three constants is relevant because, under the logarithm, only the product $\tilde{u}_0^{3/2}\tilde{v}_0$ matters, and a multiplicative change of this product can be compensated by an additive change of $\tilde{s}_0$.

where $\tilde{\mu}^0(T,p)$ is the chemical potential in a reference state. Note that the expressions in (4.22) can also be written as $U = \frac{3}{2}\tilde{N}k_{\mathrm{B}}T$ and $V = \tilde{N}k_{\mathrm{B}}T/p$, where $\tilde{N}$ is the number of molecules, $\tilde{N} = N\tilde{N}_{\mathrm{A}}$, and $k_{\mathrm{B}} = \tilde{R}/\tilde{N}_{\mathrm{A}}$ (see Appendix A).

4.3 Thermodynamic Potentials

The fundamental equations given in (4.8) and (4.9), while able to provide a complete thermodynamic description of a system, are probably not the most useful. We say this because the variables in these expressions are not usually the ones that are controlled or measured in practice. For example, it is much more likely that a process is carried out at constant T than at constant S. Recall, however, that an equation of state such as (4.11), or $U = U(T, V, N_1, N_2, \ldots, N_k)$, contains only partial thermodynamic information about the system.

Is there a function with complete thermodynamic information that depends on $T, V, N_1, N_2, \ldots, N_k$ and, if so, how does one find it? Fortunately, the answer to the first question is yes. This function and others like it are known as *thermodynamic potentials*, which can be obtained from the fundamental equation for internal energy using a simple trick involving differentials. Our goal is to eliminate dS and replace it with dT, so we subtract $d(TS) = T\,dS + S\,dT$ from both sides of (4.7), which gives

$$dF = -S\,dT - p\,dV + \sum_{\alpha=1}^{k} \tilde{\mu}_\alpha\,dN_\alpha, \tag{4.23}$$

where

$$F = U - TS. \tag{4.24}$$

The function $F = F(T, V, N_1, N_2, \ldots, N_k)$ is a thermodynamic potential called the Helmholtz free energy. In more mathematical terms, this change of variable is known as a Legendre transformation. Hence, the Helmholtz free energy $F(T)$ is the Legendre transform of $U(S)$, where we have suppressed the dependence on variables that are not affected by the Legendre transform, $(V, N_1, N_2, \ldots, N_k)$. Whereas the thermodynamic information about a system contained in the equation of state $U(T)$ is only partial, the information contained in the thermodynamic potential $F(T)$, like the fundamental equation $U(S)$, is complete. From (4.23), we can obtain the following equations of state,

$$S = -\left(\frac{\partial F}{\partial T}\right)_{V,N_\alpha}, \tag{4.25}$$

$$p = -\left(\frac{\partial F}{\partial V}\right)_{T,N_\alpha}, \tag{4.26}$$

$$\tilde{\mu}_\alpha = \left(\frac{\partial F}{\partial N_\alpha}\right)_{T,V,N_{\beta(\beta\neq\alpha)}}. \tag{4.27}$$

If instead of taking the Legendre transform of $U(S)$, it is taken on $U(V)$, we add $d(pV)$ to both sides of (4.4) and obtain the result

$$H = U + pV, \tag{4.28}$$

where $H = H(S, p, N_1, N_2, \ldots, N_k)$ is the thermodynamic potential known as the enthalpy. Applying the same procedure, it is also possible to replace both S and V as independent variables in (4.4), which gives

$$dG = -S\,dT + V\,dp + \sum_{\alpha=1}^{k} \tilde{\mu}_\alpha\,dN_\alpha, \tag{4.29}$$

or, as before,

$$G = U - TS + pV, \tag{4.30}$$

where $G = G(T, p, N_1, N_2, \ldots, N_k)$ is known as the Gibbs potential, or Gibbs free energy. Combining (4.30) with the Euler relation (4.19) reveals the close relationship between the Gibbs free energy and chemical potential, $G = \sum_{\alpha=1}^{k} \tilde{\mu}_\alpha N_\alpha$. As noted above, the thermodynamic potentials defined in (4.24), (4.28), and (4.30) contain complete thermodynamic information about a system.

Equations of state similar to those in (4.25)–(4.27) can also be identified from the total differentials of the enthalpy and Gibbs free energy. For completeness, we mention that it is also possible to perform Legendre transformations on the entropy representation of the fundamental equation (4.8), which are known as Massieu functions.

Exercise 4.2 *Legendre Transformation for Ideal Gas*
Find the Helmholtz free energy $F(T, V, N)$ by Legendre transformation of the internal energy $U(S, V, N)$ implied by the entropy $S(U, V, N)$ given in Exercise 4.1.

4.4 Second-Order Derivatives

There is an enormous wealth of information contained in the second-order derivatives of thermodynamic potentials. In this section, we reveal some insights about thermodynamic material properties, Maxwell relations, and stability criteria. A further link to equilibrium fluctuations will be made through Einstein's fluctuation theory in Section 20.4.

In applications, one often uses (piecewise) linearizations of thermodynamic equations of state. Therefore, one is interested in derivatives of equations of state, which produce second-order derivatives of a thermodynamic potential. For example, the partial derivative of $(\partial F/\partial T)_V$ with respect to V gives

$$\left(\frac{\partial}{\partial V}\left(\frac{\partial F}{\partial T}\right)_V\right)_T = \left(\frac{\partial}{\partial T}\left(\frac{\partial F}{\partial V}\right)_T\right)_V, \tag{4.31}$$

which follows if the order of differentiation is unimportant. Note that for convenience we are suppressing the dependence on N_α of the functions in the present discussion. Substitution of (4.25) and (4.26) into (4.31) yields

$$\left(\frac{\partial S}{\partial V}\right)_T = \left(\frac{\partial p}{\partial T}\right)_V, \tag{4.32}$$

which is known as a *Maxwell relation*, named after the physicist James Clerk Maxwell (1831–1879), whose name will appear several times in this book. Using the same procedure it is straightforward to derive other Maxwell relations using U, H, and G. The equality in (4.32) is an example of what is known as an *integrability condition*[7] and is intimately related to the existence of the thermodynamic potential F, from which S and p can be obtained as derivatives with respect to T and V. Maxwell relations are useful for obtaining convenient forms of the constitutive equations that are discussed in the following section. In addition, they allow one to identify the functional dependences for fundamental equations. For example, if the mechanical equation of state has the form $p \propto T/V$, then (4.32) requires S to be a logarithmic function of V, which we saw in Exercise 4.1 is true for ideal gases.

The postulate of entropy maximization at equilibrium at fixed extensive variables can, for a thermodynamic system defined by $S = S(U, V, N)$, be expressed mathematically as

$$\left(\frac{\partial S}{\partial U}\right)_{V,N} = 0, \qquad \left(\frac{\partial^2 S}{\partial U^2}\right)_{V,N} < 0. \tag{4.33}$$

The first equation in (4.33) can be used to prove that, for two systems in thermal contact, heat flows from the system at higher temperature to the system at lower temperature. The second equation in (4.33) can be used to examine the stability of a system to remain as a single phase, or to split into different phases.

---

[7] The integrability condition is quite general; in the following chapter it will be applied to define a stream function that is useful in fluid mechanics.

Derivatives of equations of state are also important in thermodynamic applications, particularly in the identification of thermodynamic properties that will be used in the following section. Differentiation of the thermal equation of state given in (4.15) with respect to U gives

$$\left(\frac{\partial^2 S}{\partial U^2}\right)_{V,N} = -\frac{1}{T^2 \left(\frac{\partial U}{\partial T}\right)_{V,N}} = -\frac{1}{T^2 C_V}, \tag{4.34}$$

where the material property C_V is the heat capacity at constant volume,

$$C_V = \left(\frac{\partial U}{\partial T}\right)_{V,N} = T \left(\frac{\partial S}{\partial T}\right)_{V,N}. \tag{4.35}$$

From the mathematical statement of entropy maximization in (4.33) it is clear that C_V is positive, which is consistent with our experience that when heat is transferred to a material, increasing its internal energy, the temperature of the material increases. Similarly, from (4.12) one can obtain a material property relating changes in volume and pressure,

$$\kappa_S = -\frac{1}{V} \left(\frac{\partial V}{\partial p}\right)_{S,N}, \tag{4.36}$$

which is known as the isentropic compressibility. These, and similarly defined material properties, are obtained from second-order derivatives of the fundamental equations (see Appendix A).

If the inequality in the second equation of (4.33), or stability criterion, is not satisfied, then a single-phase system will form two or more phases. For a multi-phase system at equilibrium, the temperature, pressure, and chemical potential of each species are equal in each phase. This means the system is described by $k + 2$ intensive variables. For each phase the Gibbs–Duhem equation (4.20) must hold so that only $k+1$ variables are independent. For a system having π coexisting distinct phases, the number of intensive variables that can independently be specified, f, is reduced by coexistence conditions and hence given by

$$f = k - \pi + 2, \tag{4.37}$$

which is the well-known *Gibbs phase rule*.

Consider a two-phase ($\pi = 2$) system comprised of a single component ($k = 1$), where one phase is designated as phase I and the other as phase II. The equilibrium conditions stated above can be expressed as $T^{\mathrm{I}} = T^{\mathrm{II}} = T$ and $p^{\mathrm{I}} = p^{\mathrm{II}} = p$; the phase rule ($f = 1$) indicates that either T or p can be specified. The third equilibrium condition, $\tilde{\mu}^{\mathrm{I}} = \tilde{\mu}^{\mathrm{II}} = \tilde{\mu}$, implies that along

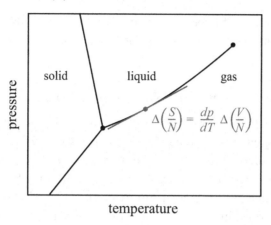

temperature

Figure 4.2 Illustration of the relevance of the Clapeyron equation.

a phase coexistence line we have $d\tilde{\mu}^{\mathrm{I}} = d\tilde{\mu}^{\mathrm{II}}$. The Gibbs–Duhem equation (4.20) then leads to the *Clapeyron equation*,

$$\frac{dp}{dT} = \frac{\Delta(S/N)}{\Delta(V/N)} = \frac{\Delta(H/N)}{T\,\Delta(V/N)}, \tag{4.38}$$

where $\Delta(.) = (.)^{\mathrm{I}} - (.)^{\mathrm{II}}$ denotes the jump in a property of coexisting phases.[8] The relationship between p and T given in (4.38) can be used to construct phase coexistence lines on p–T diagrams; from their slope one can obtain the ratio of jumps in thermodynamic quantities between different phases. For example, the phase line separating steam and water is positive because the jumps in entropy and volume have the same sign, while the phase line separating water and ice is negative because the jumps in entropy and volume have opposite signs (see Figure 4.2). The thermodynamics of interfaces plays such a critical role in transport phenomena that we give the subject its own chapter (see Chapter 13).

Exercise 4.3   Finding the Helmholtz Free Energy by Integration
Integrate (4.25) and (4.26) for the equations of state for $p(T, V, N)$ and $S(T, V, N)$ found in Exercises 4.1 and 4.2 to obtain the following expression for the Helmholtz free energy for a one-component, ideal, monatomic gas,

$$F(T, V, N) = -N\tilde{R}T \ln\left(c\frac{V}{N}T^{3/2}\right), \tag{4.39}$$

---

[8] If phase I behaves as an ideal gas, then $\Delta(V/N) \approx \tilde{R}T/p$ and (4.38) can be written as

$$\frac{dp}{dT} = \frac{p}{\tilde{R}T^2}\Delta(H/N),$$

which is known as the Clausius–Clapeyron relation.

where c is a constant of integration that can be used to make the argument of the logarithm dimensionless. Compare this with the result of Exercise 4.2.

4.5 Size Is Not Allowed to Matter

In subsequent chapters we will find it convenient to have thermodynamic constitutive equations, or equations of state, expressed entirely in terms of intensive variables. The extensive nature of the variables used to describe a thermodynamic system as expressed in (4.18) involves an arbitrary parameter λ, which is inversely proportional to the size of the system. The reader might be asking what exactly do we mean by the size of the system and how arbitrary is it? One possibility is to use the total number of moles $N = \sum_{\alpha=1}^{k} N_\alpha$ as the size. Hence, in (4.18) we set $\lambda = 1/N$,

$$U(S, V, N_1, N_2, \ldots, N_k) = N\tilde{u}(\tilde{s}, \tilde{v}, x_1, x_2, \ldots, x_{k-1}), \qquad (4.40)$$

where $\tilde{u} = U/N$, $\tilde{s} = S/N$, and $\tilde{v} = V/N$ are the molar internal energy, molar entropy, and molar volume, respectively, and $x_\alpha = N_\alpha/N$ is the mole fraction of species α. As with the extensive functions from which they are derived, the intensive fundamental equation for molar internal energy $\tilde{u}(\tilde{s}, \tilde{v}, x_1, x_2, \ldots, x_{k-1})$, or for molar entropy $\tilde{s}(\tilde{u}, \tilde{v}, x_1, x_2, \ldots, x_{k-1})$, contains complete thermodynamic information about the system. Note, however, that, because we used the total number of moles to convert from extensive to intensive variables, the intensive fundamental equation depends on $k+1$ quantities (since $\sum_{\alpha=1}^{k} x_\alpha = 1$), compared with $k+2$ for the extensive fundamental equation. This means only $k+1$ equations of state can be obtained from derivatives of the fundamental equation. The missing equation of state, however, can be obtained from the Euler equation, which is not redundant for an intensive fundamental equation.

Now that we have an idea about what type of variable may be used as a measure of the system size, we address the question of possible restrictions on the size of the system itself. In other words, does it matter how large or small the system size is? The answer is no, unless the system size, the number of molecules for example, becomes small. In such cases, the fundamental equation is no longer a homogeneous function of its arguments, and the implications of extensive and intensive variables must be abandoned. Simply put, the Gibbs–Duhem equation (4.20) no longer holds for small systems. Typically, the number of molecules in a thermodynamic system is large enough ($\approx \tilde{N}_A$) so this is not an issue. However, there are cases when small thermodynamic systems are of interest, and modifications to the structure presented here are necessary to describe them.[9] For the remainder of this chapter and throughout this book, we will assume the system under consideration is large enough that standard thermodynamics applies.

We will find it useful to consider intensive quantities that are volume densities so that the size variable is the system volume, $\lambda = 1/V$. For convenience, we consider a two-component system and our fundamental equation is $u = u(s, \rho_1, \rho_2)$, where $u = U/V$ and $s = S/V$ are the internal energy density and entropy density, respectively. The mass densities of species 1 and species 2 are defined as $\rho_1 = N_1 \tilde{M}_1/V$ and $\rho_2 = N_2 \tilde{M}_2/V$, where $\tilde{M}_1$ and $\tilde{M}_2$ are the molar masses of species 1 and 2, respectively. The total mass density of the mixture is given by $\rho = \rho_1 + \rho_2$. Our goal is to formulate expressions that relate internal energy to intensive variables such as temperature, pressure, and composition. Using the definition of temperature in (4.11) and the chain rule, we have

$$T = \left(\frac{\partial u}{\partial s} \right)_\rho, \tag{4.41}$$

and, similarly, from (4.13) we obtain

$$\hat{\mu}_1 = \left(\frac{\partial u}{\partial \rho_1} \right)_{s,\rho_2}, \qquad \hat{\mu}_2 = \left(\frac{\partial u}{\partial \rho_2} \right)_{s,\rho_1}, \tag{4.42}$$

where $\hat{\mu}_1$ and $\hat{\mu}_2$ are the chemical potential per unit mass of species 1 and 2. The total differential of $u(s, \rho_1, \rho_2)$ using (4.41) and (4.42) can be written as

$$du = T\, ds + \hat{\mu}_1\, d\rho_1 + \hat{\mu}_2\, d\rho_2. \tag{4.43}$$

The expression for pressure is obtained from (4.12) and using the chain rule on (4.40) along with (4.41) and (4.42),

$$p = -u + sT + \rho_1 \hat{\mu}_1 + \rho_2 \hat{\mu}_2, \tag{4.44}$$

which is a form of the Euler equation given in (4.19).

In some cases it is more convenient to work with intensive thermodynamic variables that are normalized by total mass rather than total volume. The total mass of a system is $N\tilde{M}$, where N is the total number of moles and $\tilde{M}$ is the molar mass of the mixture. For convenience, we again consider a two-component system and set $\lambda = 1/(N\tilde{M})$. The fundamental equation given in (4.18) then has the form $\hat{u} = \hat{u}(\hat{s}, \hat{v}, w_1)$, where $\hat{u} = U/(N\tilde{M}) = u/\rho$, $\hat{s} = S/(N\tilde{M}) = s/\rho$, and $\hat{v} = V/(N\tilde{M}) = 1/\rho$ are the specific internal energy, specific entropy, and specific volume, respectively, and $w_1 = \rho_1/\rho$ is the mass fraction of species 1.

---

[9] Hill, *Thermodynamics of Small Systems* (Dover, 1994).

In terms of specific variables we can, with the help of (4.44), write (4.43) as a differential for $\hat{u}$,

$$d\hat{u} = T\,d\hat{s} - p\,d\hat{v} + (\hat{\mu}_1 - \hat{\mu}_2)dw_1, \qquad (4.45)$$

where the equations of state now read as follows,

$$T = \left(\frac{\partial \hat{u}}{\partial \hat{s}}\right)_{\hat{v},w_1}, \qquad (4.46)$$

$$p = -\left(\frac{\partial \hat{u}}{\partial \hat{v}}\right)_{\hat{s},w_1}, \qquad (4.47)$$

$$\hat{\mu}_1 - \hat{\mu}_2 = \left(\frac{\partial \hat{u}}{\partial w_1}\right)_{\hat{s},\hat{v}}. \qquad (4.48)$$

The difference in chemical potentials in (4.48) arises from the constraint on the mass fractions, $w_1 + w_2 = 1$. In (4.45), it reflects the fact that, for a two-component system at constant temperature and volume, a change in composition results in a change in internal energy that is proportional to the difference in chemical potentials between the two components.

In practice, entropy is a quantity that can be neither measured nor controlled, so it will be useful for us to find a different way of expressing the first term in (4.45). To be more specific, we would like to find an expression for $\hat{u} = \hat{u}(T, \hat{v}, w_1)$, which is not a thermodynamic potential, but will be useful for obtaining an equation that governs temperature. This can be achieved by substituting the total differential of $\hat{s}(T, \hat{v}, w_1)$ into (4.45), which gives

$$d\hat{u} = T\left(\frac{\partial \hat{s}}{\partial T}\right)_{\hat{v},w_1} dT + \left[T\left(\frac{\partial \hat{s}}{\partial \hat{v}}\right)_{T,w_1} - p\right]d\hat{v} + \left[T\left(\frac{\partial \hat{s}}{\partial w_1}\right)_{T,\hat{v}} + (\hat{\mu}_1 - \hat{\mu}_2)\right]dw_1. \qquad (4.49)$$

Our expression for internal energy (4.49) contains derivatives of equations of state, which themselves are derivatives of fundamental equations. These second-order derivatives are material properties similar to those introduced earlier in the chapter. Also, we recognize that the second term in (4.49) can be rewritten using the intensive version of the Maxwell relation given in (4.32).

To handle the last term in (4.49), we introduce the partial specific properties associated with any specific quantity $\hat{a} = \hat{a}(T, p, w_1, \ldots, w_{k-1})$ as

$$\hat{a}_\alpha = \frac{\partial \hat{a}}{\partial w_\alpha} + \hat{a} - \sum_{\beta=1}^{k-1} w_\beta \frac{\partial \hat{a}}{\partial w_\beta}, \qquad (4.50)$$

where the following important property should be noted,

$$\sum_{\alpha=1}^{k} w_\alpha \hat{a}_\alpha = \hat{a}. \tag{4.51}$$

For a two-component system, (4.50) and (4.51) further imply

$$\hat{a}_1 - \hat{a}_2 = \left(\frac{\partial \hat{a}}{\partial w_1}\right)_{T,p}. \tag{4.52}$$

In terms of these partial specific material properties, (4.49) can (see Exercise 4.5) be written as

$$d\hat{u} = \hat{c}_{\hat{v}}\, dT + \left[T\left(\frac{\partial p}{\partial T}\right)_{\hat{v},w_1} - p\right] d\hat{v} + \left[(\hat{h}_1 - \hat{h}_2) - T\left(\frac{\partial p}{\partial T}\right)_{\hat{v},w_1}(\hat{v}_1 - \hat{v}_2)\right] dw_1,$$

$$\tag{4.53}$$

where we have used the Maxwell relation $(\partial \hat{s}/\partial \hat{v})_{T,w_\alpha} = (\partial p/\partial T)_{\hat{v},w_\alpha}$. In (4.53), $\hat{c}_{\hat{v}} = T(\partial \hat{s}/\partial T)_{\hat{v},w_\alpha}$ is the specific heat capacity at constant specific volume and composition, and $\hat{h}_\alpha$ and $\hat{v}_\alpha$ are the partial specific enthalpies and volumes, respectively. Expressions for the partial derivative in (4.53) can be obtained from the mechanical equation of state, $(\partial p/\partial T)_{\hat{v},w_1} = \alpha_p/\kappa_T$, where α_p is the isobaric thermal expansion coefficient and κ_T is the isothermal compressibility. A useful compilation of thermodynamic relations can be found in Appendix A.

We now have a convenient equation that relates changes in internal energy to changes in temperature, volume, and composition expressed entirely in terms of intensive variables. At this point, it is worthwhile to take a moment to get a feeling for (4.53) and the types of thermodynamic effects it describes. The first term shows that changes in temperature and internal energy are related by the specific heat capacity. For example, increasing the temperature of one gram of water by one degree results in an increase in internal energy that is roughly four times as large as for one gram of air. The second and third terms inside the square brackets describe the heating (cooling) that occurs when a material is compressed (expanded). The last term in (4.53) takes into account changes in internal energy due to compositional changes that might occur when mixtures are formed, or a chemical reaction takes place.

In later chapters it will be useful to have the form of the Gibbs–Duhem equation in terms of intensive variables. For a two-component system, this can be obtained from (4.44) and (4.45),

$$\hat{s}\, dT - \hat{v}\, dp + w_1\, d\hat{\mu}_1 + w_2\, d\hat{\mu}_2 = 0. \tag{4.54}$$

It will also be useful to relate changes in chemical potential to changes in composition. The total differential for $\hat{\mu}_\alpha = \hat{\mu}_\alpha(T, p, w_1)$ can be written as

$$d\hat{\mu}_\alpha = -\hat{s}_\alpha \, dT + \hat{v}_\alpha \, dp + \left(\frac{\partial \hat{\mu}_\alpha}{\partial w_1}\right)_{T,p} dw_1, \qquad (4.55)$$

where $\hat{s}_\alpha = -(\partial \hat{\mu}_\alpha / \partial T)_{p,w_\alpha}$ and $\hat{v}_\alpha = (\partial \hat{\mu}_\alpha / \partial p)_{T,w_\alpha}$ are alternative expressions for the partial specific entropy and volume, respectively.

Exercise 4.4   Hard-Sphere Gases
As the name suggests, the hard-sphere model assumes that particles (molecules) are solid spheres that occupy a finite volume. The entropy density for a hard-sphere system is given by

$$s(u, \rho) = \frac{\rho k_B}{m} \ln \left[R_0(\rho) u^{3/2} \rho^{-5/2} \right], \qquad (4.56)$$

where m is the mass of a particle and $R_0(\rho)$ is a function that takes into account the effect of the finite particle size on the density of the system. Use (4.56) to obtain equations of state $u = u(T, \rho)$ and $p = p(T, \rho)$. Also, discuss the relation between the hard-sphere and ideal gas models.

Exercise 4.5   Differentials for Specific Internal Energy
Starting with (4.49), derive the expression given in (4.53). Also, derive the alternative form for the differential of specific internal energy

$$d\hat{u} = \hat{c}_p \, dT + T \left(\frac{\partial \hat{v}}{\partial T}\right)_{p,w_1} dp - p \, d\hat{v} + (\hat{h}_1 - \hat{h}_2) dw_1, \qquad (4.57)$$

where $\hat{c}_p$ is the specific heat capacity at constant pressure and composition.

Exercise 4.6   Partial Specific Properties
Prove the following identities relating partial specific properties to chemical potentials:

$$\hat{g}_\alpha = \hat{\mu}_\alpha, \qquad \hat{s}_\alpha = -\left(\frac{\partial \hat{\mu}_\alpha}{\partial T}\right)_{p,w_\alpha}, \qquad \hat{v}_\alpha = \left(\frac{\partial \hat{\mu}_\alpha}{\partial p}\right)_{T,w_\alpha}. \qquad (4.58)$$

Exercise 4.7   Ideal Mixtures
Ideal mixtures, which are sometimes referred to as Lewis mixtures, are comprised of components that interact the same with each other as they do with themselves. As a result, the volume and internal energy are linear functions of the species mole fraction x_α,

$$\tilde{v}(T, p, x_\alpha) = \sum_{\alpha=1}^{k} x_\alpha \tilde{v}_\alpha^0(T, p), \qquad \tilde{u}(T, p, x_\alpha) = \sum_{\alpha=1}^{k} x_\alpha \tilde{u}_\alpha^0(T, p), \qquad (4.59)$$

where $\tilde{v}_\alpha^0(T, p)$ and $\tilde{u}_\alpha^0(T, p)$ are the pure component α molar volume and molar internal energy, respectively. Derive an expression for the molar entropy for a k-component ideal mixture and show that the chemical potential of species α in an ideal mixture is

$$\tilde{\mu}_\alpha = \tilde{\mu}_\alpha^0(T,p) + \tilde{R}T \ln x_\alpha, \tag{4.60}$$

where $\tilde{\mu}_\alpha^0(T,p)$ is the chemical potential of pure species α in a reference state.

Exercise 4.8   Phase Equilibria of Nonideal Mixtures

The chemical potential for a species in a nonideal mixture can be written as

$$\tilde{\mu}_\alpha = \tilde{\mu}_\alpha^{0,\mathrm{ig}}(T,p) + \tilde{R}T \ln(\mathfrak{f}_\alpha), \tag{4.61}$$

where $\tilde{\mu}_\alpha^{0,\mathrm{ig}}(T,p)$ is the chemical potential of pure species α in an ideal gas reference state, and $\mathfrak{f}_\alpha$ is the *fugacity* of species α. For gas mixtures, it is common to write the fugacity in terms of the fugacity coefficient $\bar{\phi}_\alpha$ as $\mathfrak{f}_\alpha = x_\alpha \bar{\phi}_\alpha p$, so that the chemical potential for a nonideal gas mixture is

$$\tilde{\mu}_\alpha = \tilde{\mu}_\alpha^{0,\mathrm{ig}}(T,p) + \tilde{R}T \ln(x_\alpha \bar{\phi}_\alpha p). \tag{4.62}$$

For a liquid that is a nonideal mixture, the chemical potential can be written as

$$\tilde{\mu}_\alpha^{0,\mathrm{sl}}(T, p_\alpha^{\mathrm{sat}}) + \tilde{R}T \ln(x_\alpha \bar{\gamma}_\alpha), \tag{4.63}$$

where $\tilde{\mu}_\alpha^{0,\mathrm{sl}}(T, p_\alpha^{\mathrm{sat}})$ is the chemical potential of pure species α in a saturated liquid reference state (here we neglect the Poynting correction for pressure), and $\bar{\gamma}_\alpha$ is the *activity coefficient*. For a gas phase (I) in equilibrium with a liquid phase (II), obtain the relation

$$x_\alpha^{\mathrm{I}} \bar{\phi}_\alpha p = x_\alpha^{\mathrm{II}} \bar{\gamma}_\alpha \bar{\phi}_\alpha^0 p_\alpha^{\mathrm{sat}}. \tag{4.64}$$

For an ideal gas mixture ($\bar{\phi}_\alpha = \bar{\phi}_\alpha^0 = 1$) and ideal liquid mixture ($\bar{\gamma}_\alpha = 1$) the expression above simplifies to Raoult's law $x_\alpha^{\mathrm{I}} p = x_\alpha^{\mathrm{II}} p_\alpha^{\mathrm{sat}}$.

Exercise 4.9   Henry's Law

For the special case of a gas dissolved at low concentration in a liquid, it is common to write the fugacity of the dissolved gas component as $\mathfrak{f}_\alpha = k_{\mathrm{H},\alpha} x_\alpha$, where $k_{\mathrm{H},\alpha}$ is the Henry's law constant. For a gas phase (I) in equilibrium with a liquid phase (II), obtain the relation

$$x_\alpha^{\mathrm{I}} \bar{\phi}_\alpha p = k_{\mathrm{H},\alpha} x_\alpha^{\mathrm{II}}. \tag{4.65}$$

4.6 Chemical Reactions

Chemical reactions occur in a wide range of chemical and material engineering processes. In this section we will cover only the most basic applications of equilibrium thermodynamics to chemical reaction. A chemical equation is often used to symbolically represent a chemical reaction, for example $N_2 + 3H_2 \rightleftarrows 2NH_3$, where $\rightarrow$ indicates the forward reaction and $\leftarrow$ indicates the reverse reaction. Keep in mind that the thermodynamic concepts described in this chapter apply only to systems at equilibrium. Describing the time evolution of reacting systems, or reaction kinetics, is based on governing equations that are formulated using a combination of balance equations and constitutive equations, which are discussed in subsequent chapters (see Chapters 5 and 6, in particular Section 6.6).

For a closed system, the mole number of any species N_α can change only as a result of a chemical reaction. In general, there may be multiple reactions. In each reaction, a species can be a reactant, a product, or an inert species. Here, for simplicity, we consider a single reaction and express the change in the number of moles of species α as

$$dN_\alpha = \tilde{\nu}_\alpha \, d\xi, \tag{4.66}$$

where $\tilde{\nu}_\alpha$ is the stoichiometric coefficient of species α, and ξ is the *extent of reaction*. If species α is a reactant, then $\tilde{\nu}_\alpha$ is negative; if species α is a product, then it is positive. In the example above, $\tilde{\nu}_{N_2} = -1$, $\tilde{\nu}_{H_2} = -3$, and $\tilde{\nu}_{NH_3} = 2$.

Consider a hypothetical mixture with mole numbers $N_{\alpha,0}$ that is constrained so that no reaction occurs, for which we set $\xi = 0$. The constraint is removed and the reaction is allowed to proceed. Integration of (4.66) gives $N_\alpha = N_{\alpha,0} + \tilde{\nu}_\alpha \xi$. If we multiply this result by $\tilde{M}_\alpha$ and sum over α, since the total mass $N\tilde{M} = \sum_{\alpha=1}^{k} N_\alpha \tilde{M}_\alpha$, we obtain

$$\sum_{\alpha=1}^{k} \tilde{\nu}_\alpha \tilde{M}_\alpha = \sum_{\alpha=1}^{k} \nu_\alpha = 0, \tag{4.67}$$

where $\nu_\alpha = \tilde{\nu}_\alpha \tilde{M}_\alpha / \tilde{M}_q$, and $\tilde{M}_q$ is determined from (4.67) by identifying species $\alpha = 1, \ldots, q$ as reactants, and species $\alpha = q+1, \ldots, k$ as products,

$$\tilde{M}_q = \sum_{\alpha=q+1}^{k} \tilde{\nu}_\alpha \tilde{M}_\alpha = -\sum_{\alpha=1}^{q} \tilde{\nu}_\alpha \tilde{M}_\alpha. \tag{4.68}$$

Equations (4.66) and (4.67) reflect the fact that total mass is conserved and that changes in mole numbers of the reacting species are related through the stoichiometry of the reaction.

We now look at how the energy of a system is changed by a chemical reaction. This can be done using any of the thermodynamic potentials introduced in this chapter. If we assume the reaction takes place at constant

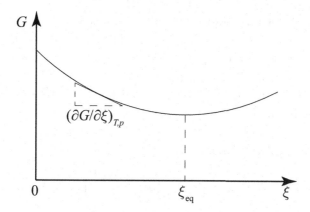

Figure 4.3 Gibbs free energy G for a system with chemical reaction as a function of the extent of reaction ξ. At equilibrium, G is minimized at ξ_{eq}.

T and p, the Gibbs free energy is the natural choice. Hence, we use (4.29) and combine this with (4.66) to obtain

$$dG = \sum_{\alpha=1}^{k} \tilde{\mu}_\alpha \, dN_\alpha = \sum_{\alpha=1}^{k} \tilde{\nu}_\alpha \tilde{\mu}_\alpha \, d\xi = \mathcal{A} \, d\xi. \tag{4.69}$$

where $\mathcal{A} = \sum_{\alpha=1}^{k} \tilde{\nu}_\alpha \tilde{\mu}_\alpha$ is the *affinity* of the chemical reaction. From (4.33) one can show that the Gibbs free energy is minimized at equilibrium (see Figure 4.3), so that, from (4.69), we have

$$\left(\frac{\partial G}{\partial \xi} \right)_{T,p} = \sum_{\alpha=1}^{k} \tilde{\nu}_\alpha \tilde{\mu}_\alpha = \mathcal{A} = 0. \tag{4.70}$$

For $\mathcal{A} < 0$, the reaction proceeds in the forward direction, and for $\mathcal{A} > 0$, the reaction proceeds in the reverse direction. At equilibrium the chemical affinity, which can be expressed as the sum of the affinities for the reaction in the forward ($\alpha = 1, \ldots, q$) and reverse ($\alpha = q+1, \ldots, k$) directions, vanishes, $\mathcal{A} = 0$. This result can be used to determine the composition of the system. In Exercise 4.7, we obtained an expression for the chemical potential for species α in an ideal mixture. Substitution of (4.60) into (4.70) gives, after some manipulation,

$$\prod_{\alpha=1}^{k} x_\alpha^{\tilde{\nu}_\alpha} = K(T), \tag{4.71}$$

which is the mass action law at equilibrium. The result in (4.71), which holds only for ideal mixtures, expresses the idea that the rates of chemical reaction in the forward and reverse directions, which are equal at equilibrium, are proportional to the product of the concentrations of the reactants and products raised to the power of their stoichiometric coefficient. If the equilibrium constant $K(T)$ is known, (4.71) can be used to determine the extent of reaction at equilibrium ξ_{eq}, and hence composition, of the reacting mixture. Note that these results can easily be generalized to multiple reactions.

The equilibrium constant $K(T)$ is given by

$$K(T) = \exp\left(-\frac{\Delta\tilde{g}^0}{\tilde{R}T}\right),\tag{4.72}$$

where $\Delta\tilde{g}^0 = \sum_{\alpha=1}^k \tilde{\nu}_\alpha\tilde{\mu}_\alpha^0$ is the standard molar Gibbs free energy for the reaction. Note that we have not explicitly indicated an additional pressure dependence of $K(T)$; in particular the equilibrium for reactions between gases can be significantly influenced by pressure. The temperature dependence of $K(T)$ in (4.72) is commonly expressed by what is known as the Van't Hoff equation,

$$\frac{d}{dT}\ln K(T) = \frac{\Delta\tilde{h}^0}{\tilde{R}T^2},\tag{4.73}$$

where $\Delta\tilde{h}^0 = \sum_{\alpha=1}^k \tilde{\nu}_\alpha\tilde{h}_\alpha^0$ is the standard molar enthalpy change of the reaction. From (4.73) we see that for endothermic reactions ($\Delta\tilde{h}^0 > 0$) $K(T)$ increases with increasing temperature, which favors the forwards reaction, while the opposite is true for exothermic reactions ($\Delta\tilde{h}^0 < 0$).

Exercise 4.10   Temperature Dependence of Equilibrium Constant
Use the equilibrium condition (4.70) and a Maxwell relation to derive (4.73).

- Thermodynamics provides the framework for the description of equilibrium systems in terms of a few macroscopic variables, which come in pairs of extensive and intensive quantities (such as volume and pressure, entropy and temperature, particle number and chemical potential ...).

- The famous first and second laws of thermodynamics can be combined into Gibbs' fundamental form, which expresses changes of internal energy in terms of changes of extensive variables, multiplied by the corresponding intensive variables.

- If we know the internal energy as a function of the extensive variables, we refer to this function as a thermodynamic potential because all equations of state for the intensive variables can be obtained as first-order derivatives with respect to the extensive variables; thermodynamic potentials for other sets of independent variables, such as Helmholtz or Gibbs free energies, can be obtained as Legendre transforms of energy.

- Second-order derivatives of thermodynamic potentials define material properties such as heat capacity or isothermal compressibility; these second-order derivatives are also related to stability and fluctuations of thermodynamic systems.

- If one is interested only in material behavior and not the size of the system, a thermodynamic description is possible entirely in terms of intensive quantities where most structural features of thermodynamics can be preserved.

5

Balance Equations

We now begin the task of formulating the equations that provide the foundation of transport phenomena. To formulate a complete set of equations, that is, a set of equations in which the number of equations equals the number of unknowns, we will need two types of equations. Equations of the first type are known as *balance equations*, which are the subject of this chapter. Equations of the second type are commonly referred to as *constitutive equations*. In the following chapter, we introduce constitutive equations of the *force–flux* type that are based on the thermodynamic quantities introduced in Chapter 4. With these equations, we will be in a position to describe a wide range of phenomena that involve the transport of mass, momentum, energy, and charge.

The formulation of balance equations is based on densities of the extensive quantities introduced in Chapter 4 such as mass and energy. These densities are position- and time-dependent quantities, or fields. For example, the mass density field is denoted by $\rho = \rho(\mathbf{r}, t)$, where we have explicitly expressed the dependence on spatial position[1] $\mathbf{r}$ and time t. For systems with flow, we will also need the *momentum density* $\mathbf{m}$, which follows from the total momentum introduced in (4.6). That is $\mathbf{m} = \mathbf{M}/V$, or

$$\mathbf{m} = \rho \mathbf{v}, \tag{5.1}$$

where $\mathbf{v}$ is the velocity field which, as discussed in Section 4.1, is an intensive thermodynamic variable. It is more common in books on transport phenomena to refer to the product $\rho \mathbf{v}$ as a mass flux, and it is rare to refer to it as momentum density and assign to it a separate variable $\mathbf{m}$ as we have done in (5.1). For our purposes, there is no difference between the momentum density $\mathbf{m}$ and the mass flux $\rho \mathbf{v}$, but we will find it convenient to use both, for example in clarifying the structure of balance equations. Note that the

---

[1] In a rectangular coordinate system (x_1, x_2, x_3) with base vectors $\boldsymbol{\delta}_i$, $i = 1, 2, 3$, the position vector is $\mathbf{r} = \sum_{i=1}^{3} x_i \boldsymbol{\delta}_i$.

velocity field is given by the ratio of the mass flux and the mass density in exactly the same way as we had previously introduced velocity as the ratio of the probability flux and the probability density after (2.6).

The remainder of this chapter is devoted to the formulation of balance equations for mass, momentum, energy, and charge. The first step is to identify the system one intends to describe. For example, in applying the law of energy conservation from equilibrium thermodynamics, the system might be the volume inside a piston–cylinder device which contains a gas. For systems described by quantities that depend on spatial position and time (that is, fields), local forms of the conservation laws that are valid at each point in space are required.

There are several different approaches one can use to formulate local forms of the balance equations that depend on the choice of the system volume. In one approach, the volume is a well-defined object, say a cube, that is fixed in space,[2] while in another, the volume has arbitrary shape and time-dependent position, but contains the same particles at all times.[3] The type of volume chosen to be the system is a matter of preference; one only needs to make sure the volume is large enough that the particles within the volume experience a large number of collisions, and small enough that the phenomena of interest are described with sufficient spatial resolution. Here, our system is fixed in space and has arbitrary shape with volume V and surface A as shown in Figure 5.1. On every surface element dA, there is an outwardly directed unit normal vector $\boldsymbol{n}$. The size and position of the volume are arbitrary. Furthermore, we allow material particles to enter and leave the volume, and allow particles within the volume to exchange energy with those outside the volume. Since our balance equations are to be formulated using volumes, we prefer to use density fields for mass $\rho(\boldsymbol{r})$, momentum $\boldsymbol{m}(\boldsymbol{r})$, and internal energy $u(\boldsymbol{r})$.

5.1 Mass Conservation

Suppose that, at differential surface element dA, fluid with density ρ crosses the surface of V with velocity $\boldsymbol{v}$. The local volumetric flow rate per unit area across dA is $\boldsymbol{n} \cdot \boldsymbol{v}\, dA$, where $\boldsymbol{n} \cdot \boldsymbol{v}$ is the component of the velocity normal to the surface of the volume.[4] If the flow is outward, then $\boldsymbol{n} \cdot \boldsymbol{v}$ is positive,

---

[2] Bird, Stewart & Lightfoot, *Transport Phenomena* (Wiley, 2001).

[3] Slattery, *Advanced Transport Phenomena* (Cambridge, 1999); Deen, *Analysis of Transport Phenomena* (Oxford, 2012).

[4] The scalar, or dot, product of the vectors $\boldsymbol{a}$ and $\boldsymbol{b}$ in a rectangular coordinate system can be written as $\boldsymbol{a} \cdot \boldsymbol{b} = \sum_{i,j=1}^{3} a_i \boldsymbol{\delta}_i \cdot b_j \boldsymbol{\delta}_j = \sum_{i,j=1}^{3} a_i b_j \delta_{ij} = \sum_{i=1}^{3} a_i b_i$, where $\boldsymbol{\delta}_i \cdot \boldsymbol{\delta}_j = \delta_{ij}$ is the Kronecker delta.

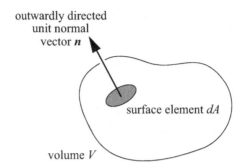

outwardly directed
unit normal
vector $\boldsymbol{n}$

surface element dA

volume V

Figure 5.1 Arbitrary volume, fixed in space, over which mass, momentum, and energy balances are made.

whereas if the flow is inward, $\boldsymbol{n} \cdot \boldsymbol{v}$ is negative. The local mass flow rate into V is given by $-\boldsymbol{n} \cdot \boldsymbol{v}\rho\,dA$.

According to the law of conservation of mass, the total mass within V will increase only as a result of a net influx of mass across the bounding surface A. Mathematically, this is expressed as

$$\frac{d}{dt} \int_V \rho\,dV = -\int_A \boldsymbol{n} \cdot \boldsymbol{v}\rho\,dA, \tag{5.2}$$

where the left-hand side is the time rate of change of mass within V and the right-hand side is the rate of mass flow into V across the surface A. When Gauss's divergence theorem[5] is used, the surface integral in (5.2) can be transformed into a volume integral,

$$\frac{d}{dt} \int_V \rho\,dV = -\int_V \boldsymbol{\nabla} \cdot (\boldsymbol{v}\rho)dV. \tag{5.3}$$

Since the volume V is fixed, we can bring the time derivative inside the integral on the left-hand side of (5.3), and then move both terms inside the same integral over V, which gives

$$\int_V \left[\frac{\partial \rho}{\partial t} + \boldsymbol{\nabla} \cdot (\boldsymbol{v}\rho)\right] dV = 0. \tag{5.4}$$

---

[5] For arbitrary vector fields $\boldsymbol{a}$ and tensor fields $\boldsymbol{\alpha}$ that are continuous and have continuous spatial derivatives in V and on A, Gauss's divergence theorem can be written as

$$\int_V \boldsymbol{\nabla} \cdot \boldsymbol{a}\,dV = \int_A \boldsymbol{n} \cdot \boldsymbol{a}\,dA, \qquad \int_V \boldsymbol{\nabla} \cdot \boldsymbol{\alpha}\,dV = \int_A \boldsymbol{n} \cdot \boldsymbol{\alpha}\,dA,$$

where $\boldsymbol{n}$ is the outward unit normal vector to V. The symbol $\boldsymbol{\nabla}$, which appears throughout this book, is the "del" or "nabla" operator. In a rectangular coordinate system $\boldsymbol{\nabla} = \sum_{i=1}^{3} \boldsymbol{\delta}_i \, \partial/\partial x_i$. Expressions for $\boldsymbol{\nabla}$ and its operation on scalars, vectors, and tensors in cylindrical and spherical coordinate systems are given in Appendix B.

We now have an integral over an arbitrary volume in space, and this integral is equal to zero. Because of the arbitrariness of V, and the assumed smoothness of the fields ρ and v, the integrand itself must be equal to zero. We thus obtain the following balance equation for mass,

$$\frac{\partial \rho}{\partial t} = -\boldsymbol{\nabla} \cdot (\boldsymbol{v}\rho), \tag{5.5}$$

which is called the *continuity equation*.

It is worthwhile to make a few observations concerning the structure of (5.5). We see that a time derivative of a density equals minus the divergence of a flux – in this case the mass density. This equation is similar in form to the Fokker–Planck equation (2.1) with a mass density that flows with velocity $\boldsymbol{v}$ and no diffusion. It is also possible to write (5.5) in the form

$$\frac{\partial \rho}{\partial t} + \boldsymbol{v} \cdot \boldsymbol{\nabla}\rho = -\rho\,\boldsymbol{\nabla} \cdot \boldsymbol{v}, \tag{5.6}$$

Equations (5.5) and (5.6) are equivalent, but can be interpreted differently. When $\boldsymbol{v}$ appears to the left of the $\boldsymbol{\nabla}$, as it does in the second term on the left-hand side of (5.6), it represents *convective* transport. The right-hand side of (5.6) expresses the change in mass density caused by a change in volume.

For flows where the left-hand side of (5.5) is zero, either because the flow is time-independent, or because the density is constant, it is sometimes convenient to replace the components of the vector field $\boldsymbol{v}$ with different variables. This is especially true when the flow field is two-dimensional or axisymmetric. As an example, consider a steady, planar flow in a rectangular coordinate system (x_1, x_2, x_3) of the form $\boldsymbol{v} = v_1\boldsymbol{\delta}_1 + v_2\boldsymbol{\delta}_2$. Here, $v_1 = v_1(x_1, x_2)$ and $v_2 = v_2(x_1, x_2)$ are the 1- and 2-components of the velocity. For this flow, (5.5) is written as

$$\frac{\partial}{\partial x_1}(\rho v_1) + \frac{\partial}{\partial x_2}(\rho v_2) = 0. \tag{5.7}$$

For a sufficiently smooth function $\psi(x_1, x_2)$, partial derivatives commute and we can write

$$\frac{\partial}{\partial x_1}\left(\frac{\partial \psi}{\partial x_2}\right) - \frac{\partial}{\partial x_2}\left(\frac{\partial \psi}{\partial x_1}\right) = 0, \tag{5.8}$$

which we use to interpret the continuity equation as an integrability condition (cf. footnote on p. 45). Comparison of (5.7) with (5.8) results in the following representation of the velocity field,

$$\rho v_1 = \frac{\partial \psi}{\partial x_2}, \qquad \rho v_2 = -\frac{\partial \psi}{\partial x_1}, \tag{5.9}$$

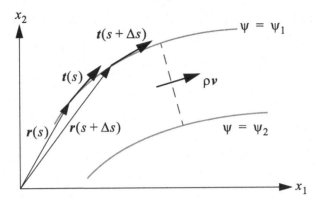

Figure 5.2 Streamline for a two-dimensional flow in the 1–2 plane showing unit tangent vector t at position r. Also shown is a second streamline connected to the first one by a dashed line at arbitrary positions.

where $\psi(x_1, x_2)$ is the *stream function* for this flow.

In addition to being a tool used to find solutions to flow problems, the stream function is also a useful way to visualize a flow field. A good way to illustrate this is to introduce the *streamline*, a curve which is tangent everywhere to the velocity at a given instant in time. Let $r(s)$ be the position of a streamline, where s is a parameter with units of length that gives the distance along the streamline as shown in Figure 5.2. The unit tangent vector to the streamline is given by $t(s) = dr/ds$. Since the velocity is tangent everywhere to a streamline, we can write $ds = v\, d\zeta$, where ζ is a parameter with units of time, and $v = \sqrt{v \cdot v}$ is the magnitude of the velocity. Hence, a streamline at a given instant in time can be obtained from the solution of

$$\frac{\partial r}{\partial \zeta} = v. \tag{5.10}$$

The stream function is useful for flow visualization because the stream function is constant along a streamline. Also, if a series of streamlines are plotted for values of the stream function that differ by a constant amount, in other words as a contour plot, it is easy to show that the mass flow rate is the same anywhere between two adjacent streamlines. Proof of these properties of the stream function is left as an exercise (see Exercise 5.3).

For a material comprised of two or more molecularly distinct species, or a *multi-component system*, the balance equation for the mass of species α is formulated using the same approach as before. However, in addition to the change in mass by flow, the mass of a species can change as the result of *homogeneous* chemical reactions.[6] In general, a given species may be involved in multiple reactions (see Exercise 5.6), but here we consider systems with a single chemical reaction. To take chemical reaction into account, we introduce the mass rate of reaction per unit volume, $\Gamma = (\tilde{M}_q/V)d\xi/dt$, where ξ is the extent of reaction introduced in (4.66), and $\tilde{M}_q$ is defined in (4.68). This leads to the continuity equation for species α,

$$\frac{\partial \rho_\alpha}{\partial t} = -\boldsymbol{\nabla} \cdot (\boldsymbol{v}_\alpha \rho_\alpha) + \nu_\alpha \Gamma, \tag{5.11}$$

where ρ_α and $\boldsymbol{v}_\alpha$ are the mass density and the velocity of species α, respectively, and ν_α, as discussed after (4.67), is related to the stoichiometric coefficient $\tilde{\nu}_\alpha$ of species α. The presence of $\nu_\alpha \Gamma$ on the right-hand side of (5.11), which appears outside the divergence term, indicates a *source* term in the structure of balance equations. Hence, the mass of a species is not a conserved quantity if chemical reactions take place.

If we sum the momenta of each species in a k-component system, we have

$$\sum_{\alpha=1}^{k} \boldsymbol{v}_\alpha \rho_\alpha = \rho \sum_{\alpha=1}^{k} \boldsymbol{v}_\alpha w_\alpha = \boldsymbol{v} \rho, \tag{5.12}$$

where $w_\alpha = \rho_\alpha/\rho$ is the mass fraction of species of α, and $\rho = \sum_{\alpha=1}^{k} \rho_\alpha$ is the mass density of the mixture. From the second equality in (5.12), we obtain the velocity $\boldsymbol{v} = \sum_{\alpha=1}^{k} w_\alpha \boldsymbol{v}_\alpha$, which is known as the mass-average, or barycentric, velocity of the mixture. Note that the addition of the momenta in (5.12) to get a properly weighted average velocity guarantees that the continuity equation (5.5) remains valid for multi-component systems, as can be seen by summing (5.11) over α and using (4.67) and (5.12). Moreover, the mass-average velocity we have just introduced leads to a natural momentum balance equation for mixtures.

The product $\boldsymbol{v}_\alpha \rho_\alpha$ is the mass flux of species α at a stationary point. It is useful to express this as the sum of two fluxes,

$$\boldsymbol{v}_\alpha \rho_\alpha = \boldsymbol{v} \rho_\alpha + \boldsymbol{j}_\alpha = \boldsymbol{n}_\alpha, \tag{5.13}$$

where $\boldsymbol{v} \rho_\alpha$ is the flux of species α from the average velocity $\boldsymbol{v}$, and $\boldsymbol{j}_\alpha$ is the flux of species α relative to the average velocity. The flux associated with $\boldsymbol{v}$ is a *convective* flux; the flux associated with $\boldsymbol{j}_\alpha$, which results from molecular interactions, is a *diffusive* flux. The two contributions to the species mass flux in (5.13) have profoundly different thermodynamic consequences: the

---

[6] Note that *heterogeneous* chemical reactions, which take place at phase interfaces, are taken into account using balance equations for interfaces, as discussed in Chapter 14.

$v\rho_\alpha$ contribution is a *reversible* transport process, while the j_α contribution is an *irreversible* transport process.

Summing (5.13) we find $\sum_{\alpha=1}^{k} j_\alpha = 0$, which follows from the constraint on the mass fractions, $\sum_{\alpha=1}^{k} w_\alpha = 1$. Substitution of (5.13) into (5.11) gives

$$\frac{\partial \rho_\alpha}{\partial t} = -\nabla \cdot (v\rho_\alpha + j_\alpha) + \nu_\alpha \Gamma. \qquad (5.14)$$

Again, we see an equation that has a structure similar to the Fokker–Planck equation (2.1). We will see in Chapter 6 that the diffusive flux j_α appearing in (5.14) is proportional to the gradient of ρ_α, so that the equation implies the presence of second-order derivatives, or diffusion.

It is also possible to express the species mass balance in terms of the mass fraction w_α. Substitution of $\rho_\alpha = w_\alpha \rho$ into (5.14) gives

$$\rho \left(\frac{\partial w_\alpha}{\partial t} + v \cdot \nabla w_\alpha \right) = -\nabla \cdot j_\alpha + \nu_\alpha \Gamma. \qquad (5.15)$$

In deriving (5.15), we have used

$$\frac{\partial}{\partial t}(\rho a) + \nabla \cdot (v\rho a) = \rho \left(\frac{\partial a}{\partial t} + v \cdot \nabla a \right) = \rho \frac{Da}{Dt}, \qquad (5.16)$$

which follows from (5.5) and is valid for any scalar or vector field a. The second equality in (5.16) introduces the *material* or *substantial derivative* of a, which gives the time rate of change of a quantity following the motion of a particle, and is defined by

$$\frac{D}{Dt} = \frac{\partial}{\partial t} + v \cdot \nabla. \qquad (5.17)$$

The substantial derivative establishes a connection between the motion of a particle and the velocity field associated with the mass flux [see discussion after (2.6)]. Equations (5.14) and (5.15) both express species mass conservation, but in different forms. When expressed in terms of a density as in (5.14), v appears to the right of ∇ and the term is interpreted as the convective flux to a stationary point. The quantity governed by (5.15) is not a density, but instead a quantity normalized by the mass density, and in this case, v appears to the left of ∇.

In some cases, those involving chemical reactions in particular, it is more convenient to work with molar, rather than mass concentration, variables. Let $c_\alpha = \rho_\alpha / \tilde{M}_\alpha$ be the molar density of species α, where $\tilde{M}_\alpha$ is the molecular weight of species α. Introducing c_α into (5.11) gives the molar form of the continuity equation for species α,

$$\frac{\partial c_\alpha}{\partial t} = -\boldsymbol{\nabla} \cdot (\boldsymbol{v}_\alpha c_\alpha) + \tilde{\nu}_\alpha \tilde{\Gamma}, \tag{5.18}$$

where $\tilde{\nu}_\alpha = \nu_\alpha \tilde{M}_q / \tilde{M}_\alpha$ is the stoichiometric coefficient of species α, which was introduced in (4.66), and $\tilde{\Gamma} = \Gamma/\tilde{M}_q = (1/V)d\xi/dt$ is the molar rate of reaction per unit volume, which is sometimes called the reaction velocity.

Now, if we sum the molar fluxes of each species in a k-component system, we have

$$\sum_{\alpha=1}^{k} \boldsymbol{v}_\alpha c_\alpha = c \sum_{\alpha=1}^{k} \boldsymbol{v}_\alpha x_\alpha = \boldsymbol{v}^* c, \tag{5.19}$$

where $x_\alpha = c_\alpha/c$ is the mole fraction of species of α, and $c = \sum_{\alpha=1}^{k} c_\alpha$ is the molar density of the mixture. From the second equality in (5.19), we obtain $\boldsymbol{v}^* = \sum_{\alpha=1}^{k} x_\alpha \boldsymbol{v}_\alpha$, which is the molar-average velocity of the mixture. This velocity is associated with the molar flux in the same spirit as we had previously associated velocities with the mass flux and with the probability flux.

Similarly to (5.13), we can express the molar flux of species α with respect to a fixed frame as the sum of convective and diffusive fluxes,[7]

$$\boldsymbol{v}_\alpha c_\alpha = \boldsymbol{v}^* c_\alpha + \boldsymbol{J}_\alpha^* = \boldsymbol{N}_\alpha, \tag{5.20}$$

where $\boldsymbol{J}_\alpha^*$ is the diffusive molar flux of species α relative to the molar average velocity $\boldsymbol{v}^*$. Summing (5.20) we find $\sum_{\alpha=1}^{k} \boldsymbol{J}_\alpha^* = \boldsymbol{0}$, which follows from the constraint on the mole fractions, $\sum_{\alpha=1}^{k} x_\alpha = 1$. Substituting (5.20) into (5.18) we obtain

$$\frac{\partial c_\alpha}{\partial t} = -\boldsymbol{\nabla} \cdot (\boldsymbol{v}^* c_\alpha + \boldsymbol{J}_\alpha^*) + \tilde{\nu}_\alpha \tilde{\Gamma}. \tag{5.21}$$

Now, if we sum (5.21) over α and use (5.19) we obtain the molar form of the continuity equation for a mixture,

$$\frac{\partial c}{\partial t} = -\boldsymbol{\nabla} \cdot (\boldsymbol{v}^* c) + \sum_{\alpha=1}^{k} \tilde{\nu}_\alpha \tilde{\Gamma}. \tag{5.22}$$

The derivation of an equation that governs the mole fraction x_α in a mixture, analogous to (5.15), from (5.21) is left as an exercise.

---

[7] We are using the common notation (see Bird, Stewart & Lightfoot, *Transport Phenomena* (Wiley, 2001)) where mass fluxes are denoted by lower-case letters ($\boldsymbol{j}_\alpha, \boldsymbol{n}_\alpha$) and molar fluxes by upper-case letters ($\boldsymbol{J}_\alpha^*, \boldsymbol{N}_\alpha$) with a superscript (or absence of one) on diffusive fluxes to indicate the relevant reference velocity ($\boldsymbol{v}, \boldsymbol{v}^*$); also see Exercise 5.5.

Clearly, the molar forms of the continuity equations are not as clean look-ing as their counterparts in terms of mass variables. More importantly, the clear separation of reversible and irreversible transport processes, which we will exploit in Chapter 6, is spoiled when a reference velocity other than $\boldsymbol{v}$ is used to define the diffusive flux. Nevertheless, the molar forms are more convenient to use in certain types of problems. It is also worth mentioning that the continuity equations can be expressed using other average velocities (see Exercise 5.5) for which different diffusive fluxes are defined.

Before proceeding to the next section, we should do some simple account-ing on the balance equations we have just derived. Let's start with the case of a one-component fluid, which means the law of mass conservation (5.5). This single equation contains two unknowns, ρ and $\boldsymbol{v}$, which means we cannot solve this equation by itself. For the case of a k-component fluid, we have an additional $k - 1$ equations given by (5.14), which have as un-knowns ρ_α, $\boldsymbol{j}_\alpha$, and Γ. (Note that the coefficients ν_α are usually known.) Clearly, to solve the balance equations introduced in this section, constitu-tive equations for $\boldsymbol{j}_\alpha$ and Γ will be required.

Exercise 5.1   *Divergence Theorem for Cubes*
Consider the following vector fields and calculate the left- and right-hand sides of Gauss's divergence theorem explicitly for a cubic volume: (i) $\boldsymbol{u}(\boldsymbol{r}) = (u_0 + u'x_1)\boldsymbol{\delta}_1$ for a cube defined by $0 \leq x_1, x_2, x_3 \leq L$ and (ii) $\boldsymbol{u}(\boldsymbol{r}) = x_2^2 x_3 \boldsymbol{\delta}_1 + x_2^3 \boldsymbol{\delta}_2 + x_1 x_3 \boldsymbol{\delta}_3$ for a cube defined by $-1 \leq x_1, x_2 \leq 1$ and $0 \leq x_3 \leq 2$.

Exercise 5.2   *Divergence Theorem for Spheres*
Consider a spherical volume of radius R centered at the origin. For the radial vector field $\boldsymbol{u}(\boldsymbol{r}) = u\,\boldsymbol{\delta}_r$, calculate the left- and right-hand sides of Gauss's divergence theorem explicitly for constant u. Repeat the calculation for the vector field $\boldsymbol{u} = u(r)\boldsymbol{\delta}_r$. Consider the special case $\boldsymbol{u} = u(r)\boldsymbol{\delta}_r$ with $u(r) = Q/r^2$. Why is $\nabla \cdot \boldsymbol{u} = 0$ but $\int_A \boldsymbol{n} \cdot \boldsymbol{u}\,dA = $ constant?

Exercise 5.3   *Properties of Stream Functions*
Since the velocity is tangent to a streamline, one has $\boldsymbol{v} \times \boldsymbol{t} = \boldsymbol{0}$.[8] Use (5.10) and the stream function defined in (5.9) to prove that the stream function is constant along a streamline, and that the mass flow rate between any two streamlines is equal to the difference between the values of the two streamlines.

---

[8] The vector, or cross, product of two vectors $\boldsymbol{a}$ and $\boldsymbol{b}$ in rectangular coordinates can be written as $\boldsymbol{a} \times \boldsymbol{b} = \sum_{i,j=1}^{3} a_i \boldsymbol{\delta}_i \times b_j \boldsymbol{\delta}_j = \sum_{i,j,k=1}^{3} a_i b_j \varepsilon_{ijk} \boldsymbol{\delta}_k$, where $\boldsymbol{\delta}_i \times \boldsymbol{\delta}_j = \sum_{k=1}^{3} \varepsilon_{ijk} \boldsymbol{\delta}_k$, and ε_{ijk} is known as the permutation symbol.

Exercise 5.4 *Species Mass Balance in Terms of Mole Fraction*

Show that the molar form of the species continuity equation (5.21) in terms of the mole fraction x_α is given by

$$c\left(\frac{\partial x_\alpha}{\partial t} + \boldsymbol{v}^* \cdot \boldsymbol{\nabla} x_\alpha\right) = -\boldsymbol{\nabla} \cdot \boldsymbol{J}_\alpha^* + \tilde{\nu}_\alpha \tilde{\Gamma} - x_\alpha \sum_{\beta=1}^{k} \tilde{\nu}_\beta \tilde{\Gamma}. \tag{5.23}$$

Exercise 5.5 *Reference Velocities*

We have just seen how different choices for reference velocities result in different definitions for the diffusive flux. For example, $\boldsymbol{j}_\alpha$ is the diffusive mass flux when $\boldsymbol{v}$ is the reference velocity, while $\boldsymbol{J}_\alpha^*$ is paired with $\boldsymbol{v}^*$. Derive the following relationships between mass- and molar-average velocities and diffusive fluxes:

$$\boldsymbol{v} - \boldsymbol{v}^* = -\frac{1}{c} \sum_{\beta=1}^{k} \frac{\boldsymbol{j}_\beta}{\tilde{M}_\beta}, \tag{5.24}$$

$$\tilde{M}_\alpha \boldsymbol{J}_\alpha^* = \boldsymbol{j}_\alpha^* = \boldsymbol{j}_\alpha - \frac{\rho_\alpha}{c} \sum_{\beta=1}^{k} \frac{\boldsymbol{j}_\beta}{\tilde{M}_\beta}. \tag{5.25}$$

Other choices for reference velocity include the solvent velocity $\boldsymbol{v}^k = \sum_{\alpha=1}^{k} \delta_{\alpha k} \boldsymbol{v}_\alpha$, which is useful for dilute mixtures. The volume-average velocity is defined as $\boldsymbol{v}^\dagger = \sum_{\alpha=1}^{k} \rho_\alpha \hat{v}_\alpha \boldsymbol{v}_\alpha$, where $\rho_\alpha \hat{v}_\alpha$ is the volume fraction of species α and $\rho_\alpha \boldsymbol{v}_\alpha = \rho_\alpha \boldsymbol{v}^\dagger + \boldsymbol{j}_\alpha^\dagger$. Substitution into (5.11) gives

$$\frac{\partial \rho_\alpha}{\partial t} = -\boldsymbol{\nabla} \cdot (\boldsymbol{v}^\dagger \rho_\alpha + \boldsymbol{j}_\alpha^\dagger) + \nu_\alpha \Gamma, \tag{5.26}$$

Derive the following relationships between mass- and volume-average velocities and diffusive fluxes:

$$\boldsymbol{v} - \boldsymbol{v}^\dagger = -\sum_{\beta=1}^{k} \hat{v}_\beta \boldsymbol{j}_\beta, \tag{5.27}$$

$$\boldsymbol{j}_\alpha^\dagger = \boldsymbol{j}_\alpha - \rho_\alpha \sum_{\beta=1}^{k} \hat{v}_\beta \boldsymbol{j}_\beta. \tag{5.28}$$

Use (5.11) to show that for ideal mixtures (constant $\hat{v}_\alpha$) without chemical reaction $\boldsymbol{\nabla} \cdot \boldsymbol{v}^\dagger = 0$.

Exercise 5.6 *Multiple Chemical Reactions*

For a system involving n independent chemical reactions, the expression in (4.66) is generalized to $dN_\alpha = \sum_{j=1}^{n} \tilde{\nu}_{\alpha,j}\, d\xi_j$, where $\tilde{\nu}_{\alpha,j}$ are the stoichiometric coefficients of species α in reaction j, and ξ_j is the extent of reaction j. Show for multiple reactions that (5.14) can be written as

$$\frac{\partial \rho_\alpha}{\partial t} = -\boldsymbol{\nabla} \cdot (\boldsymbol{v}_\alpha \rho_\alpha) + \sum_{j=1}^{n} \nu_{\alpha,j} \Gamma_j, \tag{5.29}$$

where $\Gamma_j = (\tilde{M}_{q,j}/V)d\xi_j/dt$, and $\tilde{M}_{q,j} = \sum_{\alpha=q_j+1}^{k} \tilde{\nu}_{\alpha,j}\tilde{M}_\alpha = -\sum_{\alpha=1}^{q_j} \tilde{\nu}_{\alpha,j}\tilde{M}_\alpha$.
Also, generalize (5.23) for multiple reactions.

Exercise 5.7   Plug-Flow Reactor Design Equation
The plug-flow reactor (PFR) model is commonly used by engineers to design continuous tubular reactors. The name plug flow reflects the fact that velocity is assumed to be uniform across the cross section of the tube. The PFR model also assumes that diffusive transport can be neglected. For this exercise, we consider a fluid of constant density ρ flowing through the PFR of length L with velocity $\boldsymbol{v} = v\boldsymbol{\delta}_z$. Using (5.5) and (5.14), derive the well-known design equation for a PFR,

$$\frac{L}{v} = \int_{c_\alpha(0)}^{c_\alpha(L)} \frac{dc_\alpha}{\tilde{\nu}_\alpha\tilde{\Gamma}(c_\alpha)},$$

where $c_\alpha(0)$ and $c_\alpha(L)$ are the molar densities of species α at the entrance and exit, respectively, of the PFR. Note that this expression relates the residence time in the reactor, L/v, to the reaction rate $\tilde{\Gamma}$, which can be expressed in terms of c_α.

5.2 Momentum Conservation

To formulate a balance equation based on the principle of momentum conservation, we refer to our system volume V depicted in Figure 5.1. In the previous section we noted that the local volume rate of flow across the surface element dA is $\boldsymbol{n} \cdot \boldsymbol{v}\, dA$. If this is multiplied by the momentum density of the fluid $\boldsymbol{m}$, we have $\boldsymbol{n} \cdot \boldsymbol{vm}\, dA$, which is the rate at which momentum is carried by flow across the surface element dA. The quantity $\boldsymbol{vm} = \rho\boldsymbol{vv}$ is the convective momentum flux, that is, the momentum crossing a surface per unit area and per unit time, associated with the bulk flow of fluid. This is another example of convective transport. Note the similarity between the convective transport of mass and momentum, but also that the tensorial order of the quantities involved is different. In the preceding section, the entity being transported is the mass (a scalar) and the mass flux is a vector ($\boldsymbol{v}\rho$). Here, the entity being transported is the momentum (a vector), and the momentum flux is a tensor ($\boldsymbol{vm}$).

In addition to momentum transport by flow, momentum is transferred as a result of molecular interactions. The flux of momentum associated with these intermolecular effects is denoted by $\boldsymbol{\pi}$, again a second-rank tensor. In this book, we use the convention that the jk-component π_{jk} of this tensor represents the flux of k-momentum in the j-direction. The rate of momentum flow across the surface element dA with orientation $\boldsymbol{n}$ is then $\boldsymbol{n} \cdot \boldsymbol{\pi}\, dA$.

So far, we know the momentum of our system will change as a result of the momentum fluxes vm and π. Similarly to the source term which resulted from chemical reaction in (5.11), we must allow for the possibility of momentum sources. Maybe the reader is asking the following question: what is a momentum source? External forces are sources of momentum, and a well-known source of momentum is the gravitational force. Other examples of momentum sources are forces due to the presence of electromagnetic fields, which are considered in the last section of this chapter. In this section, we consider only the gravitational force, which is the same whether the system is comprised of multiple species, or only one. Let g be the gravitational force, or the rate of momentum production per unit mass. This means the rate of momentum production per unit volume is ρg.

We are now ready to formulate a balance equation that expresses the law of momentum conservation. According to this law, the momentum of the fluid within V increases because of a net influx of momentum across A from both bulk flow and molecular interactions, and as a result of momentum sources within V. We can express this mathematically as

$$\frac{d}{dt}\int_V m\, dV = -\int_A n \cdot (vm + \pi)dA + \int_V \rho g\, dV, \qquad (5.30)$$

where the left-hand side is the rate of increase of momentum within V, and the right-hand side is the rate of momentum transport across the surface A (first term) and the total source of momentum within V (second term). When Gauss's divergence theorem (cf. footnote on p. 60) is used, the surface integral in (5.30) can be transformed into a volume integral

$$\frac{d}{dt}\int_V m\, dV = -\int_V \nabla \cdot (vm + \pi)dV + \int_V \rho g\, dV. \qquad (5.31)$$

As before, because the volume V is fixed, the order of differentiation and integration can be interchanged, which allows all terms in (5.31) to be brought inside a single integral. The integral is equal to zero, and because V is arbitrary, the integrand can be set equal to zero. This results in the balance equation for momentum,

$$\frac{\partial m}{\partial t} = -\nabla \cdot (vm + \pi) + \rho g. \qquad (5.32)$$

The presence of the source term ρg in (5.32) indicates that momentum is not a conserved quantity.

As discussed at the beginning of this chapter, the momentum density is equivalent to the mass flux as shown in (5.1). If in (5.32) we replace m with

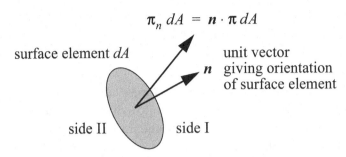

Figure 5.3 Element of surface dA across which a force $\pi_n \, dA$ is transmitted.

$\rho\boldsymbol{v}$, we can rewrite the momentum balance (5.32) as

$$\rho\frac{D\boldsymbol{v}}{Dt} = -\nabla \cdot \boldsymbol{\pi} + \rho\boldsymbol{g}, \tag{5.33}$$

where we have used (5.16). If we interpret the left-hand side of (5.33) as the product of the mass and acceleration of a fluid element, the right-hand side can be interpreted as the forces acting on the fluid element. As anticipated, the material derivative (5.17) allows us to establish a connection between the velocity associated with the mass flux and the trajectory of a fluid element. This would suggest that (5.33), which is sometimes referred to as the *equation of motion*, is an expression of Newton's second law. In fact, we could have used a somewhat different physical statement leading to (5.30): the time rate of change of the total momentum of the fluid within the volume V equals the net influx of momentum across the bounding surfaces by bulk flow plus the forces acting on the fluid within V. We then expect a surface force term of the form $\int \boldsymbol{\pi}_n \, dA$, where $\boldsymbol{\pi}_n \, dA$ is a vector describing the force exerted by the fluid on side II of dA on the fluid on side I of dA (see Figure 5.3). Comparison of the above integral with the corresponding term in (5.30) shows that $\boldsymbol{\pi}_n = \boldsymbol{n} \cdot \boldsymbol{\pi}$, which is sometimes referred to as the stress vector. That is, the force $\boldsymbol{\pi}_n \, dA$ corresponding to an orientation $\boldsymbol{n}$ of dA can be obtained from the tensor $\boldsymbol{\pi}$.

The existence of the tensor $\boldsymbol{\pi}$ and its relation to the stress vector $\boldsymbol{\pi}_n$ are sometimes referred to as the *stress principle of Cauchy*.[9] When this interpretation is used, it is more natural to refer to $\boldsymbol{\pi}$ as the *stress tensor* or, in view of the sign conventions used here,[10] as the *pressure tensor*. The

---

[9] See Section 16 of Truesdell & Noll, *Non-Linear Field Theories of Mechanics* (Springer, 1992).

[10] We have adopted the sign convention of the textbook Bird, Stewart & Lightfoot, *Transport Phenomena* (Wiley, 2001); to remind the reader that this convention is different from that of many other authors, we refer to $\boldsymbol{\pi}$ as the pressure tensor, indicating that an isotropic pressure p corresponds to $\boldsymbol{\pi} = p\boldsymbol{\delta}$. Here, $\boldsymbol{\delta}$ is the unit, or identity, tensor, which in rectangular coordinates is $\boldsymbol{\delta} = \sum_{i,j}^{3} \delta_{ij}\boldsymbol{\delta}_i\boldsymbol{\delta}_j$.

component π_{jk} is the force per unit area acting in the k-direction on a surface perpendicular to the j-direction, the force being exerted by the side II fluid on the side I fluid. The concept of force per unit area arises naturally only for short-range interactions between particles on different sides of a surface, which can be regarded as contact forces.

If one uses this viewpoint, then the integral $\int \boldsymbol{n} \cdot \boldsymbol{\pi} \, dA$ in (5.30) can be reinterpreted as the force of the fluid outside V acting on the fluid inside V across A. For some purposes, it is useful to think of $\boldsymbol{\pi}$ as a momentum flux, whereas in other situations the concept of stress is more natural. We shall use both interpretations, and the terms momentum flux tensor and pressure tensor will be used interchangeably.

Similarly, we can interpret the last term on the right-hand side of (5.32) or (5.33) as the forces acting on the fluid element that arise from the presence of external fields; for this reason they are often referred to as external forces. Conservative forces such as gravity can be expressed as the gradient of a time-independent scalar potential field. If ϕ is the gravitational potential field, then $\boldsymbol{g} = -\boldsymbol{\nabla}\phi$.

Our balance equations should also be consistent with the conservation of angular momentum, which can be expressed as a balance equation for angular momentum density $\boldsymbol{r} \times \boldsymbol{m}$. This results in

$$\boldsymbol{\pi}^{\mathrm{T}} = \boldsymbol{\pi}, \tag{5.34}$$

which means that $\boldsymbol{\pi}$ is a symmetric tensor. Proof of (5.34) is left as an exercise (see Exercise 5.8).

It is useful to consider the special case that exists when the pressure tensor is isotropic (cf. footnote on p. 70), that is $\boldsymbol{\pi} = p\boldsymbol{\delta}$, where p is the pressure. For this case, the equation of motion (5.33) simplifies to

$$\rho\frac{D\boldsymbol{v}}{Dt} = -\boldsymbol{\nabla}p + \rho\boldsymbol{g}, \tag{5.35}$$

which is known as Euler's equation, named after the mathematician Leonhard Euler (1707–1783), whose name has already appeared in Chapter 4. We will see in Chapter 6 that fluids for which the pressure tensor has the form $\boldsymbol{\pi} = p\boldsymbol{\delta}$ do not produce entropy even if they are flowing. Since no real fluid behaves this way, fluids that approach this limiting case are called *ideal fluids*. Hence, (5.35) is restricted to flows of ideal fluids. We will also see in Chapter 6 that for fluids in uniform motion ($\boldsymbol{\nabla}v = \boldsymbol{0}$), or in rigid rotation ($\boldsymbol{\nabla}v = -\boldsymbol{\nabla}v^{\mathrm{T}}$), that $\boldsymbol{\pi} = p\boldsymbol{\delta}$; in such cases (5.35) also applies.

Incompressible Flows

If the density of the fluid is independent of position and time, we have from (5.5) or (5.6) the following,

$$\nabla \cdot \boldsymbol{v} = 0, \tag{5.36}$$

which is the constraint on velocity for incompressible flow.[11] A detailed discussion of the conditions for which ρ is constant, or (5.36) holds, has been given by Batchelor,[12] and will also be discussed in Chapter 7. The scalar equation (5.36) cannot be solved for the velocity field by itself, but rather imposes a *constraint* on $\boldsymbol{v}$. It is common in the solution of fluid mechanics problems to assume incompressibility, so it is important to understand the implications of (5.36).

When the fluid density is assumed to be constant, an equation of state of the form $\rho = \rho(T, p)$ does not have meaning, because this equation cannot be inverted to express p as a function of ρ. Hence, when the constraint (5.36) is imposed, pressure in the thermodynamic sense no longer exists. The incompressibility condition (5.36) can be rewritten as

$$\int_V p^{\mathrm{L}} \, \nabla \cdot \boldsymbol{v} \, dV = 0, \tag{5.37}$$

for an arbitrary function $p^{\mathrm{L}} = p^{\mathrm{L}}(\boldsymbol{r})$. After an integration by parts and neglecting boundary terms, we have the alternative general form of the incompressibility constraint,

$$\Pi(\boldsymbol{v}) = -\int_V \boldsymbol{v} \cdot \nabla p^{\mathrm{L}} \, dV = 0. \tag{5.38}$$

Note that the functional derivative

$$\frac{\delta \Pi(\boldsymbol{v})}{\delta \boldsymbol{v}(\boldsymbol{r})} = -\nabla p^{\mathrm{L}}(\boldsymbol{r}) \tag{5.39}$$

is normal to the constraint and should be added to the equation of motion, where the Lagrange multiplier p^{L}, which we will refer to as the *pseudo-pressure*,[13] is to be determined from the condition that the constraint is to be satisfied. Adding a term normal to the constraint such that the evolution stays within the constrained manifold is nothing but a projection into the

---

[11] It may be more appropriate to describe a flow for which (5.36) holds as "noncompressing"; however, in this book we use the more common terminology "incompressible."

[12] Batchelor, *An Introduction to Fluid Dynamics* (Cambridge University Press, 1967).

[13] The term pseudo-pressure is also used in the analysis of gas flows in transmission lines and porous media, but has no relation to the function p^{L} defined in (5.37)–(5.39).

constrained manifold.[14] Note that the nabla in (5.39) is a direct consequence of the nabla in (5.36).

For the incompressible flow of an ideal fluid, the equation of motion, or Euler equation, (5.35) becomes

$$\rho \frac{D\boldsymbol{v}}{Dt} = -\boldsymbol{\nabla} p^{\mathrm{L}} + \rho \boldsymbol{g} = -\boldsymbol{\nabla}\mathcal{P}, \tag{5.40}$$

where we have introduced the *modified pressure* given by

$$\mathcal{P} = p^{\mathrm{L}} + \rho\phi, \tag{5.41}$$

which follows since $\boldsymbol{g} = -\boldsymbol{\nabla}\phi$ and ρ is constant. Taking the divergence of (5.40) and applying (5.36) gives

$$\nabla^2 \mathcal{P} = -\rho \boldsymbol{\nabla}\boldsymbol{v} : \boldsymbol{\nabla}\boldsymbol{v}, \tag{5.42}$$

The addition of $\rho\phi$ to p^{L} in (5.41), which can be done with any conservative force like gravity, further reinforces the idea that the pseudo-pressure p^{L} is an arbitrary function. Even though (5.42) was derived from (5.35), which governs the flow of ideal fluids, it turns out that this equation is valid for incompressible flow of an important class of real fluids as well.

Exercise 5.8   Symmetry of the Pressure (Stress) Tensor
Form the vector product of $\boldsymbol{r}$ with the momentum balance (5.32), to obtain[15]

$$\frac{\partial}{\partial t}(\boldsymbol{r} \times \boldsymbol{m}) = -\boldsymbol{\nabla} \cdot \boldsymbol{v}(\boldsymbol{r} \times \boldsymbol{m}) - \boldsymbol{\nabla} \cdot (\boldsymbol{r} \times \boldsymbol{\pi}^{\mathrm{T}})^{\mathrm{T}} - \boldsymbol{\varepsilon} : \boldsymbol{\pi} + \boldsymbol{r} \times \rho\boldsymbol{g},$$

which is the balance equation for angular momentum density $\boldsymbol{r} \times \boldsymbol{m}$. Show that the source associated with the pressure tensor $\boldsymbol{\varepsilon} : \boldsymbol{\pi}$ vanishes if the pressure tensor is symmetric,[16] $\boldsymbol{\pi}^{\mathrm{T}} = \boldsymbol{\pi}$.

Exercise 5.9   Irrotational Flow
A useful quantity in fluid mechanics called vorticity is defined as

$$\boldsymbol{w} = \boldsymbol{\nabla} \times \boldsymbol{v}, \tag{5.43}$$

which is a measure of the rotation of a fluid element. A flow for which the vorticity is zero, $\boldsymbol{w} = \boldsymbol{0}$, is called *irrotational flow*, or potential flow. Use the alternating identity tensor $\boldsymbol{\varepsilon}$ (see Exercise 5.8) to show for irrotational flows that the velocity gradient is symmetric: $\boldsymbol{\nabla}\boldsymbol{v} = \boldsymbol{\nabla}\boldsymbol{v}^{\mathrm{T}}$.

---

[14] As a simple example of this concept, consider a weight tied to the end of a rope traveling in a circular path. The weight is constrained to travel in a circular path because of the force exerted on it by the rope, which is perpendicular to the tangent to the circular path.

[15] The third-order tensor $\boldsymbol{\varepsilon} = \sum_{i,j,k=1}^{3} \varepsilon_{ijk}\boldsymbol{\delta}_i\boldsymbol{\delta}_j\boldsymbol{\delta}_k$ is the alternating identity tensor.

[16] Dahler & Scriven, *Nature* **192** (1961) 36.

Use the identity $\nabla \times \nabla a = \mathbf{0}$, which is valid for any scalar a, to show for an incompressible, irrotational flow that

$$\nabla^2 \Phi = 0, \tag{5.44}$$

where $\mathbf{v} = -\nabla \Phi$ and Φ is the *velocity potential*. For the two-dimensional flow defined in (5.9) with constant ρ, show that the gradients of the stream function $\psi(x_1, x_2)$ and velocity potential $\Phi(x_1, x_2)$ are orthogonal.

Exercise 5.10   Potential Flow of Ideal Fluids
Consider the steady potential flow of incompressible ideal fluids. Using the identity

$$\mathbf{a} \cdot \nabla \mathbf{a} = \frac{1}{2} \nabla a^2 - \mathbf{a} \times (\nabla \times \mathbf{a}), \tag{5.45}$$

which holds for arbitrary vector $\mathbf{a}$, show that (5.40) can be written as

$$\nabla \left(\frac{1}{2} v^2 + \frac{\mathcal{P}}{\rho} \right) = \nabla \left(\frac{1}{2} v^2 + \phi + \frac{p^{\mathrm{L}}}{\rho} \right) = \mathbf{0}. \tag{5.46}$$

Exercise 5.11   Potential Flow around a Cylinder
Consider the steady potential flow of an incompressible ideal fluid normal to an infinitely long cylinder of radius R having the velocity field $v_r = v_r(r, \theta)$, $v_\theta = v_\theta(r, \theta)$. Far from the cylinder, the fluid has uniform velocity $\mathbf{v} = V\boldsymbol{\delta}_1$, which suggests that the velocity potential has the form $\Phi = f(r)\cos\theta$. Solve (5.44) and find the velocity, and use (5.46) to find the modified pressure field. Plot the velocity potential $\Phi(r, \theta)$ and stream function $\psi(r, \theta)$.

5.3 Energy Conservation

We are now sufficiently experienced to work with local balance equations and next consider the principle of energy conservation. The total energy of a fluid element is the sum of the potential, kinetic, and internal energies. Generalizing (4.5), we write the total energy density as

$$e = u + \rho\phi + \frac{1}{2}\rho v^2. \tag{5.47}$$

The total energy is a conserved quantity, which means it increases only as a result of a net transfer of energy to the system. Balance equations for potential energy and kinetic energy, or mechanical forms of energy, are contained within the balance equations for mass and momentum formulated in the previous sections. By formulating these equations first, we can construct a balance equation for the internal energy, since the sum of these equations must result in a balance equation for a conserved quantity.

The balance equation for potential energy is formulated by multiplying (5.5) by ϕ, which, recalling that $\boldsymbol{g} = -\boldsymbol{\nabla}\phi$, can be written as

$$\frac{\partial}{\partial t}(\rho\phi) = -\boldsymbol{\nabla}\cdot(\rho\phi\boldsymbol{v}) - \rho\boldsymbol{v}\cdot\boldsymbol{g}, \tag{5.48}$$

where $\rho\phi$ is the *potential energy density* of a fluid element. The balance equation for kinetic energy can be derived by forming the scalar product of the velocity $\boldsymbol{v}$ with the momentum balance (5.32). This results, after some manipulation (see Exercise 5.12), in

$$\frac{\partial}{\partial t}\left(\frac{1}{2}\rho v^2\right) = -\boldsymbol{\nabla}\cdot\left(\boldsymbol{v}\frac{1}{2}\rho v^2 + \boldsymbol{\pi}\cdot\boldsymbol{v}\right) + \boldsymbol{\pi}:\boldsymbol{\nabla}\boldsymbol{v} + \rho\boldsymbol{v}\cdot\boldsymbol{g}, \tag{5.49}$$

where $\frac{1}{2}\rho v^2$ is the *kinetic energy density* of a fluid element. Note that the last terms on the right-hand sides of (5.48) and (5.49), which have opposite signs, are exchange terms. Adding the balance equations in (5.48) and (5.49), we obtain

$$\frac{\partial}{\partial t}\left(\frac{1}{2}\rho v^2 + \rho\phi\right) = -\boldsymbol{\nabla}\cdot\left(\boldsymbol{v}\frac{1}{2}\rho v^2 + \boldsymbol{v}\rho\phi + \boldsymbol{\pi}\cdot\boldsymbol{v}\right) + \boldsymbol{\pi}:\boldsymbol{\nabla}\boldsymbol{v}, \tag{5.50}$$

which is sometimes called the *mechanical energy balance*. The last term in (5.50) is a source term so that in general mechanical energy is not conserved.

We are now in position to formulate a balance equation for internal energy. We have just seen that a volume element in a potential field possesses potential energy, and a volume element with a velocity possesses kinetic energy. The potential and kinetic energies can thus be thought of as macroscopic forms of energy. According to (5.47), the total energy of a system equals the sum of the internal, potential, and kinetic energies. Hence, the difference between the total energy and macroscopic potential and kinetic energies equals the internal energy, which is the energy of the system associated with molecular interactions.

The total energy of the system is conserved. Hence, the balance equation for the internal energy density $u(\boldsymbol{r})$ must have source terms that cancel out the source terms in (5.50). As before, we can express the convective flux of internal energy as $\boldsymbol{v}u$. Similarly to the diffusive flux of species mass relative to the average velocity, we introduce $\boldsymbol{j}_q$, the diffusive flux of internal energy.[17] The flux $\boldsymbol{j}_q$, which is also referred to as the conductive energy flux, is a manifestation of molecular interactions analogous to the mass flux $\boldsymbol{j}_\alpha$ and momentum flux $\boldsymbol{\pi}$. Hence, the balance equation for *internal energy*

---

[17] It is more common to use $\boldsymbol{q}$, rather than $\boldsymbol{j}_q$, to denote the diffusive flux of internal energy.

density is written as

$$\frac{\partial u}{\partial t} = -\boldsymbol{\nabla} \cdot (\boldsymbol{v}u + \boldsymbol{j}_q) - \boldsymbol{\pi} : \boldsymbol{\nabla}\boldsymbol{v}, \tag{5.51}$$

with a source term for the conversion of mechanical to internal energy. The balance equation for the total energy, which is obtained by adding (5.50) and (5.51), does not have source terms since the total energy is a conserved quantity. The balance equation for the internal energy can also be expressed in terms of the specific internal energy $\hat{u} = u/\rho$, which, using (5.16), takes the form

$$\rho\frac{D\hat{u}}{Dt} = -\boldsymbol{\nabla} \cdot \boldsymbol{j}_q - \boldsymbol{\pi} : \boldsymbol{\nabla}\boldsymbol{v}. \tag{5.52}$$

We conclude this section with some additional accounting on the unknowns in our balance equations. Since our tally on the mass balance equations, we have added balance equations for momentum density (5.32) and internal energy density (5.51). The external force $\boldsymbol{g}$ is known and, since $\boldsymbol{m} = \rho\boldsymbol{v}$, the additional unknowns contained in these balance equations are $\boldsymbol{\pi}$, $\boldsymbol{j}_q$, and u. Again, we see that, on including additional balance equations, the number of unknowns in these equations exceeds the number of equations. To balance the books, now we will need to add constitutive equations for $\boldsymbol{j}_\alpha$, Γ, $\boldsymbol{\pi}$, and $\boldsymbol{j}_q$, which are the subject of Chapter 6.

Exercise 5.12   Derivation of Balance Equation for Kinetic Energy
Show that the scalar product of the velocity $\boldsymbol{v}$ with the momentum balance (5.32) results in the balance equation for kinetic energy density (5.49).

Exercise 5.13   Internal Energy Balance for a Two-Component System
Derive the following generalization of the internal energy balance (5.51) for two-component systems in the presence of external forces $\boldsymbol{f}_1$, $\boldsymbol{f}_2$ acting per unit mass of the two components:

$$\frac{\partial u}{\partial t} = -\boldsymbol{\nabla} \cdot (\boldsymbol{v}u + \boldsymbol{j}_q) - \boldsymbol{\pi} : \boldsymbol{\nabla}\boldsymbol{v} + \boldsymbol{j}_1 \cdot \boldsymbol{f}_1 + \boldsymbol{j}_2 \cdot \boldsymbol{f}_2. \tag{5.53}$$

Different external forces per unit mass of the components lead to the conversion of external into internal energy.

Exercise 5.14   Bernoulli Equation along a Streamline
Consider the steady flow of an ideal fluid so that the pressure tensor is isotropic, $\boldsymbol{\pi} = p\boldsymbol{\delta}$. Use (5.50) to derive

$$\Delta\left(\phi + \frac{1}{2}v^2\right) + \int_1^2 \frac{dp}{\rho} = 0, \tag{5.54}$$

where $\Delta(.) = (.)_2 - (.)_1$ is the difference in a quantity between points 2 and 1 along the streamline. Equation (5.54), which is known as the *Bernoulli equation along a streamline*, provides a relation among the gravitational potential, velocity, and pressure along a streamline. For example, consider a flow along a streamline at constant ϕ, in other words, at constant elevation. From (5.54), we see that, if the fluid velocity increases along the streamline, then the pressure must decrease. We benefit from this phenomenon while flying on a jet or riding on a windsurfing board.

Exercise 5.15   Speed of Sound
All sounds we hear, from the gentle trickle of a mountain stream to the music heard during a Rolling Stones concert, are the result of waves of compressed fluid that strike our eardrums. The waves of compressed fluid travel with a speed that depends on the properties of the fluid through which they travel. To derive the equation that governs the propagation of sound waves, we consider the isentropic, or constant entropy, flow of a one-component fluid. Even though the entropy balance equation will not be covered until Chapter 6, we know for the case considered here from (4.45) that $d\hat{u} = -p\,d\hat{v}$. Hence, (5.6) and (5.52) are redundant, which will only be the case if the pressure tensor is isotropic, $\boldsymbol{\pi} = p\boldsymbol{\delta}$, and the diffusive flux of internal energy is zero, $\boldsymbol{j}_q = \boldsymbol{0}$. If the fluid is otherwise at rest, and the difference between the compressed and uncompressed fluid is small, the following approximations can be used: $\rho = \rho_0 + \delta\rho$, $p = p_0 + \delta p$, and $\boldsymbol{v} = \boldsymbol{0} + \boldsymbol{v}$. Use these expansions to linearize (5.6) and (5.35) setting $\boldsymbol{g} = \boldsymbol{0}$ and derive

$$\frac{\partial^2 \delta p}{\partial t^2} = c_{\hat{s}}^2 \, \nabla^2 \delta p. \tag{5.55}$$

Notice that (5.55) is a hyperbolic, or wave, equation with wave speed $c_{\hat{s}}$,

$$c_{\hat{s}}^2 = \left(\frac{\partial p}{\partial \rho}\right)_{\hat{s}} = \frac{1}{\rho \kappa_{\hat{s}}}, \tag{5.56}$$

which is the isentropic speed of sound. The second equality in (5.56) shows the relationship between $c_{\hat{s}}$ and $\kappa_{\hat{s}}$, the isentropic compressibility (see Appendix A). The speed of sound in ordinary fluids is relatively large. For example, in air the speed of sound is about $340\,\text{m/s}$ while in water it is about $1500\,\text{m/s}$.

5.4 Charge Conservation

We now consider systems where flowing matter is comprised of charged species and hence causes and experiences electromagnetic fields. The purpose of this section is twofold: (i) we compile the basic equations describing the forces associated with moving charges, and (ii) we elaborate the deep connection between hydrodynamics and electromagnetism as field theories by establishing unified momentum and energy balances.

We denote the charge per unit mass of species α by $z_\alpha = n_\alpha \tilde{F}/\tilde{M}_\alpha$, where $\tilde{F}$ is the Faraday constant, and n_α is the elementary charge (an integer) of species α. The balance equation for the electric charge is obtained by multiplying (5.11) by z_α and summing over α. This results in

$$\frac{\partial \rho_{el}}{\partial t} = -\boldsymbol{\nabla} \cdot \boldsymbol{i} = -\boldsymbol{\nabla} \cdot (\rho_{el}\boldsymbol{v} + \boldsymbol{j}_{el}), \qquad (5.57)$$

where we have used $\sum_{\alpha=1}^{k} \tilde{M}_\alpha z_\alpha \tilde{\nu}_\alpha = 0$, which guarantees that electric charge is a conserved quantity even for chemically reacting systems. The total electric charge density ρ_{el} and the total electric flux (current) $\boldsymbol{i}$ are given, respectively, by

$$\rho_{el} = \sum_{\alpha=1}^{k} z_\alpha \rho_\alpha, \qquad \boldsymbol{i} = \sum_{\alpha=1}^{k} z_\alpha \rho_\alpha \boldsymbol{v}_\alpha. \qquad (5.58)$$

The second equality in (5.57) follows from (5.13), so that the total current density can be written as $\boldsymbol{i} = \rho_{el}\boldsymbol{v} + \boldsymbol{j}_{el}$, where $\boldsymbol{j}_{el} = \sum_{\alpha=1}^{k} z_\alpha \boldsymbol{j}_\alpha$ is the flux relative to the mass-average velocity.

The total electric charge density ρ_{el} and current density $\boldsymbol{i}$ produce electric $\boldsymbol{E}$ and magnetic $\boldsymbol{B}$ fields according to equations formulated by Maxwell, and then reformulated by Heaviside, known as *Maxwell's equations*. For a non-polarizable medium, Maxwell's equations can (in the rationalized Gauss system) be written as follows,[18]

$$\boldsymbol{\nabla} \cdot \boldsymbol{E} = \rho_{el}, \qquad (5.59)$$

$$\boldsymbol{\nabla} \cdot \boldsymbol{B} = 0, \qquad (5.60)$$

$$\frac{\partial \boldsymbol{E}}{\partial t} - c\,\boldsymbol{\nabla} \times \boldsymbol{B} = -\boldsymbol{i}, \qquad (5.61)$$

$$\frac{\partial \boldsymbol{B}}{\partial t} + c\,\boldsymbol{\nabla} \times \boldsymbol{E} = \boldsymbol{0}, \qquad (5.62)$$

where c is the speed of light in vacuum.

It is sometimes useful to express the electric and magnetic fields in terms of potential fields. From (5.60) we can write

$$\boldsymbol{B} = \boldsymbol{\nabla} \times \boldsymbol{A}, \qquad (5.63)$$

where $\boldsymbol{A}$ is the magnetic vector potential field. Using (5.62) and (5.63), we have

$$\boldsymbol{E} = -\boldsymbol{\nabla}\phi_{el} - \frac{1}{c}\frac{\partial \boldsymbol{A}}{\partial t}, \qquad (5.64)$$

where ϕ_{el} is the electric potential field. Note that $\boldsymbol{A}$ is arbitrary to the extent that (5.63) is unchanged by the addition of the gradient of some scalar field,

[18] Jackson, *Classical Electrodynamics* (Wiley, 1975).

Figure 5.4 James Clerk Maxwell, 1831–1879.

say $\boldsymbol{A} \to \boldsymbol{A} + \boldsymbol{\nabla}\Upsilon$. So that (5.64) remains unchanged, we must also transform the scalar potential, $\phi_{\text{el}} \to \phi_{\text{el}} - (1/c)\partial\Upsilon/\partial t$. These transformations allow Maxwell's equations (5.59)–(5.62) to be written as a pair of uncoupled wave equations for $\boldsymbol{A}$ and ϕ_{el}. Here we do not pursue this further, but note that these are examples of *gauge transformations* since the fields $\boldsymbol{E}$ and $\boldsymbol{B}$ are left unchanged, or are invariant under the transformations. We will see in Chapters 13 and 14 that the idea of gauge transformations provides a valuable tool also for the description of interfaces.

The Maxwell equations describe the electromagnetic fields produced by moving charged matter. How do these fields act back on matter? The answer to this question is given by the Lorentz force: the force per unit mass on the charged species α is

$$\boldsymbol{f}_\alpha = z_\alpha \left(\boldsymbol{E} + \frac{1}{c}\boldsymbol{v}_\alpha \times \boldsymbol{B} \right). \tag{5.65}$$

This equation completes our description of the mutual interactions between charged matter and electromagnetic fields. The evolving charge and current densities are the sources of electromagnetic fields and, conversely, the evolving electromagnetic fields provide sources of momentum and energy in the hydrodynamic equations. The Lorentz force in the presence of an electromagnetic field can be incorporated in exactly the same way as external forces in the momentum balance (5.32),

$$\frac{\partial \boldsymbol{m}}{\partial t} = -\boldsymbol{\nabla} \cdot (\boldsymbol{vm} + \boldsymbol{\pi}) + \sum_{\alpha=1}^{k} \rho_\alpha \boldsymbol{f}_\alpha, \qquad (5.66)$$

and (see Exercise 5.13) in the internal energy balance (5.53),

$$\frac{\partial u}{\partial t} = -\boldsymbol{\nabla} \cdot (\boldsymbol{v}u + \boldsymbol{j}_q) - \boldsymbol{\pi} : \boldsymbol{\nabla}\boldsymbol{v} + \sum_{\alpha=1}^{k} \boldsymbol{j}_\alpha \cdot \boldsymbol{f}_\alpha. \qquad (5.67)$$

These two balance equations can be combined into a balance equation for the sum of kinetic and internal energy,

$$\frac{\partial}{\partial t}\left(\frac{1}{2}\rho v^2 + u\right) = -\boldsymbol{\nabla} \cdot \left[\boldsymbol{v}\left(\frac{1}{2}\rho v^2 + u\right) + \boldsymbol{j}_q + \boldsymbol{\pi}\cdot\boldsymbol{v}\right] + \sum_{\alpha=1}^{k} \boldsymbol{j}_\alpha \cdot \boldsymbol{f}_\alpha + \boldsymbol{v}\cdot\sum_{\alpha=1}^{k} \rho_\alpha \boldsymbol{f}_\alpha. \tag{5.68}$$

We see in (5.68) source terms that represent the conversion of electromagnetic energy into internal and kinetic energy.

Our description of matter in the presence of an electromagnetic field involves the charge and current densities, ρ_{el} and $\boldsymbol{i}$. These quantities are both related through z_α to unknowns already included in the balance equations (ρ_α and $\boldsymbol{j}_\alpha$) considered in the previous sections of this chapter. Hence, ρ_{el} and $\boldsymbol{i}$ do not represent additional unknowns, and the electric $\boldsymbol{E}$ and magnetic $\boldsymbol{B}$ fields (or ϕ_{el} and $\boldsymbol{A}$) can be determined from Maxwell's equations (5.59)–(5.62).

The electromagnetic fields felt by charged matter according to (5.66) and (5.67) are produced by other flowing matter in the system and hence correspond to long-range interactions.[19] We next want to rewrite these equations as strictly local balance equations by including the momentum and energy densities of the electromagnetic fields, as well as the transport of momentum and energy implied by Maxwell's equations. In other words, we elaborate the transport aspects of electromagnetic fields.

From Maxwell's equations (5.59)–(5.62), we obtain the evolution equation (see Exercise 5.17)

$$\frac{1}{c}\frac{\partial}{\partial t}(\boldsymbol{E} \times \boldsymbol{B}) = \boldsymbol{\nabla} \cdot \boldsymbol{T} - \sum_{\alpha=1}^{k} \rho_\alpha \boldsymbol{f}_\alpha, \qquad (5.69)$$

where $\boldsymbol{T}$, which is known as Maxwell's electromagnetic stress tensor, is given

---

[19] In this discussion, we neglect the subtle problem of self-interactions for a charged particle.

by

$$T = EE + BB - \frac{1}{2}(E^2 + B^2)\,\delta. \tag{5.70}$$

By adding (5.66) and (5.69) we obtain the source-free momentum conservation law[20]

$$\frac{\partial}{\partial t}\left(m + \frac{1}{c}E \times B\right) = -\nabla \cdot (vm + \pi - T), \tag{5.71}$$

which allows us to identify $(1/c)E \times B$ as the momentum density of the electromagnetic field and T as the net momentum flux *away* from a point due to the electromagnetic field. Total momentum conservation actually dictates the form (5.65) of the Lorentz force.

Using Maxwell's equations (5.59)–(5.62) one more time, we obtain the evolution equation

$$\frac{\partial}{\partial t}\frac{1}{2}(E^2 + B^2) = E \cdot \frac{\partial E}{\partial t} + B \cdot \frac{\partial B}{\partial t} = -\nabla \cdot (cE \times B) - i \cdot E$$

$$= -\nabla \cdot (cE \times B) - \sum_{\alpha=1}^{k} j_\alpha \cdot f_\alpha - v \cdot \sum_{\alpha=1}^{k} \rho_\alpha f_\alpha, \tag{5.72}$$

where $cE \times B$ is known as the *Poynting vector*. The convective transport of electromagnetic energy happens with the speed of light and is not of the usual form. Note that there is no exchange of energy between fields and matter caused by the magnetic field. By combining (5.68) and (5.72), we obtain the source-free energy conservation law

$$\frac{\partial}{\partial t}\left[\frac{1}{2}\rho v^2 + u + \frac{1}{2}(E^2 + B^2)\right] = -\nabla \cdot \left[v\left(\frac{1}{2}\rho v^2 + u\right) + j_q + \pi \cdot v + cE \times B\right]. \tag{5.73}$$

Equations (5.71) and (5.73) unify the balance equations of hydrodynamics and electrodynamics, thus emphasizing both the field-theoretic character of hydrodynamics and the transport character of Maxwell's equations.

It is important to note that, in applying Maxwell's equations, we have neglected the electromagnetic waves generated by thermal motions. The corresponding electromagnetic energy transfer, which is also referred to as radiative heat transfer, can be taken into account by an additional source term in (5.51) and (5.67). In general, the form of this source term is extremely complicated; readers interested in this topic can refer to other sources.[21]

---

[20] de Groot & Mazur, *Non-Equilibrium Thermodynamics* (Dover, 1984).

[21] See, for example, Siegel & Howell, *Thermal Radiation Heat Transfer* (Taylor & Francis, 2002), Chapter 16 of Bird, Stewart & Lightfoot, *Transport Phenomena* (Wiley, 2001), and the references cited therein.

Exercise 5.16   *Plane Electromagnetic Waves*

For electromagnetic waves propagating in a non-conducting medium, $\rho_{\mathrm{el}} = 0$ and $i = 0$. Show, using the identity for arbitrary vector a,

$$\nabla^2 a = \nabla(\nabla \cdot a) - \nabla \times (\nabla \times a), \tag{5.74}$$

that the electric field E is governed by a wave equation,

$$\frac{1}{c^2} \frac{\partial^2 E}{\partial t^2} = \nabla^2 E. \tag{5.75}$$

Verify the plane-wave solution given by

$$E(r, t) = E_0 e^{i(k \cdot r - \omega t)}, \tag{5.76}$$

where ω is the frequency of the wave propagating with wave vector k having magnitude $k = \omega/c$, and E_0 is the (complex) amplitude of the electric field. Show that the magnetic field has the form

$$B(r, t) = B_0 e^{i(k \cdot r - \omega t)}, \tag{5.77}$$

where $B_0 = k \times E_0/k$, and that the E and B fields represent a transverse wave.

Exercise 5.17   *Maxwell's Electromagnetic Stress Tensor*

Derive the evolution equation (5.69).

- By looking at the balance of an extensive quantity in an arbitrary volume, one can derive a local balance equation for the density of the quantity in terms of the divergence of the corresponding flux, which consists of a convective and, with the exception of total mass, a non-convective or diffusive contribution.
- For the extensive quantities mass and momentum, the local balance equations are known as the continuity equation and the equation of motion, where the latter involves the pressure, or stress, tensor as an additional momentum flux.
- Balance equations for non-conserved extensive quantities, such as internal energy and species mass, have source terms along with the divergence of convective and diffusive flux terms.
- The balance equation for the electric charge density involves the electric current density, and both the electric charge density and the electric current density produce electromagnetic fields according to the Maxwell equations; in turn, electromagnetic fields produce the Lorentz force on charged matter.
- The hydrodynamic and electromagnetic momentum and energy balances can be unified into source-free total balances.
- The number of unknown quantities in the balance equations is greater than the number of equations that contain them; therefore, they cannot be solved without the addition of constitutive equations.

6
Forces and Fluxes

Where do we stand? We have the balance equations (5.5), (5.14), (5.32), and (5.51) for the densities of total mass, species mass, momentum, and internal energy which express local conservation laws. With the relationship (5.1) between momentum density and velocity, we can rewrite the momentum balance equation as an equation of motion in the form of Newton's second law (5.33). However, with every balance equation we have introduced unknown fluxes. We still need constitutive equations for the momentum and energy fluxes in order to bring all these evolution equations of the theory of transport phenomena to life. For multi-component systems, we need additional constitutive equations for the diffusive fluxes. For reasons that will soon become clear, we refer to these constitutive equations as *force–flux relations*.

Some readers, especially those with prior exposure to transport phenomena, may be familiar with Newton's law of viscosity, Fourier's law, and Fick's law. These famous constitutive equations, which were found empirically, provide expressions for the fluxes appearing in the conservation laws for mass, momentum, and energy. In this chapter, we show that these constitutive equations, as well as more general force–flux relations for mass, energy, and momentum, can be formulated using *nonequilibrium thermodynamics* – in particular, the local entropy balance equation. The field of nonequilibrium thermodynamics has deep connections with transport phenomena. Here we give an introduction to this field and make use of several of its most important concepts based on *linear irreversible thermodynamics*. Readers interested in learning more about this field should consult the classic book by de Groot and Mazur[1] or a more recent book

---

[1] de Groot & Mazur, *Non-Equilibrium Thermodynamics* (Dover, 1984).

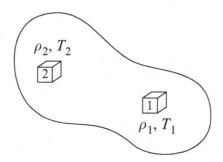

Figure 6.1 A global nonequilibrium system with local equilibrium subsystems 1 and 2.

for engineers[2] on linear irreversible thermodynamics, or a book on fully nonlinear generalizations.[3]

6.1 Local Thermodynamic Equilibrium

If we are interested in solutions in terms of quantities such as pressure, temperature, and composition, which is almost always the case, we need to incorporate the thermodynamic equations of state discussed in Chapter 4 into our set of equations. These thermodynamic quantities are applicable only to systems that are in a state of thermodynamic equilibrium. However, the phenomena we hope to describe involve time-dependent mass, momentum, and energy variables that are not spatially uniform and, therefore, are *not* at thermodynamic equilibrium.

This, at the moment at least, seems like a big problem. One way to resolve it is to divide a larger system that is not at equilibrium, say with a nonuniform mass density or temperature, into smaller systems for which a local density and temperature are defined (see Figure 6.1). The idea of a local temperature that may vary from point to point in space and with time implicitly assumes the existence of systems in *local thermodynamic equilibrium*. Without giving it much thought, we make use of this assumption in our everyday lives. For example, consider a room filled with air. If, at some position in the room, we imagine a small volume element, say of $1\,\mathrm{mm}^3$ in size, such a volume contains a large number of molecules (roughly 10^{16}), and is large compared with the typical mean free path between collisions of particles (roughly $0.1\,\mu\mathrm{m}$). For such a system, we apply equilibrium

[2] Kjelstrup *et al.*, *Non-Equilibrium Thermodynamics for Engineers* (World Scientific, 2010).
[3] Beris & Edwards, *Thermodynamics of Flowing Systems* (Oxford, 1994); Öttinger, *Beyond Equilibrium Thermodynamics* (Wiley, 2005).

thermodynamics concepts and thus assign to it a temperature, say 21 °C. We could certainly imagine the same-sized volume element at a different location in the room, say nearer to the ceiling; we would likely find this volume element to have a different temperature, maybe 25 °C. If we continue this procedure, we could identify a local temperature at each point in the room, down to a spatial resolution of 1 mm. The point of this thought experiment is that we can describe a larger system that is not at equilibrium by assuming it can be divided into smaller subsystems which are at equilibrium – in other words, by applying the concept of local thermodynamic equilibrium.

It seems we have resolved the big problem without too much effort. Note, however, that we should not take the size of the volume element used in our thought experiment $(1 \, \text{mm}^3)$ too literally. It was chosen somewhat arbitrarily, and there are situations where that particular choice would not work so well. For example, if we were trying to describe heat transfer in a gas flow through a micron-sized tube, the size of the small system would clearly be too large. We might make the volume smaller, but at some point we would approach a size where, because there are not enough molecules or they collide with the walls more frequently than with each other, we could not assume the system to be at thermodynamic equilibrium.

6.2 With a Little Help from Entropy

The local equilibrium assumption implies that the entropy density at each point can, for a one-component system, be evaluated from the densities of mass and internal energy at the same point according to the equilibrium equation of state, $s = s(u, \rho)$. The balance equations (5.5) and (5.51) for the mass and internal energy of a one-component system hence imply an evolution equation for the entropy density, which can be obtained by means of the chain rule. Neglecting external momentum sources in (5.51), we obtain

$$\frac{\partial s}{\partial t} = \left(\frac{\partial s}{\partial u}\right)_\rho \frac{\partial u}{\partial t} + \left(\frac{\partial s}{\partial \rho}\right)_u \frac{\partial \rho}{\partial t} = -\frac{1}{T}\left[\boldsymbol{\nabla} \cdot (\boldsymbol{v}u + \boldsymbol{j}_q) + \boldsymbol{\pi} : \boldsymbol{\nabla v}\right] + \frac{\hat{\mu}}{T}\,\boldsymbol{\nabla} \cdot (\boldsymbol{v}\rho)$$

$$= -\frac{1}{T}(u + p - \rho\hat{\mu})\,\boldsymbol{\nabla} \cdot \boldsymbol{v} - \boldsymbol{v} \cdot \left(\frac{1}{T}\boldsymbol{\nabla}u - \frac{\hat{\mu}}{T}\,\boldsymbol{\nabla}\rho\right) - \boldsymbol{\nabla} \cdot \frac{\boldsymbol{j}_q}{T} + \sigma$$

$$= -\boldsymbol{\nabla} \cdot \left(\boldsymbol{v}s + \frac{\boldsymbol{j}_q}{T}\right) + \sigma, \tag{6.1}$$

with

$$\sigma = \boldsymbol{j}_q \cdot \boldsymbol{\nabla}\frac{1}{T} - \frac{1}{T}\boldsymbol{\tau} : \boldsymbol{\nabla v}. \tag{6.2}$$

The second line in (6.1) is obtained by decomposing the pressure tensor in the form (cf. Exercise 5.14 on p. 76)

$$\boldsymbol{\pi} = p\boldsymbol{\delta} + \boldsymbol{\tau}, \tag{6.3}$$

and the third follows using the local Gibbs–Duhem (4.43) and Euler (4.44) relations. Note that $\boldsymbol{\tau}$, which is known as the extra pressure tensor, inherits the symmetry of $\boldsymbol{\pi}$. The final line of (6.1) is the evolution equation for the entropy density, where we see a flux term of the divergence form involving the convective entropy flux $\boldsymbol{v}s$ and the famous entropy flux $\boldsymbol{j}_s = \boldsymbol{j}_q/T$ proportional to the heat flux.

According to the entropy balance equation (6.1), the quantity σ introduced in (6.2) is recognized as the local entropy production rate. In view of the second law of thermodynamics, we must have $\sigma \geq 0$. This inequality is enormously useful in making physically meaningful choices of the irreversible contributions $\boldsymbol{\tau}$ and $\boldsymbol{j}_q$ to the fluxes of momentum and energy.

How can we guarantee that σ is always nonnegative? The simplest possibility is to make it a quadratic form by choosing the force–flux relations to be *linear*. Pre-factors, known as phenomenological coefficients, will be denoted as L with subscripts that denote the flux(es) they describe. For the energy flux we have

$$\boldsymbol{j}_q = L_q \boldsymbol{\nabla}\frac{1}{T} = -\lambda \boldsymbol{\nabla}T, \tag{6.4}$$

where $\lambda = L_q/T^2$ is the thermal conductivity, and for the extra pressure tensor which, for simplicity, we assume to be traceless, we find

$$\boldsymbol{\tau} = -\eta \left[\boldsymbol{\nabla}\boldsymbol{v} + (\boldsymbol{\nabla}\boldsymbol{v})^{\mathrm{T}} - \frac{2}{3}(\boldsymbol{\nabla} \cdot \boldsymbol{v})\boldsymbol{\delta} \right], \tag{6.5}$$

where $\eta = L_\tau/T$ is the shear viscosity. Our first guess in view of the occurrence of $-\boldsymbol{\tau} : \boldsymbol{\nabla}\boldsymbol{v}$ in (6.2) might have been $\boldsymbol{\tau} = -\eta (\boldsymbol{\nabla}\boldsymbol{v})^{\mathrm{T}}$; however, $\boldsymbol{\tau}$ needs to be symmetric. We have further assumed that $\boldsymbol{\tau}$ is traceless because the trace part requires a separate discussion. To recognize the quadratic form of the viscous entropy production, note that, if $\boldsymbol{\tau}$ is symmetric and traceless, we have $\boldsymbol{\tau} : [\boldsymbol{\nabla}\boldsymbol{v} + (\boldsymbol{\nabla}\boldsymbol{v})^{\mathrm{T}} + c\boldsymbol{\delta}] = 2\boldsymbol{\tau} : \boldsymbol{\nabla}\boldsymbol{v}$ for any scalar c.

By looking at the entropy production and turning it into a quadratic form, we are naturally led to Fourier's law (6.4) for the heat flux vector and Newton's expression (6.5) for the extra pressure tensor. The transport coefficients λ and η, which characterize the fluxes of energy and momentum, need to be nonnegative in order to guarantee a nonnegative entropy production rate. As noted earlier, the original formulation of the constitutive equations

(6.4) and (6.5) by Fourier and Newton was based on direct experimental observations rather than such theoretical considerations. A separate isotropic contribution can be added to Newton's expression for the pressure tensor,

$$\boldsymbol{\tau} = -\eta \left[\boldsymbol{\nabla v} + (\boldsymbol{\nabla v})^{\mathrm{T}} - \frac{2}{3}(\boldsymbol{\nabla} \cdot \boldsymbol{v})\boldsymbol{\delta} \right] - \eta_{\mathrm{d}}(\boldsymbol{\nabla} \cdot \boldsymbol{v})\boldsymbol{\delta}, \qquad (6.6)$$

where the nonnegative transport coefficient η_{d} is known as the dilatational or bulk viscosity leading to an additional contribution $\eta_{\mathrm{d}}(\boldsymbol{\nabla} \cdot \boldsymbol{v})^2/T$ to the entropy production rate (6.2).

A temperature gradient leads to a heat flux; a velocity gradient leads to a momentum flux. The gradients of temperature and velocity are hence referred to as the *thermodynamic driving forces* leading to the *dissipative fluxes* of energy and momentum. The introduction of pairs of forces and fluxes arises quite naturally and very generally from the analysis of the entropy production. At equilibrium, temperature and velocity are uniform so that the corresponding forces, fluxes, and entropy production vanish. Could the gradient of pressure or chemical potential in the same way act as a force? To answer this question, consider a system at equilibrium performing a uniform rigid-body rotation ($\boldsymbol{v} = \Omega r \boldsymbol{\delta}_\theta$). The symmetric velocity gradient tensor, and the divergence of the velocity, both vanish so that the extra stress vanishes. If the system has uniform temperature, the energy flux also vanishes. To compensate for the centrifugal force, the pressure has to increase radially, $\boldsymbol{\nabla}p = \rho\Omega^2 r \boldsymbol{\delta}_r$. According to the Gibbs–Duhem relation, there must also be a gradient of chemical potential $\boldsymbol{\nabla}\hat{\mu} = \Omega^2 r \boldsymbol{\delta}_r$. However, the gradient of chemical potential (or pressure) is not allowed to induce a mass flux because the system is at equilibrium. This is the basic reason why we can have the forces and fluxes in (6.4)–(6.6), whereas a diffusive mass flux in one-component systems is not allowed.[4] Note also that the entropy production terms in (6.2), which involve driving forces that are of different tensorial order, scalar ($\boldsymbol{\nabla} \cdot \boldsymbol{v}$), vector ($\boldsymbol{\nabla}(1/T)$), and second-order tensor ($\boldsymbol{\nabla v}$), are handled separately. The reason for doing this is discussed in the following section.

Will the procedure by which we have obtained Fourier's and Newton's famous constitutive equations also work so elegantly in more general situations, for example, for multi-component or complex fluids? In these more general situations, the entropy density depends on additional species densities or structural variables. As the evolution equation for the entropy density is obtained by means of the chain rule, further partial derivatives of the entropy density would occur in the entropy production rate. To

[4] Öttinger, Struchtrup & Liu, *Phys. Rev.* E **80** (2009) 056303.

guarantee a nonnegative entropy production in the most general case, the entropy production rate should be chosen as a quadratic form of the derivatives of entropy with respect to all independent system variables. This idea plays an essential role both in linear irreversible thermodynamics and in the nonlinear GENERIC framework of nonequilibrium thermodynamics.[5] In an even more general approach, quadratic forms of the derivatives of entropy can be replaced by convex positive functions.

Exercise 6.1 Forms of the Temperature Equation
Combine (4.53) with (5.52) to obtain the temperature equation

$$\rho \hat{c}_{\hat{v}} \left(\frac{\partial T}{\partial t} + \boldsymbol{v} \cdot \boldsymbol{\nabla} T \right) = -T \left(\frac{\partial p}{\partial T} \right)_{\hat{v}, w_\alpha} (\boldsymbol{\nabla} \cdot \boldsymbol{v}) - \boldsymbol{\nabla} \cdot \boldsymbol{j}_q - \boldsymbol{\tau} : \boldsymbol{\nabla} \boldsymbol{v}$$

$$+ \sum_{\alpha=1}^{k} \left[\hat{h}_\alpha - T \left(\frac{\partial p}{\partial T} \right)_{\hat{v}, w_\alpha} \hat{v}_\alpha \right] (\boldsymbol{\nabla} \cdot \boldsymbol{j}_\alpha - \nu_\alpha \Gamma). \quad (6.7)$$

Similarly, combine (4.57) with (5.52) to obtain an alternative form

$$\rho \hat{c}_p \left(\frac{\partial T}{\partial t} + \boldsymbol{v} \cdot \boldsymbol{\nabla} T \right) = -\frac{T}{\rho} \left(\frac{\partial \rho}{\partial T} \right)_{p, w_\alpha} \left(\frac{\partial p}{\partial t} + \boldsymbol{v} \cdot \boldsymbol{\nabla} p \right) - \boldsymbol{\nabla} \cdot \boldsymbol{j}_q - \boldsymbol{\tau} : \boldsymbol{\nabla} \boldsymbol{v}$$

$$+ \sum_{\alpha=1}^{k} \hat{h}_\alpha (\boldsymbol{\nabla} \cdot \boldsymbol{j}_\alpha - \nu_\alpha \Gamma). \quad (6.8)$$

6.3 Cross Effects

For a two-component system, the entropy density depends on two density variables and the internal energy density, $s = s(\rho_1, \rho_2, u)$. With (4.43), (5.14), and (5.51), we now obtain the generalized irreversible contribution to the entropy flux,

$$\boldsymbol{j}_s = \frac{1}{T} \left(\boldsymbol{j}_q - \hat{\mu}_1 \boldsymbol{j}_1 - \hat{\mu}_2 \boldsymbol{j}_2 \right), \quad (6.9)$$

and the generalized entropy production rate (see Exercise 6.2)

$$\sigma = \boldsymbol{j}_q \cdot \boldsymbol{\nabla} \frac{1}{T} - \boldsymbol{j}_1 \cdot \boldsymbol{\nabla} \frac{\hat{\mu}_1}{T} - \boldsymbol{j}_2 \cdot \boldsymbol{\nabla} \frac{\hat{\mu}_2}{T} - \frac{1}{T} \boldsymbol{\tau} : \boldsymbol{\nabla} \boldsymbol{v} - \frac{1}{T} (\nu_1 \hat{\mu}_1 + \nu_2 \hat{\mu}_2) \Gamma. \quad (6.10)$$

In view of the condition $\boldsymbol{j}_1 + \boldsymbol{j}_2 = \boldsymbol{0}$, we write the simple constitutive equation

$$\boldsymbol{j}_1 = L_{11} \boldsymbol{\nabla} \frac{\hat{\mu}_2 - \hat{\mu}_1}{T}, \quad (6.11)$$

---

[5] Grmela & Öttinger, *Phys. Rev. E* **56** (1997) 6620; Öttinger & Grmela, *Phys. Rev. E* **56** (1997) 6633; Öttinger, *Beyond Equilibrium Thermodynamics* (Wiley, 2005).

with a nonnegative transport coefficient L_{11} to supplement (6.4) and (6.6). Note, however, that we could introduce a more general quadratic form to guarantee a nonnegative entropy production rate,

$$
\sigma = \left(\begin{array}{c} \boldsymbol{\nabla}(1/T) \\ \boldsymbol{\nabla}(\hat{\mu}_2 - \hat{\mu}_1)/T \end{array} \right)^{\mathrm{T}} \cdot \left(\begin{array}{cc} L_{qq} & L_{q1} \\ L_{1q} & L_{11} \end{array} \right) \cdot \left(\begin{array}{c} \boldsymbol{\nabla}(1/T) \\ \boldsymbol{\nabla}(\hat{\mu}_2 - \hat{\mu}_1)/T \end{array} \right), \quad (6.12)
$$

where positive semidefiniteness (cf. footnote on p. 19), in addition to the conditions $L_{qq} \geq 0$ and $L_{11} \geq 0$, implies the inequality $L_{qq}L_{11} \geq L_{1q}L_{q1}$. As with (6.2), force–flux pairs having different tensorial order must be handled separately and this justifies the coupling of only vector forces and fluxes in (6.12). For this reason in (6.12) we have neglected entropy production associated with flow, which was considered in Section 6.2, and with chemical reaction, which will be discussed in Section 6.6.

By comparing the expressions (6.10) and (6.12) for the local entropy production rate, we obtain the following generalizations of the constitutive relations (6.4) and (6.11),

$$
\boldsymbol{j}_q = L_{qq} \boldsymbol{\nabla} \frac{1}{T} + L_{q1} \boldsymbol{\nabla} \left(\frac{\hat{\mu}_2 - \hat{\mu}_1}{T} \right), \quad (6.13)
$$

and

$$
\boldsymbol{j}_1 = L_{1q} \boldsymbol{\nabla} \frac{1}{T} + L_{11} \boldsymbol{\nabla} \left(\frac{\hat{\mu}_2 - \hat{\mu}_1}{T} \right). \quad (6.14)
$$

A detailed treatment of isothermal, three-component systems is given in Section 6.5; the general case with k components can be found elsewhere.[6]

The story around the matrix representation (6.12), or the constitutive equations (6.13) and (6.14), is a really big one. It shows the richness of transport phenomena and it contains the foundations of linear irreversible thermodynamics. For the theory of transport phenomena, thermodynamic driving forces may cause unexpected fluxes via so-called *cross effects*. According to (6.13), a heat flux can arise even in the absence of a temperature gradient as a consequence of a chemical potential gradient, that is, from inhomogeneities in composition. This phenomenon is known as the *Dufour effect*. Conversely, according to (6.14), a temperature gradient can produce a diffusive mass flux and hence changes in composition. This phenomenon is known as the *Soret effect*.

These Dufour and Soret effects are related because their strengths are found experimentally to be characterized by a single parameter $L_{1q} = L_{q1}$.

[6] Öttinger, *J. Chem. Phys.* **130** (2009) 021124.

Figure 6.2 Lars Onsager, 1903–1976.

The theoretical derivation of this symmetry property of the matrix of phenomenological coefficients occurring in (6.12) is associated with the name of Onsager. Onsager had related this symmetry to microscopic time-reversal properties.[7] Casimir[8] generalized Onsager's argument to show that also antisymmetric matrices of phenomenological coefficients are possible. Such antisymmetric matrices correspond to irreversible phenomena that do not produce entropy because only the symmetric part of a matrix of phenomenological coefficients contributes to entropy production; an intuitive example is provided by slip phenomena.

As elegant as (6.13) and (6.14) are, they can in practice be awkward to use. A primary goal of the study of transport phenomena is to formulate evolution equations for measurable fields like temperature and concentration. These evolution equations are derived by combining the balance equations presented in Chapter 5 with the constitutive equations discussed in this chapter. For example, the force–flux relation (6.11) or (6.14) is to be substituted into (5.14). Hence, we need to express the chemical potential gradient in terms of concentration and temperature gradients. To make this transformation, we begin with (4.55), which implies

$$\boldsymbol{\nabla}\hat{\mu}_\alpha = -\hat{s}_\alpha\,\boldsymbol{\nabla}T + \hat{v}_\alpha\,\boldsymbol{\nabla}p + \left(\frac{\partial\hat{\mu}_\alpha}{\partial w_1}\right)_{T,p}\boldsymbol{\nabla}w_1, \qquad (6.15)$$

[7] Onsager, *Phys. Rev.* **37** (1931) 405; Onsager, *Phys. Rev.* **38** (1931) 2265.
[8] Casimir, *Rev. Mod. Phys.* **17** (1945) 343.

and the Gibbs–Duhem equation (4.54) that leads to

$$\left(\frac{\partial \hat{\mu}_2}{\partial w_1}\right)_{T,p} = -\frac{w_1}{w_2}\left(\frac{\partial \hat{\mu}_1}{\partial w_1}\right)_{T,p}. \tag{6.16}$$

With these results, we obtain the useful expression

$$\boldsymbol{\nabla}\left(\frac{\hat{\mu}_2 - \hat{\mu}_1}{T}\right) = -\frac{\hat{h}_2 - \hat{h}_1}{T^2}\boldsymbol{\nabla}T + \frac{\hat{v}_2 - \hat{v}_1}{T}\boldsymbol{\nabla}p - \frac{1}{Tw_2}\left(\frac{\partial \hat{\mu}_1}{\partial w_1}\right)_{T,p}\boldsymbol{\nabla}w_1. \tag{6.17}$$

Substitution of (6.17), taking the pressure to be uniform, into (6.13) and (6.14) leads to

$$\boldsymbol{j}_q' = -\lambda'\,\boldsymbol{\nabla}T - \rho w_1\left(\frac{\partial \hat{\mu}_1}{\partial w_1}\right)_{T,p}D_{q1}\boldsymbol{\nabla}w_1, \tag{6.18}$$

$$\boldsymbol{j}_1 = -\rho w_1 w_2\frac{D_{q1}}{T}\,\boldsymbol{\nabla}T - \rho D_{12}\,\boldsymbol{\nabla}w_1, \tag{6.19}$$

where we have introduced

$$\boldsymbol{j}_q' = \boldsymbol{j}_q - (\hat{h}_1 - \hat{h}_2)\boldsymbol{j}_1. \tag{6.20}$$

Note that this modified heat flux comes with a nice expression for the entropy flux in (6.9),

$$\boldsymbol{j}_s = \frac{\boldsymbol{j}_q'}{T} + (\hat{s}_1 - \hat{s}_2)\boldsymbol{j}_1. \tag{6.21}$$

The properties of the phenomenological matrix ($L_{qq} \geq 0, L_{11} \geq 0$, and $L_{q1}^2 \leq L_{qq}L_{11}$) ensure that the modified thermal conductivity $\lambda' \geq 0$ and the diffusion coefficient $D_{12} \geq 0$; for the *thermal diffusion coefficient*[9] D_{q1} these properties lead to the inequality (see Exercise 6.5)

$$\rho w_1^2 w_2\left(\frac{\partial \hat{\mu}_1}{\partial w_1}\right)_{T,p}D_{q1}^2 \leq T\lambda'D_{12}. \tag{6.22}$$

The ratio D_{q1}/D_{12}, which is related to the Soret coefficient,[10] is usually positive and for typical liquid mixtures $D_{q1}/D_{12} \approx 0.1$–10. Note that the prefactors in front of the driving forces in (6.18) and (6.19) are no longer symmetric, which is a consequence of replacing gradients of thermodynamic potentials with gradients of their derivatives (note, however, that the two cross effect contributions to the entropy production in (6.10) are equal).

The possibility of cross effects plays a fundamental role also in the general story of linear irreversible thermodynamics. This thermodynamic framework

---

[9] It is more common to define the thermal diffusion coefficient as $D_T = D_{q1}/T$ (note that D_{q1} has units of length^2/time). The Soret coefficient is defined as $S_T = D_T/D_{12}$.

[10] Köhler & Wiegand, *Thermal Nonequilibrium Phenomena in Fluid Mixtures* (Springer, 2002).

is based on the most general linear relationship between driving forces and fluxes and their products contributing to the local entropy production. The 2×2 matrix in (6.12) is an example of a matrix of phenomenological coefficients. In (6.12), we have coupled the dissipative fluxes of mass and energy, but we have not included the momentum flux. There is a simple reason for that. If we have an isotropic material, all the transport coefficients in the matrix of phenomenological coefficients should be scalars and all the entries in the column vectors should hence be of the same vector or tensor type so that they can be superimposed in a meaningful way. This observation is known as the *Curie principle* asserting that, for isotropic systems, thermodynamic forces and fluxes having different tensorial order cannot couple.[11] For anisotropic systems, this conclusion does not hold. In particular, in the presence of interfaces, the vector normal to the interface can be used to couple forces and fluxes of different tensorial character. This possibility is important for biological systems, where a chemical reaction providing energy can, for example, drive an ion flux through a membrane (see Section 24.2). Note further that, for such a tensorial analysis, one should reduce the tensorial objects with respect to the group of rotations. For example, a second-rank tensor should be reduced according to $9 = 5 + 3 + 1$, that is, into its traceless symmetric part, its antisymmetric part, and its scalar trace. This reduction is actually the reason for the separate occurrence of the shear and bulk viscosities. Note that the Curie principle does not rule out the coupling of the scalar driving force $\nabla \cdot v$ to the chemical reactions discussed in Section 6.6. However, as we will need to go beyond quadratic forms for chemical reactions, the coupling is neither obvious nor unambiguous.

Some materials, crystals and oriented polymers for example, have an inherent anisotropy that results in anisotropic transport coefficients. One of the earliest observations of this type of behavior was anisotropic thermal conductivity in gypsum (calcium sulfate dihydrate) crystals.[12] To describe such materials, Fourier's law (6.4) is generalized to[13]

$$j_q = L_q \cdot \nabla \frac{1}{T} = -\lambda \cdot \nabla T, \tag{6.23}$$

where $\lambda = L_q/T^2$ is the thermal conductivity tensor. In this case, nonnegative entropy production in (6.2) requires the symmetric part of λ to be positive-semidefinite (cf. footnote on p. 19). For polymer liquids and

[11] Curie, *J. de Physique* **3** (1894) 393.
[12] de Sénarmont, *Pogg. Ann. Phys. Chem.* **73** (1848) 191; de Sénarmont, *Pogg. Ann. Phys. Chem.* **76** (1849) 119.
[13] Duhamel, *J. de l'École Polytech.* **13** (1832) 356.

cross-linked polymers, anisotropic thermal conductivity is the result of deformation-induced molecular orientation, which has been investigated both theoretically[14] and experimentally.[15]

Exercise 6.2   Entropy Balance for a Two-Component System
Derive the expressions for the entropy flux in (6.9) and for the entropy production rate in (6.10) for a two-component system.

Exercise 6.3   Fick's Law
Express the driving force in the constitutive equation (6.11) at constant temperature and pressure in terms of the species mass fraction to obtain Fick's law for the diffusive flux,

$$\boldsymbol{j}_1 = -\rho D_{12}\,\boldsymbol{\nabla} w_1 \tag{6.24}$$

where $D_{12} = [L_{11}/(\rho w_2 T)](\partial \hat{\mu}_1/\partial w_1)_{T,p}$. Show for an ideal mixture (see Exercise 4.7) that the mass diffusivity is given by

$$D_{12} = \frac{\tilde{R}L_{11}}{\rho w_1 w_2 (w_1 \tilde{M}_2 + w_2 \tilde{M}_1)}. \tag{6.25}$$

For constant ρ and in the absence of chemical reactions, the balance equation (5.14) for the mass density of the first species becomes

$$\frac{\partial \rho_1}{\partial t} = -\boldsymbol{\nabla}\cdot(\boldsymbol{v}\rho_1) + \boldsymbol{\nabla}\cdot(D_{12}\,\boldsymbol{\nabla}\rho_1). \tag{6.26}$$

This is our first transport equation obtained from a balance equation combined with a constitutive equation for a flux. Note that this transport equation for the species mass density is of the form of the Fokker–Planck, or diffusion, equations considered in great detail in Chapter 2.

Exercise 6.4   Alternative Forms of Fick's Law
As discussed in Section 5.1, it is sometimes more convenient to use molar concentrations and fluxes. The diffusive molar flux relative to the molar average velocity $\boldsymbol{J}_1^*$ is defined in (5.20). Using (5.24) and (6.24) show that $\boldsymbol{J}_1^*$ is given by

$$\boldsymbol{J}_1^* = -cD_{12}\,\boldsymbol{\nabla} x_1. \tag{6.27}$$

Also, using (5.27) and (6.24), show that the diffusive mass flux relative to the volume average velocity $\boldsymbol{j}_1^\dagger$ is given by,

$$\boldsymbol{j}_1^\dagger = -D_{12}\,\boldsymbol{\nabla}\rho_1, \tag{6.28}$$

---

[14] van den Brule, *Rheol. Acta* **28** (1989) 257; Öttinger & Petrillo, *J. Rheol.* **40** (1996) 857; Curtiss & Bird, *J. Chem. Phys.* **107** (1997) 5254.
[15] Venerus *et al.*, *Phys. Rev. Lett.* **82** (1999) 366; Schieber *et al.*, *Proc. Nat. Acad. Sci.* **36** (2004) 13142; Balasubramanian *et al.*, *Macromolecules* **38** (2005) 6210; Nieto Simavilla *et al.*, *J. Polym. Sci. Part B* **50** (2012) 1638.

Note the diffusivity D_{12} appearing in (6.27) and (6.28) is the same as that in (6.24).

Exercise 6.5   Thermal Diffusion Coefficient
Derive the following expressions for the transport coefficients appearing in (6.18) and (6.19),

$$\lambda' = \frac{L_{qq} + 2L_{q1}(\hat{h}_2 - \hat{h}_1) + L_{11}(\hat{h}_2 - \hat{h}_1)^2}{T^2},\tag{6.29}$$

$$D_{12} = \frac{L_{11}}{\rho w_2 T}\left(\frac{\partial \hat{\mu}_1}{\partial w_1}\right)_{T,p}, \qquad D_{q1} = \frac{L_{q1} + L_{11}(\hat{h}_2 - \hat{h}_1)}{\rho w_1 w_2 T}.\tag{6.30}$$

Verify the inequality given in (6.22) and evaluate the upper bound for the thermal diffusion coefficient D_{q1} appearing in (6.18) and (6.19) for an ideal mixture.

Exercise 6.6   Friction Coefficient and Mobility
Equations (2.24) and (6.26) suggest that a chemical potential $\tilde{\mu}_1/N_A$ and an external potential Φ^e per particle of species 1 contribute to a diffusive flux in exactly the same way. Use this observation to obtain the particle mobility and to find the Nernst–Einstein equation for the friction coefficient ζ,

$$D_{12} = \frac{k_B T}{\zeta}.\tag{6.31}$$

Exercise 6.7   Pressure Diffusion in Isothermal Systems
All of the expressions for the diffusive mass flux developed in this section have been based on the assumption of uniform pressure. Relax this assumption and show that the diffusive flux in an isothermal, two-component system can be written as

$$\boldsymbol{j}_1 = -\rho D_{12}\boldsymbol{\nabla} w_1 - \rho w_2 D_{12}\left(\frac{\partial \hat{\mu}_1}{\partial w_1}\right)_{T,p}^{-1}(\hat{v}_1 - \hat{v}_2)\boldsymbol{\nabla} p.\tag{6.32}$$

Note that the magnitude and sign of the pressure contribution depends on the difference in the partial specific volumes of the two components. Also, find the expression for the special case of an ideal fluid.

6.4 Thermoelectric Effects

Transport in materials with charged molecules, or ions, in the presence of electromagnetic fields occurs in biological systems, semiconductors, and electrochemical devices. Thermoelectric effects, which govern the transport of electrons in metals, have important applications in devices such as thermocouples used to measure and control temperatures through electric potential

differences. Advances in materials have in recent years lead to a surge in research and development on thermoelectric devices for power generation.[16]
In this section we consider thermoelectric effects as another example of cross
effects and to further appreciate the significance of Onsager's symmetry considerations.

We start our analysis by considering the entropy production rate for a
two-component system with $s = s(u, \rho_1, \rho_2)$. Using the balance equation for
internal energy given in (5.67), we obtain

$$\sigma = \boldsymbol{j}_q \cdot \boldsymbol{\nabla}\frac{1}{T} - \frac{1}{T}\boldsymbol{j}_1 \cdot \left(T\boldsymbol{\nabla}\frac{\hat{\mu}_1}{T} - \boldsymbol{f}_1\right) - \frac{1}{T}\boldsymbol{j}_2 \cdot \left(T\boldsymbol{\nabla}\frac{\hat{\mu}_2}{T} - \boldsymbol{f}_2\right), \qquad (6.33)$$

where we have excluded flow and chemical reactions. Using (5.65) and noting
that in the absence of flow $\boldsymbol{j}_{\mathrm{el}} = \boldsymbol{i}$, we can rewrite this as

$$\sigma = \frac{1}{T}\left[-\boldsymbol{j}_s \cdot \boldsymbol{\nabla}T - \boldsymbol{j}_1 \cdot \boldsymbol{\nabla}(\hat{\mu}_1 - \hat{\mu}_2) + \boldsymbol{i} \cdot \boldsymbol{E}\right], \qquad (6.34)$$

where $\boldsymbol{j}_s$ is given by (6.9).

We first consider metals, which can be thought of as two-component systems. The first component is the electrons and the second is the positive-ion
lattices of the metal. We assume that a heavy-ion lattice is immobile, that
is $\boldsymbol{j}_2 \approx \boldsymbol{0}$. Both the diffusion flow and the electric current density are determined entirely by the motion of the electrons so that

$$\boldsymbol{j}_1 = -\frac{m_{\mathrm{el}}}{e}\boldsymbol{i}, \qquad (6.35)$$

where m_{el} is the mass of an electron and e is the elementary charge. In terms
of independent force–flux pairs, (6.34) becomes

$$\sigma = \frac{1}{T}\left[-\boldsymbol{j}_s \cdot \boldsymbol{\nabla}T + \boldsymbol{i} \cdot \left(\boldsymbol{E} + \frac{m_{\mathrm{el}}}{e}\boldsymbol{\nabla}\hat{\mu}_{\mathrm{el}}\right)\right], \qquad (6.36)$$

where $T\boldsymbol{j}_s = \boldsymbol{j}_q + (m_{\mathrm{el}}/e)\hat{\mu}_{\mathrm{el}}\boldsymbol{i}$, and $\hat{\mu}_{\mathrm{el}}$ is the chemical potential of the electrons. Similarly to (6.13) and (6.14), we write the fluxes as

$$\boldsymbol{j}_s = -L_{ss}\frac{1}{T}\boldsymbol{\nabla}T + L_{se}\frac{1}{T}\left(\boldsymbol{E} + \frac{m_{\mathrm{el}}}{e}\boldsymbol{\nabla}\hat{\mu}_{\mathrm{el}}\right), \qquad (6.37)$$

$$\boldsymbol{i} = -L_{se}\frac{1}{T}\boldsymbol{\nabla}T + L_{ee}\frac{1}{T}\left(\boldsymbol{E} + \frac{m_{\mathrm{el}}}{e}\boldsymbol{\nabla}\hat{\mu}_{\mathrm{el}}\right), \qquad (6.38)$$

where we have assumed Onsager symmetry. For a system having uniform

[16] Mahan *et al.*, *Physics Today* **50** (1997) 42; Shakouri, *Annu. Rev. Mater. Sci.* **41** (2011) 399.

temperature, note that (6.38) is Ohm's law with isothermal electric conductivity $\sigma_{el} = L_{ee}/T$. Also, for $i = 0$, we can rewrite (6.37) as

$$j_s = -\left(L_{ss} - \frac{L_{se}^2}{L_{ee}}\right)\frac{1}{T}\,\boldsymbol{\nabla}T, \tag{6.39}$$

so that (6.4) suggests that $L_{ss} - L_{se}^2/L_{ee} = \lambda$ is the thermal conductivity in the absence of an electric current.

The phenomenological coefficient L_{se} appearing in (6.37) and (6.38) describes cross effects between thermal and electric phenomena. For example, for a system having a temperature gradient, we see from (6.38) that a voltage $\boldsymbol{E}$ is required to suppress an electric current ($i = 0$). This is known as the *Seebeck effect*, and is the principle upon which thermocouples are based. Similarly, for a system maintained at uniform temperature, we see from (6.37) and (6.38) that an electric field induces both an electric current and an entropy flux. This is known as the *Peltier effect*, and is used to make heating and cooling devices. More detailed discussion of these effects and the analysis of discontinuous systems (thermocouples for example) can be found elsewhere.[17] It should also be mentioned that, by a careful quantitative investigation of the Seebeck and Peltier effects, one can test Onsager's symmetry. Many experimental tests, in particular for thermoelectric effects in metallic and electrolytic thermocouples, were collected by D. G. Miller,[18] who concluded that "the experimental evidence is overwhelmingly in favor of the validity of the Onsager reciprocal relations."

We now consider a two-component system containing charged molecular species (ions). As we saw in Section 6.3, it is useful to express the gradient in chemical potential in terms of gradients in temperature and composition. Hence, we write (6.34) as

$$\sigma = -\frac{1}{T^2}j_q' \cdot \boldsymbol{\nabla}T - j_1 \cdot \left[\frac{1}{Tw_2}\left(\frac{\partial \hat{\mu}_1}{\partial w_1}\right)_{T,p}\boldsymbol{\nabla}w_1 + \frac{z_1 - z_2}{T}\boldsymbol{\nabla}\phi_{el}\right]. \tag{6.40}$$

As before, writing the fluxes as linear functions of the forces, we obtain

$$j_q' = -\lambda'\,\boldsymbol{\nabla}T - \rho w_1\left(\frac{\partial \hat{\mu}_1}{\partial w_1}\right)_{T,p}D_{q1}\,\boldsymbol{\nabla}w_1 - \rho w_1 D_{q1}z_1\,\boldsymbol{\nabla}\phi_{el}, \tag{6.41}$$

$$j_1 = -\rho w_1 w_2\frac{D_{q1}}{T}\,\boldsymbol{\nabla}T - \rho D_{12}\,\boldsymbol{\nabla}w_1 - \rho D_{12}\left(\frac{\partial \hat{\mu}_1}{\partial w_1}\right)_{T,p}^{-1}z_1\,\boldsymbol{\nabla}\phi_{el}, \tag{6.42}$$

---

[17] Chapter XIII, §6 of de Groot & Mazur, *Non-Equilibrium Thermodynamics* (Dover, 1984), Section 3.2.2 of Öttinger, *Beyond Equilibrium Thermodynamics* (Wiley, 2005).
[18] Miller, *Chem. Rev.* **60** (1960) 15.

where we have taken electroneutrality ($\rho_{\text{el}} = 0$) into account, that is $z_2 - z_1 = z_1/w_2$. Note that the λ', D_{12}, and D_{q1} appearing in (6.41) and (6.42) are the same as those in (6.18) and (6.19). Hence, for a system containing charged species, an additional contribution to the fluxes of energy and mass arises from the presence of a gradient in the electric potential, or an electric field. Detailed discussions of transport in electrolyte solutions, which includes electrokinetic phenomena such as electroosmosis and electrophoresis, can be found elsewhere.[19]

Exercise 6.8   Entropy Production with Thermoelectric Effects
Derive the expressions in (6.33) and (6.34) for the rate of entropy production. Also, show that the latter can be written as

$$\sigma = \frac{1}{T}\left[-\boldsymbol{j}_s \cdot \boldsymbol{\nabla} T - \boldsymbol{j}_1 \cdot \boldsymbol{\nabla}(\hat{\mu}_1' - \hat{\mu}_2')\right], \tag{6.43}$$

where $\hat{\mu}_\alpha' = \hat{\mu}_\alpha + z_\alpha \phi_{\text{el}}$ is the electrochemical potential, with $\boldsymbol{E} = -\boldsymbol{\nabla}\phi_{\text{el}}$. As we saw in Section 6.3, (6.43) involves coupled fluxes of energy (entropy) and mass arising from gradients of temperature and chemical potential, here modified by the presence of an electric field.

Exercise 6.9   Thermoelectric Power Generation
Thermoelectric devices for power generation exploit the Seebeck effect wherein an applied temperature gradient is used to produce an electric field that can be used to induce an electric current. From (6.37) and (6.38), derive the expressions

$$\boldsymbol{E} + \frac{m_{\text{el}}}{e}\boldsymbol{\nabla}\hat{\mu}_{\text{el}} = \frac{\boldsymbol{i}}{\sigma_{\text{el}}} + \varepsilon_{\text{el}}\boldsymbol{\nabla} T, \qquad \boldsymbol{j}_q = -\lambda\,\boldsymbol{\nabla} T + \left(\pi_{\text{el}} - \frac{m_{\text{el}}}{e}\hat{\mu}_{\text{el}}\right)\boldsymbol{i}, \tag{6.44}$$

where $\varepsilon_{\text{el}} = L_{se}/L_{ee}$ is the Seebeck coefficient and $\pi_{\text{el}} = TL_{se}/L_{ee} = \varepsilon_{\text{el}}T$ is the Peltier coefficient. The efficiency of thermoelectric devices is often characterized by the figure of merit $Z = \sigma_{\text{el}}\varepsilon_{\text{el}}^2/\lambda$, which depends only on material properties. Briefly discuss how Z is related to the efficiency of thermoelectric power generation.

Exercise 6.10   The Nernst–Planck Equation
A widely used expression for the mass flux of charged species in isothermal systems that was developed around 1890, known as the *Nernst–Planck equation*, can be expressed as

$$\boldsymbol{n}_1 = \rho_1 \boldsymbol{v} + \boldsymbol{j}_1 = \rho_1 \boldsymbol{v} - \rho D_{12}\,\boldsymbol{\nabla} w_1 - \rho_1 \frac{D_{12}\tilde{M}_1}{\tilde{R}T} z_1 \boldsymbol{\nabla}\,\phi_{\text{el}}. \tag{6.45}$$

Show for the case of a dilute, ideal system that (6.45) follows from (6.42). Briefly

---

[19] Newman & Thomas-Alyea, *Electrochemical Systems* (Wiley, 2004); Chapter 15 of Deen, *Analysis of Transport Phenomena* (Oxford, 2012).

discuss the physics behind the last term in (6.45) in terms of the Nernst–Einstein equation (see Exercise 6.6).

6.5 Diffusion in Multi-Component Systems

In the preceding sections we limited our discussion to two-component systems. There are numerous problems of importance that involve more than two components. For example, most chemical reactions involve at least three components. It turns out that going from two to three or more components leads to somewhat more complicated constitutive expressions for mass and energy fluxes. In particular, the number of mass diffusivities required increases and their dependence on composition can become significantly more complicated.

To get a feeling for these issues while keeping things relatively simple we will consider a three-component system. We focus entirely on concentration effects and hence consider a system with uniform temperature T and pressure p. We then obtain for the gradient of the chemical potentials $\hat{\mu}_\alpha(T, p, w_1, w_2)$

$$\boldsymbol{\nabla}\hat{\mu}_\alpha = \left(\frac{\partial\hat{\mu}_\alpha}{\partial w_1}\right)_{T,p,w_2} \boldsymbol{\nabla} w_1 + \left(\frac{\partial\hat{\mu}_\alpha}{\partial w_2}\right)_{T,p,w_1} \boldsymbol{\nabla} w_2, \tag{6.46}$$

where $\alpha = 1, 2, 3$. Throughout this section, we assume the simplified form of the chemical potential gradient in (6.46). For nonuniform temperature and pressure, one would need to go back to the three-component version of (4.55).

For the entropy production rate due to mass diffusion in an isothermal three-component system, we have

$$\sigma = -\boldsymbol{j}_1 \cdot \frac{\boldsymbol{\nabla}\hat{\mu}_1}{T} - \boldsymbol{j}_2 \cdot \frac{\boldsymbol{\nabla}\hat{\mu}_2}{T} - \boldsymbol{j}_3 \cdot \frac{\boldsymbol{\nabla}\hat{\mu}_3}{T}, \tag{6.47}$$

where we have also excluded entropy production terms for flow and chemical reaction. Since $\boldsymbol{j}_1 + \boldsymbol{j}_2 + \boldsymbol{j}_3 = \boldsymbol{0}$, we can write (6.47) as

$$\sigma = \boldsymbol{j}_1 \cdot \frac{\boldsymbol{\nabla}(\hat{\mu}_3 - \hat{\mu}_1)}{T} + \boldsymbol{j}_2 \cdot \frac{\boldsymbol{\nabla}(\hat{\mu}_3 - \hat{\mu}_2)}{T}. \tag{6.48}$$

We hence write the following force–flux relations,

$$\boldsymbol{j}_1 = L_{11}\frac{\boldsymbol{\nabla}(\hat{\mu}_3 - \hat{\mu}_1)}{T} + L_{12}\frac{\boldsymbol{\nabla}(\hat{\mu}_3 - \hat{\mu}_2)}{T},$$

$$\boldsymbol{j}_2 = L_{12}\frac{\boldsymbol{\nabla}(\hat{\mu}_3 - \hat{\mu}_1)}{T} + L_{22}\frac{\boldsymbol{\nabla}(\hat{\mu}_3 - \hat{\mu}_2)}{T}, \tag{6.49}$$

where the Onsager symmetry of the matrix of phenomenological coefficients has been taken into account. Positive semidefiniteness (cf. footnote on p. 19) of this matrix requires $L_{11} \geq 0$, $L_{22} \geq 0$, and $L_{11}L_{22} \geq L_{12}^2$.

The three-component version of the Gibbs–Duhem equation (4.54) can be used to eliminate $\nabla \hat{\mu}_3$; using (6.46), we can write (6.49) as

$$\boldsymbol{j}_1 = -\rho D_{13} \, \nabla w_1 - \rho D_{12} \, \nabla w_2,$$
$$\boldsymbol{j}_2 = -\rho D_{21} \, \nabla w_1 - \rho D_{23} \, \nabla w_2. \tag{6.50}$$

The four diffusivities in (6.50) are complicated functions of composition (see Exercise 6.12). For a k-component system, $(k-1)^2$ diffusivities are required, but only $(k-1)k/2$ of them are independent because the underlying matrix of phenomenological coefficients is symmetric. Positive semidefiniteness of the matrix of phenomenological coefficients implies inequalities involving diffusivities. It should also be noted that, unlike the two-component case considered in the previous section (see Exercise 6.4), the diffusivities appearing in the expressions for the mass $\boldsymbol{j}_\alpha$ and molar $\boldsymbol{J}_\alpha^*$ fluxes are not the same.

We have elaborated how the second law of thermodynamics suggests that one should introduce linear laws for fluxes in terms of forces and what kind of restrictions apply to these linear laws. Instead of expressing the fluxes as linear combinations of forces, we could alternatively express the forces as linear combinations of the fluxes. If we avoid redundant relationships, the two matrices characterizing the respective linear relationships are inverse to each other. The discussion of symmetry and positive semidefiniteness can be done equally well for either of these two representations. Even intermediate versions of establishing linear constitutive relations would be possible.

Could there be a reason to prefer a particular representation? As the inverse of a simple matrix may have a very complicated structure, or vice versa, it is definitely worthwhile to consider this possibility of obtaining more practical or even more illuminating constitutive relationships. We here consider the case of multi-component diffusion because, as we have just seen, even for a three-component system, four diffusion coefficients having complicated composition dependence are required.

In their classical work on multi-component diffusion, Maxwell (1866) and Stefan (1871) independently introduced the driving forces as linear combinations of the species velocities, that is, of the diffusive fluxes. Here we follow the approach used by Kjelstrup *et al.*,[20] who also include temperature gradients in their discussion (actually, they use an intermediate version where a combination of forces and fluxes is expressed in terms of the corresponding fluxes and forces). To illustrate how this alternative approach works, we

---

[20] See Chapter 5 of Kjelstrup *et al.*, *Non-Equilibrium Thermodynamics for Engineers* (World Scientific, 2010).

again consider an isothermal, three-component system. To express (6.47) as a quadratic form, we now write

$$-\frac{\nabla \hat{\mu}_\alpha}{T} = \sum_{\beta=1}^{3} R_{\alpha\beta} \boldsymbol{j}_\beta = R_{\alpha 1}\boldsymbol{j}_1 + R_{\alpha 2}\boldsymbol{j}_2 + R_{\alpha 3}\boldsymbol{j}_3, \qquad (6.51)$$

where the $R_{\alpha\beta}$ are elements of a symmetric matrix of phenomenological coefficients (resistivities) with zero determinant. Taking the Gibbs–Duhem equation (4.54) into account, we impose the conditions

$$\sum_{\alpha=1}^{3} \rho_\alpha R_{\alpha\beta} = \rho_1 R_{1\beta} + \rho_2 R_{2\beta} + \rho_3 R_{3\beta} = 0. \qquad (6.52)$$

With these conditions, we can write (6.51) in the reduced form

$$\begin{aligned}
\frac{\nabla \hat{\mu}_1}{T} &= R_{12}\rho_2 \left(\frac{\boldsymbol{j}_1}{\rho_1} - \frac{\boldsymbol{j}_2}{\rho_2} \right) + R_{13}\rho_3 \left(\frac{\boldsymbol{j}_1}{\rho_1} - \frac{\boldsymbol{j}_3}{\rho_3} \right), \\
\frac{\nabla \hat{\mu}_2}{T} &= R_{12}\rho_1 \left(\frac{\boldsymbol{j}_2}{\rho_2} - \frac{\boldsymbol{j}_1}{\rho_1} \right) + R_{23}\rho_3 \left(\frac{\boldsymbol{j}_2}{\rho_2} - \frac{\boldsymbol{j}_3}{\rho_3} \right).
\end{aligned} \qquad (6.53)$$

Note that, like (6.49), the flux–force relations in (6.53) involve three phenomenological coefficients. To put (6.53) on the same level as (6.50), we would need to use (6.46) to express the left-hand side of (6.53) in terms of concentration gradients.

More importantly, we can express (6.53) in terms of differences in species velocities by using (5.13). Making this substitution, and converting from mass to molar densities, we can write (6.53) as

$$\begin{aligned}
-\frac{\tilde{M}_1}{\tilde{R}T} \nabla \hat{\mu}_1 &= \frac{x_2}{\mathfrak{D}_{12}}(\boldsymbol{v}_1 - \boldsymbol{v}_2) + \frac{x_3}{\mathfrak{D}_{13}}(\boldsymbol{v}_1 - \boldsymbol{v}_3), \\
-\frac{\tilde{M}_2}{\tilde{R}T} \nabla \hat{\mu}_2 &= \frac{x_1}{\mathfrak{D}_{12}}(\boldsymbol{v}_2 - \boldsymbol{v}_1) + \frac{x_3}{\mathfrak{D}_{23}}(\boldsymbol{v}_2 - \boldsymbol{v}_3).
\end{aligned} \qquad (6.54)$$

The expressions in (6.54) are the Maxwell–Stefan equations for an isothermal, three-component system, and the $\mathfrak{D}_{\alpha\beta} = -\tilde{R}/(c\tilde{M}_\alpha \tilde{M}_\beta R_{\alpha\beta})$ are the Maxwell–Stefan diffusion coefficients. One advantage of the nicely symmetric Maxwell–Stefan formulation is that the diffusion coefficients are only weakly dependent on concentrations so that they can be approximated reasonably well by constants, and for a k-component system, there are $k(k-1)/2$ independent $\mathfrak{D}_{\alpha\beta}$. The generalization of (6.54) to non-isothermal, k-component systems is straightforward. A second advantage is that, since only species velocity differences occur in (6.54), the formulation is independent of any reference velocity.

Exercise 6.11   *Maxwell–Stefan Equations for Ideal Two-Component System*
For an ideal, two-component mixture, show using (6.54) that the results of Exercise
6.3 are recovered with $\mathfrak{D}_{12} = D_{12}$.

Exercise 6.12   *Diffusivities for a Three-Component System*
Derive expressions for the four diffusivities in (6.50) for an ideal mixture.

Exercise 6.13   *Entropy Production for the Maxwell–Stefan Equations*
Use the expression for the rate of entropy production in (6.47) to obtain

$$\sigma = -R_{12}\rho_1\rho_2(\boldsymbol{v}_1 - \boldsymbol{v}_2)^2 - R_{13}\rho_1\rho_3(\boldsymbol{v}_1 - \boldsymbol{v}_3)^2 - R_{23}\rho_2\rho_3(\boldsymbol{v}_2 - \boldsymbol{v}_3)^2 .$$

Hence, nonnegative entropy production requires the matrix of phenomenological
coefficients $R_{\alpha\beta}$ to be negative, which gives positive Maxwell–Stefan diffusion coefficients $\mathfrak{D}_{\alpha\beta} = -\tilde{R}/(c\tilde{M}_\alpha\tilde{M}_\beta R_{\alpha\beta})$.

6.6 Chemical Reaction Kinetics

The entropy production due to a chemical reaction in a k-component system
can be found as a natural generalization of the result for the two-component
system in (6.10),

$$\sigma_{\text{react}} = -\frac{\Gamma}{T}\sum_{\alpha=1}^{k}\nu_\alpha\hat{\mu}_\alpha = -\frac{\tilde{\Gamma}}{T}\sum_{\alpha=1}^{k}\tilde{\nu}_\alpha\tilde{\mu}_\alpha = -\frac{\tilde{\Gamma}}{T}\mathcal{A}, \qquad (6.55)$$

where it is convenient to make use of the molar quantities introduced in
(5.18). It is important to recognize that, while the expression in (6.55) has
the form of a force–flux pair product, unlike the previously considered cases,
the thermodynamic force here does not involve spatial derivatives ($\boldsymbol{\nabla}$). The
reason for this is that chemical reactions are not transport processes, but
rather are processes that evolve in time towards equilibrium, or *relaxation*
processes (see Exercise 6.14). Also note that, for an isolated system, the
internal energy is not affected by chemical reactions because it comprises
both the chemical energy and the energy associated with thermal motions.

The discussion of chemical reactions far from chemical equilibrium requires a generalization of our previous argument based on quadratic forms.
For this highly nonlinear relaxation process, the second law of thermodynamics can be implemented by means of the inequality

$$(e^x - e^y)(x - y) \geq 0, \qquad (6.56)$$

which can easily be verified by distinguishing the cases $x > y$ and $x <$
y. Of course, this kind of inequality can be generalized to any monotonic

function in the first factor or even in both factors. Chemical reactions are associated with exponential nonlinearity. The exponential function results from Boltzmann factors, which are related to Arrhenius' law.

The splitting of the chemical affinity $\mathcal{A} = \sum_{\alpha=1}^{k} \tilde{\nu}_\alpha \tilde{\mu}_\alpha$ occurring in (6.55) as a difference is actually very natural because we have terms with positive and negative stoichiometric coefficients $\tilde{\nu}_\alpha$ corresponding to products and reactants. Writing $\mathcal{A} = (x - y)\tilde{R}T$, we define the dimensionless variables

$$x = \frac{1}{\tilde{R}T} \sum_{\alpha=q+1}^{k} \tilde{\nu}_\alpha \tilde{\mu}_\alpha, \qquad y = -\frac{1}{\tilde{R}T} \sum_{\alpha=1}^{q} \tilde{\nu}_\alpha \tilde{\mu}_\alpha = \frac{1}{\tilde{R}T} \sum_{\alpha=1}^{q} |\tilde{\nu}_\alpha| \tilde{\mu}_\alpha, \quad (6.57)$$

which are the dimensionless affinities of the forward and reverse reactions. According to (6.55) and (6.56), we can write the nonlinear force–flux relation

$$\tilde{\Gamma} = -L_\Gamma (e^x - e^y) = L_\Gamma e^y (1 - e^{\mathcal{A}/\tilde{R}T}), \quad (6.58)$$

where the coefficient $L_\Gamma \geq 0$ is related to the *reaction rate constant*. For ideal mixtures, (6.58) produces the well-known *mass action law* where the reaction rate is given in terms of powers of reactant and product species concentrations (see Exercise 6.15). Near equilibrium $|\mathcal{A}/\tilde{R}T| \ll 1$, and (6.58) reduces to the linear force–flux relation $\tilde{\Gamma} = -L_\Gamma e^y/(\tilde{R}T)\mathcal{A}$.

At this point we feel it is necessary to comment on a commonly used terminology associated with chemical reactions. The expression for the reaction flux $\tilde{\Gamma}$ in (6.58) gives the net rate of reaction, that is, the difference between the rates of the forward and reverse reactions. A reaction where both forward and reverse reactions occur ($\rightleftharpoons$) is often termed "reversible." Similarly, a reaction where the reverse reaction is suppressed ($\rightarrow$), which occurs in the limit $\mathcal{A} \rightarrow -\infty$, is typically referred to as "irreversible." This chemical (ir)reversibility should not be confused with thermodynamic (ir)reversibility since all chemical reactions produce entropy and are therefore thermodynamically irreversible.

In general, the extension of irreversible thermodynamics from linear to nonlinear force–flux relations is a major challenge, in particular, the proper generalization of Onsager's symmetry relations. For example, consider a system with multiple chemical reactions. It is straightforward to show that the entropy production for a system with n chemical reactions (see Exercise 5.6) can be written as

$$\sigma_{\text{react}} = -\frac{1}{T} \sum_{j=1}^{n} \Gamma_j \sum_{\alpha=1}^{k} \nu_{\alpha,j} \hat{\mu}_\alpha = -\frac{1}{T} \sum_{j=1}^{n} \tilde{\Gamma}_j \sum_{\alpha=1}^{k} \tilde{\nu}_{\alpha,j} \tilde{\mu}_\alpha = -\frac{1}{T} \sum_{j=1}^{n} \tilde{\Gamma}_j \mathcal{A}_j,$$

$$(6.59)$$

where the affinity for reaction j is given by $\mathcal{A}_j = \sum_{\alpha=1}^{k} \tilde{\nu}_{\alpha,j}\tilde{\mu}_\alpha$. Nonnegative entropy production in (6.59) is ensured by writing

$$\tilde{\Gamma}_i = -\sum_{j=1}^{n} L_{ij}\mathcal{A}_j, \tag{6.60}$$

which apply to systems near equilibrium $\mathcal{A}_j/(\tilde{R}T) \ll 1$. Chemical reactions are typically coupled massively through the involvement of some species in several reactions. While (6.60) formally allows for an additional coupling of the chemical reaction rates $\tilde{\Gamma}_i$ through off-diagonal components of the matrix of phenomenological coefficients L_{ij}, there does not appear to be convincing evidence that such coupling exists.

Various thermodynamic approaches to nonlinear constitutive equations have been considered in the context of chemical reactions.[21] Chemical reactions are often governed by activated processes where a complex is formed that is at an energy state higher than the energy of both reactant and product states. An interesting approach that can handle activated processes involves the introduction of an internal variable that indicates the state system.[22] Using this approach (see Section 24.3) the reaction is described in terms of a diffusion equation for the probability that the system is in a particular state, which leads to an expression of the form in (6.58), but with a temperature-dependent prefactor having the Arrhenius form.

Exercise 6.14   Chemical Reaction as a Relaxation Process
Consider a single reaction near equilibrium with $\tilde{\Gamma} = -L_\Gamma\mathcal{A}$. Show that the extent of reaction ξ defined in (4.66) decays to its equilibrium value ξ_{eq} according to

$$\xi = \xi_{\text{eq}}[1 - \exp(-t/\tau_\Gamma)],$$

where $\tau_\Gamma^{-1} = L_\Gamma V \tilde{\nu}_\alpha^2 (\partial^2 F/\partial N_\alpha^2)_{T,V}$. Verify that the relaxation time τ_Γ is positive.

Exercise 6.15   Mass Action Law
For ideal mixtures (see Exercise 4.7), show that the molar rate of reaction per unit volume given by (6.57) and (6.58) can be written as,

$$\tilde{\Gamma} = L_\Gamma e^{-\sum_{\alpha=1}^{q} \frac{\tilde{\nu}_\alpha \tilde{\mu}_\alpha^0}{\tilde{R}T}} \left[\prod_{\alpha=1}^{q} x_\alpha^{-\tilde{\nu}_\alpha} - e^{\frac{\Delta \tilde{g}^0}{\tilde{R}T}} \prod_{\alpha=q+1}^{k} x_\alpha^{\tilde{\nu}_\alpha} \right], \tag{6.61}$$

[21] See Bataille *et al.*, *J. Non-Equilib. Thermodyn.* **3** (1978) 153; Section 12.4 of Beris & Edwards, *Thermodynamics of Flowing Systems* (Oxford, 1994); Section V.B of Öttinger & Grmela, *Phys. Rev. E* **56** (1997) 6633.

[22] See Reguera *et al.*, *J. Phys. Chem. B* **109** (2005) 21502; Chapter 7 of Kjelstrup *et al.*, *Non-Equilibrium Thermodynamics for Engineers* (World Scientific, 2010); Rubi, *Physica Scripta* **T151** (2012) 014027.

which is known as the mass action law. Identify the forward and reverse reaction rate constants, and recover (4.71) for equilibrium. Note that, for $-\Delta \tilde{g}^0/(RT) \gg 1$, the reaction can be described as chemically "irreversible."

- Assuming local equilibrium, the balance equations for the total mass, species mass, and internal energy densities can be used to derive an entropy balance equation; in particular, the entropy flux and the local entropy production rate can then be identified.
- The assumption of linear force–flux relations ensures nonnegative entropy production (the second law) and leads to Fourier's law for the heat flux, Newton's expression for the momentum flux, Fick's law for the diffusive mass flux in two-component systems, and the Maxwell–Stefan equations for multi-component diffusion.
- More generally, analysis of the entropy production rate can most naturally be performed in terms of thermodynamic force–flux pairs; in classical irreversible thermodynamics, the most general linear relation between forces and fluxes with Onsager–Casimir symmetry is assumed, where cross effects can occur whenever they do not violate a healthy vector/tensor structure of the equations (the Curie principle).
- In multi-component systems, one can identify the entropy production rate associated with chemical reactions; a properly chosen nonlinear force–flux relation guarantees the validity of the second law.

7

Measuring Transport Coefficients

We have just seen in Chapter 6 how thermodynamics, more specifically the entropy balance, guided us in the formulation of force–flux relations. Taken together with the balance equations presented in Chapter 5, we now have a set of equations that can be solved for fields such as velocity, temperature, and composition. In other words, the unknowns and equations are now equal in number. Each of the force–flux relations contains one or more parameters called *transport coefficients* (e.g., thermal conductivity λ, viscosity η). These coefficients carry information about the molecular interactions within the material we wish to describe, and therefore they should be thought of as material properties. In later chapters, we will see how the transport coefficients can be predicted using molecular models based on kinetic theory (see Chapters 21–24). Alternatively, one can find tabulated values for the transport coefficients for a large number of substances in the literature.

Suppose the system we wish to describe is too complex for kinetic theory to handle, and a value for the desired transport coefficient cannot be found. In such cases, one must go to the laboratory and use an experiment to determine the material property of interest. The process of measuring a transport coefficient requires two things. The first is an apparatus that allows one to manipulate and measure (not surprisingly) forces and fluxes. The second is a physical model that describes the expected relationship between the measured quantities and the material property to be measured. These models are formulated using the equations presented in the three preceding chapters. In other words, the model is formulated using the equations of transport phenomena. Hence, our first applications of transport phenomena will be for the analysis of experiments designed to obtain transport coefficients.

7.1 The Hydrodynamic Equations

To prepare for the remainder of this chapter, let us formulate evolution equations for velocity, temperature, and composition, or the *hydrodynamic equations*. First, we recall the mass balance (5.6),

$$\frac{\partial \rho}{\partial t} + \boldsymbol{v} \cdot \boldsymbol{\nabla} \rho = -\rho \boldsymbol{\nabla} \cdot \boldsymbol{v}. \tag{7.1}$$

The momentum balance equation for a Newtonian fluid is obtained by substitution of (6.3) with (6.6) into (5.33), which gives[1]

$$\rho \left(\frac{\partial \boldsymbol{v}}{\partial t} + \boldsymbol{v} \cdot \boldsymbol{\nabla} \boldsymbol{v} \right) = \eta \nabla^2 \boldsymbol{v} + \left(\eta_d + \frac{1}{3}\eta \right) \boldsymbol{\nabla}(\boldsymbol{\nabla} \cdot \boldsymbol{v}) - \boldsymbol{\nabla} p + \rho \boldsymbol{g}, \tag{7.2}$$

where the viscosities η and η_d are taken as constants. For the isothermal flow of a one-component fluid with an equation of state $\rho = \rho(p)$, we have with (7.1) and (7.2) a set of equations that can be solved for ρ, p, and $\boldsymbol{v}$.

To keep things simple, we now consider two special cases. One is a non-isothermal, one-component fluid, and the other is a two-component, isothermal fluid. We have already taken the first step towards formulating an evolution equation for the temperature (see Exercise 6.1). For a one-component fluid we need only the first line of (6.7) so that substitution of Fourier's law (6.4) and Newton's law of viscosity (6.6) leads to

$$\rho \hat{c}_{\hat{v}} \left(\frac{\partial T}{\partial t} + \boldsymbol{v} \cdot \boldsymbol{\nabla} T \right) = \lambda \nabla^2 T - \frac{\rho \hat{c}_{\hat{v}}(\gamma - 1)}{\alpha_p}(\boldsymbol{\nabla} \cdot \boldsymbol{v})$$

$$+ \eta [\boldsymbol{\nabla}\boldsymbol{v} + (\boldsymbol{\nabla}\boldsymbol{v})^{\mathrm{T}}] : \boldsymbol{\nabla}\boldsymbol{v} + \left(\eta_d - \frac{2}{3}\eta \right)(\boldsymbol{\nabla} \cdot \boldsymbol{v})^2, \tag{7.3}$$

where we have taken the thermal conductivity λ to be constant. For a one-component fluid with an equation of state $\rho = \rho(p, T)$, we have with (7.1), (7.2), and (7.3) a set of equations that can be solved for ρ, p, $\boldsymbol{v}$, and T. For an isothermal, two-component fluid, substitution of Fick's law (6.24) into the species continuity equation (5.15), assuming the mass diffusivity D_{AB} is constant, gives

$$\rho \left(\frac{\partial w_{\mathrm{A}}}{\partial t} + \boldsymbol{v} \cdot \boldsymbol{\nabla} w_{\mathrm{A}} \right) = \rho D_{\mathrm{AB}} \nabla^2 w_{\mathrm{A}} - \rho^2 D_{\mathrm{AB}}(\hat{v}_{\mathrm{A}} - \hat{v}_{\mathrm{B}})(\boldsymbol{\nabla} w_{\mathrm{A}})^2 + \nu_{\mathrm{A}}\Gamma. \tag{7.4}$$

The source term involving Γ in (7.4) is due to homogeneous chemical reaction, which can be expressed using a force–flux relation involving

---

[1] The symbol $\nabla^2 = \boldsymbol{\nabla} \cdot \boldsymbol{\nabla}$, which appears throughout this book, is the Laplacian operator. In a rectangular coordinate system $\nabla^2 = \sum_{i=1}^{3} \partial^2/\partial x_i \, \partial x_i$.

concentration (see Section 6.6). For an isothermal, two-component fluid with an equation of state $\rho = \rho(p, w_A)$, we have with (7.1), (7.2), and (7.4) a set of equations that can be solved for ρ, p, v, and w_A.

It should be noted that by separately considering the special cases of non-isothermal flow of a one-component fluid, and isothermal flow of a two-component fluid, we have considerably reduced the complexity of the hydrodynamic equations in (7.3) and (7.4). To describe the non-isothermal flow of a two-component fluid, we would have to account for cross effects (see Section 6.3), which would lead to modified forms of (7.3) and (7.4). For a k-component fluid, instead of (7.4) we would need $k - 1$ evolution equations for the species mass fractions and would have to allow for coupling between the species mass fluxes (see Section 6.5). Note, however, that (7.2) would not have to be modified unless a system with charged species was considered. More general forms of the hydrodynamic equations are applied in later chapters.

Notice the similarity of (7.2), (7.3), and (7.4). Each evolution equation has an accumulation term and a reversible convection $(v \cdot \nabla)$ term on the left-hand side, balanced on the right-hand side by an irreversible diffusion (∇^2) term and source terms. Note also the appearance of density ρ and velocity v both in the evolution equation for temperature (7.3) and in that for concentration (7.4), which means these equations are coupled to (7.1) and (7.2). In the examples that follow, which are our first applications of transport phenomena, we consider problems where the evolution equations can be decoupled and simplified considerably.

A significant simplification results when density variations can be neglected, which means the velocity is constrained by (5.36), that is $\nabla \cdot v = 0$. To identify conditions for which this holds, we consider a fluid composed of species A and B with $\rho = \rho(p, T, w_A)$ and write (7.1) as

$$\nabla \cdot v = -\kappa_T \frac{Dp}{Dt} + \alpha_p \frac{DT}{Dt} + \rho(\hat{v}_A - \hat{v}_B)\frac{Dw_A}{Dt}, \qquad (7.5)$$

where κ_T is the isothermal compressibility, α_p is the isobaric thermal expansion coefficient, and $\hat{v}_A, \hat{v}_B$ are the partial specific volumes of A and B.

Exercise 7.1   Viscous Dissipation for Compressible Newtonian Fluids
Prove that the viscous dissipation terms in the temperature evolution equation for a compressible Newtonian fluid (7.3) are nonnegative.

Exercise 7.2   Alternative Form of Temperature Equation for Pure Newtonian Fluids

Derive the following alternative to (7.3) for one-component Newtonian fluids:

$$\rho \hat{c}_p \left(\frac{\partial T}{\partial t} + \boldsymbol{v} \cdot \boldsymbol{\nabla} T \right) = \lambda \, \nabla^2 T + \frac{\rho \hat{c}_p (\gamma - 1)}{\gamma \alpha_p} \kappa_T \frac{Dp}{Dt}$$

$$+ \eta [\boldsymbol{\nabla} \boldsymbol{v} + (\boldsymbol{\nabla} \boldsymbol{v})^{\mathrm{T}}] : \boldsymbol{\nabla} \boldsymbol{v} + \left(\eta_{\mathrm{d}} - \frac{2}{3} \eta \right) (\boldsymbol{\nabla} \cdot \boldsymbol{v})^2. \quad (7.6)$$

Exercise 7.3   Alternative Form of Equation of Motion for Newtonian Fluids

Using the identities in (5.45) and (5.74), show that (7.2) can be written as

$$\rho \left(\frac{\partial \boldsymbol{v}}{\partial t} + \frac{1}{2} \boldsymbol{\nabla} v^2 - \boldsymbol{v} \times \boldsymbol{w} \right) = -\eta \, \boldsymbol{\nabla} \times \boldsymbol{w} + \left(\eta_{\mathrm{d}} + \frac{4}{3} \eta \right) \boldsymbol{\nabla} (\boldsymbol{\nabla} \cdot \boldsymbol{v}) - \boldsymbol{\nabla} p + \rho \boldsymbol{g}, \quad (7.7)$$

where $\boldsymbol{w} = \boldsymbol{\nabla} \times \boldsymbol{v}$ is the vorticity (see Exercise 5.9). We see for irrotational ($\boldsymbol{w} = \boldsymbol{0}$) and incompressible ($\boldsymbol{\nabla} \cdot \boldsymbol{v} = 0$) flow that viscous effects do not alter the pressure field. For applications of (7.7), see Sections 18.1 and 26.3.

7.2 Viscosity

Viscosity is a material property with which we all have some familiarity. Fluids we frequently encounter have a wide range of viscosity values. For example, the air we breathe and the maple syrup we pour on pancakes have viscosities that differ by four orders of magnitude, while water falls somewhere in between. We often use the words "thick" and "thin" to express the level of viscosity a fluid has. At some point, the reader has probably used "paint thinner," which simply reduces the viscosity of the paint, to make it easier to apply or to clean a paintbrush. The reader has probably also noticed the difficulty in getting a milkshake to flow through a straw.

The constitutive equation for stress given in (6.5) for Newtonian fluids contains the shear and bulk viscosities, η and η_{d}, respectively. Numerous methods have been developed over the years to measure η, and a device used for this purpose is called a *viscometer*. Measuring the bulk viscosity, which is related to volumetric deformations, is more difficult and will be considered in Chapter 26. In this section, we analyze a simple viscometer and develop the physical model that can be used to obtain the shear viscosity of a Newtonian liquid.

In this section we consider a liquid composed of a single component and assume the flow is isothermal. Hence, the second and third terms on the right-hand side of (7.5) vanish. This leaves only the first term, which describes changes in density due to pressure variations. This term can be

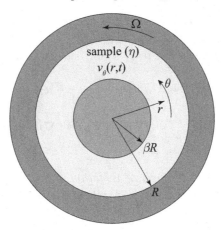

Figure 7.1 Schematic diagram of Couette flow geometry with a sample having viscosity η between a stationary inner cylinder and an outer cylinder rotating with angular velocity Ω.

neglected when $\kappa_T \Delta p \ll 1$, where Δp is a characteristic pressure difference. The assumption of incompressible flow typically fails for high-speed and/or rapidly changing flows of gases (see Section 7.5 for a more detailed discussion). We assume this commonly used assumption holds for the flow considered here, which means (7.5) reduces to $\boldsymbol{\nabla} \cdot \boldsymbol{v} = 0$. In such cases, (7.2) simplifies to the *Navier–Stokes equation*

$$\rho \left(\frac{\partial \boldsymbol{v}}{\partial t} + \boldsymbol{v} \cdot \boldsymbol{\nabla} \boldsymbol{v} \right) = \eta \nabla^2 \boldsymbol{v} - \boldsymbol{\nabla} \mathcal{P}, \qquad (7.8)$$

where we have introduced the modified pressure defined in (5.41), that is $\mathcal{P} = p^{\mathrm{L}} + \rho \phi$. The properties and solutions of the famous equation in (7.8) have been studied for more than a century,[2] but interpretation of the terms in it is straightforward. The left-hand side of (7.8) represents the mass times acceleration of a fluid element, and the right-hand side contains the forces acting on the fluid element. The modified pressure term on the right-hand side of (7.8) is a source term. Taking the divergence of (7.8) and using (5.36) produces (5.42), which we noted earlier would be the case for incompressible flows of Newtonian fluids.

A Couette viscometer consists of concentric cylinders in relative motion between which a liquid with unknown viscosity is contained. As shown in Figure 7.1, the larger cylinder with radius R rotates with angular velocity

[2] Lamb, *Hydrodynamics* (Dover, 1932); Schlichting, *Boundary Layer Theory* (McGraw-Hill, 1955); Batchelor, *An Introduction to Fluid Dynamics* (Cambridge, 1967); Happel & Brenner, *Low Reynolds Number Hydrodynamics* (Martinus Nijhoff, 1973); Landau & Lifshitz, *Fluid Mechanics* (Pergamon, 1987).

Ω, and the smaller cylinder with radius βR, where $\beta < 1$, is stationary. The tangential flow in the annular region between the cylinders is an example of Couette, or drag, flow. This device becomes a viscometer when it is possible to impose a known angular velocity Ω on the outer cylinder and measure the torque $\mathcal{M}_s$ on the (solid) inner cylinder. We assume laminar flow in the θ-direction and that the length of cylinders L is much larger than their radii $(R/L \ll 1)$ so that end effects can be neglected.

Based on the assumptions given above, the velocity field in cylindrical coordinates has the form $v_r = 0$, $v_\theta = v_\theta(r, \theta, t)$, $v_z = 0$. First, we write (5.36) in cylindrical coordinates [using (B.8)], which, after eliminating terms based on the assumed velocity field, gives

$$\frac{\partial v_\theta}{\partial \theta} = 0, \tag{7.9}$$

which further restricts the velocity field so that $v_\theta = v_\theta(r, t)$. Next, we write the r-, θ-, and z-components of (7.8) in cylindrical coordinates [using (B.10), (B.12), and (B.14)]. Eliminating terms based on the velocity field required by (7.9), we have

$$\frac{\partial \mathcal{P}}{\partial r} = \rho \frac{v_\theta^2}{r}, \qquad \rho \frac{\partial v_\theta}{\partial t} = \eta \frac{\partial}{\partial r} \left[\frac{1}{r} \frac{\partial}{\partial r}(r v_\theta) \right], \qquad \frac{\partial \mathcal{P}}{\partial z} = 0. \tag{7.10}$$

In writing the second equation in (7.10) we have used the fact that the θ-direction is periodic.[3] The θ-component of the momentum balance here has the form of a diffusion equation that we promised would appear frequently in transport phenomena. In this case, the second equation in (7.10) governs the diffusion of θ-direction momentum in the r-direction. The solution of this equation requires initial and boundary conditions, which are given by

$$v_\theta(r, 0) = 0, \qquad v_\theta(\beta R, t) = 0, \qquad v_\theta(R, t) = \Omega R, \tag{7.11}$$

where we have used the no-slip assumption to formulate boundary conditions on the surfaces of the rotating and stationary cylinders. The no-slip assumption used in the second and third equations in (7.11) is commonly used in fluid dynamics and implies that the tangential components of velocity are continuous across an interface. The specification of boundary conditions at phase interfaces, a solid–liquid interface for example, can be a delicate issue.

---

[3] The θ-component of the Navier–Stokes equation can be written as

$$\frac{\partial \mathcal{P}}{\partial \theta} = -r \left(\rho \frac{\partial v_\theta}{\partial t} - \eta \frac{\partial}{\partial r} \left[\frac{1}{r} \frac{\partial}{\partial r}(r v_\theta) \right] \right) = f(r, t).$$

Integration with respect to θ gives $\mathcal{P}(r, \theta, t) = f(r, t)\theta + g(r, t)$. Since $\mathcal{P}(r, \theta, t) = \mathcal{P}(r, \theta + 2\pi, t)$, then $f(r, t) = 0$, and $\mathcal{P} = \mathcal{P}(r, t)$.

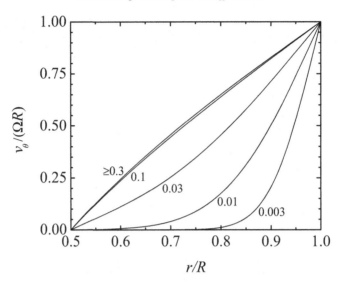

Figure 7.2 Evolution of the velocity field for Couette flow of a Newtonian fluid in an annulus with $\beta = 0.5$ given by (7.12) for $\nu t/R^2 = 0.003, 0.01, 0.03, 0.1, \geq 0.3$ (from bottom to top).

Detailed discussions of interfaces and boundary conditions can be found in later chapters.

The solution of the second equation in (7.10) subject to (7.11) can be found by the method of separation of variables, which is a special case of the eigenfunction methods described in Section 2.5. Details of the straightforward, but rather lengthy, procedure can be found elsewhere.[4] The expression for the tangential velocity can be expressed as

$$
\frac{v_\theta}{\Omega R} = -\pi \sum_{n=1}^{\infty} \frac{J_1(\beta b_n) J_1(b_n)[Y_1(\beta b_n) J_1(b_n r/R) - J_1(\beta b_n) Y_1(b_n r/R)]}{J_1{}^2(\beta b_n) - J_1{}^2(b_n)}
$$

$$
\times e^{-b_n{}^2 \nu t/R^2} + \frac{\beta}{1-\beta^2}\left(\frac{r}{\beta R} - \frac{\beta R}{r}\right), \tag{7.12}
$$

where J_1 and Y_1 are Bessel functions of the first and second kind of order one, and the b_n are roots of the equation $Y_1(\beta b_n) J_1(b_n) - J_1(\beta b_n) Y_1(b_n) = 0$. The evolution of the velocity field with time is shown in Figure 7.2. Note that the time dependence in (7.12) is an exponential decay with a characteristic time that has a quadratic dependence on R and is inversely proportional to the kinematic viscosity $\nu = \eta/\rho$, or momentum diffusivity.

---

[4] Bird & Curtiss, *Chem. Eng. Sci.* **11** (1959) 108.

For sufficiently large times compared with the time scale for momentum diffusion, $t \gg R^2/\nu$, a condition of steady state will be achieved. If our sample were a typical olive oil, which has kinematic viscosity $\nu \approx 1 \text{ cm}^2/\text{s}$, using a viscometer having a cup radius $R \approx 1$ cm, the time taken to reach a steady state would be several seconds. At steady state, (7.12) simplifies to

$$\frac{v_\theta}{\Omega R} = \frac{\beta}{1 - \beta^2} \left(\frac{r}{\beta R} - \frac{\beta R}{r} \right). \tag{7.13}$$

Notice that the velocity field in (7.13) does not depend on the viscosity of the fluid, which is a common feature of steady flows of incompressible Newtonian fluids. To find the velocity field we did not use the r- and z-components of the Navier–Stokes equation (7.8). However, the assumed form of the solution is justified by its consistency with the first and second equations in (7.10).

To find the steady-state, pseudo-pressure field $p^L = \mathcal{P} - \rho\phi$, we integrate the first and third equations in (7.10) using (7.13). The orientation of the cylindrical annulus is vertical, that is, gravity acts in the minus z-direction. For convenience we specify a reference pressure $p^L(R, h(R)) = 0$, where $h(R)$ is the height of the free surface of the liquid at the outer cylinder. To eliminate $h(R)$, we recognize that, neglecting interfacial tension, $p^L(r, h(r)) = 0$, and that the liquid volume $\pi(1 - \beta^2)R^2L$ remains constant. This calculation (see Exercise 7.4) results in the pseudo-pressure field

$$\frac{p^L}{\rho\Omega^2 R^2} = \frac{1}{N_{\text{Fr}}} \left(1 - \frac{z}{L} \right)$$

$$+ \frac{\beta^2}{2(1 - \beta^2)^2} \left[\left(\frac{r}{\beta R} \right)^2 - \left(\frac{\beta R}{r} \right)^2 - 4\ln\left(\frac{r}{\beta R} \right) \right.$$

$$\left. - \frac{3}{4\beta^2} - \frac{3}{2} + \frac{7}{4}\beta^2 + \frac{\beta^4}{2} - 3\ln(\beta) \right], \tag{7.14}$$

where $N_{\text{Fr}} = (\Omega R)^2/(gL)$ is a dimensionless quantity known as the Froude number. The first term in (7.14) is simply the hydrostatic pressure. The remaining terms in (7.14) represent the force needed to balance the centrifugal force so that fluid elements travel in circular paths. Note that the pseudo-pressure field, which is shown as a contour plot in Figure 7.3, does not depend on the fluid viscosity η. The contour for $p^L = 0$ is the height of the liquid free surface $h(r)$, which has a shape consistent with what we might expect to see in the laboratory. How would the shape of the free surface change if the lab were in a low-gravity environment like the International Space Station?

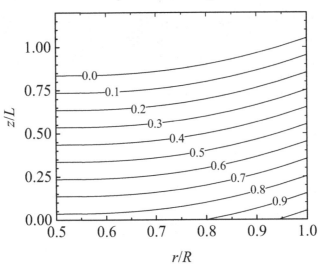

Figure 7.3 Steady-state pseudo-pressure $p^{\mathrm{L}}/(\rho\Omega^2 R^2)$ contours for Couette flow of a Newtonian fluid in an annulus given by (7.14) with $\beta = 0.5$ and $N_{\mathrm{Fr}} = 1$.

In order to use our rotating cylinder device as a viscometer, we need to derive an expression that relates η to the measurable quantities. The torque exerted by a fluid on a solid surface about the origin is the vector product of the position vector and force:

$$\mathcal{M}_{\mathrm{s}} = \int_{A_{\mathrm{s}}} \boldsymbol{r} \times (\boldsymbol{n} \cdot \boldsymbol{\pi}) dA, \qquad (7.15)$$

where $\boldsymbol{n}$ is the unit normal to A_{s} pointing into the solid. The position vector for the inner cylinder is simply $\boldsymbol{r} = \beta R \boldsymbol{\delta}_r$ and the force per unit area exerted on the cylinder by the fluid $-\boldsymbol{\delta}_r \cdot \boldsymbol{\pi}(\beta R)$. Using (7.13) and (7.14) in (7.15) gives

$$\mathcal{M}_{\mathrm{s}} = \frac{4\pi\beta^2 R^2 L\Omega\eta}{1 - \beta^2}, \qquad (7.16)$$

which provides the desired relationship from which the fluid viscosity can be determined.

We should keep in mind that we assumed the velocity field has a simple azimuthal form, and that more complicated flow fields are possible. It is easy to envision that, if the cylinder were rotated at a high rate, the liquid would not just flow in a circular pattern as indicated by (7.13), but would also have components in the r- and z-directions. The more complicated flow is induced when the centrifugal force that pushes the fluid element in the radial direction exceeds the viscous force that keeps the fluid moving in a circular path. The condition at which this transition occurs can be estimated by computing the value of a dimensionless quantity known as the Reynolds number, $N_{Re} = \rho V L/\eta$, where V and L are a characteristic velocity and length, respectively. For Couette flow between concentric cylinders we have $V = \Omega R$ and $L = R$ so that $N_{Re} = \rho \Omega R^2/\eta$. For $\beta = 0.5$, the critical Reynolds number[5] above which more complicated flow fields are observed is $N_{Re} \approx 10^5$. More complex flows such as axisymmetric toroidal (Taylor) vortices are the result of an inertially driven flow instability. Investigation of the complex transitions from the purely azimuthal base flow to more complex flows and their dependence on geometric factors for Couette flow began with the seminal work of Taylor,[6] and continues to be an active area of investigation.[7]

In addition to secondary flows caused by inertial effects, there are several other factors that may complicate viscosity measurement using a Couette viscometer. These include friction in the rotating cup, slip at the solid–fluid interfaces, end effects due to the finite length of cylinders, and viscous dissipation, which is discussed in the following section. Other commonly used geometries for viscosity measurement include flow between parallel disks (see Exercise 7.14) and flow through a tube (see Section 8.1).

The measurement of viscosity is one aspect of a larger subject known as rheometry, which itself falls within the field known as *rheology* – the study of the flow behavior of materials. Newtonian fluids are the simplest rheological fluids, and their flow behavior can be described using the rheological constitutive equation given in (6.5) and knowledge of the transport coefficient(s) η (and η_d). While many fluids we encounter on a daily basis are Newtonian, for example air and water, there are numerous examples of fluids that are non-Newtonian. Molten plastics, toothpaste, peanut butter, and blood are all non-Newtonian, or complex, fluids that display flow behavior that cannot be described by (6.5). In spite of their more complex behavior, a Couette viscometer of the type described in this section is often used to study the rheological properties of non-Newtonian fluids. However, the analysis presented above would need to be modified[8] and the result would be a nonlinear relationship between $\mathcal{M}_s$ and Ω. The fascinating subject of complex fluids is discussed in Chapters 12 and 22.

---

[5] Schlichting, *Boundary Layer Theory* (McGraw-Hill, 1955).
[6] Taylor, *Phil. Trans. R. Soc. Lond. A* **223** (1923) 289.
[7] Dutcher & Muller, *J. Fluid Mech.* **641** (2009) 85.
[8] Macosko, *Rheology* (VCH, 1994).

Exercise 7.4   Pseudo-pressure Field for Couette Flow
Derive the expression in (7.14) for the pseudo-pressure field in steady Couette flow.

Exercise 7.5   Brownian Dynamics Simulation of Couette Flow
Express the second equation in (7.10) as a Fokker–Planck equation and design a simulation to produce the behavior shown in Figure 7.2 for the initial and boundary conditions (7.11).

Exercise 7.6   Vorticity and the Stream Function
The vorticity vector $\boldsymbol{w}$ (see Exercise 5.10) is defined in (5.43). For the axisymmetric flow in spherical coordinates $v_r = v_r(r, \theta)$, $v_\theta = v_\theta(r, \theta)$, show that the following relation between the vorticity $w_\phi = w_\phi(r, \theta)$ and the stream function $\psi = \psi(r, \theta)$ holds:

$$w_\phi = -\frac{1}{r \sin \theta} E^2 \psi, \tag{7.17}$$

where the operator E^2 is given by

$$E^2 = \frac{\partial^2}{\partial r^2} + \frac{\sin \theta}{r^2} \frac{\partial}{\partial \theta} \left(\frac{1}{\sin \theta} \frac{\partial}{\partial \theta} \right). \tag{7.18}$$

Exercise 7.7   The Vorticity Transport Equation
Take the curl of (7.7) to obtain the following form of the vorticity transport equation for incompressible flows:

$$\frac{\partial \boldsymbol{w}}{\partial t} + \boldsymbol{v} \cdot \boldsymbol{\nabla} \boldsymbol{w} - \boldsymbol{w} \cdot \boldsymbol{\nabla} \boldsymbol{v} = \nu \nabla^2 \boldsymbol{w}, \tag{7.19}$$

where $\boldsymbol{w}$ is the vorticity defined in (5.43). Note the absence of pressure (and gravitational force) in (7.19). For the axisymmetric flow given by $v_r = v_r(r, z, t), v_z = v_z(r, z, t)$, show that the differential governing $w_\theta = w_\theta(r, z, t)$ is given by

$$\frac{\partial w_\theta}{\partial t} + v_r \frac{\partial w_\theta}{\partial r} + v_z \frac{\partial w_\theta}{\partial z} - \frac{v_r w_\theta}{r} = \nu \left[\frac{\partial}{\partial r} \left(\frac{1}{r} \frac{\partial}{\partial r} (r w_\theta) \right) + \frac{\partial^2 w_\theta}{\partial z^2} \right]. \tag{7.20}$$

7.3 Thermal Conductivity

Thermal conductivity is a material property that reflects the transport of internal energy due to molecular interactions. From everyday experiences, we know metals are good conductors of heat, while ceramics are good insulators. The teapot we might use to boil water when making tea is made of metal to allow heat from the stove to be transferred to the water, while the ceramic teacup we drink from allows us to hold it without burning our fingers. Numerous methods are available to measure λ, and all involve imposing a temperature gradient and measuring the heat flux, or *vice versa*.

As in the previous section, our goal is to develop a transport model that can be used to relate measurable quantities to the material property that we wish to determine.

As noted in Section 7.1, in general, the temperature equation is coupled to the equation of motion (7.2) along with the continuity equation (7.5). In practice, however, the temperature equation is usually solved using the velocity field obtained from the solution of (7.8), which is subject to the constraint $\boldsymbol{\nabla} \cdot \boldsymbol{v} = 0$. This constraint requires that the first and second terms on the right-hand side of (7.5) vanish (here we consider a single component fluid). In this case, it would be natural to use the temperature equation in (7.3) since the second and fourth terms on the right-hand side can be eliminated. However, as a vanishing right-hand side of (7.5) requires a subtle cancelation of pressure and temperature effects on density (which we cannot control), and since data for $\hat{c}_p$ are more readily available than for $\hat{c}_{\hat{v}}$, it is more common to use (7.6) and simply neglect the second term on the right-hand side. This term vanishes if $\kappa_T \Delta p \ll 1$ and/or if $\gamma \approx 1$. The latter is often assumed to be true for liquids (and solids), although in some cases this is a poor approximation[9] (for ideal gases $\gamma = 5/3$). Here, we use this approximation and write (7.6) as

$$\rho \hat{c}_p \left(\frac{\partial T}{\partial t} + \boldsymbol{v} \cdot \boldsymbol{\nabla} T \right) = \lambda \boldsymbol{\nabla}^2 T + \eta [\boldsymbol{\nabla} \boldsymbol{v} + (\boldsymbol{\nabla} \boldsymbol{v})^{\mathrm{T}}] : \boldsymbol{\nabla} \boldsymbol{v}, \qquad (7.21)$$

where the last term represents the conversion of mechanical energy to thermal energy, or viscous dissipation, which occurs for the flow of any real ($\eta > 0$) fluid. Hence, a velocity field $\boldsymbol{v}$ found by solving (7.8) can be substituted into (7.21), which can then be solved to find the temperature field.[10]

A relatively simple and widely used technique for measuring thermal conductivity is known as the *transient hot-wire* (THW) method. The device consists of a cylindrical chamber inside which the sample is placed and through which a thin metal wire passes (see Figure 7.4). A known electric field (voltage/length) with magnitude E is applied across the wire, and the electric current with magnitude i passing through the wire is measured so that the resistance (per unit length) of the wire can be determined as $R_{\mathrm{el}} = E/i$. Electromagnetic energy in the wire is converted to internal energy (see Section 5.4) at a rate per unit length given by $P_{\mathrm{el}} = i^2 R_{\mathrm{el}}$, which is transferred to the sample in contact with the wire. If R_{el} is a known

---

[9] Smith, *J. Chem. Ed.* **42** (1965) 654. For common organic liquids $\gamma = 1.2$–1.3, whereas for water one finds $\gamma = 1.01$.

[10] This is possible because in deriving (7.8) we assumed the viscosity η was constant. For cases when $\eta = \eta(T)$, the $\eta \boldsymbol{\nabla}^2 \boldsymbol{v}$ term in (7.8) must be replaced by $\boldsymbol{\nabla} \cdot [\eta(T) \boldsymbol{\nabla} \boldsymbol{v}]$. In this case, the modified form of (7.8) and (7.21) must be solved simultaneously (see Exercise 8.6).

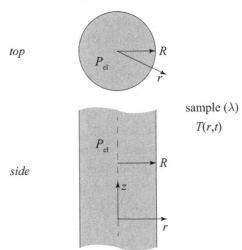

Figure 7.4 Schematic diagram of transient hot-wire (THW) experiment with wire of radius R that uniformly generates energy at a rate per unit length of P_{el} surrounded by a sample with thermal conductivity λ.

function of temperature, then the measured resistance can be used to determine the temperature of the wire. Hence, the wire acts as both a heater and a temperature sensor.

As we did in our analysis of the Couette viscometer, we now introduce several assumptions to formulate a model of the THW method that will allow the measurement of thermal conductivity. We assume the flow (if one exists) is incompressible and $\alpha_p \Delta T \ll 1$, where ΔT is a characteristic temperature difference, so that (7.5) reduces to $\boldsymbol{\nabla} \cdot \boldsymbol{v} = 0$. The wire radius R (≈ 100 μm) is much smaller than the sample chamber radius so that the sample can be considered infinitely large. The wire length L (≈ 10 mm) is much larger than its radius ($R/L \ll 1$) so that we assume end effects can be neglected. There is symmetry about the z-axis of a cylindrical coordinate system that coincides with the center of the wire.

Since both the wire and the chamber are stationary, and we have argued that flow due to thermal expansion can be neglected, it would seem reasonable to assume the fluid sample is at rest. However, the gravitational force in (7.2) can induce flow even if small density variations exist, which is known as *natural convection* (see Exercises 7.10 and 7.13). Here, we assume flow due to natural convection does not exist.

The wire and sample are different phases, and it is therefore necessary to write and solve temperature evolution equations for each. Since the wire is thin and metallic, we can make additional assumptions that will allow us to treat the wire as an ideal heater. This means that the wire temperature is uniform and increases instantaneously in response to an energy source (see Exercise 7.8). The former will be true if the thermal conductivity of the wire is large compared with that of the fluid $\bar{\lambda} \gg \lambda$, which is typically the case for a metal wire. The latter will be true if $\bar{\rho}\hat{c}_p \ll \rho\hat{c}_p$, which is a good approximation if the sample is a gas.

Based on the assumptions given above, the temperature field in the sample has the form $T = T(r,t)$. To proceed, we need (7.21) expressed in the cylindrical coordinate system. Eliminating terms based on the assumed form of the temperature field [and using (B.9)], (7.21) can be written as,

$$\rho\hat{c}_p \frac{\partial T}{\partial t} = \lambda \frac{1}{r}\frac{\partial}{\partial r}\left(r\frac{\partial T}{\partial r}\right). \tag{7.22}$$

The evolution equation for the sample temperature (7.22) is a diffusion equation (again) and, in this case, for the diffusion of internal energy in the radial direction. The solution of (7.22) requires initial and boundary conditions, which are

$$T(r,0) = T_0, \qquad \frac{\partial T}{\partial r}(R,t) = -\frac{P_{el}}{2\pi R\lambda}, \qquad T(\infty,t) = T_0, \tag{7.23}$$

where T_0 is the temperature of the wire and sample before the voltage is applied to the wire. Derivation of the boundary condition in the second equation of (7.23), which is based on the balance between the rate of energy produced in the wire and transferred to the fluid, is given as Exercise 7.8.

The solution of (7.22) and (7.23) can be obtained using the Laplace transform.[11] Alternatively, one can often find the solution to problems of this type in the literature. An excellent source is the classic book by Carslaw and Jaeger,[12] which has a large number of solutions to heat transfer problems for different boundary and initial conditions and different coordinate systems. While this and other books are excellent resources, we encourage the reader to develop the mathematical skills needed to solve partial differential equations. The solution of (7.22) and (7.23) can be expressed as

$$\frac{T(r,t) - T_0}{P_{el}/(4\pi\lambda)} = -\frac{4}{\pi}\int_0^\infty \frac{J_0(ur/R)Y_1(u) - Y_0(ur/R)J_1(u)}{u^2[J_1{}^2(u) - Y_1{}^2(u)]}(1 - e^{-u^2\chi t/R^2})du, \tag{7.24}$$

where $\chi = \lambda/(\rho\hat{c}_p)$ is the thermal diffusivity of the sample. The evolution

---

[11] The Laplace transform (cf. footnote on p. 32) of the derivative of a function $f(x,t)$ is given by

$$\int_0^\infty e^{-st}\frac{\partial f}{\partial t}\,dt = s\bar{f}(x,s) - f(x,0),$$

which is obtained using integration by parts.

[12] Carslaw and Jaeger, *Conduction of Heat in Solids* (Oxford, 1959).

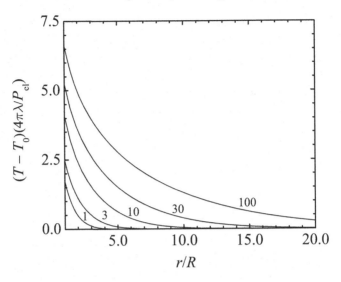

Figure 7.5 Evolution of the temperature field for a THW experiment given by (7.24) for $\chi t/R^2 = 1, 3, 10, 30, 100$ (from bottom to top).

of the temperature field given by (7.24), which has an exponential time dependence with time constant R^2/χ, is shown in Figure 7.5.

For $t \gg R^2/(4\chi)$, (7.24) gives the following expression for the temperature at the wire surface,

$$\frac{T(R,t) - T_0}{P_{\mathrm{el}}/(4\pi\lambda)} = \ln\left(\frac{4\chi t}{CR^2}\right) + \frac{R^2}{4\chi t} + O\left(\frac{R^2}{4\chi t}\right)^2, \tag{7.25}$$

where $C \approx 1.781$ is Euler's constant. For a measurement on air at room temperature and pressure ($\chi \approx 10^{-1}\,\mathrm{cm}^2/\mathrm{s}$) using a typical THW instrument ($R \approx 0.2$ mm), (7.25) would be valid roughly one millisecond after the test started. The dependence of the wire temperature rise on time predicted by (7.25) is shown in Figure 7.6. As noted above, the wire temperature is obtained from the measured resistance $R_{\mathrm{el}}(T)$, and the thermal conductivity of the sample λ is obtained from the slope and known rate of energy dissipation P_{el}.

In principle, the THW method is simple to implement. However, in practice there are a number of factors that can complicate the measurement. For example, the finite heat capacity of the wire will, for $t \lesssim R^2/(4\chi)$, cause the wire temperature rise to deviate from the logarithmic time dependence given in (7.25). For long times, the finite size of the sample chamber will affect energy transport through the sample, again causing a deviation from (7.25). The effects of these non-idealities on the temperature rise of the wire

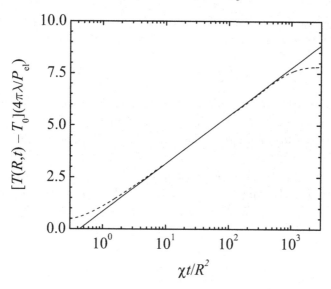

Figure 7.6 Temperature rise of wire in THW technique given by (7.25). Dashed curves at small and large times indicate nonidealities associated with the finite heat capacity of the wire and the finite sample radius.

are shown in Figure 7.6. Other effects that can complicate the THW method include flow induced by thermal expansion of the sample, natural convection (if the sample is a fluid), radiative heat transfer, and end effects associated with the finite length of the wire.[13]

Exercise 7.8   THW Temperature Equation for Wire
In the analysis above, the presence of the wire in a THW experiment is reflected only in the boundary condition given in the second equation in (7.23). This boundary condition follows from the equation governing the wire temperature $\bar{T}$. Starting with (5.67), show that the evolution equation for the average wire temperature is given by,

$$\bar{\rho}\bar{\hat{c}}_p \frac{d\langle \bar{T} \rangle}{dt} = \frac{2}{R}\bar{\lambda}\frac{\partial \bar{T}}{\partial r}(R,t) + \frac{P_{el}}{\pi R^2}, \tag{7.26}$$

where the radial-average wire temperature is defined as

$$\langle \bar{T} \rangle = \frac{2}{R^2}\int_0^R \bar{T}r\,dr. \tag{7.27}$$

---

[13] Healy *et al.*, *Physica C* **82** (1976) 392.

If we assume that the energy flux is continuous at the interface between the wire and sample, that is $\boldsymbol{n} \cdot \bar{\boldsymbol{j}}_q(R,t) = \boldsymbol{n} \cdot \boldsymbol{j}_q(R,t)$, then (7.26) can be written as

$$R^2 \frac{\bar{\rho}\bar{c}_p}{\lambda} \frac{d\langle \bar{T} \rangle}{dt} = 2R \frac{\partial T}{\partial r}(R,t) + \frac{P_{el}}{\pi\lambda}. \tag{7.28}$$

Note that the boundary condition in the second equation in (7.23) is obtained when the left-hand side of (7.28) is zero (for a more general discussion, see the remarks on the *quasi-steady-state approximation* at the end of this chapter).

Exercise 7.9   *THW Line Source Limit*
In the limit that the wire radius goes to zero ($R \to 0$), the boundary condition in the second equation in (7.23) becomes

$$\lim_{r\to 0} \left(r \frac{\partial T}{\partial r} \right) = -\frac{P_{el}}{2\pi\lambda}, \tag{7.29}$$

which is known as a line source. In this case, it is possible to convert (7.22) to an ordinary differential equation using the change of variable $\xi = r/\sqrt{4\chi t}$. Use this approach to show the solution to (7.22) with the first and third equations in (7.23) and (7.29) is

$$\frac{T(r,t) - T_0}{P_{el}/(4\pi\lambda)} = E_1 \left(\frac{r^2}{4\chi t} \right), \tag{7.30}$$

where $E_1(x)$ is the exponential integral function given by

$$E_1(x) = \int_x^\infty \frac{e^{-u}}{u} \, du = -\ln(C) - \ln(x) + x + O(x^2). \tag{7.31}$$

Show that this solution is consistent with the one given in (7.25).

Exercise 7.10   *The Boussinesq Approximation*
As discussed earlier in this section, density changes from temperature gradients can induce flow through the gravitational force in (7.2). The most general way of handling this would require solving (7.1)–(7.3) with an equation of state $\rho = \rho(T,p)$. A considerably simpler approach is to use the well-known Boussinesq approximation, which consists of two parts.[14] The first is to write the density as a linear function of temperature,

$$\rho = \rho_0[1 - \alpha_p(T - T_0)],$$

where ρ_0 and α_p are constants. The second is to replace ρ with ρ_0 in all terms in (7.1)–(7.3), with the exception of the gravitational term in (7.2). Show that these steps lead to the following forms for (7.2) and (7.3):

$$\frac{\partial \boldsymbol{v}}{\partial t} + \boldsymbol{v} \cdot \boldsymbol{\nabla}\boldsymbol{v} = \nu \nabla^2 \boldsymbol{v} - \frac{\boldsymbol{\nabla}P_0}{\rho_0} - \alpha_p(T - T_0)\boldsymbol{g}, \tag{7.32}$$

[14] Spiegel & Veronis, *Astrophys. J.* **131** (1960) 442; Mihaljan, *Astrophys. J.* **136** (1962) 1126.

where $\mathcal{P}_0 = p^L + \rho_0\phi$, and

$$\frac{\partial T}{\partial t} + \boldsymbol{v}\cdot\boldsymbol{\nabla}T = \chi\nabla^2 T, \qquad (7.33)$$

where viscous heating has been neglected. Necessary conditions for the Boussinesq approximation to be valid are $\alpha_p\Delta T \ll 1$ and $\chi^2/(L^2\hat{c}_p\Delta T) \ll 1$, where L and ΔT are the characteristic length and temperature difference, respectively.

7.4 Diffusivity

Despite the fact that the transport of mass is conceptually easier to understand than momentum or energy, our awareness of how ubiquitous and important mass transport is in our everyday lives is relatively limited. All living things, from single-cell organisms to human beings, require mass transport to exist. For example, oxygen in the air we breathe would never reach our blood without mass transport. On the other hand, the transport of carbon dioxide through the wall of a plastic bottle is what causes soda to go flat over time. Of the three transport coefficients we discuss in this chapter, the diffusivity, or diffusion coefficient, is the most difficult to measure.

A simple method for measuring the mass diffusivity D_{AB} for two-component systems is known as the *sorption experiment*. As shown schematically in Figure 7.7, the sample is a thin film of a non-volatile liquid (or solid), which we label species B, that sits on a stationary and impermeable solid surface. The other side of the film is exposed to a gas containing species A, which is soluble in species B. The solid surface is mounted on a balance

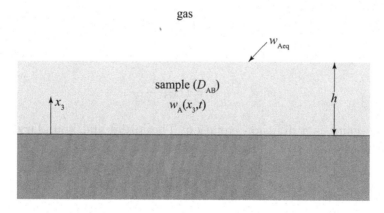

Figure 7.7 Schematic diagram of a sorption experiment with a sample having mass diffusivity D_{AB} in contact with a gas that maintains the concentration w_{Aeq} at the surface $x_3 = h$.

so that the mass of the film can be monitored as species A absorbs on and diffuses through the film.

As before, we make several assumptions that will allow us to analyze the experiment and extract the desired transport coefficient. First, we assume the film thickness h is much smaller than its lateral dimensions, so that edge effects can be neglected. The experiment is isothermal and the flow (if one exists) is incompressible so that the first and second terms on the right-hand side of (7.5) vanish. We further assume $\rho(\hat{v}_A - \hat{v}_B) \ll 1$ so that density variations due to compositional changes can be neglected. Hence, (7.5) reduces to $\nabla \cdot v = 0$ and, in the absence of chemical reaction, (7.4) simplifies to

$$\frac{\partial w_A}{\partial t} + v \cdot \nabla w_A = D_{AB} \nabla^2 w_A. \tag{7.34}$$

Based on our assumptions, the velocity and concentration fields have forms $v_1 = v_2 = 0$, $v_3 = v_3(x_3, t)$, $w_A = w_A(x_3, t)$.

For a two-component system we can write two independent mass balances; here we choose $\nabla \cdot v = 0$ and the species A continuity equation in (7.34). For the assumed velocity and concentration fields, these simplify to

$$\frac{\partial v_3}{\partial x_3} = 0, \tag{7.35}$$

$$\frac{\partial w_A}{\partial t} + v_3 \frac{\partial w_A}{\partial x_3} = D_{AB} \frac{\partial^2 w_A}{\partial x_3^2}. \tag{7.36}$$

Since the solid surface is stationary and impermeable, we have the boundary condition

$$v_3(0, t) =, 0, \tag{7.37}$$

so that the solution of (7.35) is $v_3(x_3, t) = 0$. Hence, (7.36) simplifies to

$$\frac{\partial w_A}{\partial t} = D_{AB} \frac{\partial^2 w_A}{\partial x_3^2}, \tag{7.38}$$

which (again) is a diffusion equation. The initial and boundary conditions for (7.38) are

$$w_A(x_3, 0) = w_{A0}, \qquad \frac{\partial w_A}{\partial x_3}(0, t) = 0, \qquad w_A(h, t) = w_{Aeq}, \tag{7.39}$$

where w_{A0} is the initial mass fraction of A in the film, and w_{Aeq} is the mass fraction of A at the solid(liquid)–gas interface. In writing the second equation of (7.39) we have used the fact that the solid surface is impermeable,

$n \cdot j_A(0, t) = 0$. For a sorption experiment $w_{Aeq} > w_{A0}$, and *vice versa* for a desorption experiment.

The significance of the simplification that results when $v_3(x_3, t) = 0$, which follows only if ρ is constant, cannot be overstated. If changes in ρ due to composition variations cannot be neglected, then $v_3(x_3, t) \neq 0$, which means that the diffusion of mass can induce a velocity in an otherwise stationary system. In other words, in the analysis of diffusive mass transport, one should not assume that the absence of external driving forces (pressure gradients, gravity forces, moving surfaces, for example) is a sufficient condition for the velocity field to vanish. If we had not assumed ρ to be constant, $v_3(x_3, t)$ would depend on w_A, and our analysis would require the solution of the nonlinear convection–diffusion equation in (7.36). See Exercise 14.11 for a more general analysis of sorption experiments.

In a typical sorption experiment the total mass of the film M_{tot} is measured as a function of time. The change in mass of the film $\Delta M(t) = M_{tot}(t) - M_{tot}(0)$, since ρ is constant, is given by

$$\frac{\Delta M(t)}{A\rho} = h(t) - h_0 = \int_0^{h(t)} w_A(x_3, t)dx_3 - h_0 w_{A0}, \qquad (7.40)$$

where A is the area of the film. Differentiation of (7.40) with respect to time gives, using (7.38) and the second and third equations in (7.39),

$$\frac{dh}{dt} = \frac{D_{AB}}{1 - w_{Aeq}} \frac{\partial w_A}{\partial x_3}(h, t). \qquad (7.41)$$

Hence, the boundary condition in (7.39c) is imposed at a position that is a function of time, $h = h(t)$, making this a *moving-boundary problem*. Moving-boundary problems are difficult to solve since the position at which the boundary condition is prescribed depends on the solution of the differential equation.

At equilibrium ($t \to \infty$), the concentration in the film becomes uniform, $w_A(x_3, \infty) = w_{Aeq}$, so that from (7.40) we have

$$\frac{h(\infty) - h_0}{h_0} = \frac{w_{Aeq} - w_{A0}}{1 - w_{Aeq}}, \qquad (7.42)$$

which shows that, if $|w_{Aeq} - w_{A0}|/(1 - w_{Aeq}) \ll 1$, then the change in film thickness can be neglected. Here, we will assume this condition is satisfied so that the film thickness h is effectively constant.

It is straightforward to find the solution of (7.38) using the eigenfunction method (see Section 2.5) or separation of variables. As discussed in the previous section, solutions to a large number of diffusion problems can be

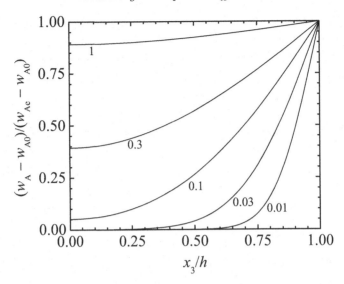

Figure 7.8 Evolution of concentration profile in sorption experiment given by (7.43) for $tD_{AB}/h^2 = 0.01, 0.03, 0.1, 0.3, 1.0$ (from bottom to top).

found in the literature. We refer the reader to the classic book by Crank.[15] The solution of (7.38) and (7.39) is given by

$$\frac{w_A - w_{A0}}{w_{Aeq} - w_{A0}} = 1 - \frac{4}{\pi} \sum_{n=0}^{\infty} \frac{(-1)^n}{2n+1} \cos\left(\frac{(2n+1)\pi x_3}{2h}\right) \exp\left(-\frac{(2n+1)^2 \pi^2 D_{AB} t}{4h^2}\right).$$

$$(7.43)$$

As we saw in the previous two sections, a time dependence of the exponential decay form is observed, with the rate of decay given by D_{AB}/h^2. The evolution of the concentration field within the film is plotted in Figure 7.8.

Substitution of (7.43) in (7.40) gives the following expression for the normalized change in mass of the film,

$$\frac{\Delta M(t)}{\Delta M(\infty)} = 1 - \frac{8}{\pi^2} \sum_{n=0}^{\infty} \frac{1}{(2n+1)^2} \exp\left(-\frac{(2n+1)^2 \pi^2 D_{AB} t}{4h^2}\right), \qquad (7.44)$$

which is plotted in Figure 7.9. From this figure, we see that the initial uptake of mass has a linear dependence on $\sqrt{tD_{AB}/h^2}$, or the square root of time normalized by the characteristic diffusion time. In practice, the value of D_{AB} is determined from the initial slope of mass uptake versus square root of time curve. As an example, let us consider a sorption experiment with carbon dioxide (species A) diffusing in a polyethylene terephthalate (species B)

[15] Crank, *Mathematics of Diffusion* (Oxford, 1975).

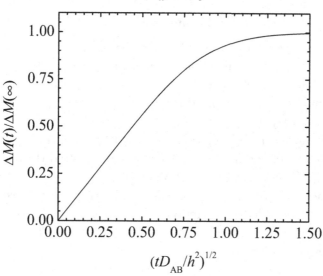

Figure 7.9 Normalized mass uptake in sorption experiment given by (7.44).
The slope of the line for $tD_{AB}/h^2 \lesssim 0.5$ is $2/\sqrt{\pi}$.

film of thickness 1 mm. At room temperature, $D_{AB} \approx 10^{-9}\,\mathrm{cm^2/s}$, so that
the characteristic diffusion time for this experiment is roughly four months.
That is why soda in a plastic bottle goes flat after several months.

Exercise 7.11   Sorption at Small Times
At small times in a sorption experiment, the concentration profile in the film remains
uniform at w_{A0} except near the surface where it is w_{Aeq} (see Figure 7.8). In this case,
it is possible to treat the film as a semi-infinite domain and replace the boundary
conditions given in the second and third equations in (7.39) by $w_A(0,t) = w_{Aeq}$ and
$w_A(\infty, t) = w_{A0}$. Use the change of variables $\xi = x_3/\sqrt{4D_{AB}t}$ to transform (7.38)
into an ordinary differential equation with the following solution that is valid at
small times,

$$\frac{w_A - w_{A0}}{w_{Aeq} - w_{A0}} = 1 - \mathrm{erf}\left(\frac{x_3}{\sqrt{4D_{AB}t}}\right), \tag{7.45}$$

where $\mathrm{erf}(x)$ is the error function given by

$$\mathrm{erf}(x) = \frac{2}{\sqrt{\pi}} \int_0^x e^{-u^2}\, du. \tag{7.46}$$

Use this result to derive an expression for the mass uptake that is consistent with
(7.44) for $t \ll h^2/D_{AB}$.

7.5 Dimensional Analysis

We have just seen how material properties like viscosity, thermal conductivity, and diffusivity can be obtained from experiments that are modeled using the equations of transport phenomena. In each of the three preceding sections, we used assumptions to simplify general forms of equations governing velocity, temperature, and concentration fields, and to formulate the initial and boundary conditions needed to solve them. This led in each case to a *linear* boundary-value problem that required the solution of a one-dimensional diffusion equation.[16] We did not spend much time discussing solution methods, but it is important to note that the linearity of the problems made it possible to find analytic solutions. For example, the solution to the boundary-value problem defined by (7.38) and (7.39) is given in (7.43).

The analytic solutions given in the preceding sections express how the dependent variable (e.g., w_A) depends on the independent variables (e.g., x_3, t). If we continue to use (7.43) as an example, notice that the dependence of w_A on spatial position involves x_3/h and on time involves $t/(h^2/D_{AB})$. These ratios are dimensionless, which is necessary since the evaluation of nonlinear mathematical functions like $\sin x$ and e^x is meaningful only if the argument x is dimensionless. The quantities used to make the independent variables dimensionless (h and h^2/D_{AB}) are characteristic quantities. Hence, we can rewrite (7.38) in the *dimensionless* form

$$\frac{\partial w_A}{\partial t} = \frac{\partial^2 w_A}{\partial x_3^2}. \tag{7.47}$$

Setting $x_3/h \to x_3$, $t/(h^2/D_{AB}) \to t$ is equivalent to setting the characteristic quantities equal to unity, that is, setting $h \to 1$, $h^2/D_{AB} \to 1$. In other words, we can transform equations from dimensional form to dimensionless form by setting the dimensional quantities, such as the characteristic length and time scales, equal to unity. Or, we measure length in units of h and time in units of h^2/D_{AB}, thus reducing the number of parameters in our problem. Once a solution has been found, the dimensional quantities can be re-introduced if desired (see Exercise 7.12).

In addition to *dimensional analysis*, there is another method for reducing the number of quantities in a problem: *normalization*. For a linear differential equation, we can divide a solution by a constant. The normalized solution satisfies the same differential equation but, typically, we can change initial and/or boundary conditions to a particular value, say zero or unity. If a

---

[16] Whereas in this chapter we have made assumptions to decouple momentum, energy, and mass transport, we shall see in Chapter 26 that the inherently coupled nature of these processes can be exploited to measure both thermodynamic and transport properties.

differential equation contains only derivatives of the dependent variable, we can add a constant to change an initial or boundary condition to zero. For example, if the variable w_A (which is already dimensionless) in (7.43) is normalized by subtracting w_{A0} and dividing by the difference $w_{Aeq} - w_{A0}$, the initial and boundary conditions in (7.39) can be written as

$$w_A(x_3, 0) = 0, \qquad \frac{\partial w_A}{\partial x_3}(0, t) = 0, \qquad w_A(1, t) = 1, \qquad (7.48)$$

which nicely take values of zero and one. Combining dimensional analysis and normalization, that is, setting $x_3/h \to x_3$, $t/(h^2/D_{AB}) \to t$, and $(w_A - w_{A0})/(w_{Aeq} - w_{A0}) \to w_A$ in (7.43), gives the solution to (7.47) and (7.48).

The above discussion about dimensional analysis and normalization for the mass diffusion problem also holds for the fluid mechanics and heat transfer problems presented in earlier sections of this chapter. Let us try to turn the ideas into a procedure to identify the number of independent dimensionless quantities that describe a given problem. For a problem that, after appropriate normalization, is described by n *independent* physical quantities involving m fundamental dimensions (e.g., mass, length, time, and temperature), the problem can be reformulated in terms of $n - m$ dimensionless quantities. The key to applying the procedure is the identification of independent quantities.

To see how this procedure works we use the heat transfer problem in Section 7.3. The problem described by (7.22) and (7.23) involves the seven quantities T, r, t, $\chi = \lambda/(\rho \hat{c}_p)$, R, T_0, P_{el}/λ. As (7.22) contains only derivatives of T, additive normalization leads to the quantities $T - T_0$, r, t, χ, R, P_{el}/λ, giving $n = 6$, which involve the dimensions length, time, and temperature so that $m = 3$. This gives $n - m = 3$, which is the number of dimensionless quantities by which the problem can be reformulated. The solution given in (7.24) expresses the dimensionless and normalized temperature $(T - T_0)/(P_{el}/4\pi\lambda)$ as a function of dimensionless independent variables r/R and $\chi t/R^2$.

Applying the same procedure to the fluid mechanics analysis in Section 7.2, the problem given by the second equation in (7.10) and (7.11), after multiplicative renormalization, involves $n = 6$ independent quantities $(v_\theta/\Omega, r, t, \nu, R, \beta)$, with $m = 2$ (length and time). The solution given in (7.12) expresses the dimensionless velocity $v_\theta/(\Omega R)$ as a function of dimensionless independent variables r/R and $\nu t/R^2$, and the dimensionless parameter β, or a relation among four $(= 6 - 2)$ dimensionless quantities. For the fluid mechanics problem at steady state, t and ν drop out of the second equation in (7.10), so that $n - m = 4 - 1 = 3$, which is consistent with the solution given in (7.13).

Dimensional analysis and normalization are powerful tools that are useful in the formulation, solution, and interpretation of problems in all branches of science and engineering, and especially in transport phenomena. They are also essential when evolution equations are solved by numerical methods. The basic concepts we have just discussed will be used throughout this book.

Once we consider only dimensionless quantities, we can try to make sure that the main quantities of interest are of order unity. For example, for the dimensionless mass diffusion problem (7.47) and (7.48), the ratios x_3/h and $t/(h^2/D_{AB})$ are normalized so that they range from zero to (approximately) unity. Of course, t goes to infinity, but when $x = t/(h^2/D_{AB}) = 3$, e^{-x} is roughly (within five percent of) the value it would have for $x \to \infty$. In this context, the numbers one and three are effectively the same, that is, they have the same order of magnitude. This is the idea of proper *scaling*. It is important to note that the "$n - m$ procedure" gives the number of independent quantities required to describe a physical problem, but does not tell us what the most appropriate dimensionless quantities are. For the simple problems considered in this chapter this did not present a problem since it was relatively easy to identify a dimensionless dependent variable and two dimensionless independent variables. If all the terms are kept in the hydrodynamic equations, in particular, nonlinear terms and all the different transport coefficients, scaling can be handled in terms of the dimensionless quantities considered next.

Dimensionless Groups

Most applications of transport phenomena do not involve the simple one-dimensional diffusion equations we have encountered in this chapter. In many cases, the phenomena we wish to describe require more than one spatial variable, and the evolution equations have (nonlinear) convective and source terms, and these equations may also be coupled. The simplification and decoupling of the evolution equations in applications of transport phenomena to more complicated problems are often based on the magnitude of dimensionless quantities, or *dimensionless groups* (e.g., $N_{Re} \ll 1$). It should be emphasized, however, that care should be taken in simplifying evolution equations based solely on the magnitude of dimensionless groups that may appear in them. For example, the more complicated flow fields that exist in Taylor–Couette flow when inertial effects become significant occur at large values of N_{Re} as is discussed in Section 7.2. On the other hand, a dimensionless group that is small (or large) may be a necessary, but not sufficient, condition for neglecting a given term appearing in an evolution equation cast in dimensionless form. Nevertheless, it is worthwhile to examine how several dimensionless groups commonly encountered in transport phenomena arise and identify their physical significance.

We would like to apply dimensional analysis (the $n-m$ procedure) to more general forms of the transport equations. All transport equations involve $\boldsymbol{v}$ and $\boldsymbol{\nabla}$, so we include the quantities V and L for velocity and length (possibly arising from boundary conditions) in our count for n.

We first consider the Navier–Stokes equation (7.8) that is solved subject to the constraint on velocity $\boldsymbol{\nabla} \cdot \boldsymbol{v} = 0$, which applies to incompressible flows. Keeping only the first term on the right-hand side of (7.5) and treating $\kappa_T^{-1} = \rho_0 c_T^2$ as a constant, we have $n = 7$ $(\boldsymbol{v}, p, \boldsymbol{r}, t, \rho_0 c_T^2, V, L)$ and $n-m = 4$. If we choose L, V, and ρ_0 as the basic units of length, velocity, and density $(L = V = \rho_0 = 1)$, we can write (7.5) as

$$\boldsymbol{\nabla} \cdot \boldsymbol{v} = -N_{\mathrm{Ma}}^2 \left(\frac{\partial p}{\partial t} + \boldsymbol{v} \cdot \boldsymbol{\nabla} p \right), \tag{7.49}$$

where $N_{\mathrm{Ma}} = V/c_T$ is the Mach number. Hence, for $N_{\mathrm{Ma}}^2 \ll 1$, a flow can be considered incompressible and (7.49) gives $\boldsymbol{\nabla} \cdot \boldsymbol{v} = 0$. In (7.8), where ρ is constant, we identify the quantities $\boldsymbol{v}, \rho, \mathcal{P}, \boldsymbol{r}, t, \nu, V, L$, so $n - m = 5$. As before, we use L, V, and ρ to fix the basic units $(L = V = \rho = 1)$ so that (7.8) can be written as

$$\frac{\partial \boldsymbol{v}}{\partial t} + \boldsymbol{v} \cdot \boldsymbol{\nabla} \boldsymbol{v} = \frac{1}{N_{\mathrm{Re}}} \nabla^2 \boldsymbol{v} - \boldsymbol{\nabla} \mathcal{P}, \tag{7.50}$$

where $N_{\mathrm{Re}} = VL/\nu$ is the Reynolds number indicating the importance of momentum convection (inertial force) relative to momentum diffusion (viscous force). For $N_{\mathrm{Re}} \gg 1$, the $\nabla^2 \boldsymbol{v}$ term in (7.50) can be neglected, giving Euler's equation (5.35) for incompressible flow. It is important to recognize that alternative choices for the basic units are possible. For example, if we instead choose L, V, and η to fix the basic units $(L = V = \eta = 1)$, (7.8) can be written as

$$N_{\mathrm{Re}} \left(\frac{\partial \boldsymbol{v}}{\partial t} + \boldsymbol{v} \cdot \boldsymbol{\nabla} \boldsymbol{v} \right) = \nabla^2 \boldsymbol{v} - \boldsymbol{\nabla} \mathcal{P}. \tag{7.51}$$

The dimensionless equations in (7.49)–(7.51) were obtained entirely by applying dimensional analysis.

We now proceed to a scaling of time as an additional step. If we introduce the scaled time $\bar{t} = t/N_{\mathrm{Re}}$, which reflects the idea that L^2/ν is the more appropriate unit of time, we can rewrite (7.51) as

$$\frac{\partial \boldsymbol{v}}{\partial \bar{t}} + N_{\mathrm{Re}} \boldsymbol{v} \cdot \boldsymbol{\nabla} \boldsymbol{v} = \nabla^2 \boldsymbol{v} - \boldsymbol{\nabla} \mathcal{P}. \tag{7.52}$$

An important special case of (7.52) is obtained when $N_{\mathrm{Re}} \ll 1$ so that the nonlinear inertial term can be neglected, which is known as *creeping flow* (see Exercises 7.14 and 7.15 and Section 17.3). For the temperature equation (7.21), we have $n = 9$ ($\boldsymbol{v}, T, \boldsymbol{r}, t, \chi, \lambda/\eta, \Delta T, V, L$) and we can fix $m = 3$ basic units by setting $L = V = \Delta T = 1$. If we introduce the scaled time $\bar{t} = t/N'_{\mathrm{Pe}}$, then (7.21) can be written as

$$\frac{\partial T}{\partial \bar{t}} + N'_{\mathrm{Pe}} \boldsymbol{v} \cdot \boldsymbol{\nabla} T = \nabla^2 T + N_{\mathrm{Br}}[\boldsymbol{\nabla} \boldsymbol{v} + (\boldsymbol{\nabla} \boldsymbol{v})^{\mathrm{T}}] : \boldsymbol{\nabla} \boldsymbol{v}, \tag{7.53}$$

where the Péclet number $N'_{\mathrm{Pe}} = VL/\chi$ indicates the importance of energy transport by convection relative to diffusion (conduction). The Brinkman number $N_{\mathrm{Br}} = \eta V^2/(\lambda \Delta T)$ indicates the importance of viscous dissipation relative to conduction. Finally, for the species mass balance (7.34) we have $n = 7$ ($\boldsymbol{v}, w_{\mathrm{A}}, \boldsymbol{r}, t, D_{\mathrm{AB}}, V, L$) and $m = 2$. If we fix $L = V = 1$ and introduce the scaled time $\bar{t} = t/N_{\mathrm{Pe}}$, then we can write (7.34) as

$$\frac{\partial w_{\mathrm{A}}}{\partial \bar{t}} + N_{\mathrm{Pe}} \boldsymbol{v} \cdot \boldsymbol{\nabla} w_{\mathrm{A}} = \nabla^2 w_{\mathrm{A}}, \tag{7.54}$$

where the Péclet number $N_{\mathrm{Pe}} = VL/D_{\mathrm{AB}}$ indicates the importance of mass transport by convection relative to diffusion. For a system with chemical reaction, (7.54) would have a source term accompanied by an additional dimensionless quantity.

The Péclet numbers appearing in (7.53) and (7.54) can also be written as $N'_{\mathrm{Pe}} = N_{\mathrm{Re}} N_{\mathrm{Pr}}$ and $N_{\mathrm{Pe}} = N_{\mathrm{Re}} N_{\mathrm{Sc}}$, where the Prandtl number is $N_{\mathrm{Pr}} = \nu/\chi$ and the Schmidt number is $N_{\mathrm{Sc}} = \nu/D_{\mathrm{AB}}$. Also note that we have used different characteristic times involving different diffusivities to normalize evolution equations for velocity, temperature, and composition. In cases where these evolution equations are coupled, ratios of these diffusivities will arise. For example, in situations involving both energy and mass transfer, the ratio of the thermal and mass diffusivities, known as the Lewis number, $N_{\mathrm{Le}} = \chi/D_{\mathrm{AB}}$, is relevant.

We have just introduced several dimensionless groups ($N_{\mathrm{Pr}}, N_{\mathrm{Sc}}, N_{\mathrm{Le}}$) that involve ratios of diffusivities for momentum, energy, and mass, and therefore depend only on material properties. Rough estimates for these dimensionless groups for broad classes of materials are given in Table 7.1. For gases, these

Table 7.1 *Typical values for Prandtl, Schmidt, and Lewis numbers*

Substance	$N_{Pr} = \nu/\chi$	$N_{Sc} = \nu/D_{AB}$	$N_{Le} = \chi/D_{AB}$
Gases	≈ 1	≈ 1	≈ 1
Low mol. wt. liquids	$1–10$	$10^2–10^3$	$10–10^3$
Liquid metals	10^{-2}	$10–10^2$	$10^3–10^4$
Complex fluids	$\gg 1$	$\gg 1$	$\gg 1$

quantities can be predicted using kinetic theory (see Table 21.1). Entries for low-molecular-weight liquids include ordinary liquids such as water and organic solvents. Liquid metals (e.g., Hg, Pb, Na) are unique in that they have unusually large thermal conductivities, due to energy transport by free electrons, which leads to small values of N_{Pr}. Complex fluids (e.g., colloidal suspensions, polymer liquids, see also Chapter 12) display a wide range of property values, but typically have relatively large viscosities and small diffusion coefficients, with thermal conductivities that are similar to those for low-molecular-weight liquids.

Similarity Transforms

The original problem defined in (7.22) and (7.23) for analysis of the THW experiment contains $T - T_0, r, t, \chi, R, P_{el}/\lambda$, so $n - m = 3$. In Exercise 7.9 the line-source limit where $R \to 0$ is considered. Since R is no longer relevant $n - m = 2$ for this case, and there is no way of forming a constant characteristic length scale from the parameters of the problem. Hence, time evolution takes place by rescaling position with the time-dependent length $\sqrt{4\chi t}$. The loss of one dimensionless quantity is reflected in the solution given in (7.30) where $(T - T_0)/(P_{el}/4\pi\lambda)$ is a function of $r/\sqrt{4\chi t}$. Combining variables in this way is known as a *similarity transform* and allows the transformation of a partial differential into an ordinary differential equation. The method is so named because partial differential equations for which such a transform goes through govern phenomena that exhibit self-similar behavior.

An example of self-similar behavior is a solution for which the spatial dependence of the solution has the same shape for all times. The solution given in (7.30) shows that the spatial dependence of the normalized temperature is the same for all times when scaled by the factor $\sqrt{4\chi t}$. We have already seen an example of self-similar behavior in the broadening of Gaussian distributions (2.16). The similarity transform method often, but not always, works for problems involving diffusive transport in infinite, or semi-infinite, domains. Exercise 7.11 is a second example where the geometric length (h) was dropped resulting in a diffusion equation on a semi-infinite domain having a similarity solution with position rescaled by $\sqrt{4D_{AB}t}$.

Quasi-steady-State Approximation

The final topic in this chapter illustrates how a useful approximation can be introduced using dimensional analysis. As discussed in Exercise 7.8, the boundary condition at the wire surface follows from the equation governing the average wire temperature $\langle \bar{T} \rangle$. In this more general case, the problem involves the time evolution of two variables T and $\langle \bar{T} \rangle$ that are coupled through (7.28). The characteristic time in this problem is R^2/χ, so we rewrite (7.28) as

$$\frac{\bar{\rho}\bar{\hat{c}}_p}{\rho \hat{c}_p} \frac{1}{\chi/R^2} \frac{d\langle \bar{T} \rangle}{dt} = 2R \frac{\partial T}{\partial r}(R,t) + \frac{P_{el}}{\pi \lambda}. \tag{7.55}$$

As discussed in this section, we can simply absorb the factor χ/R^2 into the time variable t (or set it equal to one), which leaves the ratio $\bar{\rho}\bar{\hat{c}}_p/\rho\hat{c}_p$ in front of the time derivative. For cases when $\bar{\rho}\bar{\hat{c}}_p \ll \rho\hat{c}_p$, we can set the left-hand side of (7.55) equal to zero and we recover the boundary condition in the second equation in (7.23). This means that, on the time scale of the thermal transport in the sample, the wire responds instantaneously to the heat source within it. The decoupling of time-dependent variables because their dynamics occur on different time scales is known as the *quasi-steady-state approximation*.

Exercise 7.12   Diffusion in a Sphere
Consider the (de)sorption process analyzed in Section 7.4 in a spherical sample of radius R. Initially, the sphere has uniform species A mass fraction w_{A0} and radius R_0; for $t > 0$, the species A mass fraction at the surface of the sphere is maintained at w_{Aeq}. Assuming ρ is constant and that R is approximately constant, show that the dimensionless equations governing the normalized mass fraction of species A are given by

$$\frac{\partial w_A}{\partial t} = \frac{\partial^2 w_A}{\partial r^2} + \frac{2}{r} \frac{\partial w_A}{\partial r}, \tag{7.56}$$

$$w_A(r,0) = 0, \qquad \frac{\partial w_A}{\partial r}(0,t) = 0, \qquad w_A(1,t) = 1, \tag{7.57}$$

Use the change of variable $u = r w_A$ to solve (7.56) and (7.57) and obtain

$$\frac{w_A - w_{A0}}{w_{Aeq} - w_{A0}} = 1 + \frac{2}{\pi} \frac{R}{r} \sum_{n=1}^{\infty} \frac{(-1)^n}{n} \sin\left(\frac{n\pi r}{R}\right) \exp\left(-\frac{n^2\pi^2 D_{AB}t}{R^2}\right). \tag{7.58}$$

Plot this expression as in Figure 7.8 and comment on the similarities and differences observed between rectangular and spherical geometries.

Exercise 7.13   Dimensional Analysis for Natural Convection
For more complicated problems involving coupled evolution equations, the choice of characteristic quantities can be a subtle issue. This is especially true for natural convection and the evolution equations for velocity and temperature developed in Exercise 7.10. It is natural to use L, V, and $\eta V/L$ to scale position, velocity, and pressure, respectively, and to scale temperature relative to T_0, by ΔT. Using L^2/χ to scale time, show that by a particular choice for the characteristic velocity V (7.32) and (7.33) can be written in dimensionless form

$$\frac{\partial T}{\partial t} + \boldsymbol{v} \cdot \boldsymbol{\nabla} T = \nabla^2 T, \tag{7.59}$$

$$\frac{1}{N_{\text{Pr}}} \left(\frac{\partial \boldsymbol{v}}{\partial t} + \boldsymbol{v} \cdot \boldsymbol{\nabla} \boldsymbol{v} \right) = \nabla^2 \boldsymbol{v} - \boldsymbol{\nabla} \mathcal{P}_0 - N_{\text{Ra}} \, T \frac{\boldsymbol{g}}{g}, \tag{7.60}$$

where $N_{\text{Ra}} = gL^3 \alpha_p \, \Delta T/(\nu \chi)$ is the Rayleigh number, which can also be expressed $N_{\text{Ra}} = N_{\text{Gr}} N_{\text{Pr}}$, where the Grashof number is given by $N_{\text{Gr}} = gL^3 \alpha_p \, \Delta T/\nu^2$.

Exercise 7.14   Torsional Flow between Parallel Disks
Consider the flow of an incompressible Newtonian fluid between parallel disks of radius R separated by the distance H, such that $H/R \ll 1$. The lower disk is stationary and the upper disk rotates with constant angular velocity Ω. The steady flow is laminar with a velocity field having the form $v_r = 0$, $v_\theta(r, z) = rw(z)$, $v_z = 0$. Solve for the function $w(z)$ and show that the torque on the stationary disk is given by the expression

$$\mathcal{M}_{\text{s}} = \frac{\pi R^4 \Omega \eta}{2H}. \tag{7.61}$$

Also, give the criterion that is necessary for the assumed velocity field to be consistent with *all* components of the Navier–Stokes equation.

Exercise 7.15   A Partially Filled Horizontal Couette Viscometer
A liquid half fills the annular region between horizontal concentric cylinders with length L (in the z-direction) as shown schematically in Figure 7.10. The upper half of the annular region is filled with a gas having uniform pressure p_0. The outer cylinder with radius $R \ll L$ rotates at constant angular velocity Ω and the inner cylinder of radius βR is stationary. Rotation of the outer cylinder causes the liquid to flow so that the difference in height between the exit ($\theta = \alpha$) and entrance ($\theta = \alpha - \pi$) of the liquid is $2h$. The gap between the cylinders is small, $1 - \beta \ll 1$, so that complications at the liquid–gas interfaces can be neglected. The liquid is an incompressible Newtonian fluid with constant density ρ and viscosity η. The flow is

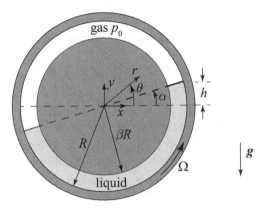

Figure 7.10 Schematic diagram showing flow of a liquid (light shaded region) that partially fills the annular region between horizontal concentric cylinders.

isothermal and inertial forces can be neglected (i.e., we have creeping flow). Derive expressions for v_θ and for h, which show how this device could be used as a crude viscometer.

- The hydrodynamic equations are obtained by combing balance equations and force–flux relations, and contain transport coefficients (diffusion coefficients, viscosities, thermal conductivity) as material properties.

- From the relation between the torque on a stationary inner cylinder and the angular velocity of a rotating coaxial outer cylinder under steady-state conditions, the viscosity of the fluid in the gap between the two cylinders can be determined (Couette viscometer).

- When an electric current is passing through a straight cylindrical wire, the time-dependence of the wire's increasing temperature is a measure of the efficiency with which the heat produced by the electric current is removed through the surrounding medium, so that the thermal conductivity of the medium can be determined (transient hot-wire method).

- When a film on an impermeable substrate is exposed to a gas containing a species that can diffuse into the film, the mass of the film increases with time; the binary diffusion coefficient for the dissolved species in the film can be determined from the initial increase of mass, which is proportional to the square root of time (sorption experiment).

- The analysis of relations between the various physical quantities involved in a transport problem (such as hydrodynamic fields, space, time, geometric or material parameters, boundary or initial values) can be simplified by introducing dimensionless variables.

8
Pressure-Driven Flow

In the preceding chapter we found a number of exact solutions to problems involving the transport of momentum, energy, and mass that are useful for the measurement of transport coefficients. This was possible because we considered idealized geometries and made further simplifying assumptions such that the transport equations were linear. We now consider a technologically important problem, namely pressure-driven flow of a compressible fluid in a tube, in which nonlinearity matters. Pressure-driven, or Poiseuille, flow in a tube is perhaps the most commonly encountered and widely studied problem in fluid mechanics. Experimental[1] and theoretical[2] studies of flow in tubes or pipes have a long history dating back to the middle of the nineteenth century. Establishing the relationship between the difference in pressure between the tube entrance and tube exit, or pressure drop, and the mass flow rate is important for applications ranging from the Trans-Alaskan Pipeline (see Exercise 1.1) to the human circulatory system. In this chapter, we use a perturbation method to solve the nonlinear equations governing weakly compressible flow in a tube. We also exploit the small aspect ratio of a tube to apply the so-called lubrication approximation.

8.1 Hagen–Poiseuille Equation

First, we consider the incompressible laminar flow of a Newtonian fluid having constant viscosity η and density ρ_0 with average velocity V in a tube of length L and radius R (see Figure 8.1). Consider a force balance in the z-direction on a cylindrical fluid element having radius r. Let p_0 and p_L be the uniform (pseudo-)pressures acting on the fluid at the surfaces where the fluid enters and exits the tube, respectively (by this convention $p_0 > p_L$).

---

[1] Hagen, *Pogg. Ann. Phys. Chem.* **46** (1839) 423; Poiseuille, *Comptes Rendus* **11** (1840) 961.
[2] Hagenbach, *Pogg. Ann. Phys. Chem.* **108** (1860) 385.

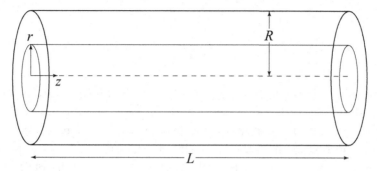

Figure 8.1 Flow in a tube showing cylindrical fluid element.

The net force on the fluid from this pressure difference is simply $\pi r^2(p_0 - p_L)$. This force is balanced by the force exerted on the cylindrical surface of the fluid element, which can be expressed as the product of the force per unit area, or stress, $\pi_{rz} = \tau_{rz}$ and the surface area $2\pi r L$. Hence, the force balance is given by

$$\frac{p_0 - p_L}{L} = 2\frac{\tau_{rz}}{r}. \tag{8.1}$$

From Section 6.2 we know for Newtonian fluids that the stress is the product of the viscosity and velocity gradient. If we assume the velocity field has the form $v_r = v_\theta = 0, v_z = v_z(r)$, then from (6.5) we can write $\tau_{rz} = -\eta\, dv_z/dr$, so that the force balance becomes

$$2\frac{\eta}{r}\frac{dv_z}{dr} = -\frac{p_0 - p_L}{L}. \tag{8.2}$$

Integrating (8.2) with respect to r, assuming no slip (as we did in Section 7.2) at the tube wall, $v_z(R) = 0$, gives

$$v_z = \frac{(p_0 - p_L)R^2}{4\eta L}\left[1 - \left(\frac{r}{R}\right)^2\right], \tag{8.3}$$

which shows the velocity has a parabolic profile. Note that since the pseudo-pressure p^L is known only to within an arbitrary constant, only differences (and derivatives) of pressure are meaningful for incompressible flows.

The mass flow rate $\mathcal{W}$ is found by integrating the mass flux ρv_z over the tube cross section

$$\mathcal{W} = 2\pi \int_0^R \rho v_z r\, dr. \tag{8.4}$$

In this section, we assume the fluid has constant density ρ_0, so that substitution of (8.3) in (8.4) gives

$$W = \frac{\rho_0 \pi R^4 (p_0 - p_L)}{8 \eta L}. \tag{8.5}$$

Equation (8.5) is the famous *Hagen–Poiseuille equation*,[3] an invaluable tool for the design of pipeline systems and the analysis of capillary viscometer experiments. The mass flow rate is proportional to the pressure drop and inversely proportional to the viscosity; more striking is its dependence on the tube radius, which enters with the fourth power. The effect of gravity, which has been neglected here, can easily be included (see Exercise 8.1).

In our derivation of the Hagen–Poiseuille equation (8.5) we assumed fully developed, laminar flow. For flow in a tube, the Reynolds number is defined as $N_{Re} = 2\rho_0 V R / \eta$; the criterion for laminar flow is $N_{Re} \lesssim 2100$, which is based on experiments. Our assumption of fully developed flow, which is the basis for setting $v_r = 0$, is valid only at some distance L_v from the entrance of the tube. A good estimate of the entrance length is given by $L_v / R \approx 1.18 + 0.112 N_{Re}$, which is based on a combination of experiments and numerical solutions of the Navier–Stokes equations.[4] We also assumed incompressible flow, which is typically justified for flows of water and other low-molecular-weight liquids in ordinary-sized pipes and tubes. For flows in micron-sized, or *microfluidic*, tubes and channels, in particular gas flows in these devices, the incompressibility assumption may no longer be valid.[5] For example, Figure 8.2 shows the normalized pressure drop for gas flows in microchannels. From this figure, we see that compressibility can reduce the pressure drop by a factor of ten relative to the incompressible case.

Exercise 8.1   Hagen–Poiseuille Equation with Gravity
For laminar flow in a tube, the velocity field has the form $v_r = v_\theta = 0, v_z = v_z(r, z)$. Use (5.36) and (7.8) to show

$$\eta \frac{1}{r} \frac{d}{dr} \left(r \frac{dv_z}{dr} \right) = \frac{d\mathcal{P}}{dz} = \frac{dp^L}{dz} - \rho g_z. \tag{8.6}$$

Integrate this result to obtain

$$v_z = 2V \left[1 - \left(\frac{r}{R} \right)^2 \right], \qquad \mathcal{P} = \mathcal{P}_L + \frac{8 \eta V L}{R^2} \left(1 - \frac{z}{L} \right), \tag{8.7}$$

---

[3] Poiseuille (1797–1869), a physician interested in the flow of blood in capillaries, carried out a careful series of experiments giving remarkably accurate values for the viscosity of water. A detailed history of (8.5) can be found in Sutera & Skalak, *Annu. Rev. Fluid Mech.* **25** (1993) 1.

[4] Atkinson *et al.*, *AIChE J.* **13** (1967) 17; Atkinson *et al.*, *AIChE J.* **15** (1969) 548.

[5] Karniadakis *et al.*, *Microflows and Nanoflows* (Springer, 2005); Kirby, *Micro- and Nanoscale Fluid Mechanics* (Cambridge, 2010).

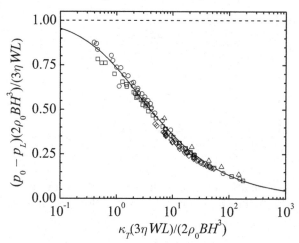

Figure 8.2 Experimental pressure drop data for gas flows in microchannels with height $2H$ and width B normalized by pressure drop for incompressible flow versus normalized isothermal compressibility. Different symbols correspond to different gases: N_2 □, He ○, Ar △, Air ◇ (see Venerus & Bugajsky, *Phys. Fluids* **22** (2010) 046101). The dashed line is the incompressible case (see Exercise 8.2) and the solid curve is given by (8.36), which is based on the lubrication approximation.

where the average velocity $V = (\mathcal{P}_0 - \mathcal{P}_L)R^2/(8\eta L)$. Also, find an expression similar to (8.5) that includes gravity.

Exercise 8.2   Pressure-Driven Flow in a Channel
Consider the pressure-driven, laminar flow of a Newtonian fluid having constant viscosity η and density ρ_0 in a channel of height $2H$, length L, and width B. The average velocity is V, and, since $H \ll B \ll L$, the velocity field has the form $v_1 = v_1(x_1, x_2), v_2 = v_3 = 0$. Use (5.36) and (7.8) to obtain the parabolic velocity field

$$v_1 = \frac{(p_0 - p_L)H^2}{2\eta L}\left[1 - \left(\frac{x_2}{H}\right)^2\right],\tag{8.8}$$

with $V = (p_0 - p_L)H^2/(3\eta L)$, which is $3/2$ the maximum velocity. Also, derive the expression

$$\mathcal{W} = \frac{2\rho_0 BH^3(p_0 - p_L)}{3\eta L}.\tag{8.9}$$

Exercise 8.3   Vorticity Transport in Incompressible Flow in a Tube
Use the vorticity transport equation (7.20) to obtain the velocity given in (8.3).

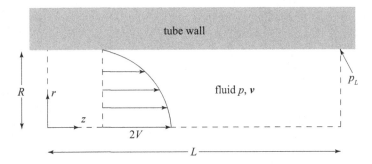

Figure 8.3 Schematic diagram of flow in a tube of radius R and length L. Note that $2V$ is the maximum velocity only for incompressible flow.

8.2 Compressible Flow in a Tube

We now consider the steady laminar flow of a compressible Newtonian fluid with constant viscosity η as shown in Figure 8.3. For convenience, we set the dilatational viscosity $\eta_d = 0$. The fluid density is taken to be a linear function of pressure, $\rho = \rho_0[1 + \kappa_T(p - p_L)]$, where κ_T is the isothermal compressibility and the exit pressure p_L at the tube wall is used as a reference pressure, with the corresponding reference density ρ_0. We assume there is no circumferential flow, that there is symmetry about the axial coordinate, and that gravitational effects are negligible.

The equations governing this flow are the steady-state forms of the continuity equation (7.1) and the equation of motion for a compressible Newtonian fluid (7.2). For this problem these equations, along with the equation of state, involve the independent quantities $\rho, \boldsymbol{v}, p - p_L, \boldsymbol{r}, R, L, V, \eta, \rho_0, \kappa_T$, so that $n = 10$ with $m = 3$ (mass, length, and time), and $n - m = 7$. Note that the average velocity $V = \mathcal{W}/(\rho_0 \pi R^2)$ is defined in terms of the mass flow rate, which is a solution feature and hence needs to be identified as part of the solution procedure. If we solve for ρ, $\boldsymbol{v}$, and $p - p_L$ as functions of $\boldsymbol{r}$, we anticipate that the solutions can depend on three dimensionless parameters. These can be taken as

$$\epsilon = \frac{\kappa_T \eta V}{R}, \qquad N_{\text{Re}} = \frac{2\rho_0 V R}{\eta}, \qquad \beta = \frac{R}{L}. \qquad (8.10)$$

As R, V, and η appear at least twice in (8.10), it is convenient to choose these parameters as the units of length, velocity, and viscosity. The corresponding units of time, mass, and pressure are R/V, $\eta R^2/V$, and $\eta V/R$, respectively. In these convenient units ($R = V = \eta = 1$), we simply have $\kappa_T = \epsilon$, $\rho_0 = N_{\text{Re}}/2$, and $L = 1/\beta$. In the solution procedure, we drop factors of R, V, and η whenever it is convenient, but we restore all the proper factors in the final results so that they hold independently of the choice of units. According to Exercise 5.15, the compressibility is related to the speed of sound. If we use the identity $\kappa_T \rho_0 = 1/c_T^2$, where c_T is the isothermal speed of sound (see Table A.1), we recognize the identity $\epsilon N_{\mathrm{Re}}/2 = \kappa_T \rho_0 = 1/c_T^2 = N_{\mathrm{Ma}}^2$, where c_T is measured in units of V. As we are interested in *subsonic flow*, we should keep both ϵ and ϵN_{Re} small.

The equation of state in dimensionless form can be written as

$$\rho = \rho_0(1 + \epsilon p) = N_{\mathrm{Re}}(1 + \epsilon p)/2, \tag{8.11}$$

where it is convenient to write p instead of $p - p_L$. Based on the assumptions stated above, we postulate the velocity field to have the form $v_r = v_r(r, z)$, $v_\theta = 0$, $v_z = v_z(r, z)$. Expressing (7.1) and (7.2) in cylindrical coordinates [with the help of (B.8), (B.10), (B.12), and (B.14)], we have[6]

$$\frac{1}{r}\frac{\partial}{\partial r}(\rho r v_r) + \frac{\partial}{\partial z}(\rho v_z) = 0, \tag{8.12}$$

$$\rho\left(v_r\frac{\partial v_r}{\partial r} + v_z\frac{\partial v_r}{\partial z}\right) = -\frac{\partial p}{\partial r} + \eta\frac{\partial}{\partial r}\left(\frac{1}{r}\frac{\partial}{\partial r}(r v_r)\right) + \eta\frac{\partial^2 v_r}{\partial z^2}$$
$$+ \frac{1}{3}\eta\left[\frac{\partial}{\partial r}\left(\frac{1}{r}\frac{\partial}{\partial r}(r v_r)\right) + \frac{\partial^2 v_z}{\partial r \partial z}\right], \tag{8.13}$$

$$\rho\left(v_r\frac{\partial v_z}{\partial r} + v_z\frac{\partial v_z}{\partial z}\right) = -\frac{\partial p}{\partial z} + \eta\frac{1}{r}\frac{\partial}{\partial r}\left(r\frac{\partial v_z}{\partial r}\right) + \eta\frac{\partial^2 v_z}{\partial z^2}$$
$$+ \frac{1}{3}\eta\left[\frac{\partial}{\partial z}\left(\frac{1}{r}\frac{\partial}{\partial r}(r v_r)\right) + \frac{\partial^2 v_z}{\partial z^2}\right]. \tag{8.14}$$

The terms with factors $1/3$ in the second lines of (8.13) and (8.14) describe the resistance to volume or density changes which, for vanishing dilatational viscosity, are governed by the shear viscosity. Velocity boundary conditions along the tube axis are from symmetry given by

$$v_r(0, z) = 0, \quad \frac{\partial v_z}{\partial r}(0, z) = 0, \tag{8.15}$$

and those at the impermeable tube wall, assuming no slip, are given by

$$v_r(R, z) = 0, \quad v_z(R, z) = 0. \tag{8.16}$$

The boundary condition for pressure is

$$p(R, L) = 0. \tag{8.17}$$

---

[6] Note that, from symmetry considerations, the θ-component of (7.2) gives $\partial p/\partial\theta = 0$.

Since we have used a characteristic velocity based on the mass flow rate $V = \mathcal{W}/(\rho_0 \pi R^2)$, it is useful to rewrite (8.4) in the form

$$\frac{2}{R^2} \int_0^R \rho v_z r \, dr = \rho_0 V. \tag{8.18}$$

Equations (8.11)–(8.17) govern the pressure-driven flow of a compressible Newtonian fluid in a tube. In our convenient units, we are free to replace η in (8.13) and (8.14), R in (8.16)–(8.18), and V in (8.18) by unity, as well as L in (8.17) by $1/\beta$, but we kept them for reasons of clarity. As anticipated, our problem involves the three dimensionless parameters ϵ, N_{Re}, and β.

Note that we have not specified the velocity boundary conditions at the entrance or exit of the tube that would be required in order to obtain a unique solution of the elliptic partial differential equations (8.13) and (8.14). Unfortunately, it is not obvious what these conditions should be. This issue will be addressed in the following section.

Exercise 8.4   *Stream Function for Compressible Flow in a Tube*
Use (8.12) to express the velocity field in terms of the stream function $\psi(r, z)$,

$$v_r = \frac{1}{\rho r} \frac{\partial \psi}{\partial z}, \qquad v_z = -\frac{1}{\rho r} \frac{\partial \psi}{\partial r},$$

and, using (8.18), show that $\psi(0, z) - \psi(1, z) = 1/2$.

8.3 A Perturbation Method

Equation (8.12) and the left-hand sides of the momentum balances (8.13) and (8.14) make these equations nonlinear. Consequently, it is not possible to find an exact analytical solution to this set of equations. From Section 8.1, we know that finding a solution to the problem of incompressible flow in a tube is relatively simple. Let us suppose we are interested in small deviations from the incompressible case. For our problem, this corresponds to $\epsilon \ll 1$; the familiar incompressible case is recovered when $\epsilon = 0$ and setting $v_r = 0$.

Techniques for finding an approximate solution to a nonlinear problem using the solution to the related linear problem and *small* corrections to it are known as *perturbation methods*. These methods are widely used in engineering and physics,[7] and here we implement the most basic version. The first step is to express the dependent variables (ρ, v_r, v_z, p) in the form of power series expansions in the *perturbation parameter* (ϵ). For the dependent

---

[7] Van Dyke, *Perturbation Methods in Fluid Mechanics* (Academic, 1964); Nayfeh, *Perturbation Methods* (Wiley, 1973); Hinch, *Perturbation Methods* (Cambridge, 1991).

variable a, we write the power series,

$$a = a^{(0)} + \epsilon a^{(1)} + \epsilon^2 a^{(2)} + \cdots, \tag{8.19}$$

where $a^{(1)}, a^{(2)}, \ldots$ are the first-, second-, ... order corrections, respectively, to the zeroth-order solution $a^{(0)}$. The second step is to substitute the expansions for all of the dependent variables into the governing equations and then collect all the terms having the same power of the perturbation parameter. If we do this for (8.11), we have $\rho^{(0)} + \epsilon\rho^{(1)} + \cdots = N_{\mathrm{Re}}[1 + \epsilon(p^{(0)} + \epsilon p^{(1)} + \cdots)]/2$, which leads to $\rho^{(0)} = N_{\mathrm{Re}}/2$, $\rho^{(1)} = N_{\mathrm{Re}}\, p^{(0)}/2$, and so on. The reader should note the similarity between this procedure and the one used to linearize the equations for sound wave propagation (see Exercise 5.15).

We have already solved the zeroth-order problem corresponding to the incompressible case in Section 8.1. By assuming $v_r^{(0)} = 0$, we found the axial velocity in the first equation in (8.7), which can be written as

$$v_z^{(0)} = 2(1 - r^2). \tag{8.20}$$

From the second equation in (8.7), we write the pressure field at zeroth order as

$$p^{(0)} = 8(L - z). \tag{8.21}$$

Note that the zeroth-order solution predicts a pressure drop $p^{(0)}(0) - p^{(0)}(L) = 8L$, which, since the pressure is measured in units of $\eta V/R$, is the dimensionless form of the Hagen–Poiseuille equation (8.5).

Now we consider the first-order problem, which involves collecting all the terms from the second step containing the perturbation parameter ϵ to the first power. This results, after substitution for the terms containing the zeroth-order solution, in the following set of dimensionless equations,

$$\frac{1}{r}\frac{\partial}{\partial r}(rv_r^{(1)}) + \frac{\partial v_z^{(1)}}{\partial z} = 16(1 - r^2), \tag{8.22}$$

$$N_{\mathrm{Re}}(1 - r^2)\frac{\partial v_r^{(1)}}{\partial z} = -\frac{\partial p^{(1)}}{\partial r} + \frac{\partial}{\partial r}\left(\frac{1}{r}\frac{\partial}{\partial r}(rv_r^{(1)})\right) + \frac{\partial^2 v_r^{(1)}}{\partial z^2}$$
$$+ \frac{1}{3}\left[\frac{\partial}{\partial r}\left(\frac{1}{r}\frac{\partial}{\partial r}(rv_r^{(1)})\right) + \frac{\partial^2 v_z^{(1)}}{\partial r\partial z}\right], \tag{8.23}$$

$$N_{\mathrm{Re}}\left(-2rv_r^{(1)} + (1 - r^2)\frac{\partial v_z^{(1)}}{\partial z}\right) = -\frac{\partial p^{(1)}}{\partial z} + \frac{1}{r}\frac{\partial}{\partial r}\left(r\frac{\partial v_z^{(1)}}{\partial r}\right) + \frac{\partial^2 v_z^{(1)}}{\partial z^2}$$
$$+ \frac{1}{3}\left[\frac{\partial}{\partial z}\left(\frac{1}{r}\frac{\partial}{\partial r}(rv_r^{(1)})\right) + \frac{\partial^2 v_z^{(1)}}{\partial z^2}\right]. \tag{8.24}$$

The equations still look a bit complicated, but notice that they are now *linear*. If, for simplicity, we again *assume* that the radial velocity is zero, $v_r^{(1)} = 0$, then all of the underlined terms in (8.22)–(8.24) are eliminated. Dropping the underlined terms, from (8.22) we see that $v_z^{(1)}$ is a linear function of z, and from (8.23) we see that pressure depends on radial position. Here we describe the basic approach used to find the first-order velocity and pressure fields from the simplified forms of (8.22)–(8.24) (details of this procedure are left as Exercise 8.5). Substitution of (8.22) into (8.23) and (8.24) leads to expressions for the radial and axial pressure gradients. These can be combined by applying the compatibility condition, which leads to an equation governing $v_z^{(1)}$. Integrating this equation subject to the boundary conditions in (8.15) and (8.16), and enforcing the proper definition of the average velocity V according to (8.18), leads to

$$v_z^{(1)} = 2(1 - r^2)\left[-8(L - z) - \frac{1}{9}N_{\mathrm{Re}}\left(2 - 7r^2 + 2r^4\right)\right]. \tag{8.25}$$

To obtain the pressure, we integrate the differential $dp^{(1)} = (\partial p^{(1)}/\partial r)dr + (\partial p^{(1)}/\partial z)dz$ subject to the boundary condition (8.17), which gives

$$p^{(1)} = 8N_{\mathrm{Re}}(L - z) - 32(L - z)^2 + \frac{16}{3}(1 - r^2). \tag{8.26}$$

We could continue this procedure to second order, but the level of complexity increases substantially, even though only simple integrations are required, while the accuracy increases only slightly. Readers interested in seeing the method carried out to second-order in ϵ, for which the radial velocity is no longer zero ($v_r^{(2)} \neq 0$), can find this elsewhere.[8]

Since we have decided to stop here, we now carry out the final step in the perturbation method: combining the different-order solutions in the series expansions with which we initiated the method. This leads to the following first-order results in ϵ in a fully restored dimensional form,

$$v_r = 0, \tag{8.27}$$

$$v_z = 2V\left(1 - \frac{r^2}{R^2}\right)\left[1 - 8\epsilon\frac{L - z}{R} - \frac{1}{9}\epsilon N_{\mathrm{Re}}\left(2 - 7\frac{r^2}{R^2} + 2\frac{r^4}{R^4}\right)\right], \tag{8.28}$$

$$p - p_L = \frac{8\eta V}{R}\left[\frac{L - z}{R} + \epsilon N_{\mathrm{Re}}\frac{L - z}{R} - 4\epsilon\frac{(L - z)^2}{R^2} + \frac{2}{3}\epsilon\left(1 - \frac{r^2}{R^2}\right)\right]. \tag{8.29}$$

Note that $8\epsilon L/R = 8\epsilon/\beta = 8\kappa_T\eta V L/R^2$, which is the isothermal compressibility scaled by the pressure drop for incompressible flow.

[8] Venerus, *J. Fluid Mech.* **555** (2006) 59–80.

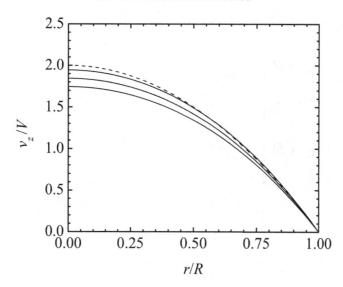

Figure 8.4 Axial velocity v_z/V profile for compressible flow in a tube from (8.28) for $\epsilon = \beta/80$ and $\epsilon N_{\mathrm{Re}} = 1/8$ or $N_{\mathrm{Ma}} = 1/4$ (solid lines) for $z/L = 0, 0.5, 1$ (bottom to top). The dashed line is for the incompressible case, $\epsilon = 0$.

Recall that we solved this flow problem without specifying velocity boundary conditions at the tube entrance and exit. There is no question that (8.27)–(8.29) satisfy the governing equations to first order in ϵ and are, therefore, a solution to compressible laminar flow in a tube. Instead of imposing complicated physical boundary conditions, however, we made a particularly simple *Ansatz* for finding our solution (see Exercise 8.5). The main purpose of this section is to illustrate in detail how perturbation theory works, not to produce physical results. Although there is no reason to believe that our solution is close to the physical solution, we nevertheless briefly discuss some implications of it.

We first examine the effect of compressibility on the velocity and pressure fields. From (8.28), we see that compressibility causes the axial velocity to increase along the tube, and that inertial effects lead to a deviation from a parabolic profile. These features are shown in Figure 8.4, where the velocity profile is flattened as a result of inertial effects. As shown by (8.29), fluid compressibility results in a nonlinear dependence of the pressure on axial position, and weak variations in the radial direction, which are caused by volumetric deformations. The pressure profile along the axial direction is shown in Figure 8.5.

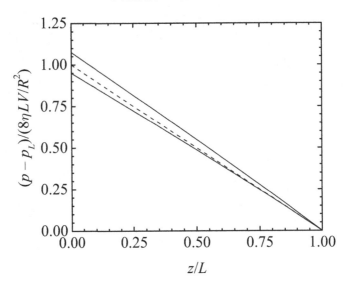

Figure 8.5 Pressure $(p - p_L)/(8\eta V L/R^2)$ profile for compressible flow in a tube from (8.29) for $\epsilon = \beta/80$ and $\epsilon N_{\text{Re}} = 0$ (bottom curve) and $\epsilon N_{\text{Re}} = 1/8$ (top curve). The dashed line is for the incompressible case, $\epsilon = 0$.

Let us now look at how compressibility affects the pressure drop for flow in a tube (that's how we got started). The pressure drop $-\Delta p = p(r,0) - p(r,L)$ is independent of r and, using (8.29), we obtain

$$-\Delta p = \frac{8\eta V L}{R^2} \left(1 - \frac{4}{\beta}\epsilon + \epsilon N_{\text{Re}}\right). \tag{8.30}$$

Recall that, for incompressible flow, $-\Delta p = 8\eta V L/R^2$, which corresponds to the Hagen–Poiseuille equation. The second term in (8.30) tells us that compressibility reduces the pressure drop relative to the incompressible case. To explain this we look at (8.28), which shows the axial velocity is reduced by compressibility to compensate for the increase in density (to satisfy mass conservation). The decreased velocity leads to a reduced shear stress, and hence reduces the drag force on the tube wall. Not so fast though. The third term in (8.30) tells us that compressibility also increases the pressure drop when inertial effects are important. The decrease in pressure drop from reduced viscous forces, and the increase due to inertia, are shown in Figure 8.6. The latter effect is explained by the acceleration of the fluid along the tube, which requires an additional pressure force (note that the mass flux $\rho \boldsymbol{v}$ is constant, but the momentum flux $\rho \boldsymbol{v} \boldsymbol{v}$ increases, along the tube). Hence, compressibility results in a nonlinear dependence of pressure drop on flow

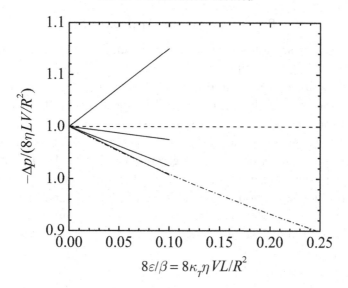

Figure 8.6 Normalized pressure drop for compressible flow in a tube from (8.30) for four values of $\beta N_{\mathrm{Re}}/8 = 0.3, 1, 3, 10$ (solid lines from bottom to top). The dash–dotted line is from the lubrication approximation (8.36). The dashed line is for the incompressible case, $\epsilon = 0$.

rate that is the manifestation of both irreversible and reversible transport of momentum.

Exercise 8.5   First-Order Perturbation Solution
Set $v_r^{(1)} = 0$ and solve (8.22)–(8.24) to obtain the first-order solutions in (8.25) and (8.26). Hint: use the compatibility condition to eliminate $p^{(1)}$.

Exercise 8.6   Viscous Heating in Couette Flow
In our analysis of Couette flow in Section 7.2, we assumed the flow was isothermal. However, for fluids with large viscosity, or for flows with large shear rate, viscous dissipation may invalidate the isothermal flow assumption. Here, we assume the thermal properties of the fluid $(\rho, \hat{c}_p, \lambda)$ are constant, and that viscosity is a linear function of temperature, $\eta(T) = \eta_0[1 + A(T - T_0)]$, where η_0 and A are constants. If both cylinders are maintained at constant temperature T_0, use a perturbation method with $N_{\mathrm{Na}} = \eta_0 \Omega^2 R^2/(\lambda/A)$ as the perturbation parameter to find the following expression for the torque exerted on the inner cylinder,

$$\mathcal{M}_s = \frac{4\pi \beta^2 R^2 L \Omega \eta_0}{1 - \beta^2} \left[1 - 2N_{\mathrm{Na}} \frac{\beta^2}{(1 - \beta^2)^2} \left(\frac{1 - \beta^2}{\ln \beta} - \frac{4\beta^2 \ln \beta}{1 - \beta^2} \right) \right].$$

Confirm that, for $A < 0$, the correction factor due to viscous dissipation (the quantity inside the square brackets) is less than one.

8.4 Lubrication Approximation

In the previous section we exploited the smallness of a dimensionless parameter (ϵ) to find an approximate solution to the problem of compressible laminar flow in a tube. There is a second parameter we have only briefly mentioned so far, which is also small: the aspect ratio of the tube, $\beta = R/L$. Typically, the length of a tube is at least 100 times larger than its radius, so that $\beta \approx 10^{-2}$, or smaller. Is there a way to exploit this and simplify the rather complicated set of equations given in (8.11)–(8.14)? The answer is yes, and it is the basis for an important method for the simplification of fluid mechanics problems. Although not necessary, for convenience we assume that inertial effects can be neglected, or apply the creeping flow approximation (see Section 7.5 and Exercise 7.14), which will be true if $\beta N_{\mathrm{Re}} \ll 1$. Recall that in the previous section we assumed $v_r = 0$, which led to the result in (8.29) that shows that the pressure is weakly dependent on radial position. Setting $v_r = 0$, from (8.13) we obtain $\partial p/\partial r \approx 0$, and (8.14) simplifies to

$$\frac{1}{r}\frac{\partial}{\partial r}\left(r\frac{\partial v_z}{\partial r}\right) = \frac{1}{\eta}\frac{dp}{dz}, \tag{8.31}$$

where we have also neglected $\partial^2 v_z/\partial z^2$ terms since these are $\sim \beta^2$ smaller than terms that have been retained. Note the similarity of (8.31) to the expression in (8.6) for incompressible flow (see Exercise 8.1) where $p = p(z)$. However, in (8.31) we note that $v_z = v_z(r, z)$.

We have just invoked what is known in fluid mechanics as the *lubrication approximation*,[9] which can be applied when the geometry of a flow contains one dimension that is much smaller than other dimensions. The name originates from journal–bearing and piston–cylinder geometries, which have one dimension that is small compared with another (e.g., the gap between the piston and cylinder is much smaller than the length of the cylinder), and lubricants are used to prevent the contact of solid surfaces in relative motion. The basic idea is that pressure variations in the direction of the small dimension can be neglected relative to those in the direction of the large dimension.

Our solution with the lubrication approximation is obtained by integrating (8.31) with respect to r subject to the boundary conditions in the second equation in (8.15) and the second equation in (8.16) to obtain an expression for v_z in terms of dp/dz. If we substitute this result into (8.12), and integrate with respect to r using the boundary conditions in the first equation in (8.15)

---

[9] Reynolds, *Phil. Trans. R. Soc. Lond. A* **177** (1886) 157.

and the first equation in (8.16), we obtain

$$\frac{d}{dz}\left(\rho\frac{dp}{dz}\right) = 0. \tag{8.32}$$

Equation (8.32) is the *Reynolds equation* for this flow, and is an important result obtained when the lubrication approximation is used. Note that, in deriving it, we satisfied the mass balance exactly, and momentum balance approximately. From (8.32), we have $dp/dz \propto 1/\rho$, which can be combined with the expression for v_z. This leads to

$$v_z = 2\rho_0/\rho(1 - r^2), \tag{8.33}$$

where we have also used (8.18). Substitution of the expression for ρ from (8.11) into the Reynolds equation (8.32), and integration subject to the boundary condition (8.17), gives

$$p - p_L = \frac{\eta V}{R}\frac{\sqrt{1 + 16\epsilon(L - z)/R} - 1}{\epsilon} = \frac{8\eta V}{R}\left[\frac{L - z}{R} - 4\epsilon\frac{(L - z)^2}{R^2}\right] + O(\epsilon^2), \tag{8.34}$$

where we have re-introduced dimensional quantities. Finally, using (8.11) and (8.34) in (8.33) gives

$$v_z = \frac{2V(1 - r^2/R^2)}{\sqrt{1 + 16\epsilon(L - z)/R}} = 2V\left(1 - \frac{r^2}{R^2}\right)\left[1 - 8\epsilon\frac{L - z}{R}\right] + O(\epsilon^2). \tag{8.35}$$

Notice that the second equations in (8.35) and (8.34) agree with the perturbation solutions (8.28) and (8.29), if in the latter we neglect the terms containing N_{Re} and radial pressure variations. The pressure drop from (8.34) is simply

$$-\Delta p = \frac{8\eta VL}{R^2}\frac{\sqrt{1 + 16\epsilon/\beta} - 1}{8\epsilon/\beta}, \tag{8.36}$$

which has been derived using both the lubrication approximation and the creeping flow approximation. The dependence of $-\Delta p$ on $8\epsilon/\beta$ from (8.36) is shown in Figure 8.6, which shows deviations from (8.30) when $\beta N_{Re}/8 \gtrsim 1$. From Figure 8.2 we see that the prediction of (8.36) is in remarkably good agreement with experimental data for gas flows in microchannels.

Exercise 8.7   Lubrication Approximation for Tube Flow
Starting from (8.31), derive (8.32), and use this to find the expressions given in (8.33)–(8.35).

Exercise 8.8   Squeezing Flow between Parallel Disks

Consider the squeezing flow of an incompressible Newtonian fluid between parallel disks of radius R separated by distance $H(t)$. Initially, the distance between the plates is $H_0 = H(0)$, and $\bar{\beta} = H_0/R \ll 1$. For $t > 0$ the upper plate moves towards the stationary lower plate with constant velocity V, which induced a flow of the form $v_r = v_r(r, z, t), v_\theta = 0, v_z = v_z(r, z, t)$. We seek dimensionless forms of the continuity and Navier–Stokes (r- and z-components, with gravity neglected) equations, and have characteristic length H_0, time H_0/V, and stress $\eta V/H_0$. Identify the characteristic quantities that lead to properly scaled variables given the smallness of the aspect ratio $\bar{\beta} = H_0/R$, and use these to put the governing equations in dimensionless form. Simplify these by invoking the lubrication ($\bar{\beta} \ll 1$) and creeping flow ($N_{\mathrm{Re}} = \rho V H_0/\eta \ll 1$) approximations. Solve the resulting equations for velocity and pressure fields. Also, using (8.39), derive the following expression for the force required to move the upper plate with velocity V:

$$\mathcal{F}_{\mathrm{s}} = \frac{3\eta\pi R^4 V}{2H^3},$$

which is known as the Stefan equation. Briefly discuss the similarities and differences between the scaling used in this problem and that used in modeling flow through a tube.

Exercise 8.9   Incompressible Flow in a Tapered Tube

Consider the steady, creeping flow of an incompressible Newtonian fluid in a slightly tapered tube with a radius that varies as follows: $R(z) = R_0 + \tan(\varphi)z$, where $|\tan(\varphi)| \ll \beta = R_0/L$ is the taper angle of the tube. Use the lubrication approximation to find expressions for the velocity and pseudo-pressure fields.

8.5 Friction Factor

There is a useful quantity for obtaining the pressure drop for flow inside conduits in more complicated situations – the so-called *friction factor*. The idea is based on finding an estimate for the drag force per unit area of the wetted surface of the conduit, which is made dimensionless with a characteristic kinetic energy density. Here we use the Fanning friction factor,[10] which is defined as

$$f_{\mathrm{s}} = \frac{\mathcal{F}_{\mathrm{s}}/A_{\mathrm{s}}}{\rho_0 \langle v \rangle^2/2}, \tag{8.37}$$

where $\langle v \rangle = \int_A v\, dA/A$ is the average velocity over the conduit cross-sectional area A, and $\mathcal{F}_{\mathrm{s}}$ is the force exerted by the fluid on the wetted

---

[10] In addition to the Fanning friction factor defined in (8.37), there are several other commonly used definitions for the friction factor, which differ from each other only by a numerical factor. For example, the Darcy friction factor is four times larger than the Fanning friction factor.

conduit surface with area A_s. The arguments that led us to the Hagen–Poiseuille equation (8.5) for laminar flow in a tube imply

$$f_s = \frac{16}{N_{Re}}. \tag{8.38}$$

In general, the force in (8.37) is defined by

$$\mathcal{F}_s = \int_{A_s} \boldsymbol{n} \cdot \boldsymbol{\pi} \, dA, \tag{8.39}$$

which appears in the macroscopic form of the momentum balance (see Exercise 8.12). It is important to recognize that the force $\mathcal{F}_s$, and hence f_s, represent the transport of momentum across an interface between a fluid and solid.

For steady flow of an incompressible fluid with density ρ_0 in a conduit with constant cross-sectional area A, the mass flow rate can be written as $\mathcal{W} = \rho_0 \langle v \rangle A$. If we further consider a straight conduit of length L (with entrance surface labeled 1 and exit surface labeled 2), the macroscopic momentum balance (see Exercise 8.12) in (8.45) gives

$$\mathcal{F}_s = -(p_2^L - p_1^L)A - \rho_0(\phi_2 - \phi_1)A = -\Delta \mathcal{P} A, \tag{8.40}$$

where $\mathcal{P}$ is the modified pressure introduced in (5.41). Combining (8.37) and (8.40), we obtain

$$-\Delta \mathcal{P} = \frac{1}{2}\rho_0 \langle v \rangle^2 \frac{L}{R_{hyd}} f_s, \tag{8.41}$$

where $R_{hyd} \, (= AL/A_s)$ is the hydraulic radius of the conduit. It is easy to verify that substitution of (8.38) into (8.41) gives (8.5), the Hagen–Poiseuille equation, which shows that the pressure drop increases linearly with flow rate.

As mentioned at the beginning of this chapter, flow in a tube is laminar for $N_{Re} \lesssim 2100$. For larger Reynolds number, say $N_{Re} \gtrsim 10^4$, the flow is no longer stable, but becomes *turbulent*. In contrast to the smooth motion of fluid elements seen in laminar flow, turbulent flow is characterized by the chaotic motion of fluid pockets called eddies. Even for flow in a tube subject to a constant pressure drop, the motion of these turbulent eddies can be time-dependent and asymmetric. The origin of these more complex flow features is the nonlinear yet reversible terms in the momentum balance, which are no longer smoothed out by the irreversible terms when N_{Re} becomes large. Despite this increased complexity, turbulent flows are still described by the same balance and constitutive equations that govern laminar flows. However, they can be solved only for relatively restricted turbulent flows, using sophisticated numerical methods on powerful computers. To quote the famous physicist Richard Feynman,[11] "Finally, there is a physical problem that is common to many fields, that is very old, and that has not been solved. It is not the problem of finding new fundamental particles, but something left over from a long time ago – over a hundred years. Nobody in physics has really been able to analyze it mathematically satisfactorily in spite of its importance to the sister sciences. It is the analysis of *circulating or turbulent fluids*."

An alternative and widely used approach to determine the relation between the pressure drop and the flow rate is to use dimensional analysis in combination with experimental data. As an example, we give the famous Blasius[12] formula,

$$f_{\rm s} \approx \frac{0.0791}{N_{\rm Re}^{1/4}}, \tag{8.42}$$

which is valid for turbulent flow in smooth pipes for $3000 \lesssim N_{\rm Re} \lesssim 10^5$. A somewhat more accurate expression is known as the Prandtl formula,[13]

$$\frac{1}{\sqrt{f_{\rm s}}} \approx 4 \log(N_{\rm Re} \sqrt{f_{\rm s}}) - 0.4, \tag{8.43}$$

which is valid for turbulent flow in smooth pipes for $3000 \lesssim N_{\rm Re} \lesssim 10^6$. A plot of the friction factor $f_{\rm s}$ versus Reynolds number $N_{\rm Re}$, which is often referred to as a Moody chart,[14] is given in Figure 8.7. For turbulent flow in tubes that are not hydraulically smooth, the friction factor increases with increasing relative roughness.

Once the friction factor is known, (8.41) allows prediction of the pressure drop for turbulent flow in a tube, which has a stronger (than linear) dependence on flow rate. This stronger dependence is the result of the increase in energy dissipation from the turbulent eddies. Readers interested in a more detailed discussion of turbulent flow should refer to other books.[15]

For incompressible laminar flows in conduits of constant cross-sectional area, it is relatively easy to derive the relationship between the pressure drop and the flow rate, and express this in terms of a friction factor. In most cases, the level of effort is comparable to finding the incompressible-flow solution given in Section 8.1. Here we give a few examples. For flow in a channel of length L, height $2H$, and width B, where $B \gg H$, the friction factor is

[11] Feynman *et al.*, *The Feynman Lectures on Physics: Volume 1* (Addison-Wesley, 1964).
[12] Blasius, *Mitteil. Forschungsarb. Ingenieurwes.* **131** (1913) 1.
[13] Prandtl, *Essentials of Fluid Mechanics* (Hafner, 1952).
[14] Moody, *Trans. Am. Soc. Mech. Eng.* **66** (1944) 671.
[15] See, for example, Frisch, *Turbulence* (Cambridge, 1995) and Chapter 5 of Bird, Stewart & Lightfoot, *Transport Phenomena* (Wiley, 2001) and the references cited therein.

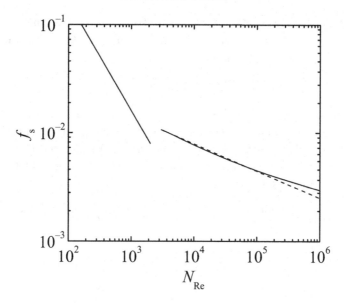

Figure 8.7 Friction factor f_s versus Reynolds number N_{Re} for incompressible flow in a smooth tube. The solid line for laminar flow ($N_{Re} \lesssim 2000$) is from (8.38). For turbulent flow ($N_{Re} \gtrsim 3000$), the solid line is from (8.42) and the dashed line is from (8.43).

given by $f_s = 24/N_{Re}$, where $N_{Re} = 2\rho_0 V H/\eta$. A perturbation solution for compressible flow in a channel, along with comparisons with experimental data, can be found elsewhere.[16] For flow in a conduit having a square cross section of $2H \times 2H$, which is a somewhat more complicated problem, the friction factor is given by $f_s \approx 14.123/N_{Re}$, where $N_{Re} = 2\rho_0 V H/\eta$.[17] The simple relations between f_s and Re for these flows are possible because the velocity profile, and hence shear stress at the solid–fluid interface, are constant along the flow direction. In other words, the pressure gradient is uniform along the conduit for these flows.

There are pressure-driven flows for which the velocity profile changes along the flow direction so that the pressure gradient is not constant. In such cases, the friction factor f_s depends on axial position, and the product $f_s N_{Re}$ is no longer equal to a constant. As we have just seen, the dependence of pressure on axial position is nonlinear for compressible laminar flow in a tube (see Exercise 8.11). A second example is incompressible flow in a conduit with increasing (or decreasing) cross section (see Exercise 8.9).

[16] Venerus & Bugajsky, *Phys. Fluids* **22** (2010) 046101.
[17] Boussinesq, *Comptes Rendus* **158** (1914) 1743; Shah & London, *Laminar Flow Forced Convection* (Academic, 1978).

Exercise 8.10   Pressure Drop along the Trans-Alaska Pipeline
Determine the pressure drop in a section of the Trans-Alaska Pipeline discussed in
Exercise 1.1. Consider a 100 km section of pipeline with a diameter of 1.22 m. Use
the following properties of crude oil: $\rho = 900$ kg/m^3, $\eta = 0.01$ Pa s.

Exercise 8.11   Friction Factor for Compressible Flow in a Tube
Consider the governing equations for compressible flow in a tube for the lubrication
approximation with inertia. Setting $v_r = 0$ in (8.13) leads to $\partial p/\partial r \approx 0$. In (8.14),
further neglect $\partial^2 v_z/\partial z^2$ and derive the following expression relating the friction
factor to the Pressure Drop:

$$\bar{f}_s N_{\mathrm{Re}} = 16\left[-\Delta p\left(1 - \frac{4\epsilon}{\beta}\Delta p\right) - \frac{\varphi}{16}\beta N_{\mathrm{Re}}\ln\left(1 - \frac{8\epsilon}{\beta}\Delta p\right)\right],$$

where $\bar{f}_s$ is the average friction factor over the length of the tube, and $\varphi = \langle v_z^2\rangle/\langle v_z\rangle^2$. Use the perturbation solution given in (8.28) and (8.29) to evaluate
this expression and comment on why it is not satisfied.

Exercise 8.12   Macroscopic Mass and Momentum Balances
Consider flow within a volume V that is bound by surface A with outward unit
normal vector n. The surface $A = A_1 + A_2 + A_s$ has an entrance with area A_1, an
exit with area A_2, and an impermeable, non-entrance–exit surface with area A_s.
At the entrance and exit surfaces, the mass flux can be written as $(\rho v)_i = \rho_i v_i n_i$
$(i = 1, 2)$, where v_i is the magnitude of the velocity and $n_1 = -n$ and $n_2 = n$.
Assume the density is uniform over A_i and use this to show from (5.5) that the
macroscopic mass balance takes the form

$$\frac{d}{dt}M_{\mathrm{tot}} = -\Delta[\rho\langle v\rangle A], \tag{8.44}$$

where $M_{\mathrm{tot}} = \int_V \rho\, dV$ is the total mass. The difference on the right-hand side is
$\Delta[.] = [.]_2 - [.]_1$, and the average is defined by $\langle(.)\rangle_i = \int_{A_i}(.)_i\, dA/A_i$. Now, assume
that extra stress $\langle\tau\rangle$ contributions to the total stress can be neglected and pressure
is uniform over A_i, to show from (5.32) that the macroscopic momentum balance
takes the form

$$\frac{d}{dt}M_{\mathrm{tot}} = -\Delta\left[(\rho\langle v^2\rangle + p)An\right] - \mathcal{F}_s + M_{\mathrm{tot}}g, \tag{8.45}$$

where the total momentum is $M_{\mathrm{tot}} = \int_V \rho v\, dV$, and $\mathcal{F}_s$ is defined in (8.39). Notice
the similarities (and differences) between (5.5) and (8.44) and between (5.32) and
(8.45).

Exercise 8.13   Macroscopic Mechanical Energy Balance
The macroscopic energy balance is useful for the design of fluid-transfer equipment.
As in Exercise 8.12 we consider a volume V bound by surface A with outward unit

normal vector $\boldsymbol{n}$. As before, $A = A_1 + A_2 + A_s$, but now for A_s we write $A_s = A_{ss} + A_{sm}$, where A_{ss} and A_{sm} are the areas of stationary and moving, respectively, non-entrance–exit surfaces. From the differential balance for mechanical energy (5.50), show that the macroscopic mechanical energy balance takes the form

$$\frac{d}{dt}(K_{tot} + \Phi_{tot}) = -\Delta \left[\rho \langle v \rangle \left(\frac{1}{2} \frac{\langle v^3 \rangle}{\langle v \rangle} + \phi + \frac{p}{\rho} \right) A \right] + \int_V p \boldsymbol{\nabla} \cdot \boldsymbol{v} \, dV + \int_V \boldsymbol{\tau} : \boldsymbol{\nabla} \boldsymbol{v} \, dV + \dot{W},$$

$$(8.46)$$

where $K_{tot} = \int_V \frac{1}{2} \rho v^2 \, dV$ and $\Phi_{tot} = \int_V \rho \phi \, dV$ are the total kinetic and potential energies, respectively, and

$$\dot{W} = -\int_{A_{sm}} (\boldsymbol{\pi} \cdot \boldsymbol{v}) \cdot \boldsymbol{n} \, dA, \qquad\qquad (8.47)$$

which is sometimes called the "shaft work," is the rate of work done on the fluid at moving surfaces (e.g. for pumps or compressors). The first and second integral terms are minus the reversible and irreversible, respectively, rates of conversion of mechanical energy to internal energy. Use (8.41) and (8.46) to show, for steady flow in straight pipes having constant cross section, that the latter can be determined by

$$-\int_V \boldsymbol{\tau} : \boldsymbol{\nabla} \boldsymbol{v} \, dV = \frac{1}{2} \rho \langle v \rangle^3 \frac{L}{R_{hyd}} f_s. \qquad\qquad (8.48)$$

Note the similarities and differences between (8.46) and the Bernoulli equation (5.54).

Exercise 8.14   Pump Power for the Trans-Alaska Pipeline
Use the macroscopic mechanical energy balance (8.46) to determine the rate of work required $\dot{W}$ in the Trans-Alaska Pipeline for the case described in Exercise 8.10. Assume the oil enters and exits the pipeline at atmospheric pressure and that changes in elevation can be neglected.

- The equations governing pressure-driven, incompressible, laminar flow in a tube are linear and their solution leads to the Hagen–Poiseuille equation, which relates flow rate and pressure drop.
- The transport equations governing pressure-driven, laminar flow of compressible fluids are a coupled system of nonlinear partial differential equations where both viscous and inertial effects are important.
- Perturbation methods provide a powerful tool for obtaining solutions to nonlinear equations expressed as power series expansions in terms of a small dimensionless parameter.
- The lubrication approximation exploits the smallness of the aspect ratio for a flow domain and leads, by neglecting pressure variations in the direction of the small dimension, to a significant simplification of the equations governing the flow.
- The friction factor, or dimensionless drag force exerted by a fluid on the wetted surface of the conduit through which it flows, provides a convenient method for determining the relationship between flow rate and pressure drop for flows in complex geometries and for turbulent flow.

9

Heat Exchangers

A primary purpose of the pressure-driven flows considered in the preceding chapter is to move fluid from one place to another. Pressure-driven flows that circulate blood through the body and that move crude oil from a remote oil field to a refinery are common examples. In such cases thermal effects are relatively unimportant. On the other hand, there are numerous situations where it is desirable to heat or cool a fluid. An efficient and commonly used method of achieving this is by having heat transferred to or from a fluid as it flows through a tube. A device where heat is transferred to a fluid as it flows through a tube is known as a *heat exchanger*. The efficient design of heat exchangers relies on models from transport phenomena that predict the temperature distribution of the fluid as it passes through the tube. Heat exchangers come in many configurations and have numerous applications. For example, heat exchangers are used in power plants where heat from a chemical or nuclear reaction is converted to electricity. Heat exchangers are also critical components in devices we use every day, ranging from refrigerators to automobiles.

The configuration of a heat exchanger depends on the means by which heat is added to or removed from the fluid passing through the tube. The most common way of achieving this is by having a second fluid in contact with the outside surface of the tube. This fluid may undergo a phase change, in which case the temperature of the tube is approximately uniform. Alternatively, the fluid outside the tube may undergo a temperature change as it flows co-currently, or counter-currently, with the fluid inside the tube. In this configuration, the temperature of the wall varies with position along the tube. It is also common to have multiple tubes inside a shell, which allows one to obtain a compact heat exchanger with a large area for heat transfer. In this chapter we consider the simplest type of heat exchanger, where a fluid flows through a heated circular tube.

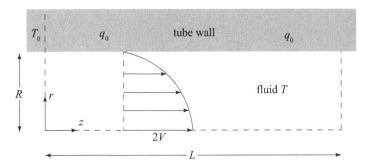

Figure 9.1 Schematic diagram of forced convection in a heated tube.

9.1 Flow in a Heated Tube

We consider the steady, incompressible, laminar flow of a Newtonian fluid with constant density ρ and viscosity η in an impermeable tube of radius R and length L (see Figure 9.1). As noted in Section 7.3 (cf. footnote on p. 117), the assumption of constant viscosity and density means that the velocity and pressure fields are decoupled from the temperature field. Hence, we take the velocity field from (8.3), which can be written as $v_z = 2V[1 - (r/R)^2]$, where $V = (p_0 - p_L)R^2/(8\eta L)$.

As shown in Figure 9.1, fluid enters the heat exchanger from a tube with constant wall temperature T_0. For $z \geq 0$, there is a constant heat flux q_0 at the tube wall. This can be achieved in practice by wrapping the tube with a heating element with known power output; q_0 is the power divided by the surface area of the tube. Since the energy flux is continuous at the tube wall, we have $\boldsymbol{n} \cdot \boldsymbol{j}_q(R, z) = -q_0$, where $\boldsymbol{n} = \boldsymbol{\delta}_r$. Assuming symmetry about the z-axis, the temperature field will have the form $T = T(r, z)$. The temperature equation for an incompressible Newtonian fluid with constant thermal conductivity λ is given in (7.21). As in the previous chapter, we put our equations in dimensionless form so that we can identify prefactors that allow the transport equations to be simplified. As a first step, we use R and R/V as units of length and time ($R = V = 1$), and we employ additive and multiplicative normalization of the temperature, $\lambda(T - T_0)/(q_0 R)$, to avoid the occurrence of parameters in the initial and boundary conditions. This allows us to write (7.21) as

$$N'_{\text{Pe}}(1 - r^2)\frac{\partial T}{\partial z} = \frac{1}{r}\frac{\partial}{\partial r}\left(r\frac{\partial T}{\partial r}\right) + \frac{\partial^2 T}{\partial z^2} + 4N_{\text{Br}}r^2, \qquad (9.1)$$

where $N'_{\mathrm{Pe}} = 2VR/\chi$ is the Péclet number and $N_{\mathrm{Br}} = 4\eta V^2/(q_0 R)$ is the Brinkman number (see Section 7.5). We note that (9.1) is an elliptic partial differential equation that, while linear, is difficult to solve analytically.

As discussed in Section 7.5, the Brinkman number indicates the importance of viscous dissipation relative to conduction. Here, viscous dissipation is scaled relative to the heat flux at the tube wall. For a heat exchanger, where heat is added or removed to affect a temperature change in the fluid, it is reasonable to expect that $N_{\mathrm{Br}} \ll 1$. Hence, in what follows we shall assume the last term on the right-hand side of (9.1) can be neglected. As discussed in Section 7.5, the Péclet number indicates the importance of convective relative to diffusive transport, and can be expressed as $N'_{\mathrm{Pe}} = N_{\mathrm{Re}}N_{\mathrm{Pr}}$, where $N_{\mathrm{Pr}} = \nu/\chi$. For gases $N_{\mathrm{Pr}} \approx 1$, and for typical low-viscosity liquids $N_{\mathrm{Pr}} \approx 1\text{--}10$ (see Table 7.1).

A typical heat exchanger is operated such that $N'_{\mathrm{Pe}} \gg 1$, and for laminar flow we require $N'_{\mathrm{Pe}} \lesssim 10^3$. This would suggest that the conduction terms on the right-hand side of (9.1) might be unimportant. That, however, is only partially true for reasons that may not be immediately obvious. Recall that both spatial variables (r, z) are given in units of R. Hence, while the dimensionless radial coordinate r in (9.1) is of order unity, the same is not true of z. For a typical tube with $L/R \approx 10^2$, for the axial coordinate in (9.1) we have $z \gg 1$. To resolve this problem, and incorporate $N'_{\mathrm{Pe}} \gg 1$ in our analysis, we rescale the axial coordinate and introduce $\bar{z} = z/N'_{\mathrm{Pe}}$. With this change of variable, (9.1) can be written as

$$(1 - r^2)\frac{\partial T}{\partial \bar{z}} = \frac{1}{r}\frac{\partial}{\partial r}\left(r\frac{\partial T}{\partial r}\right), \tag{9.2}$$

where we have taken into account the fact that $N'_{\mathrm{Pe}} \gg 1$ and $N_{\mathrm{Br}} \ll 1$. On introducing the rescaled axial coordinate we see that the axial conduction drops out, which is consistent with $N'_{\mathrm{Pe}} \gg 1$, so that axial convection is balanced by radial conduction. Also note that increasing (decreasing) the flow rate has the effect of decreasing (increasing) the effective length of the tube.

The temperature equation in (9.2) is now a parabolic partial differential equation. We assume the fluid enters the heated section of the tube with uniform temperature T_0, which is possible only if viscous dissipation is negligible. Hence, the initial and boundary conditions for the normalized temperature in (9.2) take the form

$$T(r, 0) = 0, \tag{9.3}$$

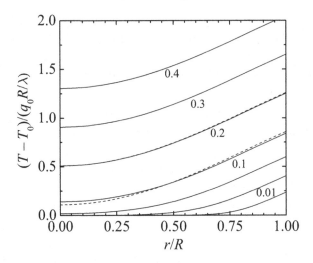

Figure 9.2 Temperature field $T(r, \bar{z})$ for laminar flow in a heated tube for $\bar{z} = z/(RN'_{\mathrm{Pe}}) = 0.005, 0.01, 0.05, 0.1, 0.2, 0.3, 0.4$ (from bottom to top). Also shown for $\bar{z} \geq 0.1$ (dashed lines) is the temperature from (9.10).

$$\frac{\partial T}{\partial r}(0, \bar{z}) = 0, \qquad \frac{\partial T}{\partial r}(1, \bar{z}) = 1. \tag{9.4}$$

While much simpler than (9.1), solving (9.2) subject to (9.3) and (9.4) is still a challenge. A closely related and widely studied problem is known as the Graetz–Nusselt[1] problem, where the wall temperature for $\bar{z} \geq 0$ is prescribed, say T_1. In this case temperature would be normalized by $T_1 - T_0$, and the second boundary condition in (9.4) is replaced by $T(1, \bar{z}) = 1$.

9.2 Solution of the Temperature Equation

Before we try to study the problem given in (9.2)–(9.4) by analytic tools, it is illustrative to look at numerical solutions first. A plot of the temperature field is shown in Figure 9.2, which was obtained using a finite difference method implemented using MATLAB® (see Exercise 9.1). Notice that for $\bar{z} \gtrsim 0.1$ the radial dependence of $T(r, \bar{z})$ is independent of $\bar{z}$, and that $T(r, \bar{z})$ increases linearly with $\bar{z}$. Since $\bar{z} = z/N'_{\mathrm{Pe}}$ and z has been normalized by R, the thermal entrance length is roughly $L_T/R \approx 0.1 N'_{\mathrm{Pe}}$. Hence, for $N_{\mathrm{Pr}} \geq 1$, the thermal entrance is larger than the entrance length for fully developed flow discussed in Section 8.1.

---

[1] Graetz, *Ann. Phys. Chem.* **18** (1883) 79; Nusselt, *Z. Ver. deutsch. Ing.* **54** (1910) 1154.

One approach to obtaining an analytical solution is to separate the heated tube into entrance and downstream regions. This is done by writing

$$T(r, \bar{z}) = \tilde{T}(r, \bar{z}) - \bar{T}(r, \bar{z}), \tag{9.5}$$

where $\tilde{T}$ and $\bar{T}$ are the temperature fields in the entrance and downstream, respectively, sections of the tube. To find $\bar{T}$, we assume the constant heating rate will lead to a linear dependence on $\bar{z}$ (see Figure 9.2) and write

$$\bar{T}(r, \bar{z}) = c_0 \bar{z} + f(r), \tag{9.6}$$

where c_0 is a constant. Note that the assumed form for $\bar{T}$ in (9.6) is not consistent with the initial condition in (9.3), which is expected since $\bar{T}$ is applicable only for $\bar{z} > 0$. To resolve this, we formulate a modified initial condition as follows. First, we multiply (9.2) by r and integrate over the interval $0 \le r \le 1$. Next we use the boundary condition in the second equation in (9.4) and integrate with respect to $\bar{z}$. This amounts to writing an energy balance on the fluid at the entrance to the heated tube, and leads to

$$\int_0^1 r(1 - r^2)\bar{T}\, dr = \bar{z}. \tag{9.7}$$

Substitution of (9.6) into (9.2) gives

$$\frac{1}{r}\frac{d}{dr}\left(r\frac{df}{dr}\right) = c_0(1 - r^2). \tag{9.8}$$

Integration of (9.8) subject to $df/dr(0) = 0$ and $df/dr(1) = 1$ gives $c_0 = 4$ and

$$f = r^2 - \frac{r^4}{4} + c_2. \tag{9.9}$$

Substitution of (9.9) into (9.6) and using the condition in (9.7) gives

$$\bar{T} = 4\bar{z} + r^2 - \frac{r^4}{4} - \frac{7}{24}. \tag{9.10}$$

From Figure 9.2 it is clear that the simple expression in (9.10) is in good agreement with the numerical solution for $\bar{z} \gtrsim 0.1$.

To find the equation that governs $\tilde{T}$, we combine (9.5) and (9.10) and substitute into (9.2)–(9.4), which leads to

$$(1 - r^2)\frac{\partial \tilde{T}}{\partial \bar{z}} = \frac{1}{r}\frac{\partial}{\partial r}\left(r\frac{\partial \tilde{T}}{\partial r}\right), \tag{9.11}$$

$$\tilde{T}(r, 0) = \bar{T}(r, 0), \tag{9.12}$$

$$\frac{\partial \tilde{T}}{\partial r}(0, \bar{z}) = 0, \qquad \frac{\partial \tilde{T}}{\partial r}(1, \bar{z}) = 0. \tag{9.13}$$

Note that by splitting the solution according to (9.5) we have moved the non-homogeneity from the boundary condition for T in the second equation in (9.4) to the initial condition for $\tilde{T}$ in the second equation in (9.13).

The solution[2] of (9.11)–(9.13) can be found by separation of variables by writing

$$\tilde{T} = \tilde{R}(r)\tilde{Z}(\bar{z}), \tag{9.14}$$

which, when substituted into (9.11), gives

$$\frac{1}{\tilde{Z}}\frac{d\tilde{Z}}{d\bar{z}} = \frac{1}{1-r^2}\frac{1}{\tilde{R}}\frac{1}{r}\frac{d}{dr}\left(r\frac{d\tilde{R}}{dr}\right) = -\beta^2. \tag{9.15}$$

It is straightforward to find $\tilde{Z}$ so that we can write (9.14) as

$$\tilde{T} = C\exp(-\beta^2\bar{z})\tilde{R}(r). \tag{9.16}$$

To find $\tilde{R}$ we must solve the Sturm–Liouville problem given by

$$\frac{d^2\tilde{R}}{dr^2} + \frac{1}{r}\frac{d\tilde{R}}{dr} + \beta^2(1-r^2)\tilde{R} = 0, \tag{9.17}$$

$$\frac{d\tilde{R}}{dr}(0) = 0, \qquad \frac{d\tilde{R}}{dr}(1) = 0. \tag{9.18}$$

The form of the differential equation in (9.17) suggests a power series for $\tilde{R}$ having the form

$$\tilde{R}(\beta, r) = \sum_{i=0}^{\infty} a_i r^i, \quad a_i = \begin{cases} 1 & \text{for} \quad i = 0, \\ (\beta/2)^2 & \text{for} \quad i = 2, \\ -(\beta/i)^2(a_{i-2} - a_{i-4}) & \text{for} \quad i = 4, 6, \dots \end{cases} \tag{9.19}$$

The $1/r$ coefficient in front of $d\tilde{R}/dr$ excludes odd powers of r, and the boundary condition in the first equation in (9.18) excludes negative powers of r. Applying the boundary condition in the second equation in (9.18) gives

$$\left[\frac{d}{dr}\tilde{R}(\beta_n, r)\right]_{r=1} = 0, \tag{9.20}$$

which is an equation with an infinite number of roots β_n $(n = 1, 2, 3, \dots)$; in

---

[2] Siegel *et al.*, *Appl. Sci. Res.* **7** (1958) 386.

Table 9.1 the first seven are given. Hence, the solution in (9.16) involves a superposition of solutions

$$\tilde{T} = \sum_{n=1}^{\infty} C_n \exp(-\beta_n^2 \bar{z}) \tilde{R}(\beta_n, r). \tag{9.21}$$

To determine C_n, we substitute (9.21) into (9.12) and obtain

$$\sum_{n=1}^{\infty} C_n \tilde{R}(\beta_n, r) = - \left(r^2 - \frac{r^4}{4} - \frac{7}{24} \right). \tag{9.22}$$

Multiplication by $r(1 - r^2)\tilde{R}_n$ and integrating over r gives

$$C_n = -\frac{\int_0^1 r(1 - r^2)\left(r^2 - \frac{r^4}{4}\right)\tilde{R}(\beta_n, r)dr}{\int_0^1 r(1 - r^2)\tilde{R}(\beta_n, r)^2 \, dr}, \tag{9.23}$$

which completes the solution.

Table 9.1 *Roots of (9.20).*

n	β_n
1	5.0675
2	9.1576
3	13.1972
4	17.2202
5	21.2355
6	25.2465
7	29.2549

Exercise 9.1   *Numerical Solution of Temperature Equation*
Develop and implement MATLAB® code to solve (9.2)–(9.4) and generate the results shown in Figure 9.2.

Exercise 9.2   *Sturm–Liouville Problem for Temperature Equation*
Verify that (9.19) is a solution to (9.17) and (9.18)

Exercise 9.3   *Viscous Dissipation in Adiabatic Flow through a Tube*
For adiabatic flow through a tube we have $q_0 = 0$. Identify the appropriate quantity to scale temperature so that temperature is governed by

$$(1 - r^2)\frac{\partial T}{\partial \bar{z}} = \frac{1}{r}\frac{\partial}{\partial r}\left(r\frac{\partial T}{\partial r}\right) + 4r^2,$$

$$T(r, 0) = 0,$$

$$\frac{\partial T}{\partial r}(0, \bar{z}) = 0, \qquad \frac{\partial T}{\partial r}(1, \bar{z}) = 0.$$

Use the expression in (9.6) to obtain the temperature field for the adiabatic case.

Exercise 9.4   Entropy Production in Adiabatic flow Through a Tube
Evaluate and plot the local rate of entropy production term in (6.2) using the results from Exercise 9.3 for adiabatic flow through a tube.

9.3 Heat Transfer Coefficient

Heat transfer coefficients are used to describe energy transport at solid–fluid interfaces in situations where heat transfer in the fluid phase is complex. The equation that defines the heat transfer coefficient h can be written as

$$d\dot{Q} = h[T(A_s) - T_\text{fl}]dA, \qquad (9.24)$$

where $d\dot{Q}$ is the differential rate of heat flow through differential area dA of the solid surface A_s. The temperature difference in (9.24) is between $T(A_s)$, the temperature at the solid surface, and T_fl, some average temperature of the fluid away from the surface. It is important to note that in general both $T(A_s)$ and T_fl may vary, so that the heat transfer coefficient h defined in (9.24) is local.

The macroscopic energy balance (see Exercise 9.7) has a term for the total rate of heat transfer to the system at solid surfaces given by

$$\dot{Q} = -\int_{A_s} \boldsymbol{n} \cdot \boldsymbol{j}_q \, dA, \qquad (9.25)$$

where $\boldsymbol{n}$ is the unit normal pointing away from the fluid. Integration of (9.24) and comparison with (9.25) gives

$$-\boldsymbol{n} \cdot \boldsymbol{j}_q = h(T - T_\text{fl}) \qquad \text{at } A_s, \qquad (9.26)$$

or, using Fourier's law (6.4) on the left-hand side,

$$\lambda \boldsymbol{n} \cdot \boldsymbol{\nabla} T = h(T - T_\text{fl}) \qquad \text{at } A_s. \qquad (9.27)$$

The heat transfer coefficient is used to take into account the resistance to heat transfer that results from the presence of a thin layer of nearly stagnant fluid at the surface of the solid. As shown schematically in Figure 9.3, the increased resistance to heat transfer across this fluid layer having thickness δ_T, known as a *boundary layer*, results in a large temperature change.

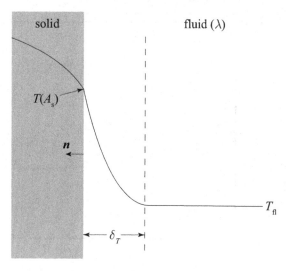

Figure 9.3 Schematic temperature profile across an interface between solid and fluid phases showing a thermal boundary layer with thickness δ_T.

Outside the boundary layer, fluid motion leads to a relatively uniform temperature. In this sense (9.26), which is sometimes referred to as Newton's law of cooling, is an approximation to Fourier's law for energy transport across the boundary layer, with h being the ratio of the fluid thermal conductivity λ to the boundary layer thickness δ_T. Hence, the heat transfer coefficient depends on the flow field and geometry on which the boundary layer thickness δ_T depends, and on fluid properties. A commonly used way of reporting heat transfer coefficients is in terms of a dimensionless group known as the Nusselt number defined as $N_{\mathrm{Nu}} = hL/\lambda \approx L/\delta_T$, where L is a characteristic dimension of the flow geometry. For complex flows, expressions for N_{Nu} are usually found from experiments.

Let us apply (9.27) to the laminar flow in a heated tube problem just considered and write

$$\frac{\partial T}{\partial r}(1, \bar{z}) = \frac{1}{2} N_{\mathrm{Nu}}[T(1, \bar{z}) - T_{\mathrm{fl}}(\bar{z})], \tag{9.28}$$

where $N_{\mathrm{Nu}} = hD/\lambda$, and $D = 2R$ is the tube diameter. The most frequently used average temperature of the fluid is known as the flow, or mixing-cup, average, which for flow through a tube is defined as

$$T_{\mathrm{fl}}(\bar{z}) = \frac{\int_0^1 v_z(r)T(r, \bar{z})r \, dr}{\int_0^1 v_z(r)r \, dr} = 4\int_0^1 (1 - r^2)T(r, \bar{z})r \, dr. \tag{9.29}$$

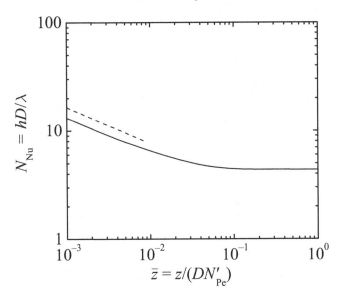

Figure 9.4 Nusselt number N_{Nu} as a function of axial position $\bar{z}$ for laminar flow in a heated tube from (9.30). The dashed line is from the asymptotic solution given in (9.31).

The second equation in (9.29) applies only to laminar flow. The solution obtained in the previous section given by (9.5) and (9.10) allows us to write (9.28) as

$$N_{Nu} = \frac{2}{11/24 + \tilde{T}(1, \bar{z})}, \tag{9.30}$$

which is plotted in Figure 9.4. For sufficiently large $\bar{z}$, the expression for $\tilde{T}$ in (9.21) vanishes, so that $N_{Nu} = 48/11 \approx 4.36$. We note for laminar flow in a tube with constant wall temperature for $\bar{z} \gg 1$ that $N_{Nu} \approx 3.66$, which is roughly 15 percent lower than for flow in a tube with constant wall heat flux.

The sum in (9.21) converges slowly for $\bar{z} \ll 1$, or the entrance region. To handle this region, a useful asymptotic solution can be found (see Exercise 9.5), from which the following expression can be obtained:

$$N_{Nu} = \frac{2\Gamma(2/3)}{(9/2)^{1/3}} \bar{z}^{-1/3} \approx 1.640\bar{z}^{-1/3}, \tag{9.31}$$

where Γ is the gamma function. As shown in Figure 9.4, the expression given in (9.31) is a reasonably good approximation to (9.30) for $\bar{z} \ll 1$.

Heat exchanger design typically involves finding the area $A_s = \pi D L$, where L is the length of the tube, required to impart a desired temperature change $\Delta T_{fl} = T_{fl}(L) - T_{fl}(0)$ in a fluid flowing at a given mass flow rate $W = (\pi/4) D^2 \rho V$. This calculation is done with the macroscopic energy balance (9.34) and the expression in (9.25). For flow in a tube with a prescribed heat flux, we have $n \cdot j_q(A_s) = -q_0$ so that $\dot{Q} = q_0 \pi D L$. In this case it is not necessary to use the heat transfer coefficient defined in (9.24), and the fluid temperature change is simply $\Delta T_{fl} = 4q_0 L/(\rho V \hat{c}_p D)$. For a heat exchanger with prescribed wall temperature, the expression in (9.25) cannot, in general, be evaluated. In such cases, the local heat transfer coefficient defined in (9.24) is replaced by an average heat transfer coefficient based on an average temperature difference between the fluid and tube wall (see Exercise 9.8).

As with the friction factor, for turbulent flow the heat transfer coefficient is determined empirically. For turbulent flow in a heated tube, correlations can be found both for local and for averaged values of N_{Nu}. There are two well-known correlations[3] for the log-mean heat transfer coefficient: $N_{Nu_{ln}} = h_{ln} D/\lambda$. These are given by

$$N_{Nu_{ln}} \approx 1.86(N_{Re} N_{Pr} L/D)^{1/3}, \tag{9.32}$$

for laminar flow, and by

$$N_{Nu_{ln}} \approx 0.026 N_{Re}^{0.8} N_{Pr}^{1/3}, \tag{9.33}$$

which is valid for $N_{Re} > 20\,000$, for turbulent flow. Note that, like (9.31), the expression in (9.32) has an $N_{Pe}'^{1/3}$ dependence and that, if (9.33) is expressed in the same form, the heat transfer coefficient has a dependence on N_{Re} similar to the friction factor in (8.42). A comprehensive tabulation of heat transfer coefficient correlations for a wide range of conditions can be found elsewhere.[4]

Exercise 9.5   Asymptotic Solution for Entrance Region
In the entrance region of a heated tube ($\bar{z} \ll 1$), the temperature change is restricted to a thin layer of fluid near the tube wall (see Figure 9.2). Within this thin layer, it is reasonable to neglect curvature effects and to take the velocity to be linear. To carry out this analysis, it is useful to re-express (9.2)–(9.4) using the change of variables $x = 1 - r$ and $\theta = \partial T/\partial x$. Derive an expression for the temperature field

---

[3] Sieder & Tate, *Ind. Eng. Chem.*, **28** (1936) 1429. We have omitted the correction factor for the dependence of viscosity on temperature.
[4] Shah & London, *Laminar Flow Forced Convection* (Academic, 1978).

and the expression for the Nusselt number given in (9.31) for the entrance region to a heated tube. (Hint: use the similarity transformation $\xi = Cx/\bar{z}^{1/3}$.)

Exercise 9.6   Heat Transfer in the Downstream Region of a Heated Channel
A Newtonian fluid with constant physical properties flows through a channel of height $2H$ with average velocity V (see Exercise 8.2). Fluid with uniform temperature T_0 enters a heated section for $x_1 \geq 0$ with constant heat flux q_0 at $x_1 \pm H$. For the downstream region, show that the Nusselt number is given by $N_{\mathrm{Nu}} = 2hH/\lambda = 70/17 \approx 4.12$.

Exercise 9.7   Macroscopic Energy Balance
The macroscopic mechanical energy balance (8.46) was found in Exercise 8.13. Here, we consider the macroscopic energy balance for a system with volume V bounded by surface A with outward unit normal vector $\boldsymbol{n}$. Make the same assumptions as in Exercises 8.12 and 8.13, that is, density and pressure are uniform over A_i, and that the $\boldsymbol{\tau}$ contribution to the total stress can be neglected at A_i. In addition, assume that $\boldsymbol{j}_q$ can be neglected at A_i. From the differential balance for total energy, which is obtained by adding (5.50) and (5.51), show that the macroscopic energy balance takes the form

$$\frac{d}{dt} E_{\mathrm{tot}} = -\Delta \left[\rho \langle v \rangle \left(\hat{h} + \frac{1}{2} \frac{\langle v^3 \rangle}{\langle v \rangle} + \phi \right) A \right] + \dot{Q} + \dot{W}, \tag{9.34}$$

where the total energy is $E_{\mathrm{tot}} = \int_V (u + \frac{1}{2}\rho v^2 + \rho\phi)dV$ and $\dot{Q}$ is defined in (9.25). Note the similarities and differences between (4.1) and (9.34).

Exercise 9.8   Heat Exchanger Design
The most commonly used average heat transfer coefficient is the "log-mean" average, in which case the heat transfer coefficient h_{ln} is defined as

$$\dot{Q} = h_{\mathrm{ln}} \frac{[T(R,0) - T_{\mathrm{fl}}(0)] - [T(R,L) - T_{\mathrm{fl}}(L)]}{\ln[T(R,0) - T_{\mathrm{fl}}(0)] - \ln[T(R,L) - T_{\mathrm{fl}}(L)]} \, \pi DL. \tag{9.35}$$

Water flowing at 4 kg/s is to be heated in a tube on which steam at $100\,^\circ\mathrm{C}$ condenses. Heat transfer resistances in the steam and tube wall can be neglected. For a tube with an inside diameter of 0.1 m, find the length of the heat exchanger required to heat the water from $25\,^\circ\mathrm{C}$ to $50\,^\circ\mathrm{C}$.

Exercise 9.9   Macroscopic Entropy Balance
Using the same approach as in Exercise 9.7, from the differential entropy balance in (6.1), show that the macroscopic entropy balance takes the form

$$\frac{d}{dt} S_{\mathrm{tot}} = -\Delta[\rho \langle v \rangle \hat{s} A] + \dot{S} + \dot{\Sigma}, \tag{9.36}$$

where the total energy is $S_{\text{tot}} = \int_V s\, dV$, and $\dot{S} = -\int_{A_s} \boldsymbol{n} \cdot \boldsymbol{j}_q/T\, dA$ is the total rate of entropy transferred to the system. The last term, $\dot{\Sigma} = \int_V \sigma\, dV$, is the total rate of entropy production.

Exercise 9.10   *Entropy Analysis for Flow in a Heated Tube*

Derive an expression for the total rate of entropy production $\dot{\Sigma}$ appearing in (9.36) for flow in a heated tube. For this analysis, neglect viscous dissipation and ignore the thermal entrance region so that the temperature field is given by (9.10).

- The evolution equation for the temperature of a fluid in laminar flow through a heated tube can be simplified to a diffusion equation where axial convection is balanced by radial conduction.
- A solution to the temperature equation is facilitated by the super-position of solutions for the entrance region and for the thermally developed region.
- Newton's law of cooling is a useful approximation that can be used to formulate boundary conditions at solid–fluid interfaces for situations where finding solutions to the temperature equation in the fluid phase is difficult or not practical.
- The heat transfer coefficient, which is defined in Newton's law of cooling, can be found either analytically for simple flows, or experimentally for more complex flows that occur in complex geometries or in turbulent flow.

10

Gas Absorption

The process known as *gas absorption*, where a component in a gas mixture is absorbed into a liquid, is a workhorse in the chemical process industry. Both absorption, and its counterpart stripping, are widely used separation processes. Reactive gas absorption, where chemical reaction takes place in concert with absorption, is used to remove unwanted substances from gas streams (e.g., NO_2 and H_2S) and in the production of basic chemicals (e.g., HNO_3 and H_2SO_4). Transport processes involving mass diffusion and chemical reaction are ubiquitous. They occur naturally in all plants and animals, and are involved in virtually every man-made substance we use in everyday life. When carried out on a large scale, the chemical reaction rate is coupled to diffusive and convective mass transport. In addition, most chemical reactions result in thermal effects, so mass and energy transport processes are coupled.

In this chapter we consider a process involving gas absorption and chemical reaction in a liquid film that flows as a result of a gravitational force. The overall efficiency of such processes is determined by a balance of the contact time between the liquid film and gas, and the rates of diffusive mass transport and chemical reaction. The example we consider illustrates how these transport processes interact, and how the complexity of the governing equations can be reduced so that useful models of these important processes can be developed.

10.1 Diffusion and Reaction in a Liquid Film

Consider a liquid film of thickness h and length $L \gg h$ flowing along a vertical planar solid surface that is in contact with a gas shown schematically in Figure 10.1. At the entrance $x_3 = 0$, the liquid film is initially pure B with uniform temperature T_0. For $x_3 > 0$, the liquid is brought into contact

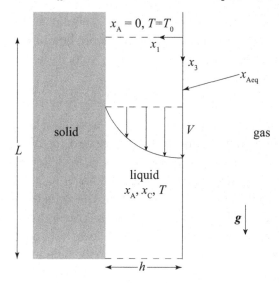

Figure 10.1 Schematic diagram of gas absorption with homogeneous chemical reaction in a liquid film flowing along a vertical solid surface.

with a gas such that at $x_1 = 0$ the concentration of species A is maintained at x_{Aeq}, and the concentration of species C is zero. As A diffuses in the liquid, it may undergo a homogeneous, "irreversible" reaction $\mathrm{A} + \mathrm{B} \to \mathrm{C}$. The liquid film has width $B \gg h$ so that all fields are independent of x_2, and we take the process to be at steady state.

Our goal is to develop a transport model that describes the concentration fields of reactants and products and allows us to examine the interplay of reaction, diffusion, and convection. We have a three-component system in which a chemical reaction takes place, so it will be convenient for us to consider molar forms of the mass balances introduced in Chapter 5. From the given reaction mechanism, we can easily identify the stoichiometric coefficients, which are $\tilde{\nu}_{\mathrm{A}} = -1$, $\tilde{\nu}_{\mathrm{B}} = -1$, $\tilde{\nu}_{\mathrm{C}} = 1$. If we assume the mixture is ideal, we can use the mass action law in (6.61) for $\tilde{\Gamma}$, which takes the form

$$\tilde{\Gamma} = L_{\Gamma} \exp\left(-\frac{\tilde{\mu}_{\mathrm{A}}^0 + \tilde{\mu}_{\mathrm{B}}^0}{\tilde{R}T}\right)\left[x_{\mathrm{A}}x_{\mathrm{B}} - \exp\left(\frac{\Delta\tilde{g}^0}{\tilde{R}T}\right)x_{\mathrm{C}}\right] = k(T)x_{\mathrm{A}}x_{\mathrm{B}}, \quad (10.1)$$

where the second equality holds for an "irreversible" reaction for which $-\Delta\tilde{g}^0/(\tilde{R}T) \gg 1$, and $k(T)$ is the reaction rate constant. Reactions involve thermal effects so that the energy balance also needs to be considered.

The diffusion and reaction process we wish to describe takes place in a liquid where there is an externally driven flow. In the absence of diffusion and reaction, the flow in a liquid film on a planar surface is driven by the gravitational force. For steady, incompressible laminar flow of a Newtonian fluid for this geometry, the (mass-average) velocity is

$$v_1 = v_2 = 0, \qquad v_3 = V\left[1 - \left(\frac{x_1}{h}\right)^2\right], \tag{10.2}$$

where $V = \rho g h^2/(2\eta)$ (see Exercise 10.1), which is sketched in Figure 10.1. Laminar flow of smooth liquid films is observed for $N_{\text{Re}} = \rho V h/\eta \lesssim 10$.

For a three-component mixture we can write three mass balances along with an energy balance, which govern the fields c, x_A, x_C (or x_B), and T. The total mass balance was used to obtain the velocity field given in (10.2), which, in general, will be modified by mass transport. Hence, we will need to either modify the velocity in (10.2), or show that the overall mass balance is consistent with (10.2). For the overall mass balance (5.22), using (10.1), we have

$$\nabla \cdot (v^* c) = -k(T) x_A x_B, \tag{10.3}$$

where v^* is the molar-average velocity. The species mass balances are obtained from (5.23) that using (10.1) can be written as

$$c v^* \cdot \nabla x_A = -\nabla \cdot J_A^* - k(T) x_A x_B (1 - x_A), \tag{10.4}$$

$$c v^* \cdot \nabla x_C = -\nabla \cdot J_C^* + k(T) x_A x_B (1 + x_C). \tag{10.5}$$

The temperature equation is obtained from (6.8). Neglecting viscous heating, and setting $\hat{h}_\alpha = \hat{h}_\alpha^0$ and $\hat{v}_\alpha = \hat{v}_\alpha^0$, we have (see Exercise 10.2)

$$\rho \hat{c}_p v \cdot \nabla T = -\nabla \cdot j_q' - k(T) x_A x_B \, \Delta \tilde{h}^0, \tag{10.6}$$

where $\Delta \tilde{h}^0 = \tilde{h}_C^0 - (\tilde{h}_A^0 + \tilde{h}_B^0)$ is the standard molar enthalpy change for the reaction (see Section 4.6). Note that (10.3)–(10.6) are coupled through the terms for chemical reaction, and that this system of equations involves two different reference velocities (v^* and v).

To proceed, we need expressions for the fluxes J_A^*, J_C^*, and j_q'. As we saw in Chapter 6, these fluxes are coupled, which would further couple the system of equations. In addition, if the velocity field is altered by the reaction and diffusion processes, the convective terms will render these evolution equations nonlinear. In the following section we consider a special case that greatly simplifies our analysis.

Exercise 10.1   Gravity-Driven Flow of Liquid Film on a Solid Surface
For the steady, laminar flow of an incompressible Newtonian liquid film in the geometry shown in Figure 10.1, show that the velocity is given by (10.2).

Exercise 10.2   *Temperature Equation for Ideal Mixtures with Reaction*

Starting with (6.8), derive the temperature equation for a k-component mixture,

$$\rho \hat{c}_p \frac{DT}{Dt} = -\frac{T}{\rho}\left(\frac{\partial \rho}{\partial T}\right)_{p,w_\alpha} \frac{Dp}{Dt} - \boldsymbol{\nabla} \cdot \boldsymbol{j}_q' - \sum_{\alpha=1}^{k} \boldsymbol{j}_\alpha \cdot \boldsymbol{\nabla}\hat{h}_\alpha - \boldsymbol{\tau} : \boldsymbol{\nabla}\boldsymbol{v} - \Gamma \sum_{\alpha=1}^{k} \nu_\alpha \hat{h}_\alpha.$$

Simplify this result for an ideal mixture to obtain (10.6).

10.2 Dilute Mixtures

We now restrict our analysis to the case of a liquid mixture that is dilute in species A ($x_A \ll 1$), and therefore is also dilute in species C ($x_C \ll 1$). This means that $x_B \approx 1$, and the molar density of the mixture c is approximately constant. Also, since $\boldsymbol{v}_B \approx \boldsymbol{v}^* \approx \boldsymbol{v}$, we can write (10.3) as

$$\boldsymbol{\nabla} \cdot \boldsymbol{v} = k'(T)x_A \approx 0, \tag{10.7}$$

where $k'(T) = k(T)/c$. Hence, for dilute mixtures, the velocity field is unaffected by diffusion and reaction and we can use the velocity given in (10.2).

To obtain expressions for the diffusive fluxes, we neglect cross effects and use the Maxwell–Stefan equations. From (6.54), we can write

$$x_A \boldsymbol{\nabla}\left(\frac{\tilde{\mu}_A}{\tilde{R}T}\right) = \frac{x_A x_B}{\mathfrak{D}_{AB}}(\boldsymbol{v}_B - \boldsymbol{v}_A) + \frac{x_A x_C}{\mathfrak{D}_{AC}}(\boldsymbol{v}_C - \boldsymbol{v}_A),$$

$$x_C \boldsymbol{\nabla}\left(\frac{\tilde{\mu}_C}{\tilde{R}T}\right) = \frac{x_C x_A}{\mathfrak{D}_{CA}}(\boldsymbol{v}_A - \boldsymbol{v}_C) + \frac{x_C x_B}{\mathfrak{D}_{CB}}(\boldsymbol{v}_B - \boldsymbol{v}_C). \tag{10.8}$$

From (5.20), we have $\boldsymbol{v}_A = \boldsymbol{N}_A/c_A$ and $\boldsymbol{v}_C = \boldsymbol{N}_C/c_C$, so that (10.8) can be written as

$$x_A \boldsymbol{\nabla}\left(\frac{\tilde{\mu}_A}{\tilde{R}T}\right) = \frac{x_B}{\mathfrak{D}_{AB}}(x_A \boldsymbol{v}_B - \boldsymbol{N}_A/c) + \frac{x_C}{\mathfrak{D}_{AC}}(x_A \boldsymbol{v}_C - \boldsymbol{N}_A/c),$$

$$x_C \boldsymbol{\nabla}\left(\frac{\tilde{\mu}_C}{\tilde{R}T}\right) = \frac{x_A}{\mathfrak{D}_{CA}}(x_C \boldsymbol{v}_A - \boldsymbol{N}_C/c) + \frac{x_B}{\mathfrak{D}_{CB}}(x_C \boldsymbol{v}_B - \boldsymbol{N}_C/c). \tag{10.9}$$

For our dilute mixture $x_A \ll 1, x_C \ll 1, x_B \approx 1$; since $\boldsymbol{v}^* \approx \boldsymbol{v}_B$, we can write $\boldsymbol{N}_\alpha/c \simeq x_\alpha \boldsymbol{v}_B + \boldsymbol{J}_\alpha^*/c$. Applying these to (10.9), we obtain

$$\boldsymbol{J}_A^* \approx -c\mathfrak{D}_{AB}x_A \boldsymbol{\nabla}\left(\frac{\tilde{\mu}_A}{\tilde{R}T}\right) = -cD_{AB}\,\boldsymbol{\nabla}x_A,$$

$$\boldsymbol{J}_C^* \approx -c\mathfrak{D}_{CB}x_C \boldsymbol{\nabla}\left(\frac{\tilde{\mu}_C}{\tilde{R}T}\right) = -cD_{CB}\,\boldsymbol{\nabla}x_C, \tag{10.10}$$

where the second equality follows for ideal mixtures and setting $\mathfrak{D}_{AB} = D_{AB}$

and $\mathfrak{D}_{CB} = D_{CB}$. The expressions in (10.10) are Fick's law for the *pseudo-binary approximation*, which can easily be generalized for an arbitrary number of dilute components (see Exercise 10.3).

Typically, the reaction rate constant $k'(T)$ is a strong function of temperature. A well-known expression to describe the temperature dependence is Arrhenius' law, which is given by

$$k'(T) = A \exp\left(-\frac{\tilde{E}_{act}}{\tilde{R}T}\right) \approx k_0'\left[1 + \frac{\tilde{E}_{act}}{\tilde{R}T_0^2}(T - T_0)\right], \qquad (10.11)$$

where $\tilde{E}_{act}$ is the activation energy, and the prefactor A is the frequency factor. The expression in (10.11) is based on the idea that a reaction passes through energy states above the energy states of both reactants and products (see Section 24.3 for a detailed discussion). The second equality introduces the constant $k_0' = k'(T_0)$ and holds for the dilute reactant limit.

We are now in a position to write down the evolution equations that govern the fields x_A, x_C, and T, which depend on x_1 and x_3. Substitution of (10.2), (10.10), and (10.11) in (10.4) and (10.5) gives

$$V\left[1 - \left(\frac{x_1}{h}\right)^2\right]\frac{\partial x_A}{\partial x_3} = D_{AB}\left[\frac{\partial^2 x_A}{\partial x_1^2} + \frac{\partial^2 x_A}{\partial x_3^2}\right] - k_0'\left[1 + \frac{\tilde{E}_{act}}{\tilde{R}T_0^2}(T - T_0)\right]x_A,$$
$$(10.12)$$

$$V\left[1 - \left(\frac{x_1}{h}\right)^2\right]\frac{\partial x_C}{\partial x_3} = D_{CB}\left[\frac{\partial^2 x_C}{\partial x_1^2} + \frac{\partial^2 x_C}{\partial x_3^2}\right] + k_0'\left[1 + \frac{\tilde{E}_{act}}{\tilde{R}T_0^2}(T - T_0)\right]x_A,$$
$$(10.13)$$

where have assumed that the diffusivities are constants. Using the expression in (6.18) for j_q', and again neglecting cross effects, along with (10.2) and (10.11), the temperature equation (10.6) can be written as

$$V\left[1 - \left(\frac{x_1}{h}\right)^2\right]\frac{\partial T}{\partial x_3} = \chi\left[\frac{\partial^2 T}{\partial x_1^2} + \frac{\partial^2 T}{\partial x_3^2}\right] - k_0'\left[1 + \frac{\tilde{E}_{act}}{\tilde{R}T_0^2}(T - T_0)\right]\frac{\Delta\tilde{h}^0}{\tilde{c}_p}x_A,$$
$$(10.14)$$

where we have assumed λ is constant and used $\tilde{c}_p = \tilde{M}\hat{c}_p$.

The solution of (10.12)–(10.14) requires the specification of boundary conditions. At $x_1 = 0$, we have

$$x_A(0, x_3) = x_{Aeq}, \qquad x_C(0, x_3) = 0, \qquad T(0, x_3) = T_0, \qquad (10.15)$$

which imply thermal equilibrium with the gas, which is free of C. If the solid at $x_1 = h$ is impermeable and an insulator, then we can write

$$\frac{\partial x_A}{\partial x_1}(h, x_3) = 0, \qquad \frac{\partial x_C}{\partial x_1}(h, x_3) = 0, \qquad \frac{\partial T}{\partial x_1}(h, x_3) = 0. \qquad (10.16)$$

The remaining boundary conditions will be given in the following section.

Exercise 10.3   Pseudo-binary Approximation for Dilute Ideal Mixtures
Use the results from Exercise 6.12 to show that, for an ideal mixture with component
3 as the solvent ($w_3 \approx 1$) and $w_\alpha \ll 1$ for $\alpha = 1, 2$, for a k-component system the
diffusive mass fluxes are given by

$$j_\alpha = -\rho D_{\alpha k} \, \boldsymbol{\nabla} w_\alpha, \qquad D_{\alpha k} = \frac{\tilde{R} L_{\alpha\alpha}}{\rho \tilde{M}_\alpha w_\alpha}. \tag{10.17}$$

These expressions, like those in (10.10), show that the diffusive fluxes are decoupled
in dilute mixtures.

Exercise 10.4   Diffusion and Chemical Reaction in Multi-component Systems
Consider an isothermal system where the "reversible" reaction $A + B \rightleftarrows 2C$ occurs
in an ideal mixture that is *not* dilute. Use the Maxwell–Stefan equations in (10.8)
to obtain the diffusive molar fluxes in (5.23). If the system is bounded by an impermeable surface where $\boldsymbol{v}^* = \boldsymbol{0}$, show that the evolution equations for the species
concentrations can be written as

$$\frac{\partial x_A}{\partial t} = \tilde{D}_{AB} \, \nabla^2 x_A + \tilde{D}_{AC} \, \nabla^2 x_C - k_f x_A (1 - x_A - x_C) + k_r x_C^2,$$

$$\frac{\partial x_C}{\partial t} = \tilde{D}_{CA} \, \nabla^2 x_A + \tilde{D}_{CB} \, \nabla^2 x_C + 2 k_f x_A (1 - x_A - x_C) - 2 k_r x_C^2,$$

where k_f, k_r are the rate constants for the forward and reverse reactions. Obtain
expressions for the diffusion coefficients $\tilde{D}_{\alpha\beta}$, which here are taken as constants, in
terms of $\mathfrak{D}_{\alpha\beta}$. Under certain conditions, equations having this form evolve towards
steady solutions where species concentrations display spatially periodic patterns
known as Turing patterns.[1] These patterns, which are thought to occur frequently in
nature, and in particular in biological systems, are the result of symmetry breaking
instabilities that lead to dissipative structures.[2]

10.3 Solution of the Diffusion–Reaction Equations

By considering the dilute limit, we have considerably simplified our system
of evolution equations. However, the source terms in (10.12)–(10.14) are
still nonlinear, so it is not possible to obtain analytic solutions. It will be
useful to put the governing equations in dimensionless form so that we can
identify prefactors that allow the problem to be further simplified. We use
h and h/V as units of length and time ($h = V = 1$), and $\Delta \tilde{h}^0 x_{\text{Aeq}}/\tilde{c}_p$ as
our unit temperature to make temperature, relative to T_0, dimensionless
($\Delta \tilde{h}^0 x_{\text{Aeq}}/\tilde{c}_p = 1$). Note that, while the dimensionless spatial coordinate x_1

[1] Turing, *Phil. Trans. R. Soc. Lond.* B **237** (1952) 37.
[2] Prigogine *et al.*, *Nature* **30** (1969) 913.

is of order unity the same is not true of x_3. As in Section 9.1, we resolve this by rescaling one of the coordinates using the Péclet number $N_{\text{Pe}} = Vh/D_{\text{AB}}$. Here, we rescale the transverse (to the flow) direction as $\bar{x}_1 = \sqrt{N_{\text{Pe}}}x_1$. For $N_{\text{Pe}} \gg 1$, this means the diffusion length is small compared with the film thickness, so the velocity within the film can be treated as approximately uniform. Also, this rescaling means that diffusion in the x_3-direction can be neglected relative to diffusion in the $\bar{x}_1$-direction (see Section 9.1). These simplifications allow (10.12)–(10.14) to be written as the following set of dimensionless equations,

$$\frac{\partial x_{\text{A}}}{\partial x_3} = \frac{\partial^2 x_{\text{A}}}{\partial \bar{x}_1^2} - N_{\text{Da}}(1 + \varepsilon T)x_{\text{A}}, \tag{10.18}$$

$$\frac{\partial x_{\text{C}}}{\partial x_3} = \beta\frac{\partial^2 x_{\text{C}}}{\partial \bar{x}_1^2} + N_{\text{Da}}(1 + \varepsilon T)x_{\text{A}}, \tag{10.19}$$

$$\frac{\partial T}{\partial x_3} = N_{\text{Le}}\frac{\partial^2 T}{\partial \bar{x}_1^2} - N_{\text{Da}}(1 + \varepsilon T)x_{\text{A}}, \tag{10.20}$$

where the Damköhler number[3] $N_{\text{Da}} = k_0'h/V$ gives the ratio of the flow-based time to the reaction time. The parameter $\varepsilon = (x_{\text{Aeq}}\Delta\tilde{h}^0\tilde{E}_{\text{act}})/(\tilde{c}_p\tilde{R}T_0^2)$, which is positive (negative) for an endothermic (exothermic) reaction, indicates the effect of temperature on the reaction rate constant. Also appearing in (10.19) and (10.20) are the ratio of diffusivities $\beta = D_{\text{CB}}/D_{\text{AB}}$ and the Lewis number $N_{\text{Le}} = \chi/D_{\text{AB}}$ (see Table 7.1). The initial conditions for (10.18)–(10.20) are given by

$$x_{\text{A}}(\bar{x}_1, 0) = 0, \qquad x_{\text{C}}(\bar{x}_1, 0) = 0, \qquad T(\bar{x}_1, 0) = 0. \tag{10.21}$$

The boundary conditions in (10.15) take the form

$$x_{\text{A}}(0, x_3) = 1, \qquad x_{\text{C}}(0, x_3) = 0, \qquad T(0, x_3) = 0, \tag{10.22}$$

and the boundary conditions in (10.16) are, since the domain can now be treated as semi-infinite, replaced by

$$x_{\text{A}}(\infty, x_3) = 0, \qquad x_{\text{C}}(\infty, x_3) = 0, \qquad T(\infty, x_3) = 0. \tag{10.23}$$

While considerably simplified, the equations in (10.18)–(10.20) are still coupled and nonlinear due to the chemical reaction terms. Also note that, in general $(\Delta\tilde{h}^0 \neq 0)$, chemical reaction results in a nonisothermal system. One

---

[3] Damköhler, Z. *Elektrochemie* **42** (1936) 846. The definition $N_{\text{Da}} = k_0'h/V$ is sometimes referred to as the "first" Damköhler number. The "second" Damköhler number is defined as $N_{\text{Da}} = k_0'h^2/D_{\text{AB}}$, which can be obtained by multiplying the first Damköhler number by the Péclet number $N_{\text{Pe}} = Vh/D_{\text{AB}}$.

approach to solving this system of equations is to use a perturbation method (see Section 8.3) taking $|\varepsilon|$ to be small. For convenience, we consider the case $\varepsilon = 0$ so that (10.18) and (10.19) are decoupled from the temperature equation (10.20) that, however, is still coupled to (10.18).

Setting $\varepsilon = 0$ and taking the Laplace transform (cf. footnote on p. 32) of (10.18) and using in the first equation in (10.21) gives

$$\frac{d^2 \bar{x}_A}{d\bar{x}_1^2} - (N_{Da} + s)\bar{x}_A = 0. \tag{10.24}$$

Taking the Laplace transform of the boundary conditions in the first equation in (10.22) and the first in (10.23) gives $\bar{x}_A(0, s) = 1/s$ and $\bar{x}_A(\infty, s) = 0$, respectively, so that the solution of (10.24) is simply

$$\bar{x}_A = \frac{1}{s} \exp\left(-\bar{x}_1 \sqrt{N_{Da} + s}\right). \tag{10.25}$$

To find the inverse Laplace transform of (10.25), we use the convolution theorem (cf. footnote on p. 32), which gives

$$\begin{aligned}
x_A &= \int_0^{x_3} \frac{\bar{x}_1}{\sqrt{4\pi u^3}} \exp\left(-N_{Da} u - \frac{\bar{x}_1^2}{4u}\right) du \\
&= \frac{2}{\sqrt{\pi}} \int_{\bar{x}_1/\sqrt{4x_3}}^{\infty} \exp\left(-\xi^2 - N_{Da} \frac{\bar{x}_1^2}{4\xi^2}\right) d\xi,
\end{aligned} \tag{10.26}$$

where the second equality follows from the change of variables $\xi = \bar{x}_1/\sqrt{4u}$. Evaluation of the integral gives

$$\begin{aligned}
x_A &= \frac{e^{\sqrt{N_{Da}}\bar{x}_1}}{2} \operatorname{erfc}\left(\frac{\bar{x}_1}{\sqrt{4x_3}} + \sqrt{N_{Da} x_3}\right) \\
&\quad + \frac{e^{-\sqrt{N_{Da}}\bar{x}_1}}{2} \operatorname{erfc}\left(\frac{\bar{x}_1}{\sqrt{4x_3}} - \sqrt{N_{Da} x_3}\right).
\end{aligned} \tag{10.27}$$

The evolution of the concentration field for species A is shown in Figure 10.2. From this figure we see that chemical reaction, as one would expect, leads to more rapid decrease of the reactant species concentration. We also see that the reactant concentration profile becomes independent of position along the film for sufficiently large x_3/h when chemical reaction occurs. Recall that we have stretched the x_1-coordinate so that the profiles shown in Figure 10.2 correspond to a thin layer of the liquid film in contact with the gas.

The same procedure can be used to find x_C. Taking the Laplace transform of (10.19) and using the second equation in (10.21) gives

$$\frac{d^2 \bar{x}_C}{d\bar{x}_1^2} - \frac{s}{\beta} \bar{x}_C = -\frac{N_{Da}}{\beta s} \exp\left(-\bar{x}_1 \sqrt{N_{Da} + s}\right). \tag{10.28}$$

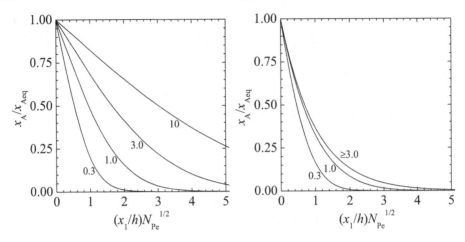

Figure 10.2 Species A concentration profile from (10.27) for gas absorption without $N_{Da} = 0$ (left), and with $N_{Da} = 1$ (right), homogeneous chemical reaction at several locations along the film: $x_3/h = 0.3, 1, 3, 10$.

The Laplace transform of the boundary conditions in the second equation in (10.22) and the second in (10.23) gives $\bar{x}_C(0, s) = \bar{x}_C(\infty, s) = 0$. Hence, the solution of (10.28) is given by

$$\bar{x}_C = \frac{N_{Da}}{s[(1 - \beta)s - \beta N_{Da}]} \left[\exp\left(-\bar{x}_1\sqrt{N_{Da} + s}\right) - \exp\left(-\bar{x}_1\sqrt{s/\beta}\right)\right].$$

$$(10.29)$$

Since $\beta \lesssim 1$, for convenience we set $\beta = 1$; using the convolution theorem to invert (10.29), we find

$$x_C = \text{erfc}\left(\frac{\bar{x}_1}{\sqrt{4x_3}}\right) - \frac{e^{\sqrt{N_{Da}}\bar{x}_1}}{2}\text{erfc}\left(\frac{\bar{x}_1}{\sqrt{4x_3}} + \sqrt{N_{Da}x_3}\right)$$

$$- \frac{e^{-\sqrt{N_{Da}}\bar{x}_1}}{2}\text{erfc}\left(\frac{\bar{x}_1}{\sqrt{4x_3}} - \sqrt{N_{Da}x_3}\right),$$

$$(10.30)$$

which is plotted in Figure 10.3. From this figure we see that the concentration of the product species increases along the film, leading to a larger gradient, and hence rate of desorption, at the gas–liquid interface $\bar{x}_1 = 0$.

Finally, to find the temperature field, we take the Laplace transform of (10.20) and using the third equation in (10.21) we obtain

$$\frac{d^2\bar{T}}{d\bar{x}_1^2} - \frac{s}{N_{Le}}\bar{T} = -\frac{N_{Da}}{N_{Le}s}\exp\left(-\bar{x}_1\sqrt{N_{Da} + s}\right).$$

$$(10.31)$$

Transforming boundary conditions in the third equation in (10.22) and the third in (10.23) gives $\bar{T}(0, s) = \bar{T}(\infty, s) = 0$; the solution of (10.31) can be

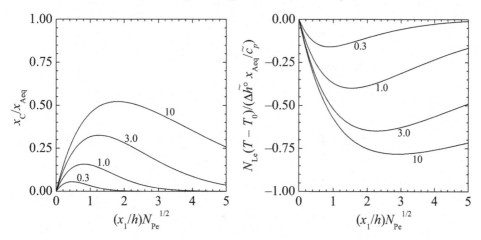

Figure 10.3 Species C concentration profile (left) from (10.30) and temperature profile (right) from (10.33) with $N_{Le} = 10$ for gas absorption with homogeneous chemical reaction with $N_{Da} = 1$ at several locations along the film: $x_3/L = 0.3, 1, 3, 10$.

written as

$$\bar{T} = \frac{N_{Da}}{s[(N_{Le} - 1)s + N_{Le}N_{Da}]} \left[\exp\left(-\bar{x}_1\sqrt{N_{Da} + s}\right) - \exp\left(-\bar{x}_1\sqrt{s/N_{Le}}\right) \right].$$
(10.32)

For typical liquid mixtures $N_{Le} \gg 1$ (see Table 7.1), which facilitates inversion of (10.32) giving

$$
\begin{aligned}
N_{Le}\bar{T} &= \frac{e^{\sqrt{N_{Da}}\bar{x}_1}}{2} \operatorname{erfc}\left(\frac{\bar{x}_1}{\sqrt{4x_3}} + \sqrt{N_{Da}x_3} \right) \\
&+ \frac{e^{-\sqrt{N_{Da}}\bar{x}_1}}{2} \operatorname{erfc}\left(\frac{\bar{x}_1}{\sqrt{4x_3}} - \sqrt{N_{Da}x_3} \right) - e^{-N_{Da}x_3}\operatorname{erfc}\left(\frac{\bar{x}_1}{\sqrt{4x_3}} \right) \\
&- \left(1 - e^{-N_{Da}x_3}\right)\operatorname{erfc}\left(\frac{\bar{x}_1}{\sqrt{4N_{Le}x_3}} \right),
\end{aligned}
$$
(10.33)

where we have also used $N_{Le} \gg N_{Da}$ to further simplify the expression. The temperature field given by (10.33) is plotted in Figure 10.3. Since the normalized temperature is negative, for an endothermic reaction ($\Delta\tilde{h}^0 > 0$) the temperature decreases (relative to T_0), while for an exothermic reaction ($\Delta\tilde{h}^0 < 0$) the temperature increases (relative to T_0). If the reaction rate constant were allowed to depend on temperature, this would alter the concentration fields of reactant and product species. For example, with an exothermic reaction we have $1 + \varepsilon T > 0$, which would increase the reaction term (10.18), leading to a decrease in x_A relative to the $\varepsilon = 0$ case.

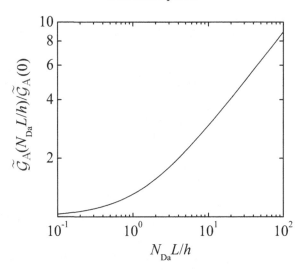

Figure 10.4 The dependence of the molar rate of absorption $\widetilde{\mathcal{G}}_A$ on the reaction rate and film length $N_{\mathrm{Da}}L/h$ as given by (10.35).

The effectiveness of the gas absorption process is related to the rate at which species A is absorbed by a liquid film. Let $\widetilde{\mathcal{G}}_A$ be the total molar rate of absorption of A, which is found using the expression in (10.38) and $\widetilde{\mathcal{G}}_A = \mathcal{G}_A/\tilde{M}_A$. When applied at the surface of the liquid film $\bar{x}_1 = 0$, we have

$$\frac{\widetilde{\mathcal{G}}_A}{c_{\mathrm{Aeq}}B\sqrt{D_{\mathrm{AB}}VL}} = -\sqrt{\frac{h}{L}}\int_0^{L/h}\frac{\partial x_A}{\partial \bar{x}_1}(0, x_3)dx_3$$
$$= \left[\frac{\exp(-N_{\mathrm{Da}}L/h)}{\sqrt{\pi}} + \frac{(1+2N_{\mathrm{Da}}L/h)\mathrm{erf}(\sqrt{N_{\mathrm{Da}}L/h})}{2\sqrt{N_{\mathrm{Da}}L/h}}\right],$$

$$\tag{10.34}$$

where the second line is obtained by substitution of (10.27). The effect chemical reaction has on the absorption process can be ascertained by the ratio

$$\frac{\widetilde{\mathcal{G}}_A(N_{\mathrm{Da}}L/h)}{\widetilde{\mathcal{G}}_A(0)} = \frac{1}{2}\left[\exp(-N_{\mathrm{Da}}L/h) + \frac{\sqrt{\pi}(1+2N_{\mathrm{Da}}L/h)\mathrm{erf}(\sqrt{N_{\mathrm{Da}}L/h})}{2\sqrt{N_{\mathrm{Da}}L/h}}\right].$$

$$\tag{10.35}$$

The dependence of the normalized molar rate of absorption given by (10.35) on N_{Da} is shown in Figure 10.4. From this figure, we see that, for $N_{\mathrm{Da}}L/h = 10$, the rate of absorption is increased by roughly a factor of three when a chemical reaction takes place within the film.

Exercise 10.5 *Solution of Reactive Gas Absorption Equations*
Use Laplace transforms to solve the coupled system of equations in (10.18)–(10.23) for $\varepsilon = 0$ to obtain the expressions (10.27), (10.30), and (10.33).

Exercise 10.6 *Solution of Diffusion–Reaction Equations*
In many cases it is possible to find the solution to diffusion–reaction equations of the type given in (10.18) by solving a simpler diffusion equation.[4] Consider a diffusion–reaction equation for $f(x, t)$ given by

$$\frac{\partial f}{\partial t} = \frac{\partial^2 f}{\partial x^2} - Cf,$$

$$f(x, 0) = 0, \qquad f(0, t) = 1, \qquad f(\infty, t) = 0,$$

where C is a constant. Show that the solution f is given by

$$f = \int_0^t \frac{\partial \hat{f}}{\partial t} \exp(-Ct')dt',$$

where $\hat{f}$ is the solution to the problem with $C = 0$.

Exercise 10.7 *Diffusion and Reaction in a Stationary Liquid*
Consider a stationary, semi-infinite liquid in the region $x_1 > 0$. Initially, the liquid is pure B, and, for $t > 0$, the liquid is brought into contact with a gas such that, at $x_1 = 0$, the concentration of species A is maintained at x_{Aeq}. As A diffuses in the liquid, an "irreversible" reaction $A + B \to C$ takes place with reaction rate constant k. The liquid mixture is dilute in species A ($x_A \ll 1$), and therefore is also dilute in species C ($x_C \ll 1$) so that $x_B \approx 1$. In this case, the molar density of the mixture c and the diffusivity D_{AB} can be taken as constants, and the process can be assumed to be isothermal. Show that the mole fraction of species A is given by

$$\frac{x_A}{x_{\text{Aeq}}} = \frac{e^{x_1/\sqrt{D_{AB}/k'}}}{2} \operatorname{erfc}\left(\frac{x_1}{\sqrt{4D_{AB}t}}\right) + \frac{e^{-x_1/\sqrt{D_{AB}/k'}}}{2} \operatorname{erfc}\left(\frac{x_1}{\sqrt{4D_{AB}t}}\right).$$

Exercise 10.8 *Mass Transfer Coefficients*
Analogously to the heat transfer coefficient introduced in Section 9.3, a mass transfer coefficient can be used to describe mass transport at solid–fluid (or liquid–gas) interfaces in situations where mass transfer in a fluid phase is complex. Similarly to (9.26), the equation defining the mass transfer coefficient k_m is for dilute systems given by

$$-\boldsymbol{n} \cdot (\rho_\alpha \boldsymbol{v} + \boldsymbol{j}_\alpha) = -\boldsymbol{n} \cdot \boldsymbol{n}_\alpha = k_m(w_\alpha - w_{\alpha\text{fl}}) \qquad \text{at } A_s, \tag{10.36}$$

where $\boldsymbol{n}$ points away from the fluid (or gas) phase, and $w_{\alpha\text{fl}}$ is an average mass

[4] Danckwerts, *Trans. Faraday Soc.* **46** (1950) 300.

fraction for species α of the fluid away from the surface. Formulate an alternative boundary condition for the first equation in (10.15) in terms of the mass transfer coefficient.

Exercise 10.9   Macroscopic Species Mass Balance
Consider the macroscopic balance (see Exercises 8.12 and 9.7) for species mass. As before, the system has volume V and is bounded by surface A with outward unit normal vector n. For the non-entrance–exit surface A_s, which is assumed to be stationary, we write $A_s = A_{si} + A_{sp}$, where A_{si} is the area of impermeable and inert surfaces, and A_{sp} is the area of permeable and/or reactive surfaces. Assume the species mass density ρ_α is uniform over A_i, and that j_α can be neglected at entrance and exit surfaces A_i. From the differential species mass balance (5.14), show that the macroscopic species mass balance takes the form

$$\frac{d}{dt} M_{\alpha,\text{tot}} = -\Delta[\rho_\alpha \langle v \rangle A] + \mathcal{G}_\alpha + \mathcal{R}_\alpha, \tag{10.37}$$

where $M_{\alpha,\text{tot}} = \int_V \rho_\alpha \, dV$ is the total species mass, $\mathcal{R}_\alpha = \int_V \nu_\alpha \Gamma \, dV$ is the species mass rate of production by homogeneous chemical reaction, and $\mathcal{G}_\alpha$ is defined as

$$\mathcal{G}_\alpha = - \int_{A_{sp}} n_\alpha \cdot n \, dA, \tag{10.38}$$

which is the rate of species mass transferred to the system at permeable and/or reactive surfaces with area A_{sp}.

Exercise 10.10   Continuous Stirred Tank Reactor Design Equation
The continuous stirred tank reactor (CSTR) model is commonly used by engineers to design chemical reactors. A typical CSTR has a feed stream and a product stream, and is operated at steady state. Perfect mixing is assumed within the tank so that the product stream has the same composition as the contents of the tank. Consider a CSTR with constant volume V and feed stream with molar densities $c_{\alpha 0}$. Use (10.37) to obtain

$$\frac{V}{\langle v \rangle A} = \frac{c_\alpha - c_{\alpha 0}}{\tilde{\nu}_\alpha \tilde{\Gamma}}.$$

The ratio of the reactor volume V to the feed (product) stream flow rate $\langle v \rangle A$ is the residence time.

- The description of multi-component systems in which flow, diffusion, and chemical reaction occur involves a coupled set of evolution equations for concentrations and temperature.

- For ideal mixtures, source terms for homogeneous chemical reaction appearing in the species mass balance equations have the form of the mass-action law.

- For cases when the total mass, or molar, density of a mixture is effectively constant, the velocity field for an externally imposed flow can be found from solution of the total mass and momentum balances and used in evolution equations for concentration and temperature.

- For dilute ideal mixtures, the pseudo-binary approximation can be used to obtain greatly simplified expressions for the diffusive fluxes having the form of Fick's law.

- Analytic solutions of the linearized and decoupled evolution equations for species concentration and temperature can be found and used to examine the effectiveness of gas absorption processes.

11

Driven Separations

In the previous chapter we considered a classical separation process in which mass diffusion and chemical reaction took place in a flowing liquid. To keep things relatively simple, we excluded the coupling of diffusive fluxes of energy and mass so that mass diffusion occurred as a result of concentration gradients alone. As we saw in Chapter 6, diffusive species mass fluxes can also result from other forces such as gradients of pressure and temperature. In this chapter, we consider examples where species mass fluxes are driven not only by gradients of concentration, but also by gradients of pressure, temperature, or electric potential. Large-scale separation processes (e.g., gas absorption) are based on equilibrium thermodynamic properties of mixtures such as solubility differences, and are typically used for mixtures of ordinary gases and liquids. By contrast, the separation processes considered in this chapter are useful for more complicated mixtures where classical processes are impractical or inadequate.

11.1 Ultracentrifugation

A centrifuge is a device that rotates a material around an axis, producing a force perpendicular to the axis of rotation, or a centrifugal force. This causes denser components in the material to move outward in the radial direction, while components that are less dense move towards the axis of rotation. Centrifuges have numerous applications that include hematocrit (volume fraction of red blood cells in whole blood) measurement, uranium enrichment (isotope separation), dairy processing, water treatment, and materials processing. An ultracentrifuge, which can achieve rotation rates of 10^5 revolutions per minute, is capable of generating accelerations in excess of $10^6 g$. The analytic ultracentrifuge, developed by the chemist Theodor Svedberg (1926 Nobel Prize in Chemistry), is equipped with an optical detection

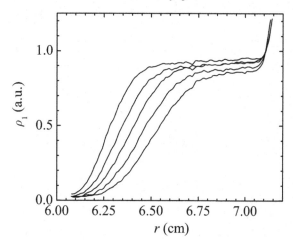

Figure 11.1 Concentration of the enzyme RNase A as function of position within sample cell for different times (increasing from left to right at 20 min intervals) for an analytic ultracentrifuge experiment run at 20 °C, at pH 6.5, and at a rotation rate of 60 000 revolutions per min. [Adapted from Laue & Stafford, *Annu. Rev. Biophys. Biomol. Struct.* **28** (1999) 75.]

system that allows one to monitor the evolution of the concentration within the sample. These experiments have been used to study the size, shape, and interactions of colloids and both biological and synthetic polymers. Analytic ultracentrifuges find important uses in molecular biology, biochemistry, and polymer science. An example of data collected from an analytic ultracentrifuge experiment is shown in Figure 11.1 for bovine pancreatic ribonuclease (RNase A),[1] which is an enzyme that cleaves single-stranded RNA. For readers interested in a detailed discussion of the analysis and applications of ultracentrifuges, we recommend the book by Fujita.[2] In this section we develop the transport equations that describe, and hence can be used to analyze, or design, an ultracentrifuge.

We consider a two-component sample comprised of a solute (species 1) and a solvent (species 2). We assume that the system is isothermal and that there are no chemical reactions. The sample occupies the truncated sector of a cylinder (see Figure 11.2) with inner radius $R - L$, outer radius $R + L$, where $L < R$, and angle $\alpha \ll 1$ between the side walls. The sample cell rotates at constant angular velocity Ω about the z-axis, which is parallel to the gravitational force $\boldsymbol{g} = -g\boldsymbol{\delta}_z$. Impermeable walls surround the sample cell on all sides.

[1] Laue & Stafford, *Annu. Rev. Biophys. Biomol. Struct.* **28** (1999) 75.
[2] Fujita, *Foundations of Ultracentrifugal Analysis* (Wiley, 1975).

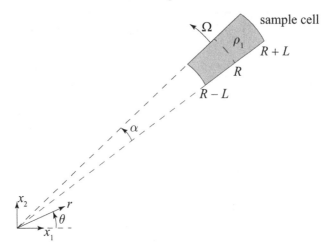

Figure 11.2 Schematic diagram of a centrifuge with a sample cell rotating with angular velocity Ω. The sample occupies the region described by $R - L \leq r \leq R + L$, and has solute concentration $\rho_1 = \rho w_1$.

Our goal is to formulate an equation that governs the concentration of the solute species within the sample. We take the sample to be a Newtonian fluid with constant viscosity η and negligible bulk viscosity ($\eta_d = 0$). Hence, we deal with the fields v and p, which are governed by (7.1) and (7.2), along with an equation of state $\rho = \rho(p, w_1)$. To complete our model, we need an evolution equation for the solute mass fraction w_1, or mass density $\rho_1 = w_1 \rho$.

The sample rotates with angular velocity Ω so that the velocity has a tangential component, $v_\theta = \Omega r$. As noted above, the centrifugal force leads to a concentration gradient in the radial direction, which means that, in general (see Section 7.4), the velocity will also have a nonzero radial component ($v_r \neq 0$). Here, we assume all fields are uniform in the z-direction and that $v_z = 0$. Since the surfaces bounding the sample have positions $\theta = \Omega t$ and $\theta = \Omega t + \alpha$, all fields depend on θ. However, in a coordinate system rotating about the z-axis with angular velocity Ω these surfaces are stationary, and, since they are impermeable, it would be reasonable to assume the concentration field is uniform in the θ-direction. This suggests that the balance equations may become simpler if expressed in a rotating coordinate system.

Balance equations are usually applied in an inertial coordinate system. By inertial coordinate system we mean a coordinate system in which only interaction forces (e.g., gravitational, electromagnetic) are present in

Newton's second law. All inertial frames are in a state of constant, rectilinear motion with respect to one another. A *non-inertial coordinate system* accelerates (by translation and/or rotation) with respect to an inertial coordinate system and, because of this, additional so-called inertial forces are present. A coordinate system that is stationary with respect to a laboratory or processing facility, which is fixed to the earth's surface, is often taken to be an approximately inertial coordinate system. The surface of the earth is, of course, accelerating because of the earth's rotation. In most cases, however, we do not notice this motion, that is, we perceive only the interaction forces. By contrast, in a coordinate system fixed to a racing car, the driver is stationary yet still experiences additional forces as the car accelerates or makes a sharp turn. Similarly to the racing car driver, the material in a centrifuge sample cell is essentially at rest in a coordinate system fixed to the sample cell, yet the material experiences a centrifugal force.

The transformation of evolution equations from inertial to non-inertial frames involves some delicate issues. Details of this transformation are given at the end of this section. Here, we give results for a non-inertial coordinate system with position vector $c(t)$ and rotation matrix $Q_{jk}(t)$ relative to an inertial coordinate system. When appropriate, primed quantities refer to the non-inertial coordinate system. For example, r' is the position vector in a non-inertial coordinate system. The velocity vector v' of a particle following a trajectory described by the time-dependent coordinates x'_k is defined as

$$v' = \sum_{k=1}^{3} \frac{dx'_k}{dt} \delta'_k,\qquad(11.1)$$

where the base vectors δ'_k are treated as time-independent (see discussion on p. 196). This neglect of the time dependence of δ'_k is the reason why inertial forces need to be introduced.

The species continuity equation takes the form

$$\frac{\partial' \rho'_\alpha}{\partial t} = -\nabla' \cdot (v' \rho'_\alpha + j'_\alpha),\qquad(11.2)$$

which we see has the same (neglecting chemical reaction) structure as (5.14). The notation $\partial'/\partial t$ indicates time differentiation at fixed position in the non-inertial frame. The equation of motion for a Newtonian fluid is given by

$$\rho'\left(\frac{\partial' v'}{\partial t} + v' \cdot \nabla' v'\right) = \eta \nabla'^2 v' + \frac{1}{3}\eta \nabla'(\nabla' \cdot v') - \nabla' p' + \rho' g$$

$$+ \rho'\left[\frac{d^2 c}{dt^2} - \frac{d\omega}{dt} \times r' - 2\omega \times v' - \omega \times (\omega \times r')\right],$$

$$(11.3)$$

where $\boldsymbol{\omega}$ is the angular velocity vector,

$$\boldsymbol{\omega} = \frac{1}{2} \sum_{i,j,k=1}^{3} \varepsilon_{ijk} A_{kj} \boldsymbol{\delta}'_i, \tag{11.4}$$

and where the A_{kj} are the components of the angular velocity matrix,

$$A_{kj} = \sum_{i=1}^{3} Q_{ki} \frac{dQ_{ji}}{dt}. \tag{11.5}$$

The first line of (11.3) has the same structure as (7.2). The second line of (11.3), however, contains various additional forces that arise in the non-inertial coordinate system. The first and second terms are due to translational and rotational acceleration, respectively; the third term is the Coriolis acceleration, and the fourth is the centrifugal acceleration.

Returning to our analysis of the ultracentrifuge, we have pure rotation about the z-axis so that $\boldsymbol{c} = \boldsymbol{0}$, and Q_{jk} is given by

$$Q_{jk} = \begin{pmatrix} \cos\phi & \sin\phi & 0 \\ -\sin\phi & \cos\phi & 0 \\ 0 & 0 & 1 \end{pmatrix}, \tag{11.6}$$

where $\phi = \Omega t$. From (11.4) and (11.6) it is straightforward to show that $\boldsymbol{\omega} = \Omega \boldsymbol{\delta}_z$. Since $v' \sim v_r$, which is the velocity induced by mass diffusion, it is reasonable to assume that inertial and viscous forces are negligible. Hence, we neglect all terms involving $\boldsymbol{v}'$ in (11.3). For the case of steady rotation, (11.3) then simplifies to

$$\boldsymbol{0} = -\boldsymbol{\nabla}'p' + \rho'\boldsymbol{g} + \rho'\Omega^2 r'\boldsymbol{\delta}'_r, \tag{11.7}$$

or, in component form,

$$\frac{\partial p'}{\partial r'} = \rho'\Omega^2 r', \qquad \frac{\partial p'}{\partial \theta'} = 0, \qquad \frac{\partial p'}{\partial z'} = -\rho'g. \tag{11.8}$$

Since $\Omega^2 R \gg g$, we neglect pressure variations in the z-direction. Hence, by writing the momentum balance in a coordinate system that rotates with the sample cell, and assuming the system is at mechanical equilibrium, we find a pressure gradient in the radial direction in the first equation in (11.8), which is needed to compensate for the centrifugal force.

We now consider the mass balances for the two-component system within the sample cell. Since the form of the mass balance is unchanged in the non-inertial coordinate system, in the remainder of this discussion we drop the primes. As noted above, separation in an ultracentrifuge is based on density differences of the different components. This means the density ρ must depend on composition and therefore *cannot* be constant. As discussed in Section 7.4, density variations can result in a diffusion-induced convective velocity, which renders the evolution equation for the species mass density nonlinear. This complication can be avoided by the choice of reference velocity. Here, we use the volume-average velocity $v^\dagger$, and from (5.27) we write

$$v_r = v_r^\dagger + (\hat{v}_2 - \hat{v}_1)(j_1)_r.\tag{11.9}$$

If we assume the partial specific volumes $\hat{v}_\alpha = \hat{v}_\alpha^0$ are constant, that is, independent of both composition (ideal mixture) and pressure, then the total mass balance takes the form $\nabla \cdot v^\dagger = 0$ (see Exercise 5.5). Now, since $v_r^\dagger(R \pm L, t) = 0$, we have $v_r^\dagger(r, t) = 0$. Hence, combining (11.9) and (11.2), we obtain [using (B.8)]

$$\frac{\partial \rho_1}{\partial t} = -\frac{1}{r}\frac{\partial}{\partial r}\left[r\rho\hat{v}_2^0(j_1)_r\right].\tag{11.10}$$

Note that, using $v^\dagger$ as the reference velocity, and being able to argue that it vanishes, convective transport is taken into account by the factor $\rho\hat{v}_2^0$ modifying the diffusive mass flux.

The diffusive mass flux of solute in (11.10) is the result of both concentration and pressure gradients, which is a case considered in Exercise 6.7. Multiplying (6.32) by $\rho\hat{v}_2$, we can write

$$\rho\hat{v}_2 j_1 = -D_{12}\nabla\rho_1 - \frac{L_{11}}{T}\rho\hat{v}_2(\hat{v}_1 - \hat{v}_2)\nabla p,\tag{11.11}$$

where $D_{12} = L_{11}(\partial\hat{\mu}_1/\partial w_1)_{T,p}/(\rho_2 T)$. The r-component of (11.11), using the first equation in (11.8), can be written as

$$\rho\hat{v}_2^0(j_1)_r = -D_{12}\frac{\partial\rho_1}{\partial r} + s_1\rho_1\Omega^2 r,\tag{11.12}$$

where we have introduced the *sedimentation coefficient* s_1. This parameter, which is defined as

$$s_1 = \frac{L_{11}\rho^2\hat{v}_2^0(\hat{v}_2^0 - \hat{v}_1^0)}{\rho_1 T},\tag{11.13}$$

is positive if the solute is denser than the solvent, and *vice versa*. Substitution of (11.12) into (11.10) leads to

$$\frac{\partial\rho_1}{\partial t} = \frac{1}{r}\frac{\partial}{\partial r}\left(rD_{12}\frac{\partial\rho_1}{\partial r} - s_1\Omega^2 r^2\rho_1\right),\tag{11.14}$$

which is known as the Lamm equation.[3]

In general, both D_{12} and s_1 are functions of ρ_1, making (11.14) nonlinear.

[3] Lamm, *Ark. Math. Astr. Fys.* **21B**(2) (1929) 1.

However, if we consider a dilute system ($w_1 \ll 1$), both D_{12} and s_1 can be treated as constants, so that (11.14) can be written as

$$\frac{\partial \rho_1}{\partial t} = D_{12} \frac{1}{r} \frac{\partial}{\partial r} \left(r \frac{\partial \rho_1}{\partial r} \right) - s_1 \Omega^2 \frac{1}{r} \frac{\partial}{\partial r} (r^2 \rho_1). \tag{11.15}$$

The initial condition for (11.15) is taken to be

$$\rho_1(r, 0) = \rho_{10}, \tag{11.16}$$

although the initial concentration need not be uniform. The walls of the sample cell are impermeable, $(j_1)_r(R \pm L, t) = 0$, so that from (11.12) the following boundary conditions are obtained,

$$D_{12} \frac{\partial \rho_1}{\partial r}(R \pm L, t) = s_1 \Omega^2 (R \pm L) \rho_1(R \pm L, t). \tag{11.17}$$

Equations (11.15)–(11.17) describe the evolution of the solute concentration in an ultracentrifuge. Applying the $n - m$ procedure (see Section 7.5), we have the quantities $\rho_1, r, t, D_{12}, s_1 \Omega^2, \rho_{10}, R, L$, so that $n - m = 8 - 3 = 5$. Hence, the solute concentration ρ_1/ρ_{10} depends on the radial position $(r - R)/L$, the time $t L^2/D_{12}$, and two dimensionless parameters. Here, we choose $\Lambda_p = s_1 \Omega^2 R^2/D_{12}$, which does not depend on transport coefficients (see Exercise 11.1), and $\beta = L/R$. Typical values for an ultracentrifuge are $R \sim 10^{-1}$ m and $\Omega \sim 10^4$ s^{-1}, and for macromolecular solutes $s_1 \sim 10^{-13}$ s and $D_{12} \sim 10^{-8} - 10^{-10}$ m^2/s, so that $\Lambda_p \sim 10^2$–10^3.

The equilibrium solute concentration (see Exercise 11.1) is plotted in Figure 11.3, which shows that, for large values of Λ_p, the solute concentration at the outer wall of the cell ($r = R + L$) is increased by more than a factor of 10 relative to the initial value. On the other hand, the solute concentration can be reduced by more than a factor of 10^3 within the inner half of the cell for large values of Λ_p. Analysis of an equilibrium experiment allows the determination of Λ_p, or the ratio of s_1 to D_{12}. For a dilute ideal mixture, $D_{12} = \tilde{R} L_{11}/(\rho_1 \tilde{M}_1)$, which, when substituted into (11.13), gives

$$\tilde{M}_1 = \frac{\tilde{R} T s_1}{\rho^2 \hat{v}_2^0 (\hat{v}_2^0 - \hat{v}_1^0) D_{12}} = \frac{\tilde{R} T s_1}{(1 - \rho \hat{v}_1^0) D_{12}}, \tag{11.18}$$

where the second equality follows since $\rho_2 \approx \rho \approx 1/\hat{v}_2^0$. This expression, which is known as the Svedberg relation, allows the determination of the molecular weight of the solute from measurements of Λ_p.

Although it is linear, obtaining an analytic solution to the problem given by (11.15)–(11.17) presents a significant challenge. An approximate solution was obtained by Faxén,[4] and an exact (but somewhat cumbersome) solution

---

[4] Faxén, *Ark. Math. Astr. Fys.* **21B** (1929) 1.

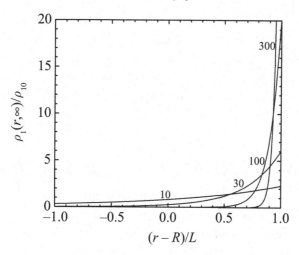

Figure 11.3 Steady-state concentration field for an ultracentrifuge from (11.30) with $\beta = 0.1$ for $\Lambda_p = 10, 30, 100, 300$.

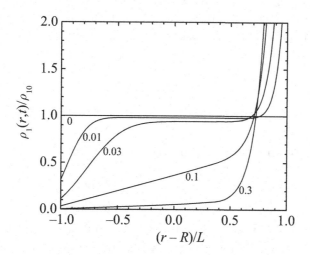

Figure 11.4 Time evolution of the solute concentration field for an ultra-centrifuge obtained from numerical solution of (11.15)–(11.17) with $\beta = 0.1$ and $\Lambda_p = 100$ for different times $tD_{12}/L^2 = 0, 0.01, 0.03, 0.1, 0.3$.

was found by Archibald.[5] A detailed discussion of these and other solutions can be found elsewhere in Fujita's book. Figure 11.4 shows the time evolution of the solute concentration field obtained from numerical solution of (11.15)–(11.17). From this figure we see, as for the experimental results in

---

[5] Archibald, *Phys. Rev.* **53** (1938) 746; Archibald, *Phys. Rev.* **54** (1938) 371.

Figure 11.1, that a concentration front moves from the inner wall of the cell $(r = R - L)$ to the outer wall of the cell $(r = R + L)$, where a steep concentration gradient develops. The large concentration gradient at the outer wall of the cell can create both experimental and numerical difficulties. Measurement of the concentration-front propagation allows separate determination of s_1 and D_{12}.

Transport in Non-inertial Coordinate Systems

Our goal here is to express the hydrodynamic equations in a non-inertial coordinate system, which means we will need to derive expressions for fields and their derivatives in a non-inertial coordinate system from those in an inertial coordinate system. Alternative approaches and more detailed discussions on changes in coordinate system can be found elsewhere.[6] Consider an inertial coordinate system with origin O and a non-inertial coordinate system with origin O' as shown in Figure 11.5. The position of point P is $\boldsymbol{r}$ in the inertial coordinate system, and $\boldsymbol{r}'$ in the non-inertial coordinate system, and $\boldsymbol{c} = \boldsymbol{c}(t)$ is the position of O relative to O'. From Figure 11.5 it is clear that

$$\boldsymbol{r}' = \boldsymbol{r} + \boldsymbol{c}. \tag{11.19}$$

The relationship between the base vectors in the non-inertial and inertial coordinate systems is given by $\boldsymbol{\delta}'_j = \sum_k Q_{jk}\boldsymbol{\delta}_k$, where the $Q_{jk}(t)$ are the components of an orthogonal matrix,[7]

$$Q_{jk} = \boldsymbol{\delta}'_j \cdot \boldsymbol{\delta}_k, \qquad \sum_k Q_{ik}Q_{jk} = \delta_{ij}. \tag{11.20}$$

Writing $\boldsymbol{r} = \sum_j x_j\boldsymbol{\delta}_j$ (cf. footnote on p. 58) and $\boldsymbol{r}' = \sum_j x'_j\boldsymbol{\delta}'_j$, it is straightforward to obtain the following relations between the coordinate positions in the two coordinate systems,[8]

$$x'_j = \sum_k Q_{jk}x_k + c'_j, \qquad x_j = \sum_k Q_{kj}x'_k - c_j, \tag{11.21}$$

where we have used $\boldsymbol{c} = \sum_j c_j\boldsymbol{\delta}_j = \sum_j c'_j\boldsymbol{\delta}'_j$.

---

[6] Slattery, *Advanced Transport Phenomena* (Cambridge, 1999); Vrentas & Vrentas, *Diffusion and Mass Transfer* (CRC Press, 2012).

[7] Here, $\sum_k$ implies summation with $k = 1, 2, 3$.

[8] An important special case of (11.21) known as a Galilean transformation is when $\boldsymbol{Q}$ is independent of time, and the non-stationary coordinate system moves with constant velocity relative to the stationary coordinate system, that is, $\boldsymbol{c}$ is linear in time.

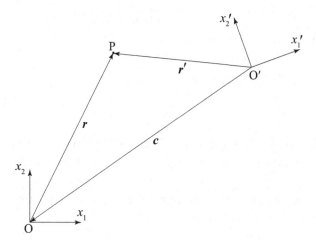

Figure 11.5 Two-dimensional representation of a non-inertial coordinate system with origin O' relative to an inertial coordinate system with origin O. Point P is an arbitrary point in space.

To obtain a transformation for the velocity field, we invoke the concept of particle trajectories where the $x_j = x_j(t)$ represent the time-dependent coordinates of a particle in an inertial coordinate system.[9] On taking the time derivative of the second equation in (11.21), we obtain the following definition for velocity:

$$\frac{dx_j}{dt} = v_j = \sum_k Q_{kj}\frac{dx'_k}{dt} + \sum_k \frac{dQ_{kj}}{dt}x'_k - \frac{dc_j}{dt}. \tag{11.22}$$

Multiplication of this result by Q_{ij} and summing over j gives, after rearranging terms,

$$\frac{dx'_i}{dt} = v'_i = \sum_j Q_{ij}v_j - \sum_k A_{ik}x'_k + \sum_j Q_{ij}\frac{dc_j}{dt}, \tag{11.23}$$

where the A_{ik} ($= -A_{ki}$) are the components of the skew-symmetric angular velocity matrix

$$A_{ik} = \sum_j Q_{ij}\frac{dQ_{kj}}{dt} = -\sum_j \frac{dQ_{ij}}{dt}Q_{kj}, \tag{11.24}$$

where the second equality follows from (11.20). We now have rules for transforming velocities in the different coordinate systems.

---

[9] That is, we make use of the connection between the particle velocity associated with a trajectory and the velocity field associated with the mass flux first motivated after (2.6) and repeatedly discussed in Chapter 5.

Let $a = a(x_1, x_2, x_3, t)$ be any function where the x_j now resume the role of independent variables. We assume $a' = a$, where $a' = a'(x'_1, x'_2, x'_3, t)$ is the function in the non-inertial coordinate system. Using the chain rule and (11.21), it can be shown that

$$\frac{\partial a}{\partial x_j} = \sum_k Q_{kj} \frac{\partial a'}{\partial x'_k}, \qquad \frac{\partial a}{\partial t} = \frac{\partial' a'}{\partial t} - \sum_{k,i} \left(A_{ki} x'_i - Q_{ki} \frac{dc_i}{dt} \right) \frac{\partial a'}{\partial x'_k}, \quad (11.25)$$

where, as usual, $\partial/\partial t$ is the partial time derivative at fixed position in the inertial coordinate system (x_1, x_2, x_3), and $\partial'/\partial t$ is the time derivative at fixed position in the non-inertial coordinate system (x'_1, x'_2, x'_3).

The results we have just obtained can be used to express the hydrodynamic equations in a non-inertial coordinate system. Writing the continuity equation (5.14) in component form and using (11.22) and (11.25), we obtain

$$\frac{\partial' \rho'_\alpha}{\partial t} = -\sum_k \frac{\partial}{\partial x'_k} [v'_k \rho'_\alpha + (j'_\alpha)_k], \quad (11.26)$$

where we have assumed the transformation $(j_\alpha)_j = \sum_k Q_{kj} (j'_\alpha)_k$.[10] With this assumption, the form of the continuity equation is unchanged in the non-inertial coordinate system. Using the same approach, we write (7.2) in component form (setting $\eta_d = 0$ for convenience) and use (11.22) and (11.25). This derivation (see Exercise 11.5) leads to

$$\rho' \left(\frac{\partial' v'_i}{\partial t} + \sum_k v'_k \frac{\partial v'_i}{\partial x'_k} \right) = \eta \sum_k \frac{\partial^2 v'_i}{\partial x'_k \partial x'_k} + \frac{1}{3} \eta \frac{\partial}{\partial x'_i} \sum_k \left(\frac{\partial v'_k}{\partial x'_k} \right) - \frac{\partial p'}{\partial x'_i} + \rho' g'_i$$

$$+ \rho' \sum_k \left(Q_{ik} \frac{d^2 c_k}{dt^2} - \frac{dA_{ik}}{dt} x'_k - 2 A_{ik} v'_k \right.$$

$$\left. - \sum_j A_{ij} A_{jk} x'_k \right). \quad (11.27)$$

The first line of (11.27) has the same structure as (7.2); the second and third lines contain terms resulting from transformation to the non-inertial coordinate system. Note that, for a Galilean transformation (cf. footnote on p. 194), the second and third lines of (11.27) vanish, leaving the equations of motion unchanged, or Galilean invariant. Since A_{ik} is skew-symmetric, it can be written as

$$A_{ik} = \begin{pmatrix} 0 & -\omega_3 & \omega_2 \\ \omega_3 & 0 & -\omega_1 \\ -\omega_2 & \omega_1 & 0 \end{pmatrix}, \qquad \omega_i = \frac{1}{2} \sum_{j,k} \varepsilon_{ijk} A_{kj}, \quad (11.28)$$

---

[10] This transformation for $(j_\alpha)_j$, which from Chapter 6 we know is a linear function of the gradient of a scalar, is consistent with the transformation in the first equation in (11.25).

where the ω_i are the components of the angular velocity vector $\boldsymbol{\omega}$. This allows the last three terms in the second line of (11.27) to be expressed in terms of cross products (cf. footnote on p. 67) as follows: $\sum_k A_{ik} v_k = \sum_{j,k} \epsilon_{ijk} \omega_j v_k$.

As a final step, we express our results in vector form. To do so, we multiply the results expressed in terms of components by the base vectors $\boldsymbol{\delta}_i'$. For example, this leads to the definition in (11.1). Similarly, (11.26) can be written as (11.2) and (11.27) as (11.3); these can be expressed in any other coordinate system (e.g., cylindrical) treating the base vectors as time independent.

Exercise 11.1   Ultracentrifuge at Equilibrium
Use (4.55) and the first equation in (11.8) to obtain the following equation that governs the steady-state solute concentration,

$$\frac{d\rho_1}{dr} = \left(\frac{\partial \hat{\mu}_1}{\partial w_1}\right)_{T,p}^{-1} \rho^2 \rho_2 \hat{v}_2 (\hat{v}_2 - \hat{v}_1)\Omega^2 r = \Lambda_p \frac{r}{R^2}\rho_1, \qquad (11.29)$$

where the second equality follows for a dilute, ideal mixture with $\Lambda_p = s_1\Omega^2 R^2/D_{12}$. Show that this is equivalent to the steady-state form of (11.15). Solve (11.29) to obtain

$$\frac{\rho_1}{\rho_{10}} = \frac{2\Lambda_p \beta \exp\left(\dfrac{\Lambda_p r^2}{2R^2}\right)}{\exp\left(\dfrac{\Lambda_p(1+\beta)^2}{2}\right) - \exp\left(\dfrac{\Lambda_p(1-\beta)^2}{2}\right)}. \qquad (11.30)$$

Exercise 11.2   Ultracentrifuge without Diffusion
Use the method of characteristics to show that the solution of (11.15)–(11.17) for the case $D_{12} = 0$ is given by

$$\frac{\rho_1}{\rho_{10}} = \begin{cases} 0 & \text{for} \quad R-L \le r < (R-L)\exp(s_1\Omega^2 t), \\ \exp(-2s_1\Omega^2 t) & \text{for} \quad (R-L)\exp(s_1\Omega^2 t) < r \le R+L. \end{cases}$$

Note that, because the order of (11.15) is reduced for this case, only the boundary condition given in (11.17) at $r = R - L$ is satisfied.

Exercise 11.3   Ultracentrifuge Neglecting Curvature
Use $r \to (r-R)/L$, $t \to D_{12}t/L^2$, and $\rho_1 \to \rho_1/\rho_{10}$ to rescale the equations in (11.15)–(11.17) and show that, for $\beta \ll 1$, the solute concentration field neglecting curvature is governed by

$$\frac{\partial \rho_1}{\partial t} = \frac{\partial^2 \rho_1}{\partial r^2} - \Lambda_p \beta \frac{\partial \rho_1}{\partial r},$$

$$\rho_1(r,0) = 1, \qquad \frac{\partial \rho_1}{\partial r}(\pm 1, t) = \Lambda_p \beta \rho_1(\pm 1, t).$$

Express this evolution equation for ρ_1 as a Fokker–Planck equation, develop a simulation, and compare the results with those given in Figure 11.4.

Exercise 11.4   *Sedimentation Coefficient and Mobility*
Use (11.18) to find a relation between the sedimentation coefficient and the mobility of the solute (see Exercise 6.6). Briefly discuss the physical origin of small values of the sedimentation coefficient s_1.

Exercise 11.5   *Equations of Motion in a Non-inertial Coordinate System*
Write (7.2) in component form and use (11.22) and (11.25) to obtain (11.27).

11.2 Field-Flow Fractionation

Field-flow fractionation is a family of techniques developed by the chemist J. Calvin Giddings[11] (1930–1996) used to separate and characterize macromolecular and colloidal fluids. All field-flow fractionation (F^3) techniques involve the flow of a fluid mixture through a channel inside which an external field is applied perpendicular to the main flow direction. The transverse field exerts a force that depends on the properties of the different components in the flowing mixture. The applied field can be gravitational or centrifugal, but more common versions of F^3 involve the application of electric or thermal fields. The separation results from the interaction of the nonuniformity that exists in the main flow with the transverse external forces on the individual components. An example of results obtained from an electric F^3 experiment is given in Figure 11.6, which shows elution curves for the separation of proteins found in blood.[12] Detailed discussions of the numerous types and applications of F^3 can be found elsewhere.[13] In this section we develop the equations that describe the thermal field-flow fractionation technique. The analysis of electric F^3 is given as an exercise at the end of this section.

We consider a two-component liquid comprised of a solute (species 1) and a solvent (species 2). Fluid flows through a rectangular channel with height $2H$, width B, and length $L \gg H$ (see Figure 11.7). Typically $B \gg H$, so we assume all fields are uniform in the x_3-direction. For $t < 0$, the liquid flowing through the channel is pure solvent ($\rho_1 = 0$) and has uniform temperature T_0. At $t = 0$, a liquid plug having concentration ρ_{10} and length L_p is injected at the channel entrance ($x_1 = 0$), and for $t \geq 0$ a temperature difference ΔT is maintained between the channel walls. For a typical thermal

[11] Giddings, *Sep. Sci.* **1** (1966) 123; Giddings *et al.*, *Science* **193** (1976) 1244; Giddings *et al.*, *Science* **260** (1993) 1456.
[12] Kesner *et al.*, *Anal. Chem.* **48** (1976) 1834.
[13] Schrimpf *et al.*, *Field-Flow Fractionation Handbook* (Wiley, 2000).

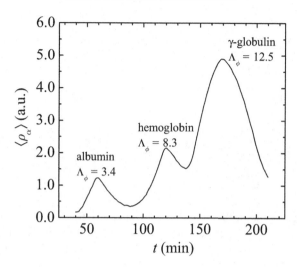

Figure 11.6 Exiting average concentrations versus time for the separation of different blood proteins (species α) in an electric F^3 experiment carried out at $25\,^{\circ}\mathrm{C}$, at pH 4.5, with an electric potential gradient of 2.95 V/cm. The parameter Λ_ϕ (see Exercise 11.9) is proportional to the molecular weight of the protein. [Adapted from Krishnamurthy & Subramanian, *Sep. Sci.* **12** (1977) 347.]

F^3 separation, $\Delta T \sim 10$–100 K, and $H \sim 0.1$–1 mm, so that temperature gradients can be in the range 10^4–10^6 K/m. The temperature gradient in the x_2-direction, because of the Soret effect (see Section 6.3), induces a diffusive flux. For convenience we consider ideal mixtures that are dilute ($w_{10} = \rho_{10}/\rho \ll 1$). There are no chemical reactions, and we neglect viscous heating and gravitational forces.

Our goal is to formulate the equations that govern the temperature and solute concentration of the fluid as it flows through the channel. The liquid is an incompressible Newtonian fluid with constant viscosity η. Since $w_{10} \ll 1$, if we further require that thermal expansion is negligible ($\alpha_p \Delta T \ll 1$), then the fluid density ρ is constant. For steady, pressure-driven, laminar flow in the channel, the velocity (see Exercise 8.2) is given by

$$v_1 = \frac{3}{2}V\left[1 - \left(\frac{x_2}{H}\right)^2\right], \qquad v_2 = 0, \qquad v_3 = 0, \qquad (11.31)$$

where V is the average velocity through the channel.

For the solute mass balance we use the species continuity equation given in (5.15) that, for constant ρ, takes the form

$$\frac{\partial \rho_1}{\partial t} + \frac{3}{2}V\left[1 - \left(\frac{x_2}{H}\right)^2\right]\frac{\partial \rho_1}{\partial x_1} = -\boldsymbol{\nabla} \cdot \boldsymbol{j}_1, \qquad (11.32)$$

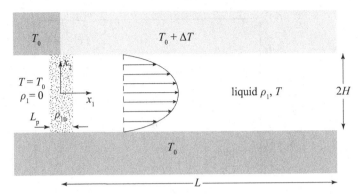

Figure 11.7 Schematic diagram of a thermal field-flow fractionation (F^3) channel. For $t < 0$, the liquid is pure solvent ($\rho_1 = 0$) and has uniform temperature T_0. At $t = 0$, a liquid plug having concentration ρ_{10} and length L_p is injected at the channel entrance ($x_1 = 0$), and for $t \geq 0$ a temperature difference ΔT is maintained between the channel walls.

where we have used (11.31). The temperature equation comes from (6.8), which, for an ideal mixture (see Exercise 10.2), can be written as

$$\frac{\partial T}{\partial t} + \frac{3}{2}V\left[1 - \left(\frac{x_2}{H}\right)^2\right]\frac{\partial T}{\partial x_1} = -\frac{1}{\rho \hat{c}_p}\boldsymbol{\nabla} \cdot \boldsymbol{j}'_q, \qquad (11.33)$$

where we have again used (11.31). As discussed in Section 6.3, the diffusive fluxes on the right-hand sides of (11.32) and (11.33) are coupled. For a dilute ideal mixture, we write (6.18) and (6.19) as

$$\boldsymbol{j}_1 = -D_{12}\boldsymbol{\nabla}\rho_1 - D_T\rho_1\boldsymbol{\nabla}T, \qquad (11.34)$$

$$\boldsymbol{j}'_q = -\frac{\tilde{R}}{\tilde{M}_1}D_T T^2\boldsymbol{\nabla}\rho_1 - \lambda\boldsymbol{\nabla}T, \qquad (11.35)$$

where $D_T = D_{q1}/T$ is the thermal diffusion coefficient (cf. footnote on p. 92).

Substitution of (11.34) into (11.32) and of (11.35) into (11.33), taking D_{12}, λ, and D_T to be constants, leads to

$$\frac{\partial \rho_1}{\partial t} + \frac{3}{2}V\left[1 - \left(\frac{x_2}{H}\right)^2\right]\frac{\partial \rho_1}{\partial x_1} = D_{12}\nabla^2\rho_1 + D_T\boldsymbol{\nabla}\cdot(\rho_1\boldsymbol{\nabla}T), \qquad (11.36)$$

$$\frac{\partial T}{\partial t} + \frac{3}{2}V\left[1 - \left(\frac{x_2}{H}\right)^2\right]\frac{\partial T}{\partial x_1} = \chi\nabla^2 T + \left(\frac{\tilde{R}D_T}{\rho\hat{c}_p\tilde{M}_1}\right)\boldsymbol{\nabla}\cdot(T^2\boldsymbol{\nabla}\rho_1). \qquad (11.37)$$

Our evolution equations for solute concentration and temperature, because of the Soret and Dufour effect (terms with D_T), are a coupled and nonlinear set of equations. As a first step towards simplifying these equations, we recognize that, for typical liquid mixtures, $N_{\text{Le}} = \chi/D_{12} \gg 1$, where N_{Le} is the Lewis number (see Table 7.1). This implies that for large times relative to the time scale for mass diffusion, and for large distances along the channel compared with the thermal entrance length, terms on the left-hand side of (11.37) can be neglected. We also recognize that the imposed temperature difference ΔT will dominate the temperature field, and that through the Soret effect term in (11.36) a temperature gradient drives diffusive mass transfer. The resulting concentration gradient, through the Dufour effect term in (11.37), drives diffusive energy transfer. However, because we consider dilute mixtures ($w_{10} \ll 1$), this is a higher-order effect that we assume to be negligible. In this case, only the conduction term in (11.37) survives, that is $\nabla^2 T = 0$. Integration subject to the boundary conditions $T(x_1, -H, t) = T_0$ and $T(x_1, H, t) = T_0 + \Delta T$ gives

$$T = T_0 + \frac{H + x_2}{2H} \Delta T. \tag{11.38}$$

Hence, the case we consider neglects the transport of energy by convection and by the Dufour effect, which leads to a uniform temperature gradient throughout the channel.

Substitution of (11.38) into (11.36) leads to

$$\frac{\partial \rho_1}{\partial t} + \frac{3}{2} V \left[1 - \left(\frac{x_2}{H}\right)^2 \right] \frac{\partial \rho_1}{\partial x_1} = D_{12} \left(\frac{\partial^2 \rho_1}{\partial x_1^2} + \frac{\partial^2 \rho_1}{\partial x_2^2} \right) + \left(\frac{D_T \, \Delta T}{2H} \right) \frac{\partial \rho_1}{\partial x_2}. \tag{11.39}$$

The initial condition for (11.39) can, for $L_{\mathrm{p}} \to 0$, be written as[14]

$$\rho_1(x_1, x_2, 0) = \frac{M_{10}}{2HB} \delta(x_1), \tag{11.40}$$

where $M_{10} = 2HBL_{\mathrm{p}} \rho_{10}$ is the initial mass of solute in the plug injected at $t = 0$. The boundary conditions for (11.39) are given by

$$\rho_1(\infty, x_2, t) = 0, \tag{11.41}$$

$$D_{12} \frac{\partial \rho_1}{\partial x_2}(x_1, \pm H, t) = - \left(\frac{D_T \, \Delta T}{2H} \right) \rho_1(x_1, \pm H, t). \tag{11.42}$$

The boundary conditions in (11.42) follow from (11.34) and (11.38) since the upper and lower surfaces of the channel are taken to be impermeable.

---

[14] As shown in Figure 11.7, $\rho_1(x_1, x_2, 0) = [\mathcal{H}(x_1 + L_{\mathrm{p}}/2) - \mathcal{H}(x_1 - L_{\mathrm{p}}/2)]\rho_{10}$, where $\mathcal{H}(x)$ is the Heaviside step function

$$\mathcal{H}(x) = \begin{cases} 0 & \text{for} \quad x < 0, \\ 1 & \text{for} \quad x \geq 0. \end{cases}$$

Note that $\lim_{\epsilon \to 0} [\mathcal{H}(x + \epsilon) - \mathcal{H}(x - \epsilon)]/(2\epsilon) = \delta(x)$, or $(d/dx)[\mathcal{H}(x)] = \delta(x)$.

Note the similarity of these boundary conditions and those in (11.17) for the ultracentrifuge analysis.

The linear system of equations (11.39)–(11.42) describes the evolution of the solute concentration $\rho_1 = \rho_1(x_1, x_2, t)$ in a thermal field-flow fractionation device. Applying the $n - m$ procedure, we have the quantities $\rho_1, x_1, x_2, t, V, D_{12}, D_T \Delta T, H, M_{10}/B$, so that $n - m = 9 - 3 = 6$, which means the model involves two dimensionless parameters. For our characteristic mass (density), length, and time, we choose $M_{10}/(2H^2B)$, H, and H^2/D_{12}, respectively. One parameter is the now-familiar Péclet number given by $N_{\text{Pe}} = 3VH/(2D_{12})$ and the other is $\Lambda_T = D_T \Delta T/(2D_{12})$, which indicates the strength of the applied thermal field. As we did in Section 9.1, we further rescale the flow-direction coordinate by N_{Pe} (indicated by an overbar). An analytic solution of (11.39)–(11.42) has been found[15] using generalized dispersion theory.[16] This solution has the form

$$\rho_1 = \sum_{j=0}^{\infty} f_j \frac{\partial^j \langle \rho_1 \rangle}{\partial \bar{x}_1^j}, \tag{11.43}$$

where $f_j = f_j(x_2, t)$, and $\langle \rho_1 \rangle = \langle \rho_1 \rangle(\bar{x}_1, t)$ is the transverse-direction-average solute concentration defined as

$$\langle \rho_1 \rangle = \frac{1}{2} \int_{-1}^{1} \rho_1 \, dx_2. \tag{11.44}$$

We see from (11.43) that the functions f_j carry information about the effect of the velocity distribution in the transverse direction on the average concentration. In typical F^3 applications, the average solute concentration $\langle \rho_1 \rangle$, which can be measured optically, is the quantity of interest, in which case there is no need to know the functions f_j. The evolution equation for $\langle \rho_1 \rangle$ is given by

$$\frac{\partial \langle \rho_1 \rangle}{\partial t} = \sum_{i=1}^{\infty} K_i \frac{\partial^i \langle \rho_1 \rangle}{\partial \bar{x}_1^i} \approx K_1 \frac{\partial \langle \rho_1 \rangle}{\partial \bar{x}_1} + K_2 \frac{\partial^2 \langle \rho_1 \rangle}{\partial \bar{x}_1^2}, \tag{11.45}$$

where the coefficients $K_1 = K_1(t)$ and $K_2 = K_2(t)$ are found from f_0 and f_1 (see Exercise 11.6). A discussion on the validity of the truncation of the sum in (11.45) after two terms can be found in the cited references. Comparing with (2.1), we see that (11.45) is a diffusion equation with drift velocity $-K_1$ and diffusion coefficient $2K_2$.

To keep things simple, here we consider the long-time solution, which gives accurate results for $t \gtrsim 1$. In this case, K_1, K_2 are independent of time, and

15 Krishnamurthy & Subramanian, *Sep. Sci.* **12** (1977) 347.
16 Gill & Sankarasubramanian, *Proc. Roy. Soc. A* **316** (1970) 341.

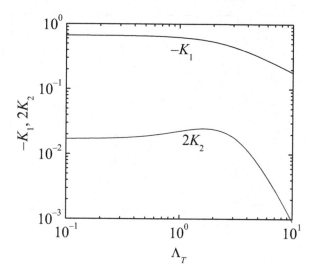

Figure 11.8 Dependence of the coefficients $-K_1$ and K_2 with $N_{\text{Pe}} \gg 1$ on $\Lambda_T = D_T\,\Delta T/(2D_{12})$ given by (11.47) and (11.48) from dispersion theory solution.

the solution of (11.45) is given by

$$\langle \rho_1 \rangle = \frac{1}{\sqrt{4\pi K_2 t}} \exp\left[-\frac{(\bar{x}_1 + K_1 t)^2}{4K_2 t}\right]. \tag{11.46}$$

The reader should recognize the expression in (11.46) as a Gaussian probability density (distribution) with time-dependent mean $-K_1 t$ and variance $2K_2 t$ (see Section 2.3). Expressions for f_0 and f_1, which are lengthy and not given here, lead to the following expressions for the coefficients:

$$-K_1 = \frac{2}{\Lambda_T^2}[\Lambda_T \coth(\Lambda_T) - 1] = \frac{2}{3} - \frac{2}{45}\Lambda_T^2 + O(\Lambda_T^4), \tag{11.47}$$

and

$$K_2 = \frac{1}{N_{\text{Pe}}^2} + \frac{4}{\Lambda_T^4 \sinh(\Lambda_T)}\left[\frac{2\Lambda_T^2}{3\sinh(\Lambda_T)} - \frac{10\cosh(\Lambda_T)}{\Lambda_T} + \frac{14\sinh(\Lambda_T)}{\Lambda_T^2}\right.$$
$$\left. + \frac{2}{\sinh(\Lambda_T)} - \frac{2\Lambda_T\cosh(\Lambda_T)}{\sinh^2(\Lambda_T)} + \frac{4\cosh^2(\Lambda_T)}{\sinh(\Lambda_T)} + 6\sinh(\Lambda_T)\right]$$
$$= \frac{1}{N_{\text{Pe}}^2} + \frac{8}{945} + \frac{4}{135}\Lambda_T + O(\Lambda_T^3). \tag{11.48}$$

The dependence of K_1 and K_2 on Λ_T is shown in Figure 11.8.

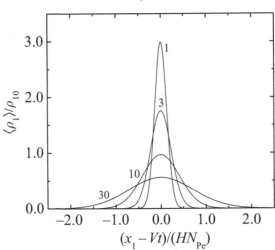

Figure 11.9 Time evolution of the radial-average solute concentration field for laminar flow in a channel without temperature gradient $\Lambda_T = 0$ from (11.46)–(11.48) with $N_{Pe} \gg 1$ in a coordinate system that moves with the average velocity of the fluid for different times $tD_{12}/H^2 = 1, 3, 10, 30$.

The evolution equation for $\langle \rho_1 \rangle$ in (11.45) is known as a *dispersion equation*, and the coefficient K_2 plays the role of a diffusion coefficient. However, from (11.48) we see that, even if diffusion in the flow direction is negligible ($N_{Pe} \gg 1$) and there is no applied field ($\Lambda_T = 0$), there is a diffusion-like process in the flow direction, that is $K_2 \neq 0$. This phenomenon, which is commonly referred to as *Taylor dispersion*, results from the interplay between transverse diffusion and a nonuniform velocity in the flow direction. Taylor found an approximate solution (see Exercise 11.8) to the problem of dispersion in a tube,[17] which was generalized by Aris.[18]

To better understand the phenomenon of dispersion, we consider flow in a channel for $N_{Pe} \gg 1$ with $\Lambda_T = 0$. This is equivalent to eliminating the terms containing $\partial^2 \rho_1 / \partial x_1^2$ and $\partial \rho_1 / \partial x_2$ in (11.39). It will be convenient to examine $\langle \rho_1 \rangle$ in a coordinate system that moves with the average velocity of the fluid. In this case, we apply (11.19) with $\boldsymbol{c} = -Vt\boldsymbol{\delta}_1$, so that $\bar{x}'_1 = \bar{x}_1 - (2/3)t$. Figure 11.9 shows the time evolution of $\langle \rho_1 \rangle$ for laminar flow in a channel. From this figure we see that the solute injected at the channel entrance at $t = 0$ has a Gaussian distribution that is stationary (in the coordinate system moving with the average fluid velocity)

[17] Taylor, *Proc. Roy. Soc. A* **219** (1953) 186.
[18] Aris, *Proc. Roy. Soc. A* **67** (1956) 235.

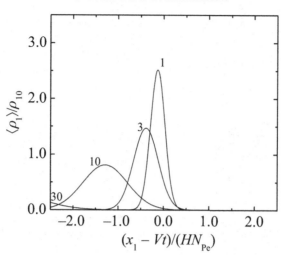

Figure 11.10 Time evolution of the radial-average solute concentration field for laminar flow in a channel with temperature gradient $\Lambda_T = 2$ from (11.46)–(11.48) with $N_{Pe} \gg 1$ in a coordinate system that moves with the average velocity of the fluid for different times $tD_{12}/H^2 = 1, 3, 10, 30$.

and becomes broader as time increases. This broadening is due entirely to dispersion since here we have neglected flow-direction diffusion. It should be noted that this interpretation requires a sufficiently long residence time in the channel for transverse diffusion to take place. As we are considering the case $tD_{12}/H^2 \gtrsim 1$, this is equivalent to $Vt/H \gtrsim N_{Pe}$.

If we re-introduce dimensional variables in (11.45), the coefficient in front of the second-order derivative, which is known as the *dispersion coefficient*, is given by $K_d = K_2 D_{12} N_{Pe}^2$. Note that K_d depends not only on the diffusion coefficient, but also on the average velocity and channel dimension. If axial diffusion is included, the expression for the dispersion coefficient is given by $K_d = D_{12}[1 + (8/945)N_{Pe}^2]$, that, since $N_{Pe} = 3VH/(2D_{12})$, has a nonmonotonic dependence on D_{12}. Dispersion experiments where $\langle \rho_1 \rangle$ is determined as a function of time or position are commonly used for the measurement of mass diffusivities.

Figure 11.10 shows the time evolution of the average solute concentration field for thermal F^3 from (11.46)–(11.48). The application of a transverse temperature gradient, which induces a solute flux towards the lower channel wall, causes the solute on average to move slower than the average velocity of the fluid. For the case shown in this figure ($\Lambda_T = 2$), we see that the transverse field has a pronounced effect on convective transport

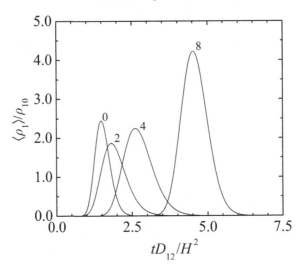

Figure 11.11 Elution curves at channel location $x_1/(HN_{\mathrm{Pe}}) = 1$ in the thermal F^3 method from (11.46)–(11.48) with $N_{\mathrm{Pe}} \gg 1$ for $\Lambda_T = 0, 2, 4, 8$.

(K_1), and a relatively small effect on dispersion (K_2). If D_{12} is known from a separate experiment (e.g., dispersion), then analysis of a thermal F^3 experiment of the type shown in Figure 11.10 with known ΔT can be used to obtain Λ_T, or the thermal diffusion coefficient $D_T = D_{q1}/T$ for the solute.

As mentioned above, F^3 techniques are often used for the separation of mixtures. In this case, liquid is collected from a downstream location in the channel as a function of time. Figure 11.11 shows the time dependence of $\langle \rho_1 \rangle$ at a particular channel location $x_1/(HN_{\mathrm{Pe}}) = 1$ for several values of Λ_T. The curves in this figure, sometimes called elution curves or fractograms, show that the time required for the solute to pass through the channel increases with increasing Λ_T. Consider the injection into the channel of a mixture having several different solutes that is sufficiently dilute that the pseudo-binary approximation can be applied (see Section 10.2). Separation of such a mixture can be achieved if each solute has a different value of Λ_T. In the case of polymers and colloids, D_{12} decreases with increasing (effective) size of the solute, while $D_T = D_{q1}/T$ is relatively insensitive to size. This means that, for a given ΔT, each solute species will have a different Λ_T, and therefore will exit the channel at a different time as shown in Figure 11.11. Based on this principle, thermal F^3 has been used to separate colloidal particles having different sizes (10–10^3 nm) and polymers having different molecular weights.

Exercise 11.6   Derivation of the Dispersion Equation

It can be shown, by integration of (11.39) over the channel thickness, that the coefficients K_i appearing in the dispersion equation (11.45) are given by

$$K_i = \frac{\delta_{i2}}{N_{\mathrm{Pe}}^2} - \frac{1}{2} \int_{-1}^{1} f_{i-1}(1 - x_2^2) dx_2.$$

Use (11.43) and (11.45) to show that the equations governing the f_j are given by

$$\frac{\partial f_j}{\partial t} + \sum_{i=1}^{j} K_i f_{j-i} + (1 - x_2^2) f_{j-1} = \frac{1}{N_{\mathrm{Pe}}^2} f_{j-2} + \frac{\partial^2 f_j}{\partial x_2^2} + \Lambda_T \frac{\partial f_j}{\partial x_2}.$$

For $t \gg 1$, solve this equation for f_0 and use this to find the expression in (11.47).

Exercise 11.7   Retention Ratio in Field-Flow Fractionation

It is common to characterize F^3 experiments by assuming the solute concentration profile becomes independent of $\bar{x}_1$ and time. The experiment is then described by the retention ratio R_{ret}, defined as

$$R_{\mathrm{ret}} = \frac{\frac{1}{2}\int_{-1}^{1} v_1 \rho_1 \, dx_2}{\frac{1}{2}\int_{-1}^{1} v_1 \, dx_2 \, \frac{1}{2}\int_{-1}^{1} \rho_1 \, dx_2} = \frac{\frac{1}{2}\int_{-1}^{1}(1 - x_2^2)\rho_1 \, dx_2}{\frac{2}{3}\langle \rho_1 \rangle},$$

which indicates the increase in residence time of the solute due to the transverse field. For this case, find the solute concentration and describe the significance of $1/\Lambda_T$, which is called the retention parameter. Also, show that $R_{\mathrm{ret}} = -2K_1$.

Exercise 11.8   Dispersion in Flow through a Tube

Dispersion of a solute in laminar flow through a tube is governed by

$$\frac{\partial \rho_1}{\partial t} + v_z \frac{\partial \rho_1}{\partial z} = D_{12} \left[\frac{1}{r}\frac{\partial}{\partial r}\left(r \frac{\partial \rho_1}{\partial r}\right) + \frac{\partial^2 \rho_1}{\partial z^2} \right],$$

where $v_z = 2V[1 - (r/R)^2]$. Taylor obtained an approximate solution by writing $\rho_1 = \langle \rho_1 \rangle(z) + \delta \rho_1(r, z)$, where $\langle \rho_1 \rangle = 2/R^2 \int_0^R \rho_1 r \, dr$. He further assumed for $t \gtrsim R^2/D_{12}$ that $|\delta \rho_1| \ll \langle \rho_1 \rangle$. Use this approach to show that the dispersion coefficient is given by $K_d = D_{12}[1 + (1/48)N_{\mathrm{Pe}}^2]$.

Exercise 11.9   Electric Field-Flow Fractionation

The experiments shown in Figure 11.6 were obtained using electric F^3. Use the Nernst–Planck equation (6.45) and (11.32) to formulate the analog to (11.39) for isothermal electric F^3. Identify the parameter $\Lambda_\phi = z_1 \Delta\phi_{\mathrm{el}} \tilde{M}_1/(2\tilde{R}T)$, where $\Delta\phi_{\mathrm{el}}$ is the applied electric field potential across the channel.

- The pressure gradient required to balance the centrifugal force resulting from large rotation rates in an ultracentrifuge leads to an additional driving force for mass diffusion in liquid mixtures.
- An ultracentrifuge can be used to separate mixtures or study their properties based on density differences between the components.
- It is sometimes advantageous to apply the hydrodynamic equations in non-inertial coordinate systems, which, for the equations of motion, involves additional inertial forces such as the centrifugal and Coriolis forces.
- Field-flow fractionation is a family of methods where a liquid mixture flowing through a channel is subjected to a transverse field that interacts with the components of the mixture.
- Thermal field-flow fractionation can be used to separate mixtures or study their properties based on the Soret effect.
- Dispersion is a diffusion-like phenomenon in the direction of flow that occurs in liquid mixtures, which results from the interaction of convective transport having a nonuniform velocity distribution with diffusive transport in the direction transverse to the flow.

12

Complex Fluids

So far we have characterized the transport properties of fluids entirely in terms of the transport coefficients, shear viscosity, bulk viscosity, thermal conductivity, and, if we are dealing with a multi-component system, diffusion coefficients. This description is remarkably universal and the corresponding Navier–Stokes–Fourier and diffusion equations turned out to be impressively rich, even when solved subject to simple boundary conditions.

However, there exist many fluids that cannot be described by the transport equations we have developed so far. For example, Newton's law of viscosity does not account for the phenomenon of viscoelasticity, which needs to be considered in a variety of different fields such as polymer processing, food technology, or biological systems. The occurrence of viscoelasticity is related to slowly relaxing structural features in complex fluids, which we describe by additional structural variables such as the average overall shape of polymer molecules. The choice of structural variables and the formulation of evolution equations for them is much less universal than for hydrodynamics with the densities of the conserved quantities mass, momentum, and energy and the corresponding balance equations. We hence develop some basic knowledge for modeling fluids with microstructure, or *complex fluids*, and solving transport problems for them. Combining this new understanding of complex fluids with our previous knowledge of transport phenomena, we develop as a modern application a set of hydrodynamic equations for relativistic fluids that is consistent with all the principles of special relativity and nonequilibrium thermodynamics.

12.1 Importance of Soft Matter

Pierre-Gilles de Gennes, who may be considered as a founding father of soft matter physics, began his Nobel Lecture (1991) with the words "What do we mean by soft matter? Americans prefer to call it 'complex fluids.' This is a rather ugly name, which tends to discourage the young students. But it does indeed bring in two of the major features [complexity, flexibility]." For the brief overview given in this section, we rely on two excellent textbooks,[1] the titles of which confirm de Gennes' observation.

Why did we choose the "discouraging" title complex fluids for the present chapter? An emphasis on fluids and hence on momentum transport seems to be appropriate for a textbook on transport phenomena. However, in the present section, we seize the opportunity to present some encouraging motivation for developing the tools needed to investigate complex fluids or soft matter.

Soft matter includes polymers, colloids, emulsions, foams, gels, liquid crystals, granular materials, surfactants, and other materials that possess flow-dependent microstructures. The unifying feature is that soft matter contains molecular or structural units much larger than atoms. In some cases we deal with very large molecules; in other cases soft matter self-organizes into large physical structures. The large structural units possess long relaxation times that lead to the characteristic slow response of soft matter. Slow structural variables are associated with nonequilibrium behavior on experimentally accessible time scales so that soft matter provides an ideal playground for nonequilibrium thermodynamics.

A further characteristic of soft matter is that small applied forces lead to large responses. Drastic changes in mechanical properties are a typical feature and, as emphasized before, soft matter shows complex behavior between fluids and solids. The energies required for deformation are of the order of the thermal energy at room temperature so that fluctuations become relevant. Large deformations lead to nonlinear behavior, and nonequilibrium thermodynamics in the nonlinear domain is needed to describe the dynamics of soft matter, including the corresponding theory of fluctuations. Enormous progress in developing the experimental and theoretical tools for investigating soft matter has been made in the last few decades. As a result, soft condensed matter can now be discussed on the same sound physical basis as solid condensed matter.

Soft materials are important in a wide range of technologies and have a promising potential in many future technologies, such as medical and environmental applications. They may be used as structural and packaging materials, adhesives, detergents and cosmetics, pharmaceuticals, paints, foods and food additives, lubricants and fuel additives, electrically or magnetically

---

[1] Larson, *Structure and Rheology of Complex Fluids* (Oxford, 1999); Doi, *Soft Matter Physics* (Oxford, 2013).

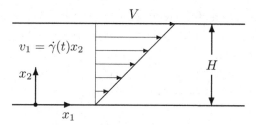

Figure 12.1 Time-dependent shear flow between parallel plates with shear rate $\dot{\gamma}(t)$.

responsive material (e.g., in liquid crystal displays), and so on. The development of smart materials is increasingly based on soft matter components, where self-healing materials provide a most prominent example. Moreover, a number of biological materials can be classified as soft matter. Even when soft matter properties are not relevant in the final product, the transport properties of soft matter may often influence the processability of materials, most importantly in the plastics and ceramics industries.

In view of the above characteristics of complex fluids, we need to discuss the material properties characterizing their flow behavior that are of interest to us. For that purpose, we first present some selected basic ideas to describe non-Newtonian flow behavior. We start with linear viscoelasticity as a first step towards understanding behavior that can range from fluid- to solid-like. In view of the large responses expected from soft matter, we then introduce some material functions that characterize nonlinear material behavior. *Rheology*,[2] the study of the flow behavior of matter, is a highly developed and still rapidly progressing research field of great practical importance. In the present chapter, we restrict ourselves to giving only a flavor of its essence, and provide some basic tools that are used to describe the rheology of soft condensed matter. In doing so, we heavily rely on the textbook *Beyond Equilibrium Thermodynamics*.[3]

12.2 Linear Viscoelasticity

Let us consider homogeneous shear flow with the velocity field given by $v_1 = \dot{\gamma}x_2$ as shown in Figure 12.1, where the shear rate $\dot{\gamma}$ is given by the ratio of the upper plate velocity V to the distance between the plates H. Here we are interested in flows with time-dependent shear rate $\dot{\gamma} = \dot{\gamma}(t)$, such as the

[2] Morrison, *Understanding Rheology* (Oxford, 2001); Barnes, Hutton & Walters, *Introduction to Rheology* (Elsevier, 1989).

[3] See Section 4.1 of Öttinger, *Beyond Equilibrium Thermodynamics* (Wiley, 2005).

start-up of steady shear flow where $\dot{\gamma}(t) = 0$ for $t < 0$, and $\dot{\gamma}(t) = \dot{\gamma}$ for $t \geq 0$. Note that, for any fluid, a finite time is required for the velocity to obtain the linear profile shown in Figure 12.1. For a Newtonian fluid this time is $t \gg H^2/\nu$ (see Section 7.2). Recall the rheological constitutive equation for the incompressible flow of a Newtonian fluid (6.6), that is $\boldsymbol{\tau} = -\eta[\boldsymbol{\nabla v} + (\boldsymbol{\nabla v})^{\mathrm{T}}]$. Hence, the shear stress $\tau_{12}(t)$ at any time would be proportional to the instantaneous shear rate $\dot{\gamma}(t)$, that is $\tau_{12}(t) = 0$ for $t < 0$ and $\tau_{12}(t) = -\eta\dot{\gamma}$ for $t \geq 0$. For a complex fluid, the measured shear stress will not jump immediately to the steady-state value at the start-up of the steady shear flow. If the shear rate $\dot{\gamma}$ is sufficiently small, then we expect an expression of the linear form

$$\tau_{12}(t) = -\int_{-\infty}^{t} G(t - t')\dot{\gamma}(t')dt', \tag{12.1}$$

where $G(t)$ is the shear relaxation modulus. Equation (12.1) is the fundamental equation of linear viscoelasticity. For the stress growth in start-up of steady shear flow, we obtain from (12.1) for $t \geq 0$

$$\tau_{12}(t) = -\dot{\gamma}\int_{0}^{t} G(t - t')dt' = -\dot{\gamma}\int_{0}^{t} G(t')dt'. \tag{12.2}$$

When, for large times t, a steady state is reached, we can identify the steady-state viscosity as

$$\eta_0 = \int_{0}^{\infty} G(t')dt'. \tag{12.3}$$

For shorter times, we would find a stress growth as sketched in Figure 12.2, where have included a Newtonian contribution with viscosity η_∞. In a stress-growth experiment, one can measure the integral of $G(t')$, or stress-growth coefficient,

$$\eta^+(t) = \int_{0}^{t} G(t')dt'. \tag{12.4}$$

In cessation of steady flow, that is $\dot{\gamma}(t) = \dot{\gamma}$ for $t < 0$ and $\dot{\gamma}(t) = 0$ for $t \geq 0$, one can measure the stress decay coefficient $\eta^-(t) = \eta_0 - \eta^+(t)$.

If we want to measure the modulus $G(t)$ directly, we need to perform a small instantaneous shear step of size γ, described by $\dot{\gamma}(t) = \gamma\delta(t)$. After the step, we find a shear stress proportional to the shear relaxation modulus

$$\tau_{12}(t) = -\gamma\int_{-\infty}^{t} G(t - t')\delta(t')dt' = -\gamma G(t). \tag{12.5}$$

Although we have shown how the material property $G(t)$ can be measured

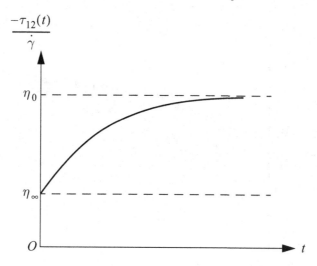

Figure 12.2 Stress growth in start-up of steady shear flow.

in principle, it is quite difficult in practice to produce steps in shear strain or strain rate. It often is advantageous to avoid abrupt steps and to perform oscillatory experiments, $\gamma(t) = \gamma_0 \sin(\omega t)$, where $\gamma_0 \ll 1$ is the strain amplitude and ω is the angular frequency. In oscillatory flow (see Exercise 12.1), one can measure the storage modulus

$$G'(\omega) = \omega \int_0^\infty G(t) \sin(\omega t) dt \qquad (12.6)$$

and loss modulus

$$G''(\omega) = \omega \int_0^\infty G(t) \cos(\omega t) dt. \qquad (12.7)$$

A *complex modulus* can be defined as

$$G^*(\omega) = G'(\omega) + iG''(\omega) = i\omega \int_0^\infty G(t) e^{-i\omega t} dt. \qquad (12.8)$$

If one prefers to leave out the factor $i\omega$ in front of the integral in (12.8), one can introduce the *complex viscosity*

$$\eta^*(\omega) = \int_0^\infty G(t) e^{-i\omega t} dt. \qquad (12.9)$$

The time integrals from 0 to ∞ in the preceding formulas can be extended to integrals from $-\infty$ to ∞ because we have $G(t) = 0$ for $t < 0$.

The dynamic modulus for three polystyrene melts having narrow molecular-weight distributions but different (weight-averaged) molecular

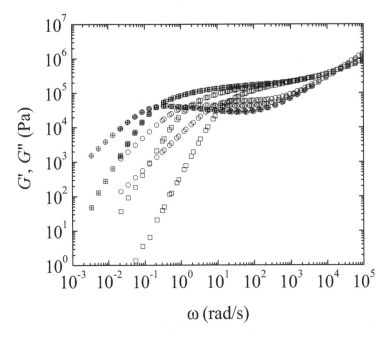

Figure 12.3 Shear storage G' (squares) and loss G'' (circles) moduli versus frequency for three polystyrene melts at $180\,^\circ$C: $\tilde{M}_{\mathrm{w}} = 100$ kDa (open); $\tilde{M}_{\mathrm{w}} = 200$ kDa (dots); $\tilde{M}_{\mathrm{w}} = 400$ kDa (crosses), from experiments performed by Teresita Kashyap at IIT, Chicago.

weights ($\tilde{M}_{\mathrm{w}}$) are shown in Figure 12.3. Notice that at large frequencies, which probe the dynamics of the polymer chain on small time and length scales, the dynamic moduli are independent of molecular weight. At intermediate frequencies, the dynamic modulus is relatively insensitive to frequency, which is known as the rubbery or elastic response. The width of the rubbery region increases with increasing molecular weight due to the increase in the density of entanglements, or temporary crosslinks, between polymer chains. The frequency at which the loss modulus equals the storage modulus is roughly equal to the inverse of the longest relaxation time of the polymer chain. In what is known as the viscous response, at smaller frequencies the dynamics of the entire chain determines the dependence of the moduli on frequency. The behavior shown in Figure 12.3 is characteristic for entangled polymer melts, and it is the basis for molecular theories for polymer dynamics.[4]

---

[4] Doi, *Soft Matter Physics* (Oxford, 2013).

From the theory of Fourier transforms, we obtain

$$G(t) = \frac{1}{2\pi} \int_{-\infty}^{\infty} \frac{G^*(\omega)}{i\omega} e^{i\omega t} \, d\omega, \tag{12.10}$$

and, in particular,

$$G(0) = \frac{1}{2\pi} \int_{-\infty}^{\infty} \frac{G^*(\omega)}{i\omega} \, d\omega = \frac{1}{\pi} \int_{0}^{\infty} \frac{G''(\omega)}{\omega} \, d\omega. \tag{12.11}$$

Note that at $t = 0$, $G(t)$ jumps from $G(0-) = 0$ to a finite value $G(0+)$, and, by Fourier transformation, we pick up the mean value $G(0+)/2$ so that we have

$$G(0+) = \frac{2}{\pi} \int_{0}^{\infty} \frac{G''(\omega)}{\omega} \, d\omega. \tag{12.12}$$

Equation (12.12) is an example of a *sum rule*. In general, sum rules express the relaxation modulus and its derivatives in the limit $t \downarrow 0$ in terms of moments of $G'(\omega)$ or $G''(\omega)$. Such sum rules are important for checking the consistency of experimental data or phenomenological models. Such consistency checks can be particularly useful when one realizes that quantities such as $G(0+)$ are amenable to equilibrium statistical mechanics.

In the start-up, step-strain, and oscillatory experiments considered so far, we specified a time-dependent shear deformation. A further option is not to prescribe the time dependence of the deformation, but it may be advantageous to control the stress and to measure the resulting deformation. Again, various types of experiments are possible, such as creep and recoil experiments.[5]

It is moreover possible to treat other types of flow than shear. For that purpose, (12.1) can be generalized to the integral constitutive equation of linear viscoelasticity by writing the symmetric pressure tensor as

$$\boldsymbol{\tau}(t) = - \int_{-\infty}^{t} G(t - t') \dot{\boldsymbol{\gamma}}(t') dt', \tag{12.13}$$

where $\dot{\boldsymbol{\gamma}}$ is the symmetrized velocity gradient tensor, that is,

$$\dot{\boldsymbol{\gamma}} = \boldsymbol{\nabla} \boldsymbol{v} + (\boldsymbol{\nabla} \boldsymbol{v})^{\mathrm{T}}. \tag{12.14}$$

Assuming incompressibility, we have here neglected the divergence of the velocity field. Using integration by parts and setting $G(\infty) = 0$, (12.13) can be expressed in the alternative form

$$\boldsymbol{\tau}(t) = -\int_{-\infty}^{t} M(t - t')\boldsymbol{\gamma}(t, t')dt', \tag{12.15}$$

where $M(t - t') = dG/dt'$ is the memory function, and $\boldsymbol{\gamma}(t, t')$ is the strain tensor, given by

$$\boldsymbol{\gamma}(t, t') = \int_{t'}^{t} \dot{\boldsymbol{\gamma}}(t'')dt''. \tag{12.16}$$

The strain tensor can also be obtained from the gradient of the displacement vector $\boldsymbol{u}(\boldsymbol{r}, t, t') = \boldsymbol{r}' - \boldsymbol{r}$, where $\boldsymbol{r}'$ and $\boldsymbol{r}$ are the positions of a particle at past time t' and present time t, respectively, as $\boldsymbol{\gamma} = \boldsymbol{\nabla u} + (\boldsymbol{\nabla u})^{\mathrm{T}}$. We see from (12.15) and (12.16) that the stress is given by past strains, relative to the strain at the present time, weighted by the memory function.

The relaxation modulus for a purely viscous fluid can be written as $G(t) = \eta_\infty \delta(t)$, in which case (12.13) gives (6.6), the constitutive equation for a Newtonian fluid. For the case of a constant relaxation modulus $G(t) = G_0$, we can integrate (12.13) to obtain

$$\boldsymbol{\tau}(t) = -G_0\boldsymbol{\gamma}(t, -\infty), \tag{12.17}$$

which is the constitutive equation for a linear elastic solid, known as Hooke's law. Hence, we see that the constitutive equation of linear viscoelasticity given in (12.13), or equivalently in (12.15), includes as limiting cases the constitutive equations for a linearly viscous (Newtonian) fluid and a linearly elastic (Hookean) solid.

Exercise 12.1   Complex Modulus from Oscillatory Shear Flow
Oscillatory shear flow is defined by $\gamma(t) = \gamma_0 \sin(\omega t)$. From (12.1), derive the expressions in (12.6) and (12.7) for G' and G''. Also, show that the shear stress in this flow can be written as

$$\tau_{12} = -\gamma_0 |G^*| \sin(\omega t + \delta), \qquad \tan\delta = \frac{G''(\omega)}{G'(\omega)}, \tag{12.18}$$

where $|G^*|$ is the magnitude of the complex modulus in (12.8), and δ is the phase angle between the stress and applied strain.

Exercise 12.2   Decomposition of Complex Viscosity
From (12.9), derive integral expressions for the real-valued, frequency-dependent viscosities $\eta'(\omega)$ and $\eta''(\omega)$ in

$$\eta^*(\omega) = \eta'(\omega) - i\eta''(\omega). \tag{12.19}$$

Exercise 12.3   Elongational Flows

When a fluid is elongated in the x_1-direction with the elongational rate $\dot{\epsilon}$, incompressibility requires that the fluid contracts in the two transverse directions with half of that rate. This is known as simple elongational flow and is defined by the symmetric velocity gradient tensor

$$\boldsymbol{\nabla v} = (\boldsymbol{\nabla v})^{\mathrm{T}} = \frac{1}{2}\dot{\boldsymbol{\gamma}} = \begin{pmatrix} \dot{\epsilon} & 0 & 0 \\ 0 & -\dot{\epsilon}/2 & 0 \\ 0 & 0 & -\dot{\epsilon}/2 \end{pmatrix}. \tag{12.20}$$

Determine the linear viscoelastic viscosity growth coefficient in simple elongation

$$\eta_{\mathrm{E}}^{+}(t) = -\frac{\tau_{11}(t) - \tau_{33}(t)}{\dot{\epsilon}} \tag{12.21}$$

for start-up of steady elongational flow at time $t = 0$. Similarly, when a fluid is elongated in the x_1- and x_2-directions with the elongational rate $\dot{\epsilon}_{\mathrm{B}}$, incompressibility requires that the fluid contracts in the x_3-direction with twice that rate. This is known as equibiaxial elongational flow and is defined by

$$\boldsymbol{\nabla v} = (\boldsymbol{\nabla v})^{\mathrm{T}} = \frac{1}{2}\dot{\boldsymbol{\gamma}} = \begin{pmatrix} \dot{\epsilon}_{\mathrm{B}} & 0 & 0 \\ 0 & \dot{\epsilon}_{\mathrm{B}} & 0 \\ 0 & 0 & -2\dot{\epsilon}_{\mathrm{B}} \end{pmatrix}. \tag{12.22}$$

Determine the linear viscoelastic viscosity growth coefficient in equibiaxial elongation

$$\eta_{\mathrm{B}}^{+}(t) = -\frac{\tau_{11}(t) - \tau_{33}(t)}{\dot{\epsilon}_{\mathrm{B}}} \tag{12.23}$$

for start-up of steady elongational flow at time $t = 0$. Briefly discuss the relationship between these two elongational deformations in terms of $\dot{\epsilon}$ and $\dot{\epsilon}_{\mathrm{B}}$.

All the physically important information is contained in the memory integral (12.1), so that all other equations of this section are about the consequences of (12.1) in various types of time-dependent flow situations. In principle, all of these equations contain equivalent information. In practice, however, different types of experiments come with different limitations. If one wants to collect information about the flow behavior of a material over the widest possible range of times or frequencies, it is important to bring together all the information available from different experiments. One hence needs to know all the different material functions of this section and how they are related.

A most convenient concept to perform the various transformations is the *relaxation spectrum*, which can be introduced through the assumption

$$G(t) = \eta_{\infty}\delta(t) + \sum_{j}\frac{\eta_j}{\lambda_j}e^{-t/\lambda_j}, \tag{12.24}$$

Table 12.1 *Material functions of linear viscoelasticity*

Linear viscoelastic material function (in shear)	Expression in terms of relaxation spectrum	General nomenclature*
Relaxation modulus $G(t)$	$\eta_\infty \delta(t) + \sum_j \frac{\eta_j}{\lambda_j} e^{-t/\lambda_j}$	Response function
Stress-growth coeff. $\eta^+(t)$	$\eta_\infty + \sum_j \eta_j (1 - e^{-t/\lambda_j})$	
Stress-decay coeff. $\eta^-(t)$	$\sum_j \eta_j e^{-t/\lambda_j}$	Relaxation function
Complex modulus $G^*(\omega)$	$i\omega\eta_\infty + \sum_j \frac{i\omega\eta_j}{1+i\omega\lambda_j}$	
Storage modulus $G'(\omega)$	$\sum_j \eta_j \frac{\omega^2\lambda_j}{1+\omega^2\lambda_j^2}$	
Loss modulus $G''(\omega)$	$\eta_\infty\omega + \sum_j \frac{\eta_j\omega}{1+\omega^2\lambda_j^2}$	
Complex viscosity $\eta^*(\omega)$	$\eta_\infty + \sum_j \frac{\eta_j}{1+i\omega\lambda_j}$	Complex admittance
Dynamic viscosity $\eta'(\omega)$	$\eta_\infty + \sum_j \frac{\eta_j}{1+\omega^2\lambda_j^2}$	
$\eta''(\omega)$	$\sum_j \eta_j \frac{\omega\lambda_j}{1+\omega^2\lambda_j^2}$	

*According to Section 3.1 of Kubo *et al.*, *Statistical Physics II* (Springer, 1991).

where $j = 1, 2, \ldots$ labels a finite or infinite number of relaxation times. The η_∞-term, which should be considered as the limit of $(\eta_\infty/\lambda_\infty)\exp(-t/\lambda_\infty)$ for $\lambda_\infty \to 0$, describes an instantaneous Newtonian viscosity contribution, as can be seen by calculating

$$\eta^+(t) = \eta_\infty + \sum_j \eta_j (1 - e^{-t/\lambda_j}). \tag{12.25}$$

In (12.25), mode j contributes with a rubber-like term $(\eta_j/\lambda_j)t$ for $t \ll \lambda_j$ and with a Newtonian liquid-like term η_j for $t \gg \lambda_j$. For steady state, we obtain

$$\eta_0 = \eta_\infty + \sum_j \eta_j. \tag{12.26}$$

If in (12.24) we allow $\lambda_j \to \infty$, while keeping $\eta_j/\lambda_j \to G_0$ finite, we obtain an elastic contribution to the stress as in (12.17).

The relationships between the various material functions of linear viscoelasticity discussed in this section in terms of the relaxation spectrum are summarized in Table 12.1. The material functions associated with stress-controlled experiments cannot be expressed directly in terms of the relaxation spectrum; rather, a retardation spectrum is needed. One should realize from Table 12.1 that all the material functions contain equivalent information, except for $\eta^-(t)$, $G'(\omega)$, and $\eta''(\omega)$, which are lacking the information about η_∞. In particular, the real and imaginary parts of the complex viscosity, or more precisely $\eta'(\omega) - \eta_\infty$ and $\eta''(\omega)$, contain equivalent information. This equivalence is a consequence of the fact that the relaxation modulus $G(t)$ is fully determined by its values on the positive real axis.[6] The explicit relationship between $\eta'(\omega) - \eta_\infty$ and $\eta''(\omega)$ is an example of the *Kramers–Kronig relations*,[7]

$$\eta'(\omega) - \eta_\infty = \frac{2}{\pi} \int_0^\infty \frac{\omega' \eta''(\omega') - \omega \eta''(\omega)}{\omega'^2 - \omega^2} d\omega', \qquad (12.27)$$

and

$$\eta''(\omega) = \frac{2\omega}{\pi} \int_0^\infty \frac{\eta'(\omega') - \eta'(\omega)}{\omega^2 - \omega'^2} d\omega'. \qquad (12.28)$$

The determination of relaxation spectra from any of the material functions in Table 12.1, which may be considered as a first step toward transformations, is much more subtle than one might expect at first sight. In mathematical terms, the determination of a spectrum is an *ill-posed problem*, that is, drastically different spectra can be used to fit the data for some linear viscoelastic material function equally well. Two extreme strategies seem to be particularly useful. One involves choosing a minimal set of modes by placing the relaxation times and selecting the weights of the corresponding modes such that a given data set can be fitted with the desired precision.[8] This approach is, for example, useful when one needs a rheological model for efficient numerical flow calculations, and experimental results for linear viscoelastic material functions should be reproduced faithfully. The other approach involves the construction of a continuous (in practice, sufficiently dense) spectrum of relaxation times, subject to certain smoothness conditions.[9] This approach is, for example, useful for theoretical investigations or for material characterization. As the determination of spectra is an ill-posed problem, a number of approximate but direct conversion formulas between the material functions in Table 12.1 can be useful for practical purposes.[10]

---

[6] $G(t)$ is identically zero on the negative axis; the existence of a relationship between $\eta'(\omega) - \eta_\infty$ and $\eta''(\omega)$ thus is a consequence of *causality*. In other words, the stress in (12.1) cannot depend on future strain rates but is fully determined by the flow history.

[7] A guided derivation of the Kramers–Kronig relations can be found in Problem 5D.2 on pp. 290–291 of Bird, Armstrong & Hassager, *Dynamics of Polymeric Liquids. Volume 1. Fluid Mechanics* (Wiley, 1987); in fact, Exercise 12.5 may be considered as an alternative derivation of a Kramers–Kronig relation.

[8] Winter, *J. Non-Newtonian Fluid Mech.* **68** (1997) 225.

[9] Honerkamp & Weese, *Rheol. Acta* **32** (1993) 65.

[10] See, for example, Chapter 4 of Ferry, *Viscoelastic Properties of Polymers* (Wiley, 1980), Chapter 8 of Tschoegl, *Linear Viscoelastic Behavior* (Springer, 1989), or Section 8.8 of Schwarzl, *Polymermechanik* (Springer, 1990).

A useful tool for the practical determination of linear viscoelastic material functions is the time–temperature-superposition principle. It is based on the idea of accounting for the effect of temperature by rescaling all relaxation times λ_j by the *same* temperature-dependent factor (see Exercise 12.6). The time–temperature superposition is frequently employed to extend the effective frequency range by varying the temperature, and it helps to reduce significantly the amount of experimental work required for a complete rheological characterization of viscoelastic materials. In a typical experiment, three to four decades of frequency (or time) can be covered. The dynamic modulus data shown in Figure 12.3, which cover roughly seven decades in frequency, were obtained from measurements made at several temperatures and shifted to a reference temperature.

The theory of linear viscoelasticity holds for small deformations $\gamma_0 \ll 1$. Different types of responses to deformation (i.e., viscous, elastic) can conveniently be classified using the famous Deborah number, which is the ratio of the (longest) relaxation time of the fluid to the characteristic time of the flow $N_{\mathrm{De}} = \lambda/\tau_{\mathrm{flow}}$. The characteristic time for a flow τ_{flow}, which may be thought of as the time of observation, for an oscillatory deformation would be ω^{-1}. For small (large) N_{De}, the response is viscous (elastic), while for $N_{\mathrm{De}} \approx 1$, the response is viscoelastic. Linear viscoelastic behavior is also observed if the product of the deformation rate and the longest relaxation time, which is known as the Weissenberg number, $N_{\mathrm{Wi}} = \dot{\gamma}\lambda$, is small. That is, a large total deformation is irrelevant since the fluid "remembers" only a small part of the deformation. The plot shown in Figure 12.4, which is known as a Pipkin diagram,[11] nicely demonstrates the relationship between different types of fluid responses and the Deborah and Weissenberg numbers.

Exercise 12.4   Material Functions in Terms of Spectra
Verify the expressions for the material functions of linear viscoelasticity given in Table 12.1.

Exercise 12.5   Application of Kramers–Kronig Relations
Calculate $\eta''(\omega)$ from $\eta'(\omega)$ in Table 12.1 by means of the Kramers–Kronig relation (12.28).

Exercise 12.6   Time–Temperature Superposition
Assume that the temperature dependence of all relation times is (approximately)

---

[11] Pipkin, *Lectures on Viscoelastic Theory* (Springer, 1972).

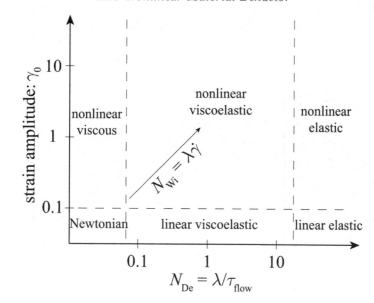

Figure 12.4 Pipkin diagram showing the dependence of the material response on the Deborah, N_{De}, and Weissenberg, N_{Wi}, numbers.

given by

$$\lambda_j(T) = a_T \lambda_j(T_0), \tag{12.29}$$

where the factor a_T depends also on the reference temperature T_0, and that the ratios $\eta_j(T)/\lambda_j(T)$ occurring as weights in the spectrum are independent of temperature. Derive the following relationships:

$$G(t;T) = G(t/a_T;T_0), \tag{12.30}$$

$$G'(\omega;T) = G'(a_T\omega;T_0), \tag{12.31}$$

and

$$G''(\omega;T) = G''(a_T\omega;T_0). \tag{12.32}$$

How does the viscosity (12.3) depend on temperature?

12.3 Nonlinear Material Behavior

For sufficiently large and rapid deformations, most complex fluids exhibit a strongly nonlinear behavior (see Figure 12.4). For example, the steady-state viscosity in shear flow depends on the shear rate, as shown in Figure 12.5 for a polystyrene melt.[12] The shear-rate dependence of the viscosity, referred to

[12] Schweizer *et al.*, *Rheol. Acta* **47** (2008) 943.

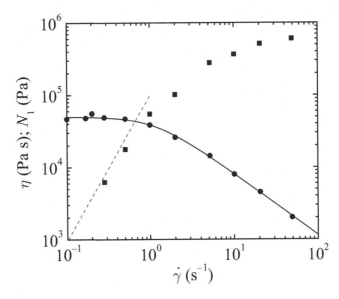

Figure 12.5 Steady-state viscosity η and first normal-stress difference N_1 versus shear rate for a polystyrene melt at 180 °C ($\tilde{M}_w$ =200 kDa), from Schweizer *et al.*, *Rheol. Acta* **47** (2008) 943. The solid curve is the fit of the expression given in (12.39).

as shear thinning when the viscosity is a decreasing function of shear rate, is a pronounced effect with important consequences. By running processing equipment at high flow rates, one not only obtains higher output rates, but also can drastically reduce the viscosity and hence the energy dissipation (at least, the energy dissipation does not increase quadratically with shear rate). For high shear rates, one often finds a power-law decay of the viscosity with an exponent smaller than unity; for typical polymeric liquids, the exponent is found experimentally to be between 0.4 and 0.9.

Another important nonlinear effect is illustrated in Figure 12.6. For a Newtonian fluid, a steady shear flow between parallel plates is maintained by a constant shear stress. For a complex fluid, one typically needs to exert an additional normal stress in order to keep the plates at the prescribed distance. The intuitive explanation for the need for such a normal force is as follows: for a polymeric fluid, the molecules in the shear flow of Figure 12.6 are more stretched in the main flow direction than in transverse directions; these macromolecules hence naturally tend to retract, thus contracting the entire material in the main flow direction. The normal pressure on the plates in Figure 12.6 is needed to prevent the material from contracting in the main flow direction by pushing the plates apart. The occurrence of

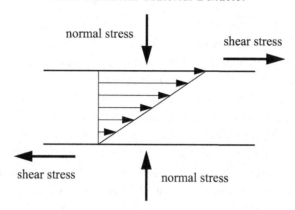

Figure 12.6 Occurrence of normal-stress differences in shear flow of viscoelastic liquids.

normal-stress differences is another characteristic of a liquid with elastic features, previously recognized in memory effects.

Going from a simple to a complex fluid, an extra pressure tensor of the more general form

$$\boldsymbol{\tau} = \begin{pmatrix} \tau_{11} & \tau_{12} & 0 \\ \tau_{12} & \tau_{22} & 0 \\ 0 & 0 & \tau_{33} \end{pmatrix}, \tag{12.33}$$

instead of the Newtonian form (6.6) with equal diagonal elements, is expected for shear flow in the x_1–x_2 plane. While no additional off-diagonal entries can occur for symmetry reasons, in general, all diagonal elements can be different. The off-diagonal entries are used to define the shear-rate dependent viscosity,

$$\tau_{12} = -\eta(\dot{\gamma})\dot{\gamma}. \tag{12.34}$$

Since we have assumed incompressibility, from (6.3) we have $\boldsymbol{\pi} = p^L\boldsymbol{\delta} + \boldsymbol{\tau}$, where p^L is the pseudo-pressure. Hence, only differences in the diagonal elements of (12.33) are of interest, which can be expressed in terms of the first and second normal-stress differences N_1 and N_2,

$$N_1 = \tau_{11} - \tau_{22} = -\Psi_1(\dot{\gamma})\dot{\gamma}^2 \tag{12.35}$$

and

$$N_2 = \tau_{22} - \tau_{33} = -\Psi_2(\dot{\gamma})\dot{\gamma}^2, \tag{12.36}$$

or, equivalently, by the first and second normal-stress coefficients Ψ_1 and Ψ_2. These definitions take into account that we have $\tau_{11} = \tau_{22} = \tau_{33}$ at equilibrium, and that, for symmetry reasons, differences between the diagonal components of the pressure tensor at small shear rates are usually quadratic in the shear rate. Therefore, the zero-shear-rate limits of Ψ_1 and Ψ_2 are finite. The rheological material properties in shear flow, η, Ψ_1, and Ψ_2, are collectively known as the *viscometric functions*.

The shear-rate dependence of the viscosity (see Exercise 12.9) and the occurrence of normal-stress differences, which also depend on the shear rate, are typical and important nonlinear effects in viscoelastic fluids.[13] We hence need to explain these nonlinear effects in addition to the enormously high viscosities and large relaxation times. We need to understand qualitatively how viscoelasticity, shear thinning, and normal-stress differences arise. As a next step, we need quantitative predictions for the rheological material properties, ideally in terms of the molecular architecture.

Exercise 12.7   Mean Relaxation Time from Viscometric Functions
Although the first normal-stress coefficient occurs as an intrinsically nonlinear material property, continuum mechanical considerations[14] show that Ψ_1 can be expressed in terms of the shear relaxation modulus,

$$\Psi_{10} = 2 \int_0^\infty t G(t) dt. \tag{12.37}$$

Discuss the significance of $\lambda_{\mathrm{avg}} = \Psi_{10}/(2\eta_0)$ as a mean relaxation time.

Exercise 12.8   Cone and Plate Rheometry
A commonly used geometry for measuring the viscometric functions η, Ψ_1, and Ψ_2 defined in (12.34)–(12.36) is known as cone and plate flow. This flow is created by placing a fluid between a flat plate and a conical surface with radius R that makes an angle β with the plate. The plate is stationary and the cone rotates with angular velocity Ω. The velocity field in cone and plate flow takes the form $v_r = v_\theta = 0, v_\phi = r\sin(\theta)w(\theta)$, where $w(\theta)$ is an unknown function. Typically $\beta \ll 1$, so that the shear rate within the fluid is approximately uniform. Derive relationships between the torque on the plate and η, between the pressure distribution on the plate and N_1 and N_2, and between the total force on the plate and N_1.

---

[13] See Bird & Curtiss, *Physics Today* **37/1** (1984) 36 and Chapter 2 of Bird, Armstrong & Hassager, *Dynamics of Polymeric Liquids. Volume 1. Fluid Mechanics* (Wiley, 1987).

[14] See Section 9.6 on the memory-integral expansion in Bird, Armstrong & Hassager, *Dynamics of Polymeric Liquids. Volume 1. Fluid Mechanics* (Wiley, 1987), in particular Example 9.6-1 and Section 6.2 on the retarded-motion expansion, giving the expression (6.2-6) for Ψ_1.

Exercise 12.9   Generalized Newtonian Fluids

For incompressible, purely viscous fluids, a commonly used expression for the extra stress tensor can be written as

$$\boldsymbol{\tau} = -\eta(\dot{\gamma})\dot{\boldsymbol{\gamma}}, \tag{12.38}$$

where $\eta(\dot{\gamma})$ is a function of $\dot{\gamma} = \sqrt{(\dot{\boldsymbol{\gamma}} : \dot{\boldsymbol{\gamma}})/2}$, which is known as the second invariant of $\dot{\boldsymbol{\gamma}}$. The rheological constitutive equation in (12.38) is a nonlinear relationship between the stress tensor and the velocity gradient. Because of its similarity to (6.5), it is called a generalized Newtonian fluid. A useful, three-parameter model for $\eta(\dot{\gamma})$ known as the Carreau model is given by[15]

$$\eta(\dot{\gamma}) = \eta_0[1 + (\lambda\dot{\gamma})^2]^{\frac{n-1}{2}}, \tag{12.39}$$

where η_0 is the zero-shear rate viscosity and n the power-law index. For $\lambda\dot{\gamma} \gg 1$, the widely used "power-law model," which can be written as $\eta(\dot{\gamma}) = K\dot{\gamma}^{n-1}$, is recovered ($K = \eta_0\lambda^{n-1}$). Find the parameters η_0, λ, and n that fit the steady-state viscosity data shown in Figure 12.5.

Exercise 12.10   Yield Stress Fluids

A simple model for a material possessing a yield stress proposed by Oldroyd[16] is given by

$$\boldsymbol{\tau} = \begin{cases} -G\boldsymbol{\gamma} & \text{for} \quad \sqrt{\text{tr}(\boldsymbol{\tau} \cdot \boldsymbol{\tau})/2} < 2\sigma, \\ -\left[\eta + \sigma/\sqrt{\text{tr}(\dot{\boldsymbol{\gamma}} \cdot \dot{\boldsymbol{\gamma}})/2}\right]\dot{\boldsymbol{\gamma}} & \text{for} \quad \sqrt{\text{tr}(\boldsymbol{\tau} \cdot \boldsymbol{\tau})/2} \geq 2\sigma, \end{cases} \tag{12.40}$$

where σ is the yield stress. In the limit $G \to \infty$, the classical Bingham fluid[17] model is recovered, so that the material remains in an undeformed state ($\boldsymbol{\gamma} = \mathbf{0}$) for $\sqrt{\text{tr}(\boldsymbol{\tau} \cdot \boldsymbol{\tau})} < 2\sigma$.

Consider the pressure-driven flow of a Bingham fluid in an infinitely wide (in the x_3-direction) channel with height $2B$ and length L. For incompressible flow, the velocity field has the form $v_1 = v_1(x_2), v_2 = v_3 = 0$, and the pressure drop over the length of the channel is $-\Delta p^L = p^L(0) - p^L(L) > 0$. For $|x_2| < b$, the stress is below the yield stress and the fluid will have a uniform velocity. Derive an expression for the velocity distribution $v_1(x_2)$.

12.4 Evolution Equations for Structural Variables

So far, all our evolution equations have been balance equations, and, as such, they had to possess a very particular structure. To capture the memory effects of viscoelastic fluids, we need evolution equations for suitable

[15] Carreau, *Trans. Soc. Rheol.* **16** (1972) 99.
[16] Oldroyd, *Proc. Cambridge Phil. Soc.* **43** (1947) 100.
[17] Bingham, *Fluidity and Plasticity* (McGraw-Hill, 1922).

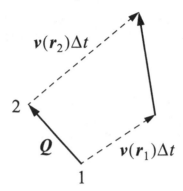

Figure 12.7 Translation and deformation of a vector by convection.

structural variables which characterize the internal state of a complex fluid undergoing flow and are not related to conserved quantities. The simplest idea is to use the pressure tensor as a structural variable. As we have no guidance from balance equations, we need to be careful with formulating proper time derivatives. As we have no conservation laws, we should expect relaxation to become important in addition to transport.

Consider Figure 12.7 to see what happens to a vector $Q = r_2 - r_1$ when convected with a velocity field $v(r)$. The figure suggests that the vector is not just transported but also deformed. More quantitatively, we have

$$Q \to r_2 + v(r_2)\Delta t - r_1 - v(r_1)\Delta t = Q + Q \cdot \nabla v\, \Delta t = Q + \kappa \cdot Q\, \Delta t, \quad (12.41)$$

with the transposed velocity gradient tensor $\kappa = (\nabla v)^{\mathrm{T}}$. As we have previously incorporated the translation effect into the concept of a material derivative, we now include also the deformational effect into a modified time derivative by defining

$$Q_{(1)} = \frac{\partial Q}{\partial t} + v \cdot \nabla Q - \kappa \cdot Q, \quad (12.42)$$

which is known as the *convected time derivative* of the vector Q. The convective time derivative of vectors can easily be generalized to tensors by considering the dyadic $\alpha = QQ$, which suggests

$$\alpha_{(1)} = Q_{(1)}Q + QQ_{(1)} = \frac{\partial \alpha}{\partial t} + v \cdot \nabla \alpha - \kappa \cdot \alpha - \alpha \cdot \kappa^{\mathrm{T}}. \quad (12.43)$$

With this kind of derivative, we can now approach the problem of formulating an evolution equation for the pressure tensor describing complex fluids undergoing flow. The analysis of the behavior of structural variables under the space transformations associated with convection is an important tool in nonequilibrium thermodynamics, which is closely related to the representation of reversible dynamics in terms of so-called Poisson brackets.[18]

In order to combine viscous and elastic effects, we write the following differential equation for the pressure tensor $\boldsymbol{\tau}$,

$$\lambda \boldsymbol{\tau}_{(1)} + \boldsymbol{\tau} = -\eta(\boldsymbol{\kappa} + \boldsymbol{\kappa}^{\mathrm{T}}), \tag{12.44}$$

where λ is the relaxation time and η is the viscosity. For $\lambda = 0$, we recover Newton's expression (6.5) for the extra pressure tensor for incompressible fluids, that is, stress is proportional to the strain rate. For $\lambda \to \infty$, the first term on the left-hand side dominates and the rate of change of stress is proportional to the rate of change of strain, or stress proportional to strain, so that we recover a Hookean solid with modulus $G = \eta/\lambda$. For intermediate λ, we have the desired combination of viscous and elastic behavior. With the definition (12.43), the rheological constitutive equation (12.44) can be rewritten as

$$\frac{\partial \boldsymbol{\tau}}{\partial t} = -\boldsymbol{v} \cdot \boldsymbol{\nabla} \boldsymbol{\tau} + \boldsymbol{\kappa} \cdot \boldsymbol{\tau} + \boldsymbol{\tau} \cdot \boldsymbol{\kappa}^{\mathrm{T}} - G(\boldsymbol{\kappa} + \boldsymbol{\kappa}^{\mathrm{T}}) - \frac{1}{\lambda} \boldsymbol{\tau}. \tag{12.45}$$

This evolution equation represents the famous *convected Maxwell model* for the pressure tensor.[19] Whereas (12.44) looks linear, the more explicit reformulation in (12.45) reveals the nonlinear effects implied by the convected time derivative, both through translation and through deformation. Therefore, models of this type are occasionally referred to as quasilinear. The last term in (12.45) represents a linear relaxation term.

As a first application of our new rheological constitutive equation (12.45), we consider the homogeneous shear flow of Figure 12.1 with a pressure tensor of the form (12.33), where the components may depend on time but not on position. In this situation, (12.45) for the components of the pressure tensor becomes

$$\lambda \frac{d\tau_{12}}{dt} = \lambda \dot{\gamma} \left(\tau_{22} - G \right) - \tau_{12}, \tag{12.46}$$

$$\lambda \frac{d\tau_{11}}{dt} = 2\lambda \dot{\gamma} \, \tau_{12} - \tau_{11}, \tag{12.47}$$

and

$$\lambda \frac{d\tau_{22}}{dt} = -\tau_{22}, \qquad \lambda \frac{d\tau_{33}}{dt} = -\tau_{33}. \tag{12.48}$$

[18] Öttinger, *Beyond Equilibrium Thermodynamics* (Wiley, 2005).
[19] More precisely, we should refer to it as the *upper convected Maxwell model*; in addition to the upper convected derivative (12.42), there exists a lower convected derivative associated with the deformation effect on the cross product of two vectors, with the corresponding generalization to tensors.

The components τ_{22} and τ_{33} relax to zero, no matter how we choose the time-dependent shear rate $\dot{\gamma}$. As these components cannot arise from any shear-flow history, we assume $\tau_{22} = \tau_{33} = 0$. Then, (12.46) simplifies to

$$\lambda \frac{d\tau_{12}}{dt} = -\lambda\dot{\gamma}\, G - \tau_{12}. \tag{12.49}$$

For the special case of shear flow, (12.49) and (12.47) of the Maxwell model for τ_{12} and τ_{11} become linear. They nevertheless go beyond linear viscoelasticity because an equation for the normal stress τ_{11} arises.

The steady-state solution of (12.47) and (12.49) for constant shear rate is given by $\tau_{12} = -\lambda\dot{\gamma}\, G$ and $\tau_{11} = -2(\lambda\dot{\gamma})^2\, G$. The definitions (12.34)–(12.36) of the viscometric functions hence imply

$$\eta = \lambda G, \qquad \Psi_1 = 2\lambda^2 G, \qquad \Psi_2 = 0. \tag{12.50}$$

Note that these results imply $\Psi_1/(2\eta) = \lambda$ and hence nicely support the ideas of Exercise 12.7. To expand on this check, we look at a time-dependent shear flow with $\dot{\gamma}(t) = \gamma\delta(t)$, which allows us to find the relaxation modulus according to (12.5). For $t > 0$, this implies free relaxation of the components τ_{12} and τ_{11} with relaxation time λ so that we need know only the initial condition. By integrating (12.49) from $0-$ to $0+$ we obtain the initial condition $\tau_{12}(0+) = -\gamma G$ and hence

$$\tau_{12}(t) = -\gamma\, G\, e^{-t/\lambda}, \tag{12.51}$$

which implies the shear relaxation modulus $G(t) = G\, e^{-t/\lambda}$. Then, the result of calculating Ψ_1 from (12.37) coincides with the result given in (12.50).

In start-up of steady shear flow, we need to solve (12.47) and (12.49) for constant shear rate $\dot{\gamma}$ with the initial conditions $\tau_{12}(0) = \tau_{22}(0) = 0$. The solutions of these linear ordinary differential equations are given by

$$\tau_{12}(t) = -\lambda\dot{\gamma}\, G\, (1 - e^{-t/\lambda}) \tag{12.52}$$

and

$$\tau_{11}(t) = -2(\lambda\dot{\gamma})^2 G\left[1 - \left(1 + \frac{t}{\lambda}\right) e^{-t/\lambda}\right]. \tag{12.53}$$

Equation (12.50) demonstrates that the convected Maxwell model can qualitatively account for effects of viscosity and the first normal-stress coefficient. However, these material properties are predicted to be independent of the shear rate, which is not realistic. A nonlinear relaxation term with a relaxation time becoming smaller for larger stress or strain rate would be required for a more realistic prediction. Of course, also the parameter η or G

can depend on stress or strain rate. To obtain a nonzero second normal-stress coefficient, an anisotropic relaxation term would be required.

For homogeneous simple elongational flow (see Exercise 12.3), we can write (12.45) for the components τ_{11} and $\tau_{22} = \tau_{33}$ of the pressure tensor as

$$\lambda \frac{d\tau_{11}}{dt} = 2\lambda\dot{\epsilon}\,(\tau_{11} - G) - \tau_{11} \tag{12.54}$$

and

$$\lambda \frac{d\tau_{22}}{dt} = -\lambda\dot{\epsilon}\,(\tau_{22} - G) - \tau_{22}. \tag{12.55}$$

The steady-state solutions to these equations are

$$\tau_{11} = -\frac{2\lambda G}{1 - 2\lambda\dot{\epsilon}}\,\dot{\epsilon} \qquad \tau_{22} = \frac{\lambda G}{1 + \lambda\dot{\epsilon}}\,\dot{\epsilon}. \tag{12.56}$$

Note that the normal-stress difference diverges at the finite dimensionless strain rates $\lambda\dot{\epsilon} = 1/2$ and $\lambda\dot{\epsilon} = -1$. The linear relaxation of the Maxwell model is too weak to avoid such singular behavior.

Exercise 12.11   Conformational Entropy Balance
Assume that the conformational contribution to the entropy per unit mass is of the form

$$\hat{s}(\boldsymbol{\tau}) = \frac{k_{\mathrm{B}}}{m}\,f(\boldsymbol{c}) \qquad \text{with} \quad \boldsymbol{c} = \boldsymbol{\delta} - \frac{1}{G}\boldsymbol{\tau}, \tag{12.57}$$

where m is the mass of a structural unit (such as a polymer molecule). Derive the conformational entropy balance equation

$$\frac{\partial \hat{s}}{\partial t} = -\boldsymbol{v} \cdot \boldsymbol{\nabla}\hat{s} + 2\frac{k_{\mathrm{B}}}{m}\left(\boldsymbol{c} \cdot \frac{\partial f}{\partial \boldsymbol{c}}\right) : \boldsymbol{\kappa} - \frac{1}{\lambda}\frac{k_{\mathrm{B}}}{m}\frac{\partial f}{\partial \boldsymbol{c}} : (\boldsymbol{c} - \boldsymbol{\delta}). \tag{12.58}$$

Discuss the three terms on the right-hand side of (12.58). Argue why $f(\boldsymbol{c})$ must be of the functional form

$$f(\boldsymbol{c}) = \frac{1}{2}\left[\mathrm{tr}(\boldsymbol{\delta} - \boldsymbol{c}) + \ln(\det \boldsymbol{c})\right], \tag{12.59}$$

and why the dimensionless conformation tensor $\boldsymbol{c}$ must be positive-definite.

Exercise 12.12   Oldroyd B Fluids
A widely used rheological constitutive equation known as the Oldroyd B fluid is obtained by the superposition of a "polymer" stress $\boldsymbol{\tau}^{\mathrm{p}}$ from the convected Maxwell model and a Newtonian "solvent" stress given by $\boldsymbol{\tau}^{\mathrm{s}} = -\eta^{\mathrm{s}}\dot{\boldsymbol{\gamma}}$. Show, using (12.44) with viscosity η^{p}, that by writing $\boldsymbol{\tau} = \boldsymbol{\tau}^{\mathrm{s}} + \boldsymbol{\tau}^{\mathrm{p}}$ the following is obtained,

$$\lambda\boldsymbol{\tau}_{(1)} + \boldsymbol{\tau} = -\eta\left[\dot{\boldsymbol{\gamma}} + \lambda\frac{\eta^{\mathrm{s}}}{\eta^{\mathrm{p}}}\dot{\boldsymbol{\gamma}}_{(1)}\right], \tag{12.60}$$

where $\eta = \eta^s + \eta^p$. We will see in Chapter 22 that a simple molecular model gives the rheological constitutive equation in (12.60).

Exercise 12.13   Second-Order Fluids

One approach to developing nonlinear rheological constitutive equations for purely viscous fluids is to write the extra stress as a polynomial expansion in terms of time derivatives of the rate of strain tensor. The zeroth-order derivative is $\dot{\gamma}$, while higher-order derivatives are obtained from the convected time derivative given in (12.43). In general, these equations are known as retarded motion expansions, and, if the expansion is truncated at second order, the following rheological constitutive equation is obtained,

$$\boldsymbol{\tau} = -[b_1 \dot{\gamma} + b_2 \dot{\gamma}_{(1)} + b_{11}(\dot{\gamma} \cdot \dot{\gamma})], \tag{12.61}$$

where b_1, b_2, and b_{11} are material parameters. For the homogeneous shear flow in Figure 12.1, obtain expressions for the viscometric functions η, Ψ_1, and Ψ_2 defined in (12.34)–(12.36) in terms of the parameters in the second-order fluid equation (12.61).

12.5 Rayleigh Problem

So far, we have not solved any transport problem for a viscoelastic fluid yet. We have merely calculated the stress response in time-dependent or steady homogeneous shear or elongational flows. As a real flow calculation, we consider the so-called Rayleigh problem. A semi-infinite domain $(x_2 \geq 0)$ filled with a viscoelastic fluid with constant density ρ is bounded by a planar surface at $x_2 = 0$. Before $t = 0$, the fluid and the surface are at rest; after $t = 0$, the surface moves with constant velocity V in the positive x_1-direction. Under isothermal conditions, we wish to find out how the momentum propagates into the viscoelastic fluid. Is the mechanism significantly different from that for a Newtonian fluid? This and a number of further flow problems are solved in a comprehensive textbook on the dynamics of polymeric liquids.[20]

We assume that the solution is given by a nonhomogeneous shear flow, that is $v_1 = v_1(x_2, t)$ and $v_2 = v_3 = 0$. This assumption implies incompressibility. The momentum balance equation amounts to

$$\rho \frac{\partial v_1}{\partial t} = -\frac{\partial \tau_{12}}{\partial x_2}, \tag{12.62}$$

and we are again faced with (12.47) and (12.49). Here, we have partial

---

[20] See Chapter 7 of Bird, Armstrong & Hassager, *Dynamics of Polymeric Liquids. Volume 1. Fluid Mechanics* (Wiley, 1987).

instead of ordinary time derivatives and a space- and time-dependent shear rate $\dot{\gamma} = \partial v_1/\partial x_2$. Hence, we write (12.49) as

$$\lambda \frac{\partial \tau_{12}}{\partial t} = -\eta \frac{\partial v_1}{\partial x_2} - \tau_{12}. \tag{12.63}$$

Combining (12.62) and (12.63), we obtain

$$\frac{\partial v_1}{\partial t} + \lambda \frac{\partial^2 v_1}{\partial t^2} = \frac{\eta}{\rho} \frac{\partial^2 v_1}{\partial x_2^2}. \tag{12.64}$$

For a Newtonian fluid with $\lambda = 0$, we have a simple diffusion equation with diffusion coefficient η/ρ; the solution satisfying the boundary conditions can be expressed in terms of the error function. For a viscoelastic fluid with $\lambda > 0$, the character of a wave equation with propagation speed $\sqrt{\eta/(\lambda\rho)}$ appears, which dominates for large λ. Our problem involves the quantities v_1, x_2, t, λ, η/ρ, V, so that $m = 6$, $n = 2$ (length and time), and $n - m = 4$. Therefore, the solution $v_1 = v_1(x_2, t)$ depends on only one parameter. If we use units of length and time such that $\lambda = \eta/\rho = 1$, (12.64) becomes

$$\frac{\partial v_1}{\partial t} + \frac{\partial^2 v_1}{\partial t^2} = \frac{\partial^2 v_1}{\partial x_2^2}. \tag{12.65}$$

The only parameter of our problem is given by the dimensionless velocity $V\sqrt{\lambda\rho/\eta}$ imposed at $x_2 = 0$ for $t > 0$. As (12.65) is linear in v_1, we can easily avoid the explicit occurrence of this parameter by looking at the normalized velocity $\phi = v_1/V$, for which the problem does not involve any parameter.

In order to obtain an ordinary differential equation, we introduce the Laplace transform $\bar{\phi}(x_2, s)$ of $\phi(x_2, t)$, as previously defined in (3.22). Equation (12.65) is then found to be equivalent to

$$(s + s^2)\, \bar{\phi} = \frac{\partial^2 \bar{\phi}}{\partial x_2^2}, \tag{12.66}$$

with the boundary condition $\bar{\phi}(0, s) = 1/s$. For $x_2 \to \infty$, the velocity field vanishes, giving the second boundary condition $\bar{\phi}(\infty, s) = 0$. The desired solution hence is

$$\bar{\phi}(x_2, s) = \frac{1}{s} e^{-\sqrt{s+s^2}\, x_2}. \tag{12.67}$$

The inverse Laplace transform can be expressed in terms of a modified Bessel function of the first kind, I_1. A nonzero velocity is obtained only for $x_2 < t$, and it is given by

$$\frac{v_1(x_2, t)}{V} = e^{-x_2/2} + \frac{x_2}{4} \int_0^t \frac{I_1\left(\frac{1}{2}\sqrt{y^2 - x_2^2}\right)}{\sqrt{y^2 - x_2^2}} e^{-y/2}\, dy. \tag{12.68}$$

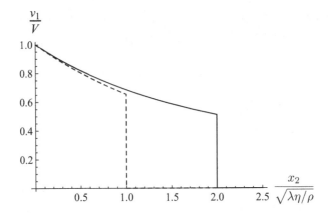

Figure 12.8 The velocity profile (12.68) for $t/\lambda = 1$ (dashed line) and $t/\lambda = 2$ (continuous line).

Note that the propagation speed in the chosen units of length and time is 1. The time-dependent velocity profiles shown in Figure 12.8 illustrate the finite propagation speed of the velocity perturbation in a Maxwell fluid.

12.6 Relativistic Hydrodynamics

In this section, we explain why relativistic hydrodynamics is related to complex fluids, and we present a full covariant set of hydrodynamic equations. Such equations are important for modeling phenomena such as the time evolution of the universe, galaxy formation, transport processes in a gravitational collapse, or processes occurring in high-energy heavy-ion collisions. The relativistic generalization turns out to be less universal than classical hydrodynamics, which is entirely based on conservation laws, so that simplicity and thermodynamic consistency become major issues in formulating proper evolution equations.

All the non-relativistic balance equations of hydrodynamics contain first-order time derivatives and second-order space derivatives associated with the diffusive smoothing of composition, velocity, or temperature fields by diffusivity, viscosity, or thermal conductivity. As time and space are recognized to be on an equal footing in relativity, the equations of relativistic hydrodynamics must be fundamentally different from the non-relativistic ones.

Note that the two space derivatives of the non-relativistic equations express very different physics: one of them is related to balancing and expresses the divergence of a flux; the other derivative results from fluxes proportional to thermodynamic driving forces, which are gradients of intensive thermo-dynamic variables. It is hence natural to avoid second-order space deriva-tives by using a different type of constitutive equations for the fluxes while keeping the divergences of fluxes. We assume that dissipation arises from the relaxation of fluxes rather than directly from transport processes. The momentum and energy flux, or closely related tensor and vector variables, should be considered as additional independent variables, just like the struc-tural variables of complex fluids. In the same way as the pressure tensor is used as an additional variable in the preceding section, here also the heat flux vector becomes an additional structural variable. We have already en-countered a finite propagation speed in Figure 12.8, which, in relativistic hydrodynamics, should be given by the speed of light.

We do not assume that our typical reader has a deep background in rela-tivity. However, the structure of the equations can be appreciated from the perspective of transport and complex fluids with little knowledge about rel-ativity. Our presentation is guided by two papers deeply rooted in nonequi-librium thermodynamics and the references given therein.[21]

We first need to introduce some basic notation of special relativity. To put space and time on an equal footing, we introduce the time coordinate $x^0 = -x_0 = ct$ in addition to the space coordinates $x^1 = x_1$, $x^2 = x_2$, $x^3 = x_3$ to obtain our first four-vectors, x^μ and x_μ ($\mu = 0, 1, 2, 3$). The speed of light c is used to give $x^0 = -x_0$ the dimension of length. Distinguishing between upper and lower indices matters only for the time component, where, as a general rule, the difference is a minus sign. Our next four-vector is given by the partial derivatives $\partial_\mu = \partial/\partial x^\mu$. In this notation, the non-relativistic continuity equation (5.5) can be rewritten as

$$\partial_0 \rho + \partial_j \left(\rho \frac{v^j}{c} \right) = 0, \qquad (12.69)$$

where we are using Einstein's summation convention.[22] Equation (12.69) motivates us to introduce a dimensionless velocity four-vector

$$u^0 = \gamma, \quad u^1 = \gamma \frac{v^1}{c}, \quad u^2 = \gamma \frac{v^2}{c}, \quad u^3 = \gamma \frac{v^3}{c}, \qquad (12.70)$$

where the proportionality constant γ, which is known as the Lorentz factor,

---

[21] Öttinger, *Physica A* **259** (1998) 24 and *Physica A* **254** (1998) 433.
[22] In the context of this section, the Einstein summation convention can be stated as follows: a repeated upper and lower index in a term implies summation over that index with values $1, 2, 3$ for a Latin index and with values $0, 1, 2, 3$ for a Greek index.

is chosen such that $u^\mu u_\mu = -1$, that is,

$$\gamma = \frac{1}{\sqrt{1 - v^2/c^2}}. \tag{12.71}$$

In terms of the velocity four-vector, we thus arrive at the elegant relativistic continuity equation

$$\partial_\mu (\rho_f u^\mu) = 0. \tag{12.72}$$

Note that we have been careful to put the subscript "f" on the mass density ρ. By this subscript, we indicate quantities evaluated in the comoving local reference frame of the fluid.[23] By being specific, we avoid possible issues arising from velocity-dependent masses or length contraction, both of which affect the definition of a mass density.

Convenient as the four-vector notation may be, it appears to be very formal: we add a zero-component to the column vector of spatial components and we associate a minus sign with it. What are the deeper reasons for doing that? The physics of the minus sign can be understood from $x^\mu x_\mu = x^2 - c^2 t^2$, which describes the propagation of crests of a spherical light wave in space and time. The experimental observation that the description of light waves coincides for all reference frames moving with a constant velocity with respect to each other leads to the postulate that $x^2 - c^2 t^2$ must be invariant. Our four-vectors are hence much more than formal column vectors: just as three-vectors have a well-defined transformation behavior under rotations and translations (leaving x^2 invariant), four-vectors have a well-defined transformation behavior under Lorentz transformations (leaving $x^2 - c^2 t^2$ invariant). Lorentz transformations mix space and time coordinates, such as for relative motion in the x_1-direction,

$$t' = \gamma(t - vx_1/c^2), \quad x_1' = \gamma(x_1 - vt), \quad x_2' = x_2, \quad x_3' = x_3. \tag{12.73}$$

In the limit $c \to \infty$, and hence $\gamma \to 1$, this transformation coincides with a Galilean transformation (see footnote associated with (11.21)). The generalization for finite c is introduced such that $x'^2 - c^2 t'^2 = x^2 - c^2 t^2$, as the reader should check.

For the transparent formulation of relativistic equations, it is convenient to introduce the Minkowski tensor with components $\eta_{00} = \eta^{00} = -1$, $\eta_{11} = \eta^{11} = \eta_{22} = \eta^{22} = \eta_{33} = \eta^{33} = 1$, $\eta_{\mu\nu} = \eta^{\mu\nu} = 0$ for $\mu \neq \nu$. This tensor allows us to write $x_\mu = \eta_{\mu\nu} x^\nu$ and $x^\mu = \eta^{\mu\nu} x_\nu$, and can more generally be used for lowering or raising indices. The Minkowski tensor is the proper

---

[23] There are different possibilities for defining the local reference frame of the fluid; we here use the Eckart convention that the spatial components of the velocity four-vector vanish in the local reference frame of the fluid, so that we have $u_f^0 = 1$, $u_f^1 = u_f^2 = u_f^3 = 0$.

relativistic generalization of the unit tensor, where $\eta_\mu{}^\nu$ actually coincides with Kronecker's δ symbol. The components of the Minkowski tensor are independent of the reference frame. We further note that the tensor $\hat{\eta}^{\mu\nu} = \eta^{\mu\nu} + u^\mu u^\nu$ can be interpreted as a spatial projector in the local reference frame, where it possesses the diagonal components are $0, 1, 1, 1$ (and all off-diagonal components are zero).

In a relativistic treatment, energy, and momentum should be recognized as the temporal and spatial components of a unified quantity. We hence combine the energy and momentum balances into the compact equation

$$\partial_\mu T^{\mu\nu} = 0\,, \tag{12.74}$$

where $T^{\mu\nu}$ is known as the energy-momentum tensor. Actually, this tensor comprises energy and momentum fluxes in addition to the energy and momentum densities. We have already noted that, as $T^{\mu\nu}$ cannot contain spatial derivatives, we need to construct it by using relaxing structural tensor and vector variables, $\alpha_{\mu\nu}$ and ω_μ, allowing us to model the momentum and energy flux, respectively. We first formulate the relaxation equations for the symmetric four-tensor $\alpha_{\mu\nu}$ and for the four-vector ω_μ, and then express $T^{\mu\nu}$ in terms of these quantities.

By imposing the constraint

$$\alpha_{\mu\lambda} u^\lambda = u_\mu\,, \tag{12.75}$$

we relate the four-tensor $\alpha_{\mu\nu}$ to velocities and restrict it to possess only six independent components; moreover, $\alpha_{\mu\nu}$ is implied to be dimensionless. By acting with ∂_ν on (12.75), we obtain

$$u^\lambda \partial_\nu \alpha_{\mu\lambda} = (\eta_{\mu\lambda} - \alpha_{\mu\lambda})\, \partial_\nu u^\lambda\,. \tag{12.76}$$

A natural constraint for $\omega_\lambda u^\lambda$ arises in the further development. This quantity will be identified with a local equilibrium variable in the rest frame and hence does not relax.

Motivated by the formulation (12.45) of the convected Maxwell model, we write

$$u^\lambda(\partial_\lambda \alpha_{\mu\nu} - \partial_\mu \alpha_{\lambda\nu} - \partial_\nu \alpha_{\mu\lambda}) = -\frac{1}{c\lambda_0}\overline{\alpha}_{\mu\nu} - \frac{1}{c\lambda_2}\overset{\circ}{\alpha}_{\mu\nu}\,. \tag{12.77}$$

The term $u^\lambda \partial_\lambda \alpha_{\mu\nu}$ contains the time derivative and the convection term. The fact that u^λ is not differentiated implies that $\alpha_{\mu\nu}$ represents a tensor, not a tensor density. The other two terms on the left-hand side describe the mixing of tensor components by rotation and deformation as well as the velocity gradients occurring in (12.45), as can be verified by means of (12.76).

In formulating the left-hand side of (12.77), we nicely benefit from our knowledge about transport and complex fluids. The result is properly expressed in terms of relativistic four-vectors and -tensors. The index structure of the terms in parentheses corresponds to a so-called Lie derivative and guarantees the Hamiltonian structure that one expects for the reversible terms representing convection that we have collected on the left-hand side of (12.77).

The right-hand side of (12.77) contains the irreversible relaxation terms with the relaxation times λ_0 and λ_2. Alerted by the experience with Newton's stress tensor in (6.6), we assume that the trace part $\overline{\alpha}_{\mu\nu}$ and the trace-free part $\overset{\circ}{\alpha}_{\mu\nu}$ of $\alpha_{\mu\nu}$ can relax independently, where we define

$$\overline{\alpha}_{\mu\nu} = \frac{1}{3}(\alpha_\lambda{}^\lambda - 1)\,\hat{\eta}_{\mu\nu}\,, \qquad \overset{\circ}{\alpha}_{\mu\nu} = \alpha_{\mu\nu} + u_\mu u_\nu - \overline{\alpha}_{\mu\nu}\,. \tag{12.78}$$

Note that, when (12.77) is contracted with u^μ, both sides vanish.

Motivated by (12.42) and (12.77), the evolution equation for the four-vector ω_μ is written as

$$u^\nu(\partial_\nu\omega_\mu - \partial_\mu\omega_\nu) = -\frac{1}{c\lambda_1}\hat{\eta}_{\mu\nu}\omega^\nu\,. \tag{12.79}$$

Again, upon contraction with u^μ, both sides of this evolution equation vanish. This is the reason why the spatial projector $\hat{\eta}_{\mu\nu}$ had to be included in the relaxation term. For the further calculations, it is convenient to introduce $\hat{\omega}_\mu = \hat{\eta}_{\mu\nu}\,\omega^\nu$.

We next need to formulate a constitutive equation for the energy-momentum tensor $T^{\mu\nu}$ so that we can benefit from the balance equation (12.74). Motivated by our experience from Section 6.2, we do this by analyzing the entropy balance equation. We assume that the entropy density in the comoving reference frame is of the form $s_f(\rho_f, \epsilon_f, \alpha_{\mu\nu}, \hat{\omega}_\mu)$, where ϵ_f is the internal energy density. We here use the spatial projection $\hat{\omega}_\mu$ because the temporal part of ω_μ contains thermodynamic information and would hence be redundant with ρ_f, ϵ_f. Of course, s_f can depend on $\alpha_{\mu\nu}$ and $\hat{\omega}_\mu$ only through scalar invariants, which can then be evaluated in the comoving frame.

We can now derive an entropy balance equation by using the chain rule for $s_f(\rho_f, \epsilon_f, \alpha_{\mu\nu}, \hat{\omega}_\mu)$, and we find

$$\partial_\mu(s_f u^\mu) = \left(s_f - \rho_f\frac{\partial s_f}{\partial\rho_f}\right)\partial_\mu u^\mu + \frac{\partial s_f}{\partial\epsilon_f}u^\mu\,\partial_\mu\epsilon_f + \frac{\partial s_f}{\partial\alpha_{\mu\nu}}u^\lambda\,\partial_\lambda\alpha_{\mu\nu} + \frac{\partial s_f}{\partial\hat{\omega}_\mu}u^\nu\,\partial_\nu\hat{\omega}_\mu\,, \tag{12.80}$$

where the continuity equation (12.72) has been used. Note that the derivatives with respect to $\alpha_{\mu\nu}$ and $\hat{\omega}_\mu$ need to be performed under constraints. As $\hat{\omega}_\mu$ is spatial, also $\partial s_{\rm f}/\partial \hat{\omega}_\mu$ can be chosen to be spatial. Similarly, $\partial s_{\rm f}/\partial \alpha_{\mu\nu}$ can be chosen as symmetric and spatial. By means of the evolution equations (12.77) and (12.79), we further obtain

$$\partial_\mu (s_{\rm f} u^\mu) = \frac{\partial s_{\rm f}}{\partial \epsilon_{\rm f}} p_{\rm f}\, \partial_\mu u^\mu + \frac{\partial s_{\rm f}}{\partial \epsilon_{\rm f}}\, \partial_\mu (\epsilon_{\rm f} u^\mu) + 2 \frac{\partial s_{\rm f}}{\partial \alpha_{\mu\lambda}}\, u^\nu\, \partial_\mu \alpha_{\lambda\nu}$$

$$+ \frac{\partial s_{\rm f}}{\partial \hat{\omega}_\mu} u^\nu\, \partial_\mu \omega_\nu + u^\lambda \omega_\lambda \frac{\partial s_{\rm f}}{\partial \hat{\omega}_\nu}\, u^\mu\, \partial_\mu u_\nu$$

$$- \frac{\partial s_{\rm f}}{\partial \alpha_{\mu\nu}} \left(\frac{1}{c\lambda_0} \overline{\alpha}_{\mu\nu} + \frac{1}{c\lambda_2} \overset{\circ}{\alpha}_{\mu\nu} \right) - \frac{1}{c\lambda_1} \frac{\partial s_{\rm f}}{\partial \hat{\omega}_\mu} \hat{\omega}_\mu, \quad (12.81)$$

where, moreover, we have introduced the pressure $p_{\rm f}$ in the comoving reference frame according to the usual local-equilibrium Euler equation

$$\frac{\partial s_{\rm f}}{\partial \epsilon_{\rm f}} p_{\rm f} = s_{\rm f} - \rho_{\rm f} \frac{\partial s_{\rm f}}{\partial \rho_{\rm f}} - \epsilon_{\rm f} \frac{\partial s_{\rm f}}{\partial \epsilon_{\rm f}}. \quad (12.82)$$

With the auxiliary equation

$$\frac{u^\nu}{u^\lambda \omega_\lambda}\, \partial_\mu \left(u^{\lambda'} \omega_{\lambda'} \frac{\partial s_{\rm f}}{\partial \hat{\omega}_\mu} \omega_\nu \right) - \partial_\mu \left(u^{\lambda'} \omega_{\lambda'} \frac{\partial s_{\rm f}}{\partial \hat{\omega}_\mu} \right) = \frac{\partial s_{\rm f}}{\partial \hat{\omega}_\mu} u^\nu\, \partial_\mu \omega_\nu \quad (12.83)$$

and the condition (12.75), we can rearrange our entropy evolution equation into the form

$$\partial_\mu \left(s_{\rm f} u^\mu + u^\nu \omega_\nu \frac{\partial s_{\rm f}}{\partial \hat{\omega}_\mu} \right) = -\frac{\partial s_{\rm f}}{\partial \epsilon_{\rm f}} u_\nu\, \partial_\mu \left\{ (\epsilon_{\rm f} + p_{\rm f}) u^\mu u^\nu + p_{\rm f} \eta^{\mu\nu} \right\}$$

$$+ \frac{1}{u^{\nu'} \omega_{\nu'}} u_\nu\, \partial_\mu \left\{ u^{\lambda'} \omega_{\lambda'} \left[2 \frac{\partial s_{\rm f}}{\partial \alpha_{\mu\lambda}} (\alpha_\lambda{}^\nu - \eta_\lambda{}^\nu) \right.\right.$$

$$\left.\left. + \frac{\partial s_{\rm f}}{\partial \hat{\omega}_\mu} \omega^\nu \right] \right\} - u_\nu\, \partial_\mu \left\{ u^\lambda \omega_\lambda \frac{\partial s_{\rm f}}{\partial \hat{\omega}_\nu} u^\mu \right\}$$

$$- \frac{\partial s_{\rm f}}{\partial \alpha_{\mu\nu}} \left(\frac{1}{c\lambda_0} \overline{\alpha}_{\mu\nu} + \frac{1}{c\lambda_2} \overset{\circ}{\alpha}_{\mu\nu} \right) - \frac{1}{c\lambda_1} \frac{\partial s_{\rm f}}{\partial \hat{\omega}_\mu} \hat{\omega}_\mu.$$

$$(12.84)$$

If we could eliminate the terms containing $u_\nu\, \partial_\mu \{\ldots\}$ from the right-hand side, we would have a nice entropy equation with an entropy-flux four-vector on the left-hand side and the entropy production on the right-hand side. To achieve this, we have to combine the first two terms into one, and we can then use the energy-momentum balance (12.74). These two terms can be combined if the prefactors differ only be a constant factor.[24] We hence

---

[24] The third term can easily be included because it can be nonzero only if the derivative ∂_μ hits $\partial s_{\rm f}/\partial \hat{\omega}_\nu$, so that any desired factor and its reciprocal can be introduced before and after the derivative.

identify both prefactors in terms of a temperature,

$$\frac{\partial s_f}{\partial \epsilon_f} = \frac{1}{T_f}, \qquad u^\nu \omega_\nu = T_f. \tag{12.85}$$

The definition of temperature in terms of the derivative of s_f with respect to ϵ_f in the first part of (12.85) is not as unambiguous as in non-relativistic hydrodynamics because, in addition to ρ_f, we keep the structural variables constant. If other combinations of variables were kept constant, then we would arrive at another definition of temperature.

Once we have identified the prefactors, we can eliminate the terms containing $u_\nu \partial_\mu \{\ldots\}$ from (12.84) by choosing the energy-momentum tensor as

$$T^{\mu\nu} = (\rho_f c^2 + \epsilon_f) u^\mu u^\nu + p_f \hat{\eta}^{\mu\nu} - 2 T_f \frac{\partial s_f}{\partial \alpha_{\mu\lambda}} (\alpha_\lambda{}^\nu - \eta_\lambda{}^\nu)$$
$$- T_f \frac{\partial s_f}{\partial \hat{\omega}_\mu} \hat{\omega}^\nu + T_f^2 \left(\frac{\partial s_f}{\partial \hat{\omega}_\mu} u^\nu + u^\mu \frac{\partial s_f}{\partial \hat{\omega}_\nu} \right). \tag{12.86}$$

The term involving $\rho_f c^2$, which, in view of the continuity equation (12.72), has no effect, has been added to $T^{\mu\nu}$ to obtain the proper identification of the internal energy density $u_\mu T^{\mu\nu} u_\nu = \rho_f c^2 + \epsilon_f$, on top of the energy density associated with the mass density according to Einstein's famous formula $E = mc^2$.

The final entropy balance equation now becomes

$$\partial_\mu \left(s_f u^\mu + T_f \frac{\partial s_f}{\partial \hat{\omega}_\mu} \right) = -\frac{\partial s_f}{\partial \alpha_{\mu\nu}} \left(\frac{1}{c\lambda_0} \overline{\alpha}_{\mu\nu} + \frac{1}{c\lambda_2} \overset{\circ}{\alpha}_{\mu\nu} \right) - \frac{1}{c\lambda_1} \frac{\partial s_f}{\partial \hat{\omega}_\mu} \hat{\omega}_\mu. \tag{12.87}$$

The entropy flux four-vector contains a contribution proportional to $\partial s_f / \partial \hat{\omega}_\mu$, which shows how $\hat{\omega}_\mu$ is indeed related to the conductive heat flux. The entropy production due to relaxation on the right-hand side of (12.87) is not automatically guaranteed to be nonnegative. The situation would have been much better if we had not formulated linear relaxation terms in (12.77) and (12.79) for the structural variables, but had instead expressed relaxation in terms of $\partial s_f / \partial \alpha_{\mu\nu}$ and $\partial s_f / \partial \hat{\omega}_\mu$, respectively. Such a construction would also be more natural according to the general formulation of nonequilibrium thermodynamics.[25] Linear relaxation would be achieved by quadratic configurational contributions to the entropy density,

$$s_f(\rho_f, \epsilon_f, \alpha_{\mu\nu}, \hat{\omega}_\mu) = s_f^{eq}(\rho_f, \epsilon_f) - \frac{1}{2}\rho_f \left[H_\alpha(\alpha_{\mu\nu}\alpha^{\mu\nu} - 1) + H_\omega \hat{\omega}_\mu \hat{\omega}^\mu \right], \tag{12.88}$$

---

[25] See Chapter 1 of Öttinger, *Beyond Equilibrium Thermodynamics* (Wiley, 2005).

with positive constants H_α and H_ω.[26] From (12.88), we obtain the constrained derivatives

$$\frac{\partial s_f}{\partial \alpha_{\mu\nu}} = -\rho_f H_\alpha(\alpha^{\mu\nu} + u^\mu u^\nu), \qquad \frac{\partial s_f}{\partial \hat{\omega}_\mu} = -\rho_f H_\omega \hat{\omega}^\mu, \qquad (12.89)$$

and hence the energy-momentum tensor

$$T^{\mu\nu} = (\rho_f c^2 + \epsilon_f)u^\mu u^\nu + p_f \hat{\eta}^{\mu\nu} + \rho_f T_f \left[2H_\alpha(\alpha^{\mu\lambda}\alpha_\lambda{}^\nu - \alpha^{\mu\nu}) + H_\omega \hat{\omega}^\mu \hat{\omega}^\nu \right]$$
$$- \rho_f T_f^2 H_\omega(\hat{\omega}^\mu u^\nu + u^\mu \hat{\omega}^\nu). \qquad (12.90)$$

Note that this tensor is symmetric. Our theory of relativistic hydrodynamics consists of this expression for $T^{\mu\nu}$, the balance equations (12.72), (12.74), and the relaxation equations (12.77), (12.79). As an additional result, we obtain a thermodynamically admissible balance equation for the entropy.

In summary, we have developed a full set of evolution equations for relativistic hydrodynamics by using our accumulated knowledge about the structure of balance equations, the relevance of the entropy balance equation in identifying constitutive equations, and the role of convected derivatives for structural variables in describing complex fluids. Our unknown variables are ρ_f, ϵ_f, u_μ, $\alpha_{\mu\nu}$, and ω_μ (1+1+3+6+3 = 14 in view of the various constraints). The count for the number of evolution equations is $1 + 4 + 6 + 3 = 14$ from (12.72), (12.74), (12.77), and (12.79), respectively. Of course, it is crucial to have the expression (12.86) for the energy-momentum tensor so that the balance equation (12.74) becomes useful. For the evaluation of the energy-momentum tensor, one needs to know the derivatives of the entropy density with respect to the structural variables, where simple examples are given in (12.89). More generally, these derivatives of entropy density should be used to describe relaxation in (12.77) and (12.79) so that the entropy production in (12.87) can be guaranteed to be nonnegative.

Exercise 12.14   Entries of the Energy-Momentum Tensor
Interpret the various entries of the energy-momentum tensor as the energy density, energy flux, momentum density, and pressure tensor. How do the structural variables affect these quantities?

Exercise 12.15   Symmetry of the Energy-Momentum Tensor
Evaluate the structural contribution to the energy-momentum tensor (12.86) for the entropy contribution $s_f = -(1/2)\rho_f H \hat{\omega}_\mu \alpha^{\mu\nu} \hat{\omega}_\nu$ and verify its symmetry.

---

[26] In order to describe energetic effects by the structural variables one should choose H_α and H_ω proportional to $1/T_f$; this would eliminate a factor T_f from the structural contributions in (12.90).

- There is a large variety of complex fluids, often referred to as forms of soft matter, which exhibit large responses to small forces and mechanical behavior between that of fluids and that of solids; they are important in a wide range of present and future technological applications.
- The linear viscoelastic behavior of complex fluids can be characterized by a number of different experiments providing time- or frequency-dependent material functions containing equivalent information; the range of accessible times or frequencies can often be extended by the idea of time–temperature superposition.
- The convected Maxwell model provides an evolution equation for the stress tensor, which may be regarded as a convenient structural variable for complex fluids; in shear flow, the model can be used to calculate the viscometric functions, where a nonzero first normal-stress coefficient provides a first signal of nonlinear behavior.
- In a convected Maxwell fluid, momentum propagates with a finite speed; this result highlights a fundamental difference between viscoelastic and Newtonian fluids.
- The accumulated knowledge on transport and complex fluids is used to develop a set of hydrodynamic equations for relativistic fluids that is consistent with all the principles of special relativity and nonequilibrium thermodynamics.

13

Thermodynamics of Interfaces

In the previous chapters, we have implicitly assumed single-phase systems, say fluids or gases, in which the transport of mass, momentum, or energy takes place. In many applications, however, we need to consider more complex, multi-phase systems. Whereas by now we know how to describe transport processes in individual bulk phases, the problem of stitching these equations together by capturing the proper physics of interfaces requires additional ingredients. Paralleling the structure of Chapters 4, 5, and 6, we devote three chapters to developing the building blocks of *interfacial transport phenomena*, beginning with thermodynamics in the present chapter. The two chapters that follow address balance equations and constitutive equations for interfaces. In these three chapters, we develop powerful instruments to treat boundary conditions of a very general form, based on the local equilibrium assumption and the second law of thermodynamics for interfaces.

13.1 Multi-Phase Systems

To motivate our interest in multi-phase systems and interfaces, we consider the cytoplasm of a cell separated from the extracellular fluid by a membrane. For the survival and functioning of a cell it is essential that the cell responds properly to external stimuli. Calcium ions are an important player in the cell signaling chain that governs the cellular activities. By binding to suitable proteins in the cell, Ca^{2+} ions released in response to chemical stimuli in the extracellular space play an important role, for example, in muscle contraction or cell migration. To fulfill this function, it is important to keep the concentration of Ca^{2+} ions inside a cell in the absence of stimuli sufficiently low and to transport Ca^{2+} ions against a concentration gradient from the cytoplasm into the extracellular fluid (see Figure 13.1). To understand what

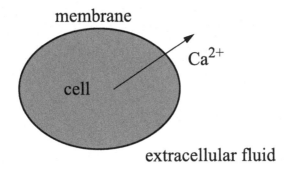

Figure 13.1 Calcium transport across a cell membrane.

renders such a transport process possible, we need to study the properties of the membrane as the interface between the cytoplasm of the cell and the extracellular fluid. Either activation from some chemical reaction (typically from the hydrolysis of adenosine triphosphate), or the concentration gradient of another ion, is required to enable Ca^{2+} ion transport against a large electrochemical gradient. From a microscopic point of view, the required cross couplings must be part of the thermodynamic properties of the interface. From a molecular point of view, the interface must contain devices functioning as ion pumps or ion exchangers.

This example from cell biology is certainly beyond our present reach (see Sections 24.2 and 24.3 for a detailed discussion). Let us consider the much simpler example of a one-component two-phase system, say coexisting gas and liquid phases in a container. By changing the temperature on one of the walls, we can initiate a heat flux (and possibly other fluxes) through the system. How can we describe the heat transfer from the gas to the liquid? What are the resulting boundary conditions for the transport problem in the bulk phases? How does the resulting temperature profile look, and will it be continuous at the interface? What kind of properties do we need to assign to the interface, in particular, when the interface might start to move?

More generally, we need to identify the relevant variables living in essentially two-dimensional interfaces and to couple their evolution to the evolution of the variables characterizing the bulk phases. The bulk phases act as reservoirs controlling the state of the interface, and the state of the interface determines the boundary conditions for the bulk processes. This mutual-coupling situation is illustrated in Figure 13.2.

Even the simple problem of heat transport through a two-phase system raises a number of interesting questions. To address them, let us start from an even more down-to-earth level: how can we describe the properties of

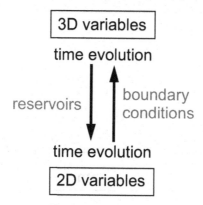

Figure 13.2 Two-way coupling between bulk phases and interfaces.

an interface between two coexisting bulk phases at equilibrium? This problem was addressed and solved by Josiah Willard Gibbs (1839–1903) in the second part of his monumental work *On the Equilibrium of Heterogeneous Substances* in 1877–1878.

13.2 Where Exactly Is the Interface?

Let us consider a two-dimensional interface between two coexisting bulk phases at equilibrium. The thickness of the interfacial region is typically on the nanometer scale so that, from a macroscopic thermodynamic perspective, it cannot be resolved. Throughout the system we have the same temperature and chemical potential but, across the interfacial region, we expect changes in the densities of mass, energy, and entropy. For simplicity, we assume that the interface is planar.[1]

Suppressing two lateral dimensions, the idealization of an interface is illustrated in Figure 13.3. Instead of a continuous density profile that changes within a few nanometers from the value of one bulk phase to the value of another bulk phase, we assume that the two bulk phases touch at the *dividing interface*. The precise choice of the location of this dividing interface within the *interfacial region* is not unique.[2] A possible procedure to fix the position is to postulate that the total masses obtained by integrating the two profiles in Figure 13.3 are exactly equal. This definition looks quite natural, but

---

[1] For comments on curved interfaces, see Section 13.4.

[2] Gibbs uses the terms "dividing surface" instead of "dividing interface" and "surface of discontinuity" instead of "interfacial region"; we find the distinction between "dividing interface" and "interfacial region" nicely intuitive and use the term "interface" to comprise all the aspects of sharp and smooth (or, diffuse) interfaces.

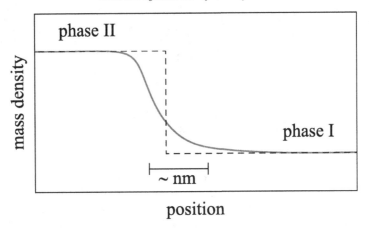

Figure 13.3 True continuous density profile (continuous line) and an idealized profile for two bulk phases touching at a dividing interface (dashed line).

things are not really so simple. For example, for a multi-component system, the proposed construction works only for one selected species of particles. For any other species, we expect an excess mass density, which is defined as the mass obtained from the true continuous profile minus the mass from the discontinuous idealization per unit area. Such excess densities, which can be associated with all extensive properties, depend on the precise location of the dividing interface. Note that these excess quantities can be attributed to the dividing interface; therefore, a corresponding interface density should be defined as the amount of the quantity per unit area instead of the amount per unit volume occurring naturally in the usual bulk density.

Whereas excess densities are ambiguous under small displacements of the dividing interface, relationships between excess densities may be unambiguous, thus containing important physical information. We call a small displacement of the dividing interface, which leads to changes of the various excess densities, a *gauge transformation*. We refer to properties that are not affected by the precise location of the dividing interface as *gauge-invariant*.

In the words of Gibbs, "It will be observed that the position of this [dividing] surface is as yet to a certain extent arbitrary."[3] Excess quantities "are determined partly by the state of the physical system which we are considering, and partly by the various imaginary surfaces by means of which these quantities have been defined."[4] Gibbs refers to the corresponding surface densities as superficial densities.

[3] See p. 219 of Gibbs, *Scientific Papers I* (Woodbridge, 1993).
[4] See p. 224 of Gibbs, *Scientific Papers I* (Woodbridge, 1993).

What happens if we move the dividing interface in Figure 13.3 by a distance ℓ to the right? Of course, the integral under the continuous curve does not change. The integral under the dashed profile increases by $\ell(\rho^{II} - \rho^{I})$, where ρ^{I} and ρ^{II} are the mass densities of the two bulk phases. For the excess mass density ρ^{s}, defined as the difference of these two integrals, we hence find the gauge transformation

$$\rho^{s} \to \rho^{s} + \ell(\rho^{I} - \rho^{II}). \tag{13.1}$$

Note that ℓ times volume density has the proper dimensions of an interfacial density. Similar gauge transformations in terms of jumps of bulk values exist for all excess densities, such as,

$$u^{s} \to u^{s} + \ell(u^{I} - u^{II}), \tag{13.2}$$

$$s^{s} \to s^{s} + \ell(s^{I} - s^{II}), \tag{13.3}$$

for the excess internal energy and entropy densities, and

$$\rho_{\alpha}^{s} \to \rho_{\alpha}^{s} + \ell(\rho_{\alpha}^{I} - \rho_{\alpha}^{II}), \tag{13.4}$$

for the excess species mass densities.

Exercise 13.1   A Jump Relation
Use the gauge transformations (13.1)–(13.3) to derive the relation

$$\frac{\partial f^{s}(T, \rho^{s})}{\partial \rho^{s}} = \frac{u^{I} - u^{II}}{\rho^{I} - \rho^{II}} - T \frac{s^{I} - s^{II}}{\rho^{I} - \rho^{II}}. \tag{13.5}$$

13.3 Fundamental Thermodynamic Equations

We next generalize the fundamental thermodynamic equations of Chapter 4 to include interfaces. For a one-component system,[5] an increase in the internal energy U of the total system can be expressed by Gibbs' fundamental form

$$dU = T\,dS - p\,dV + \tilde{\mu}\,dN + \gamma\,dA, \tag{13.6}$$

which, as a generalization of (4.4), introduces temperature T, pressure p, chemical potential $\tilde{\mu}$, and the surface or interfacial tension γ. These intensive variables relate changes of U to changes in entropy S, volume V, number of moles N, and surface area A. In particular, we have $\gamma = (\partial U/\partial A)_{S,V,N}$ in addition to (4.11)–(4.13). As we are assuming global equilibrium and a planar dividing interface, p, T, and μ are constant throughout the two-phase

[5] Multi-component systems are considered later in this section.

system, including the interface, and the interfacial tension γ is uniform over the dividing interface. Making proper use of the extensivity of U, S, V, N, and A in the spirit of Section 4.2, we further obtain the Euler equation for the entire system at equilibrium, $U = TS - pV + \tilde{\mu}N + \gamma A$, which generalizes (4.19), and hence the Gibbs–Duhem equation $S\,dT - V\,dp + N\,d\tilde{\mu} + A\,d\gamma = 0$, which generalizes (4.20). The thermodynamics for each of the two equilibrium bulk phases leads to $U^i = TS^i - pV^i + \tilde{\mu}N^i$ and $S^i\,dT - V^i\,dp + N^i\,d\tilde{\mu} = 0$, for $i = \mathrm{I,II}$. Defining the interfacial quantities $N^{\mathrm{s}} = N - N^1 - N^2$, $U^{\mathrm{s}} = U - U^1 - U^2$, and $S^{\mathrm{s}} = S - S^1 - S^2$ (note that $V = V^1 + V^2$) and taking differences of the above equations, one obtains the Euler and Gibbs–Duhem equations for the interface in global equilibrium,

$$u^{\mathrm{s}} = Ts^{\mathrm{s}} + \gamma + \hat{\mu}\rho^{\mathrm{s}}, \tag{13.7}$$

and

$$s^{\mathrm{s}}\,dT + d\gamma + \rho^{\mathrm{s}}\,d\hat{\mu} = 0. \tag{13.8}$$

Whereas these fundamental equations for interfaces look very similar to those of Section 4.5, it should be noted that they relate the well-defined intensive quantities T, $\hat{\mu}$, γ and the ambiguous or gauge-variant excess densities u^{s}, s^{s}, ρ^{s}. The ambiguity of the excess densities is related to the fact that, according to the Gibbs phase rule (4.37), one independent physical variable is lost because of the coexistence of two phases. If we still use the same set and hence the same number of variables in (13.7) and (13.8), there must exist one ambiguous gauge degree of freedom for the dividing interfaces between coexisting bulk phases.[6]

By assuming that the Gibbs–Duhem equation (13.8) still holds after introducing the transformations (13.1) and (13.3), and subtracting the original form of (13.8), we obtain the jump ratio

$$\frac{s^{\mathrm{I}} - s^{\mathrm{II}}}{\rho^{\mathrm{I}} - \rho^{\mathrm{II}}} = -\frac{d\hat{\mu}(T)}{dT}, \tag{13.9}$$

where $\hat{\mu}(T)$ is the chemical potential leading to phase coexistence at temperature T. Gauge invariance of the Euler equation (13.7) similarly implies the further jump ratio

$$\frac{u^{\mathrm{I}} - u^{\mathrm{II}}}{\rho^{\mathrm{I}} - \rho^{\mathrm{II}}} = \hat{\mu}(T) - T\frac{d\hat{\mu}(T)}{dT}. \tag{13.10}$$

The Clapeyron-type equations (13.9) and (13.10) for jump ratios (see (4.38) for comparison) are immediate consequences of the gauge invariance of the

---

[6] For three-phase coexistence at a contact line, we would have two gauge degrees of freedom.

fundamental equations (13.7) and (13.8) for interfaces. Note that these equations are consistent with (13.5) if the derivative of the excess Helmholtz free energy with respect to the excess mass density at constant temperature is given by the chemical potential $\hat{\mu}$, as we would anticipate in a global equilibrium state. By combining (13.9) and (13.10), and using the symbol Δ for the jump of a bulk quantity, for example $\Delta\rho = \rho^{\mathrm{I}} - \rho^{\mathrm{II}}$, we get an interesting Maxwell-like relation involving mixed differential and difference quotients,

$$\frac{d(\Delta f/\Delta \rho)}{dT} = \frac{\Delta(\partial f/\partial T)}{\Delta \rho}. \tag{13.11}$$

Exercise 13.2   Interfacial Euler and Gibbs–Duhem Equations
Derive the Euler equation (13.7) and the Gibbs–Duhem equation (13.8).

Exercise 13.3   Clapeyron-Type Equations
Derive the Clapeyron-type equations (13.9) and (13.10).

For the gauge $\rho^{\mathrm{s}} = 0$, (13.7) and (13.8) imply $s^{\mathrm{s}} = -d\gamma/dT$ and $u^{\mathrm{s}} = \gamma - T\, d\gamma/dT$. On moving the dividing interface by an arbitrary distance ℓ, which by means of (13.1) can be reconstructed from ρ^{s} as $\ell = \rho^{\mathrm{s}}/(\rho^{\mathrm{I}} - \rho^{\mathrm{II}})$, the transformations (13.2) and (13.3), combined with the jump ratios (13.9) and (13.10), lead to[7]

$$s^{\mathrm{s}}(T, \rho^{\mathrm{s}}) = -\frac{d\gamma(T)}{dT} - \rho^{\mathrm{s}}\frac{d\hat{\mu}(T)}{dT}, \tag{13.12}$$

and

$$u^{\mathrm{s}}(T, \rho^{\mathrm{s}}) = \gamma(T) - T\frac{d\gamma(T)}{dT} + \rho^{\mathrm{s}}\left[\hat{\mu}(T) - T\frac{d\hat{\mu}(T)}{dT}\right]. \tag{13.13}$$

The excess densities in (13.12) and (13.13) consist of a gauge-invariant part and an ambiguous part proportional to ρ^{s}. These equations illustrate nicely why Gibbs says that excess quantities are determined partly by the state of the physical system and partly by the imaginary surface.

From Chapter 4, we know that complete thermodynamic information about a bulk phase of a one-component system is contained in a single thermodynamic potential depending on two variables, say the Helmholtz free energy density $f(T, \rho)$. For an interface, however, complete thermodynamic information is contained in two functions of one variable. The function $\hat{\mu}(T)$ encodes the condition for phase coexistence, while the function $\gamma(T)$ characterizes the structure of the interfacial region. For interfaces in more general

[7] Savin *et al.*, *Europhys. Lett.* **97** (2012) 40002.

systems, we need two functions of one fewer independent variables than for a bulk phase.

The fundamental equation (13.6) would suggest using the extensive quantities S, V, N, and A as independent variables. However, it is much more natural to work with intensive variables to characterize interfaces. This is so because isolated interfaces do not exist so that interfaces are always controlled by the adjacent bulk phases acting as reservoirs. After performing all the Legendre transformations to intensive independent variables, the resulting thermodynamic potential turns out to be pressure-like. It is hence not surprising that the interfacial tension $\gamma(T)$ contains the thermodynamic information about the state of the interface.

For a complete generalization of the global equilibrium description for planar interfaces to a local equilibrium one, we still need to discuss the momentum density, which can be chosen to be absent for global equilibrium. If we assume that $\boldsymbol{v}^{\mathrm{s}} = \boldsymbol{m}^{\mathrm{s}}/\rho^{\mathrm{s}}$ is the gauge-invariant velocity of the interface, the gauge condition $\rho^{\mathrm{s}} = 0$ implies that also $\boldsymbol{m}^{\mathrm{s}}$ must vanish for this gauge. Then, moving the dividing interface by an arbitrary distance ℓ and adding the gauge transformation

$$\boldsymbol{m}^{\mathrm{s}} \to \boldsymbol{m}^{\mathrm{s}} + \ell(\boldsymbol{m}^{\mathrm{I}} - \boldsymbol{m}^{\mathrm{II}}), \tag{13.14}$$

to the list (13.1)–(13.3) and (13.4), we obtain an additional, nonequilibrium Clapeyron-type equation,

$$\frac{\boldsymbol{m}^{\mathrm{I}} - \boldsymbol{m}^{\mathrm{II}}}{\rho^{\mathrm{I}} - \rho^{\mathrm{II}}} = \boldsymbol{v}^{\mathrm{s}}, \tag{13.15}$$

for the jump of the momentum density.

We now generalize our results to multi-component systems. According to the Gibbs phase rule, for a two-phase system we lose one degree of freedom. Here, we characterize the thermodynamic state of an interface by $\hat{\mu}_k = \hat{\mu}_k(T, \hat{\mu}_1, \ldots, \hat{\mu}_{k-1})$ and $\gamma = \gamma(T, \hat{\mu}_1, \ldots, \hat{\mu}_{k-1})$. The Euler equation for multi-component systems takes the form

$$u^{\mathrm{s}} = T s^{\mathrm{s}} + \gamma + \sum_{\alpha=1}^{k-1} \hat{\mu}_\alpha \rho_\alpha^{\mathrm{s}} + \hat{\mu}_k \rho_k^{\mathrm{s}}, \tag{13.16}$$

and we write the Gibbs–Duhem equation as

$$s^{\mathrm{s}} \, dT + d\gamma + \sum_{\alpha=1}^{k-1} \rho_\alpha^{\mathrm{s}} \, d\hat{\mu}_\alpha + \rho_k^{\mathrm{s}} \, d\hat{\mu}_k = 0. \tag{13.17}$$

We begin by applying the gauge transformations in (13.3) and (13.4) to the

Gibbs–Duhem equation (13.17), which gives

$$\frac{s^{\mathrm{I}} - s^{\mathrm{II}}}{\rho_k^{\mathrm{I}} - \rho_k^{\mathrm{II}}} = -\left(\frac{\partial \hat{\mu}_k}{\partial T}\right)_{\hat{\mu}_{\alpha \neq k}}, \tag{13.18}$$

$$\frac{\rho_\alpha^{\mathrm{I}} - \rho_\alpha^{\mathrm{II}}}{\rho_k^{\mathrm{I}} - \rho_k^{\mathrm{II}}} = -\left(\frac{\partial \hat{\mu}_k}{\partial \hat{\mu}_\alpha}\right)_{T, \hat{\mu}_{\beta \neq \alpha, k}}, \tag{13.19}$$

for $\alpha = 1, \ldots, k-1$. Using the gauge transformations in (13.2), (13.3), and (13.4), along with (13.18) and (13.19), in the Euler equation (13.16), we obtain

$$\frac{u^{\mathrm{I}} - u^{\mathrm{II}}}{\rho_k^{\mathrm{I}} - \rho_k^{\mathrm{II}}} = \hat{\mu}_k - T\left(\frac{\partial \hat{\mu}_k}{\partial T}\right)_{\hat{\mu}_{\alpha \neq k}} - \sum_{\alpha=1}^{k-1} \hat{\mu}_\alpha \left(\frac{\partial \hat{\mu}_k}{\partial \hat{\mu}_\alpha}\right)_{T, \hat{\mu}_{\beta \neq \alpha, k}}. \tag{13.20}$$

Note that (13.18)–(13.20) are multi-component generalizations of the Clapeyron-type equations in (13.9) and (13.10).

Now, by using (13.17) in the gauge $\rho_k^{\mathrm{s}} = 0$ and then transforming to an arbitrary gauge, we can obtain the gauge-invariant expressions[8]

$$s^{\mathrm{s}} = -\left(\frac{\partial \gamma}{\partial T}\right)_{\hat{\mu}_{\alpha \neq k}} - \rho_k^{\mathrm{s}}\left(\frac{\partial \hat{\mu}_k}{\partial T}\right)_{\hat{\mu}_{\alpha \neq k}}, \tag{13.21}$$

where we have used (13.18), and

$$\rho_\alpha^{\mathrm{s}} = -\left(\frac{\partial \gamma}{\partial \hat{\mu}_\alpha}\right)_{T, \hat{\mu}_{\beta \neq \alpha, k}} - \rho_k^{\mathrm{s}}\left(\frac{\partial \hat{\mu}_k}{\partial \hat{\mu}_\alpha}\right)_{T, \hat{\mu}_{\beta \neq \alpha, k}}, \tag{13.22}$$

for $\alpha = 1, \ldots, k-1$, where we have used (13.19). Similarly, starting from (13.16) in the gauge $\rho_k^{\mathrm{s}} = 0$, we obtain the gauge-invariant expression

$$u^{\mathrm{s}} = \gamma - T\left(\frac{\partial \gamma}{\partial T}\right)_{\hat{\mu}_{\alpha \neq k}} - \sum_{\alpha=1}^{k-1} \hat{\mu}_\alpha \left(\frac{\partial \gamma}{\partial \hat{\mu}_\alpha}\right)_{T, \hat{\mu}_{\beta \neq \alpha, k}}$$
$$+ \rho_k^{\mathrm{s}}\left[\hat{\mu}_k - T\left(\frac{\partial \hat{\mu}_k}{\partial T}\right)_{\hat{\mu}_{\alpha \neq k}} - \sum_{\alpha=1}^{k-1} \hat{\mu}_\alpha \left(\frac{\partial \hat{\mu}_k}{\partial \hat{\mu}_\alpha}\right)_{T, \hat{\mu}_{\beta \neq \alpha, k}}\right], \tag{13.23}$$

where we have made use of (13.20)–(13.22).

13.4 Curved Interfaces

So far, we have considered only planar interfaces. A few remarks on curved interfaces would seem to be appropriate.

---

[8] The expressions in (13.21)–(13.23) differ from those given in Savin *et al.*, *Europhys. Lett.* **97** (2012) 40002, which are missing the terms multiplied by ρ_k^{s}.

A planar interface is characterized by a normal vector that is independent of position. We can hence recognize curvature from a change of the normal vector along the interface. The larger the curvature, the more rapidly the normal vector changes. Large curvature further implies a small radius of curvature. The quantitative relationship is illustrated in Figure 13.4. If R_1 and R_2 are the principal radii of curvature in perpendicular directions along a two-dimensional interface, such as an inflatable pool tube, Figure 13.4 suggests

$$\boldsymbol{\nabla}_\| \cdot \boldsymbol{n} = \frac{1}{R_1} + \frac{1}{R_2}, \qquad (13.24)$$

where $\boldsymbol{\nabla}_\|$ is the surface gradient, or surface nabla, operator.[9] If the radius of a sphere increases from R to $R + \Delta R$, the relative increase of surface area is $2\,\Delta R/R = \Delta R\,\boldsymbol{\nabla}_\| \cdot \boldsymbol{n}$, where the latter result can be generalized to normal displacements of arbitrary interfaces. In view of this change of area, the change of the surface energy density in the system upon displacing the interface by ΔR is not just given by the work against the pressure difference, $(p^I - p^{II})\Delta R$. The work against the interfacial tension associated with a change of surface area needs to be added, $(p^I - p^{II} + \gamma\,\boldsymbol{\nabla}_\| \cdot \boldsymbol{n})\Delta R$, where $\boldsymbol{n}$ points from phase II into phase I. The total force per unit area against displacements of the interface is hence given by $p^I - p^{II} + \gamma\,\boldsymbol{\nabla}_\| \cdot \boldsymbol{n}$, and equilibrium is achieved for

$$p^I = p^{II} - \gamma\,\boldsymbol{\nabla}_\| \cdot \boldsymbol{n}. \qquad (13.25)$$

This relation for the pressure jump across a curved interface at equilibrium is the Young–Laplace equation discovered in the early nineteenth century. The proportionality between the pressure difference and the mean curvature reflects our everyday experience that blowing up a balloon becomes easier when the balloon gets larger and that the pressure in tires is larger for bicycles than for cars.

Of course, the radii of curvature should be large compared with the thickness of the interfacial region. To obtain well-defined thermodynamic properties of the interface, one should be allowed to treat interfaces over sufficiently large transverse dimensions as planar objects so that they contain a large number of particles.

---

[9] $\boldsymbol{\nabla}_\|$ can formally be written as the projection of $\boldsymbol{\nabla}$ onto a surface with unit normal $\boldsymbol{n}$. Note, however, that for fields defined only in the interface, the full $\boldsymbol{\nabla}$ is actually undefined, whereas differentiations in tangential directions are possible. In subsequent chapters, we will use the surface, or tangential, projection tensor $\boldsymbol{\delta}_\| = \boldsymbol{\delta} - \boldsymbol{nn}$, which has the property $\boldsymbol{\delta}_\| \cdot \boldsymbol{a}^s = \boldsymbol{a}^s$ for any tangential vector $\boldsymbol{a}^s$.

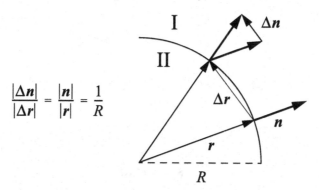

Figure 13.4 Relationship between change of normal vector and radius of curvature.

Interfaces having an irregular surface, for example a surface that is not planar, cylindrical or spherical, are common both in nature and in various processing technologies. In such cases, it is convenient to describe the surface in terms of a function defined by $\varphi^{\mathrm{s}}(\boldsymbol{r}, t) = 0$. The unit normal vector $\boldsymbol{n}$ is then given by

$$\boldsymbol{n} = \frac{\boldsymbol{\nabla}\varphi^{\mathrm{s}}}{|\boldsymbol{\nabla}\varphi^{\mathrm{s}}|}. \qquad (13.26)$$

To be consistent with our convention for choosing $\boldsymbol{n}$, the function φ^{s} is negative in phase II, zero on the surface, and positive in phase I.

Our new knowledge about the relative increase of surface area moreover allows us to generalize the gauge transformations (13.1)–(13.3), (13.4), and (13.14) to curved interfaces. If a is the density of an arbitrary extensive quantity, the generalization can be written as

$$a^{\mathrm{s}} \to a^{\mathrm{s}} + \ell(a^{\mathrm{I}} - a^{\mathrm{II}} - a^{\mathrm{s}}\,\boldsymbol{\nabla}_{\|} \cdot \boldsymbol{n}). \qquad (13.27)$$

A relative increase of the surface (corresponding to positive ℓ) leads to a decrease of the surface density, and vice versa.

Exercise 13.4   Curvature of a Sphere
For a sphere characterized by spherical coordinates (r, θ, ϕ) with constant $r = R$, calculate $\boldsymbol{\nabla}_{\|} \cdot \boldsymbol{n}$ by explicit differentiations.

Exercise 13.5   Kelvin Equation
Consider a spherical liquid drop (phase II) of radius R in equilibrium with a gas (phase I). At equilibrium, the pressures in the two phases are related by the Young–Laplace equation (13.25) with $\boldsymbol{\nabla}_{\|} \cdot \boldsymbol{n}$ determined as in Exercise 13.4. Assume ideal

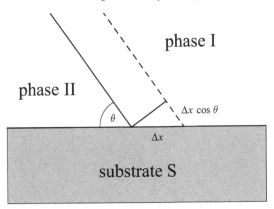

Figure 13.5 Contact angle θ for two fluid phases I and II in equilibrium with a rigid substrate.

gas behavior and that the density of liquid drop is independent of pressure to derive the following expression showing the effect of interfacial tension on vapor pressure,

$$\frac{p^{\mathrm{I}}}{p_0^{\mathrm{I}}} = \exp\left[\frac{\tilde{v}^{\mathrm{II}}}{\tilde{R}T}\left(\frac{2\gamma}{R}\right) + \frac{\tilde{v}^{\mathrm{II}}}{\tilde{R}T}(p^{\mathrm{I}} - p_0^{\mathrm{I}})\right], \tag{13.28}$$

where p_0^{I} is what the pressure in the gas phase would be if the gas–liquid interface were planar. When the second term inside the square brackets, which is usually negligible, is excluded, the expression above is known as the Kelvin equation. For a drop of water with a radius of 0.1 μm at $p_0^{\mathrm{I}} = 1$ bar and $T = 298$ K, estimate $p^{\mathrm{I}}/p_0^{\mathrm{I}}$.

Exercise 13.6   Young's Equation for the Contact Angle
Consider two fluid phases I and II in equilibrium with a relatively rigid substrate S (see Figure 13.5). How much work is required to move the dividing interface between the fluid phases I and II by the distance Δx? Use the result to derive Young's equation

$$\gamma^{\mathrm{I,S}} - \gamma^{\mathrm{II,S}} = \cos\theta\,\gamma^{\mathrm{I,II}}, \tag{13.29}$$

which characterizes the so-called contact angle θ in terms of the interfacial tensions between the various pairs of phases.

Exercise 13.7   Meniscus Shape Near a Wetted Solid
Consider a stagnant liquid with constant density ρ that wets, or forms a meniscus with, a vertical solid wall as shown in Figure 13.6. The local height of the meniscus is $x_3 = h(x_1)$, and the unit normal $\boldsymbol{n}$ points into the gas (I) and away from the liquid (II). Combine the equation governing the pseudo-pressure in the liquid $\boldsymbol{\nabla} p^{\mathrm{L}} = \rho \boldsymbol{g}$ with the Young–Laplace equation (13.25) to obtain the following equation governing

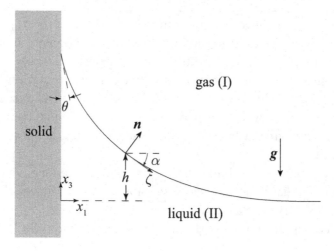

Figure 13.6 Meniscus formed by liquid in contact with a planar solid with local height h and contact angle θ.

the meniscus height

$$l_{\text{cap}}^2 \frac{d^2 h}{dx_1^2} = \left[1 + \left(\frac{dh}{dx_1} \right)^2 \right]^{3/2} x_3,$$

where $l_{\text{cap}} = \sqrt{\gamma/(\rho g)}$ is the so-called capillary length. Show, by parameterization of the meniscus $x_3 = h(x_1)$ using the angle α and distance ζ defined in Figure 13.6, that the previous result can be written as

$$\frac{dx_3}{d\zeta} = -\sin \alpha, \qquad \frac{d\alpha}{d\zeta} = -\frac{x_3}{l_{\text{cap}}^2}.$$

Use this to find the following expression for the meniscus height:

$$h(0) = l_{\text{cap}} \sqrt{2(1 - \sin \theta)}, \qquad (13.30)$$

where θ is the contact angle. Estimate $h(0)$ for a typical water–air–glass system. Finally, derive expressions that give the shape of the meniscus in terms of α and ζ.

- Interfaces can have a major impact on the transport properties of multi-phase systems; in particular, many functions of biological systems are based on active interfacial transport.

- Choosing the precise location of a dividing surface within the interfacial region can have a strong influence on the value of excess densities; to analyze observable properties, we associate gauge transformations with this degree of freedom, which are particularly simple for planar interfaces.

- At equilibrium, the bulk phases around an interface act as thermodynamic baths; the interfacial tension as a function of the intensive variables serves as a thermodynamic potential.

- In local equilibrium, the thermodynamic behavior of a bulk phase of a k-component system is characterized by a thermodynamic potential of $k+1$ variables, whereas the behavior of an interface is characterized by two functions of k variables: one function acts as a thermodynamic potential characterizing the structure of the interfacial region, and the other one encodes the condition for the coexistence of bulk phases.

- Any degree of freedom lost for coexisting phases according to the Gibbs phase rule is turned into a gauge degree of freedom; Clapeyron-type equations for the ratios of jumps between bulk phases follow from the gauge invariance of the fundamental thermodynamic relations.

14

Interfacial Balance Equations

We have just considered the thermodynamics of interfaces between bulk phases that are in equilibrium. In particular, we have seen that the principle of gauge invariance can be used to establish relationships between bulk phase thermodynamic quantities and to identify relevant thermodynamic quantities of the interface.

Our goal here is to formulate balance equations for interfaces that separate bulk phases that are not necessarily at equilibrium. In previous chapters, we considered situations in which the formulation of boundary conditions at interfaces was relatively straightforward. For example, at solid–fluid interfaces, the energy flux was taken to be continuous (e.g., transient hot wire, heated tube), or solid–fluid interfaces were taken to be impermeable so that mass fluxes vanished (e.g., sorption experiment, ultracentrifuge).

Interfacial balance equations are essential for prescribing boundary conditions at more complicated phase interfaces. A number of questions immediately come to mind. For example, how does one formulate balance equations at interfaces that are not stationary, and do interfaces themselves possess evolving fields analogous to the mass, momentum, and energy densities used to describe bulk phases? A number of approaches have been used to address some of these issues in the formulation of balance equations for interfaces. The approach we use is based on intuition and experience from Chapters 5 and 6 as well as insights from Chapter 13. More formal (and less intuitive) approaches can be found in the books by Edwards, Brenner, and Wasan[1] and by Slattery, Sagis, and Oh.[2] The novel approach presented in this chapter is the basis for a recently published paper.[3]

---

[1] Edwards *et al.*, *Interfacial Transport Processes* (Butterworth, 1991).
[2] Slattery *et al.*, *Interfacial Transport Phenomena* (Springer, 2007).
[3] Öttinger & Venerus, *AIChE J.* **60** (2014) 1424.

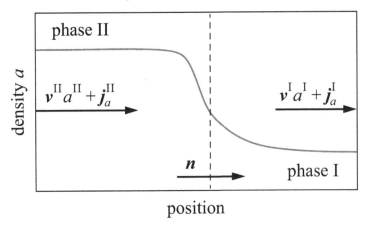

Figure 14.1 Density profile of a conserved quantity. As a consequence of a net accumulation of that quantity in the interfacial region, the dividing interface for $a^s = 0$ (dashed line) moves to the right.

14.1 Dividing Interfaces and Normal Velocities

We are interested in balance equations for the excess densities of mass ρ^s, momentum $\boldsymbol{m}^s$, internal energy u^s, and species mass ρ_α^s. However, we already know from Chapter 13 that all of these quantities are ambiguous because they depend on the choice of the dividing interface. To obtain unambiguous balance equations, we would need to look at the evolution of quantities like u^s in the gauge fixed by $\rho^s = 0$. To avoid the need to consider two excess densities simultaneously, we start with a simpler question. How can we determine the velocity of the evolving interface for which a particular excess density vanishes at all times?

For any density a of a conserved quantity, we have the corresponding ambiguous excess density a^s. We consider the interface for which the excess density vanishes, $a^s = 0$, and study the motion of this particular interface. For the special case of a planar homogeneous interface, the normal velocity v_{na}^s of the dividing interface, defined by $a^s = 0$, at all times is given by

$$v_{na}^s \left(a^{II} - a^I\right) = \boldsymbol{n} \cdot \left(\boldsymbol{v}^{II} a^{II} + \boldsymbol{j}_a^{II} - \boldsymbol{v}^I a^I - \boldsymbol{j}_a^I\right). \tag{14.1}$$

The superscripts I and II indicate the value of the quantity in the bulk phases extrapolated to the interface.[4] We arbitrarily choose the unit normal $\boldsymbol{n}$ to the interface so that it points into phase I and out of phase II (see Figure 14.1). The right-hand side of (14.1) represents the flux of a coming

---

[4] The precise location of the dividing interface does not matter for the extrapolation because bulk properties do not vary noticeably on the length scale of the order of the thickness of the interfacial region.

from the bulk phases into the interfacial region, split into the convective contribution va and the diffusive contribution j_a known from the balance equations for each bulk phase. If, as suggested in Figure 14.1, the net effect is an inflow of a into the interfacial region, the bulk phase with the higher value of a expands at the expense of that with the lower value of a, where the required positive value of the normal velocity v_{na}^{s} is given by (14.1).

For an inhomogeneous interface, the dividing interface defined by $a^{\mathrm{s}} = 0$ is no longer planar, and its normal velocity may be affected by transport processes taking place within the interface. We hence generalize (14.1) to

$$v_{na}^{\mathrm{s}} \left(a^{\mathrm{II}} - a^{\mathrm{I}} \right) = \boldsymbol{n} \cdot \left(\boldsymbol{v}^{\mathrm{II}} a^{\mathrm{II}} + \boldsymbol{j}_a^{\mathrm{II}} - \boldsymbol{v}^{\mathrm{I}} a^{\mathrm{I}} - \boldsymbol{j}_a^{\mathrm{I}} \right) - \boldsymbol{\nabla}_{\parallel} \cdot \boldsymbol{j}_a^{\mathrm{s}}, \tag{14.2}$$

where the tangential vector field $\boldsymbol{j}_a^{\mathrm{s}}$ is the interface analog of the bulk conductive fluxes $\boldsymbol{j}_a^{\mathrm{I}}$ and $\boldsymbol{j}_a^{\mathrm{II}}$. Note that a convection term in the interface is not required in the present discussion because we look at the dividing interface with $a^{\mathrm{s}} = 0$. Additional source terms can be included for densities such as entropy and species mass for multi-component systems in the presence of reactions in the interface, as we discuss below.

By applying (14.2) to the exchange of mass and momentum between the bulk and the interfacial region, we obtain

$$v_{n\rho}^{\mathrm{s}} \left(\rho^{\mathrm{II}} - \rho^{\mathrm{I}} \right) = \boldsymbol{n} \cdot \left(\boldsymbol{v}^{\mathrm{II}} \rho^{\mathrm{II}} - \boldsymbol{v}^{\mathrm{I}} \rho^{\mathrm{I}} \right), \tag{14.3}$$

and

$$v_{nm}^{\mathrm{s}} \left(\boldsymbol{m}^{\mathrm{II}} - \boldsymbol{m}^{\mathrm{I}} \right) = \boldsymbol{n} \cdot \left(\boldsymbol{v}^{\mathrm{II}} \boldsymbol{m}^{\mathrm{II}} + \boldsymbol{\pi}^{\mathrm{II}} - \boldsymbol{v}^{\mathrm{I}} \boldsymbol{m}^{\mathrm{I}} - \boldsymbol{\pi}^{\mathrm{I}} \right) - \boldsymbol{\nabla}_{\parallel} \cdot \boldsymbol{\pi}^{\mathrm{s}}, \tag{14.4}$$

where the symmetric momentum flux tensor $\boldsymbol{\pi}^{\mathrm{s}}$ is purely transverse. In writing (14.3), we have assumed a vanishing diffusive mass flux, $\boldsymbol{j}_\rho^{\mathrm{s}} = \boldsymbol{0}$ (see Section 6.2 for the corresponding discussion in the bulk case).

14.2 The Interface Velocity and Jump Balances

We have just seen that there are a number of interface normal velocities corresponding to different dividing interfaces for which one of the extensive excess densities is zero. Are any of these velocities special, and is it possible to identify $\boldsymbol{v}^{\mathrm{s}}$, the full velocity vector of the interface? To answer these questions, we note that $v_{n\rho}^{\mathrm{s}}$ in (14.3) is nicely consistent with the general definition (13.15) of the interfacial velocity $\boldsymbol{v}^{\mathrm{s}}$, which has been obtained from gauge invariance. The discussion of the gauge transformation behavior of the momentum density in Section 13.3 also implies that v_{nm}^{s} in (14.4) coincides with the normal component of $\boldsymbol{v}^{\mathrm{s}}$.

We can now rewrite (14.3) and (14.4) in terms of the velocity defined in (13.15) as

$$n \cdot (v^{\mathrm{I}} - v^{\mathrm{s}}) \rho^{\mathrm{I}} = n \cdot (v^{\mathrm{II}} - v^{\mathrm{s}}) \rho^{\mathrm{II}}, \tag{14.5}$$

and

$$n \cdot [(v^{\mathrm{I}} - v^{\mathrm{s}}) m^{\mathrm{I}} + \pi^{\mathrm{I}}] = n \cdot [(v^{\mathrm{II}} - v^{\mathrm{s}}) m^{\mathrm{II}} + \pi^{\mathrm{II}}] - \nabla_{\parallel} \cdot \pi^{\mathrm{s}}. \tag{14.6}$$

These equations are known as *jump balances* for mass and momentum at an interface. In the absence of mass transfer (phase change) across the interface, from (14.5) we have

$$v^{\mathrm{I}} \cdot n = v^{\mathrm{s}} \cdot n = v^{\mathrm{II}} \cdot n, \tag{14.7}$$

which means that the normal component of velocity is continuous across the interface. If we consider a stationary interface between two fluids at rest and assume that the pressure tensor is given by $\pi^{\mathrm{s}} = -\gamma \delta_{\parallel}$, then one can show that the normal component of (14.6) simplifies to the Young–Laplace equation (13.25). For phase coexistence at a planar interface, we recover the equality of pressure in the two phases.

From the definition of v^{s} in (13.15), it is possible to write

$$\rho^{\mathrm{I}}(v^{\mathrm{I}} - v^{\mathrm{s}}) = \left(\frac{1}{\rho^{\mathrm{II}}} - \frac{1}{\rho^{\mathrm{I}}} \right)^{-1} (v^{\mathrm{II}} - v^{\mathrm{I}}) \tag{14.8}$$

and

$$\rho^{\mathrm{II}}(v^{\mathrm{II}} - v^{\mathrm{s}}) = \left(\frac{1}{\rho^{\mathrm{II}}} - \frac{1}{\rho^{\mathrm{I}}} \right)^{-1} (v^{\mathrm{II}} - v^{\mathrm{I}}), \tag{14.9}$$

which show that the velocity differences for each phase depend on the same jumps in velocity and density. Using (14.8) and (14.9), the jump momentum balance (14.6) can be rewritten in the alternative form

$$-\left(\frac{1}{\rho^{\mathrm{II}}} - \frac{1}{\rho^{\mathrm{I}}} \right)^{-1} n \cdot (v^{\mathrm{II}} - v^{\mathrm{I}})(v^{\mathrm{II}} - v^{\mathrm{I}}) = n \cdot (\pi^{\mathrm{II}} - \pi^{\mathrm{I}}) - \nabla_{\parallel} \cdot \pi^{\mathrm{s}}. \tag{14.10}$$

Equation (14.10) relates jumps in the pressure tensor to jumps in velocity and reciprocal density (that is, specific volume).

Note that, in general, we cannot simply write a jump balance for the excess energy e^{s} that would be analogous to the jump balances (14.5) and (14.6) for mass and momentum. The reason is that v^{s}_{ne} is different from $v^{\mathrm{s}}_{n\rho} = v^{\mathrm{s}}_n = n \cdot v^{\mathrm{s}}$. Although the difference between these velocities must be small because the distance between the dividing interfaces with $\rho^{\mathrm{s}} = 0$ and $e^{\mathrm{s}} = 0$ is of the order of the thickness of the interfacial region, this difference is exactly what matters for the evolution of the interfacial energy in the gauge $\rho^{\mathrm{s}} = 0$. The physical relevance of the small difference $v^{\mathrm{s}}_{ne} - v^{\mathrm{s}}_{n\rho}$ is a consequence of the fact that we are dealing with ambiguous variables. Each v^{s}_{na} can be made to vanish by a proper choice of the dividing interface within the interfacial region but, in general, one cannot make all these small excess quantities vanish simultaneously.

We next introduce a splitting of the velocity v^{s} into translational and deformational parts, $v^{\mathrm{s}}_{\mathrm{tr}}$ and $v^{\mathrm{s}}_{\mathrm{def}}$, which is motivated by the wish to introduce partial time derivatives of fields living in the interface. We are then faced with the problem of defining a time derivative at a "fixed" position within the dividing interface although the interface is moving. Such a concept requires a given mapping of equivalent points of the interface at different times.[5] Instead of a global mapping, we equivalently introduce the velocity $v^{\mathrm{s}}_{\mathrm{tr}}$ associated with a trajectory of equivalent points at different times (see Figure 14.2) and write

$$v^{\mathrm{s}} = v^{\mathrm{s}}_{\mathrm{tr}} + v^{\mathrm{s}}_{\mathrm{def}}. \tag{14.11}$$

The conditions that equivalent points follow the interface and that the deformation takes place within the interface are expressed by

$$\boldsymbol{n} \cdot v^{\mathrm{s}}_{\mathrm{tr}} = \boldsymbol{n} \cdot v^{\mathrm{s}}, \quad \boldsymbol{n} \cdot v^{\mathrm{s}}_{\mathrm{def}} = 0. \tag{14.12}$$

The minimalist way of "following the interface" is given by the normal–parallel splitting

$$v^{\mathrm{s}}_{\mathrm{tr}} = \boldsymbol{nn} \cdot v^{\mathrm{s}} = v^{\mathrm{s}}_n \boldsymbol{n}, \quad v^{\mathrm{s}}_{\mathrm{def}} = (\boldsymbol{\delta} - \boldsymbol{nn}) \cdot v^{\mathrm{s}} = v^{\mathrm{s}}_{\parallel}. \tag{14.13}$$

Whereas we mostly assume the normal–parallel splitting (14.13), other choices are possible and may be more reasonable. A simple situation in which we would choose a different splitting is given by the uniform motion of a non-deforming interface; in that situation we would clearly choose $v^{\mathrm{s}}_{\mathrm{tr}} = v^{\mathrm{s}}$, $v^{\mathrm{s}}_{\mathrm{def}} = \boldsymbol{0}$ as an alternative solution of (14.12). An important ingredient in the general splitting expressed in (14.11) and (14.12) is that uniform motion effects only $v^{\mathrm{s}}_{\mathrm{tr}}$ so that our equations are Galilean invariant (see the end of Section 11.1).

We are now in a position to introduce an "interface partial time derivative" by following the interface without deformation, symbolically expressed as

$$\frac{\partial^{\mathrm{s}}}{\partial t} = \frac{\partial}{\partial t} + v^{\mathrm{s}}_{\mathrm{tr}} \cdot \nabla = \frac{D}{Dt} - v^{\mathrm{s}}_{\mathrm{def}} \cdot \nabla_{\parallel}. \tag{14.14}$$

Note that (14.14) is merely a formal equation because, strictly speaking, the differential operators in the intermediate expression have no meaning. This

[5] Slattery *et al.*, *Interfacial Transport Phenomena* (Springer, 2007).

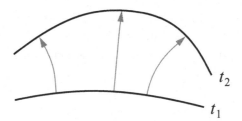

Figure 14.2 Instead of mapping equivalent points of a continuously deform-
ing interface at two times t_1 and t_2 (thick lines), we can alternatively look
at the trajectories associated with equivalent points (thin arrows) and their
time derivatives providing the translational velocity field v_{tr}^s.

heuristic equation is only meant to express the idea that for an interface
partial time derivative we need to follow the equivalent points in the spirit
of a substantial or material time derivative. For the splitting in (14.13), the
partial time derivative in (14.14) is taken at a position within the interface
that is "as constant as possible."

For interfaces having a time-dependent unit normal vector n, such as
the one shown in Figure 14.2, determining the translational part of the
interface velocity requires following the trajectories of equivalent points
on the interface. The expression for n in (13.26) follows from the equation $\varphi^s(r, t) = 0$. Taking the material derivative of this equation, we can
write

$$\frac{\partial \varphi^s}{\partial t} + (v^s \cdot n)|\nabla \varphi^s| = 0, \tag{14.15}$$

which gives the condition that a point remains on a surface at all times.

Exercise 14.1   Convective Transport within the Interface
Show that only for the normal–parallel splitting in (14.13) is the following valid,

$$\nabla_\| \cdot (v_{def}^s a^s) + a^s \, \nabla_\| \cdot v_{tr}^s = \nabla_\| \cdot (v^s a^s). \tag{14.16}$$

Exercise 14.2   Mass and Momentum Balances at a Spherical Interface
Consider a spherical interface having unit normal $n = \delta_r$ and location $r = R$
separating phases I and II that are Newtonian fluids with constant density. The
interface velocity has only a normal component $v^s \cdot n = dR/dt$. Assume spherical
symmetry so that fields depend only on radial position r and the velocities in the
bulk phases are purely radial. For convenience, use the notation $\rho^I = \rho$, $\rho^{II} = \bar{\rho}$ for
density, and $v^I = v(R)$, $v^{II} = \bar{v}(R)$ for velocity. Show that the jump mass balance

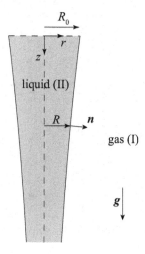

Figure 14.3 Falling liquid (II) jet surrounded by gas (I).

(14.5) takes the form

$$\rho\left[v_r(R) - \frac{dR}{dt}\right] = \bar{\rho}\left[\bar{v}_r(R) - \frac{dR}{dt}\right]. \tag{14.17}$$

Also show that the normal component of the jump momentum balance (14.6) with $\pi^s = -\gamma\boldsymbol{\delta}_\parallel$ can be written as

$$\rho[v_r(R) - \bar{v}_r(R)]\left[v_r(R) - \frac{dR}{dt}\right] + p(R) - 2\eta\frac{\partial v_r}{\partial r}(R) = \bar{p}(R) - 2\bar{\eta}\frac{\partial \bar{v}_r}{\partial r}(R) - 2\frac{\gamma}{R}, \tag{14.18}$$

where η and $\bar{\eta}$ are the viscosities of phases I and II, respectively.

Exercise 14.3   Shape of a Falling Liquid Jet
Consider a liquid jet falling under the influence of gravity through a gas as shown in Figure 14.3. At $z = 0$, the jet has radius R_0 and uniform velocity V. The mass flow rate of the jet $W = \rho\pi R_0^2 V$ is constant. Assume that $N_{\text{Re}} = W/(\eta\pi R_0)$ is sufficiently large that viscous effects can be neglected. If the gas is an inviscid fluid, it is reasonable to further assume that the velocity is independent of radial position. Combine the Bernoulli equation (5.54) with the jump balances (14.5) and (14.6) for mass and momentum to obtain the following equation for the jet radius:

$$\frac{R}{R_0} = \left[1 + \frac{2g}{V^2}z + \frac{2\gamma}{\rho V^2 R_0}\left(1 - \frac{R_0}{R}\right)\right]^{-1/4}.$$

Identify expressions for the Froude, N_{Fr}, and Weber, N_{We}, numbers.

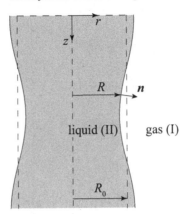

Figure 14.4 Liquid (II) cylinder with unperturbed radius R_0 surrounded by gas (I).

Exercise 14.4   The Rayleigh–Plateau Instability
The Rayleigh–Plateau instability is responsible for the breakup of liquid jets that emerge from fountains and faucets. Here, we consider the *linear stability analysis* performed by Rayleigh,[6] which excludes viscous and gravitational forces in a cylindrical liquid column surrounded by an inviscid gas as shown in Figure 14.4. Assume the radius of the perturbed surface of the liquid has the form

$$R = R_0 + \tilde{R}e^{\omega t + ikz},$$

where $\tilde{R} \ll R_0$ is the amplitude of the radius perturbation, ω is the growth rate, and k is the wave number of the disturbance. The perturbation of the surface results in small perturbations to the velocity and pressure fields in the liquid, which are assumed to have the same form as the radius perturbation. Solve the linearized fluid mechanics problem for the velocity and pressure perturbations. Show using a balance between the flow-induced pressure field and the capillary pressure that the following dispersion relation is obtained:

$$\omega^2 = \frac{\gamma}{\rho R_0^3} \frac{I_1(kR_0)}{I_0(kR_0)} kR_0 [1 - (kR_0)^2],$$

where γ is the interfacial tension of the gas–liquid interface. This result shows that the liquid cylinder is unstable to a disturbance having a wavelength that is larger than the circumference of the cylinder. Use this result to find the wave number for the most unstable (fastest growing) disturbance.

---

[6] Rayleigh, *Proc. Lond. Math. Soc.* **10** (1878) 4.

14.3 Time Evolution of Excess Densities

As pointed out before, the normal velocity v_{na}^s associated with the extensive interfacial density a^s can be different from $v_{n\rho}^s = v_{nm}^s = v_n^s$. This difference can be translated into a rate of change of the excess density for the gauge with $\rho^s = 0$. For this gauge, the rate of change of a^s at a fixed position on the interface can be expressed as

$$\frac{\partial^s a^s}{\partial t} = (v_{na}^s - v_n^s)(a^{\mathrm{II}} - a^{\mathrm{I}}) - \boldsymbol{\nabla}_{\|} \cdot (\boldsymbol{v}_{\mathrm{def}}^s\, a^s) - a^s\, \boldsymbol{\nabla}_{\|} \cdot \boldsymbol{v}_{\mathrm{tr}}^s. \tag{14.19}$$

According to (14.2), $v_{na}^s(a^{\mathrm{II}} - a^{\mathrm{I}})$ is the amount of an extensive quantity that accumulates in the interfacial region; in (14.19), the effect of the moving reference interface is subtracted. The last two terms in (14.19) correspond to convection in the interface and to a relative change of surface area, which leads to a decrease of any interfacial density. We are now in a position to obtain convenient evolution equations for various excess densities by combining our general results (14.2) and (14.19).

For the species mass balance, one should take into account a source of species mass from chemical reactions within the interface, or *heterogeneous* chemical reactions, and a diffusive mass flux j_α^s within the interface. By introducing a source term into (14.2), we obtain for the velocity $v_{n\rho_\alpha}^s$ of the dividing interface with $\rho_\alpha^s = 0$

$$v_{n\rho_\alpha}^s(\rho_\alpha^{\mathrm{II}} - \rho_\alpha^{\mathrm{I}}) = \boldsymbol{n}\cdot(\boldsymbol{v}^{\mathrm{II}}\rho_\alpha^{\mathrm{II}} + \boldsymbol{j}_\alpha^{\mathrm{II}} - \boldsymbol{v}^{\mathrm{I}}\rho_\alpha^{\mathrm{I}} - \boldsymbol{j}_\alpha^{\mathrm{I}}) - \boldsymbol{\nabla}_{\|}\cdot\boldsymbol{j}_\alpha^s + \nu_\alpha^s\Gamma^s, \tag{14.20}$$

where $\Gamma^s = (1/A)d\xi^s/dt$ is the mass rate of reaction per unit area, and ν_α^s is related to the stoichiometric coefficient of species α (see Sections 4.6 and 5.1). By combining with (14.19), we obtain evolution equations for the excess species mass densities

$$\frac{\partial^s \rho_\alpha^s}{\partial t} = -\boldsymbol{\nabla}_{\|}\cdot(\boldsymbol{v}_{\mathrm{def}}^s\rho_\alpha^s + \boldsymbol{j}_\alpha^s) - \rho_\alpha^s\,\boldsymbol{\nabla}_{\|}\cdot\boldsymbol{v}_{\mathrm{tr}}^s + \nu_\alpha^s\Gamma^s$$
$$+ \boldsymbol{n}\cdot[(\boldsymbol{v}^{\mathrm{II}} - \boldsymbol{v}^s)\rho_\alpha^{\mathrm{II}} + \boldsymbol{j}_\alpha^{\mathrm{II}} - (\boldsymbol{v}^{\mathrm{I}} - \boldsymbol{v}^s)\rho_\alpha^{\mathrm{I}} - \boldsymbol{j}_\alpha^{\mathrm{I}}], \tag{14.21}$$

where we have chosen the gauge in which the total mass density vanishes $\rho^s = 0$. For systems with chemical reaction, the species masses change in a concerted way so the total mass does not change. Moreover, we assumed $\sum_{\alpha=1}^k \boldsymbol{j}_\alpha^s = \boldsymbol{0}$.

By applying (14.2) to the excess total energy density e^s, we find

$$v_{ne}^s(e^{\mathrm{II}} - e^{\mathrm{I}}) = \boldsymbol{n}\cdot(\boldsymbol{v}^{\mathrm{II}}e^{\mathrm{II}} + \boldsymbol{j}_q^{\mathrm{II}} + \boldsymbol{\pi}^{\mathrm{II}}\cdot\boldsymbol{v}^{\mathrm{II}} - \boldsymbol{v}^{\mathrm{I}}e^{\mathrm{I}} - \boldsymbol{j}_q^{\mathrm{I}} - \boldsymbol{\pi}^{\mathrm{I}}\cdot\boldsymbol{v}^{\mathrm{I}}) - \boldsymbol{\nabla}_{\|}\cdot(\boldsymbol{j}_q^s + \boldsymbol{\pi}^s\cdot\boldsymbol{v}^s), \tag{14.22}$$

where we have introduced $\boldsymbol{j}_q^s$ for the diffusive flux of energy within the interface. By combining this with (14.19) for the total energy density, we arrive at the evolution equation for the excess total energy density

$$\frac{\partial^s e^s}{\partial t} = -\nabla_\parallel \cdot (v^s_{\text{def}} e^s + j^s_q + \pi^s \cdot v^s) - e^s \, \nabla_\parallel \cdot v^s_{\text{tr}}$$
$$+ n \cdot \left[(v^{\text{II}} - v^s) e^{\text{II}} + j^{\text{II}}_q + \pi^{\text{II}} \cdot v^{\text{II}} - (v^{\text{I}} - v^s) e^{\text{I}} - j^{\text{I}}_q - \pi^{\text{I}} \cdot v^{\text{I}} \right],$$

$$(14.23)$$

where we have again used the reference velocity v^s corresponding to the gauge $\rho^s = 0$. Hence, the evolution equation for internal energy density u^s is simply obtained by setting $e^s = u^s$ in (14.23). Using (14.6), (14.8), and (14.9), we obtain (see Exercise 14.5) the evolution equation for the excess internal energy density,

$$\frac{\partial^s u^s}{\partial t} = -\nabla_\parallel \cdot (v^s_{\text{def}} u^s + j^s_q) - u^s \, \nabla_\parallel \cdot v^s_{\text{tr}} - \pi^s : \nabla_\parallel v^s$$

$$+ n \cdot \left[(v^{\text{II}} - v^s) h^{\text{II}} + j^{\text{II}}_q - (v^{\text{I}} - v^s) h^{\text{I}} - j^{\text{I}}_q \right] + \left(\frac{1}{\rho^{\text{II}}} - \frac{1}{\rho^{\text{I}}} \right)^{-1}$$

$$\times \left\{ \left[\frac{1}{2} \frac{\rho^{\text{I}} + \rho^{\text{II}}}{\rho^{\text{I}} - \rho^{\text{II}}} (v^{\text{II}} - v^{\text{I}})^2 + n \cdot \left(\frac{\tau^{\text{II}}}{\rho^{\text{II}}} - \frac{\tau^{\text{I}}}{\rho^{\text{I}}} \right) \cdot n \right] n \cdot (v^{\text{II}} - v^{\text{I}}) \right.$$

$$\left. + n \cdot \left(\frac{\tau^{\text{II}}}{\rho^{\text{II}}} - \frac{\tau^{\text{I}}}{\rho^{\text{I}}} \right) \cdot (v^{\text{II}} - v^{\text{I}})_\parallel \right\}.$$

$$(14.24)$$

It is important to note that, in contrast to the procedure used to derive (5.51) for a bulk phase, (14.24) is not obtained by subtracting the mechanical energy from the total energy. The terms in the third line of (14.24) are a source due to a discontinuity in the normal velocity (phase change), and the fourth line is a source due to a discontinuity in the tangential velocity (slip).

In applications of interfacial balance equations it is helpful to distinguish between *active* and *passive* interfaces. The theory of this chapter is developed for the general case of an active interface, where the interface is characterized by time-evolution equations for excess densities, like the balance equations (14.21), (14.23), or (14.24). The same gauge, $\rho^s = 0$, is assumed in all these evolution equations. A much simpler situation arises if all ambiguous excess densities can simultaneously be assumed to be zero so that the various normal velocities of the interface coincide. As the time-evolution equations are then reduced to jump balances, we refer to this simpler case as that of passive interfaces. Whereas active interfaces are governed by evolution equations for interfacial variables, passive interfaces are governed by jump balances that do not involve time derivatives.

For passive interfaces, (14.21) leads to the *jump species mass balances*

$$n \cdot \left[(v^{\text{I}} - v^s) \rho^{\text{I}}_\alpha + j^{\text{I}}_\alpha \right] = n \cdot \left[(v^{\text{II}} - v^s) \rho^{\text{II}}_\alpha + j^{\text{II}}_\alpha \right] - \nabla_\parallel \cdot j^s_\alpha + \nu^s_\alpha \Gamma^s, \quad (14.25)$$

and, similarly, (14.24) simplifies to the *jump energy balance*

$$\boldsymbol{n} \cdot [(\boldsymbol{v}^{\mathrm{I}} - \boldsymbol{v}^{\mathrm{s}})h^{\mathrm{I}} + j_q^{\mathrm{I}}] = \boldsymbol{n} \cdot [(\boldsymbol{v}^{\mathrm{II}} - \boldsymbol{v}^{\mathrm{s}})h^{\mathrm{II}} + j_q^{\mathrm{II}}] + \left(\frac{1}{\rho^{\mathrm{II}}} - \frac{1}{\rho^{\mathrm{I}}} \right)^{-1}$$

$$\times \left\{ \left[\frac{1}{2} \frac{\rho^{\mathrm{I}} + \rho^{\mathrm{II}}}{\rho^{\mathrm{I}} - \rho^{\mathrm{II}}} (\boldsymbol{v}^{\mathrm{II}} - \boldsymbol{v}^{\mathrm{I}})^2 \right. \right.$$

$$\left. + \boldsymbol{n} \cdot \left(\frac{\boldsymbol{\tau}^{\mathrm{II}}}{\rho^{\mathrm{II}}} - \frac{\boldsymbol{\tau}^{\mathrm{I}}}{\rho^{\mathrm{I}}} \right) \cdot \boldsymbol{n} \right] \boldsymbol{n} \cdot (\boldsymbol{v}^{\mathrm{II}} - \boldsymbol{v}^{\mathrm{I}})$$

$$\left. + \boldsymbol{n} \cdot \left(\frac{\boldsymbol{\tau}^{\mathrm{II}}}{\rho^{\mathrm{II}}} - \frac{\boldsymbol{\tau}^{\mathrm{I}}}{\rho^{\mathrm{I}}} \right) \cdot (\boldsymbol{v}^{\mathrm{II}} - \boldsymbol{v}^{\mathrm{I}})_{\parallel} \right\}$$

$$- \boldsymbol{\nabla}_{\parallel} \cdot j_q^{\mathrm{s}} - \boldsymbol{\pi}^{\mathrm{s}} : \boldsymbol{\nabla}_{\parallel} \boldsymbol{v}^{\mathrm{s}}. \tag{14.26}$$

In most situations, the terms for kinetic energy and extra stress associated with phase change in the second and third lines of (14.26) can be neglected (see Exercise 14.7). As in the case of the jump momentum balance (14.6), some interfacial terms affect the equations relating the jumps between bulk phases in the jump balances for species mass (14.25) and energy (14.26).

It is worthwhile to make some observations about the structure of the interfacial balance equations we have just formulated. More specifically, let us compare the balance equation for species mass in the bulk (5.14) with its interfacial counterpart in (14.21). The left-hand sides express the time rate of change of a density – at a fixed point in space for the bulk, and at a fixed point on the surface for the interface. Notice that in going from the bulk to the interface, terms corresponding to divergences of fluxes $-\boldsymbol{\nabla} \cdot (.)$ appear in two forms: one is a surface divergence $-\boldsymbol{\nabla}_{\parallel} \cdot (.)$ and the second appears as a difference $\boldsymbol{n} \cdot [(.)^{\mathrm{II}} - (.)^{\mathrm{I}}]$, where motion of the interface is subtracted from the convective part. This structure is also found in the jump balances (14.5) and (14.6), which, however, are not time-evolution equations. Terms of the form $a^{\mathrm{s}} \boldsymbol{\nabla}_{\parallel} \cdot \boldsymbol{v}_{\mathrm{tr}}^{\mathrm{s}}$ in (14.21), which do not have counterparts in the bulk, describe changes in an excess density due to changes in surface area.

Exercise 14.5   Interfacial Internal Energy Balance Equation
Derive the balance equation for excess internal energy given in (14.24).

Exercise 14.6   Molar Form of Jump Balance for Species Mass
Show that the jump balance for species mass (14.25) can be written as

$$\boldsymbol{n} \cdot [(\boldsymbol{v}^{*\mathrm{I}} - \boldsymbol{v}^{\mathrm{s}})c_\alpha^{\mathrm{I}} + \boldsymbol{J}_\alpha^{*\mathrm{I}}] = \boldsymbol{n} \cdot [(\boldsymbol{v}^{*\mathrm{II}} - \boldsymbol{v}^{\mathrm{s}})c_\alpha^{\mathrm{II}} + \boldsymbol{J}_\alpha^{*\mathrm{II}}] - \boldsymbol{\nabla}_{\parallel} \cdot \boldsymbol{J}_\alpha^{\mathrm{s}} + \tilde{\nu}_\alpha^{\mathrm{s}} \tilde{\Gamma}^{\mathrm{s}}, \tag{14.27}$$

where $\boldsymbol{J}_\alpha^{\mathrm{s}} = j_\alpha^{\mathrm{s}} / \tilde{M}_\alpha$ is the molar flux within the interface relative to $\boldsymbol{v}_{\mathrm{def}}^{\mathrm{s}}$, $\tilde{\nu}_\alpha^{\mathrm{s}} = \nu_\alpha^{\mathrm{s}} \tilde{M}_q^{\mathrm{s}} / \tilde{M}_\alpha$ are the stoichiometric coefficients for the interfacial reaction, and $\tilde{\Gamma}^{\mathrm{s}} = \Gamma^{\mathrm{s}} / \tilde{M}_q^{\mathrm{s}}$ is the molar rate of reaction per unit area.

Exercise 14.7   Jump Balance for Energy with Phase Change
Consider the spherical surface with radius R described in Exercise 14.2. Show that
for this case the jump balance for energy (14.26) can be written as

$$-\lambda \frac{\partial T}{\partial r}(R) = -\bar{\lambda} \frac{\partial \bar{T}}{\partial r}(R) - \gamma \frac{2}{R} \frac{dR}{dt} + \rho \left(v_r(R) - \frac{dR}{dt} \right)$$

$$\times \left\{ \Delta \hat{h} + \frac{1}{2} \left(\frac{1}{\bar{\rho}} + \frac{1}{\rho} \right) \left(\frac{1}{\bar{\rho}} - \frac{1}{\rho} \right)^{-1} [\bar{v}_r(R) - v_r(R)]^2 \right.$$

$$\left. - 2\nu \frac{\partial v_r}{\partial r}(R) + 2\bar{\nu} \frac{\partial \bar{v}_r}{\partial r}(R) \right\}, \tag{14.28}$$

where $\Delta \hat{h} = \hat{h}^{II} - \hat{h}^{I}$ is the difference in specific enthalpy between phases I and II,
and λ and $\bar{\lambda}$ are the thermal conductivities of phases I and II, respectively.

Consider the phase change for H_2O involving a spherical particle composed of
either a solid (ice) or vapor (steam), surrounded by a liquid (water) that is otherwise
stationary. Use (14.28) to compare the magnitudes of the mechanical energy effects
with those associated with the enthalpy change. Assume that the densities of both
the particle and the surrounding liquid are constant.

Exercise 14.8   Solidification of a Stagnant Liquid
A quiescent liquid occupying the region $x_1 > 0$ has uniform initial temperature
T_0. For $t > 0$, the temperature at $x_1 = 0$ is maintained at a temperature of T_1,
where $T_1 < T_m < T_0$, and T_m is the melting temperature of the material. This
causes the liquid to solidify, and the solid–liquid interface is located at $x_1 = H(t)$.
Both the solid and liquid have constant physical properties denoted as $\bar{\rho}, \bar{\hat{c}}_p, \bar{\lambda}$ and
$\rho, \hat{c}_p, \lambda$, respectively. Assume that $\bar{\rho} \approx \rho$, that the interface can be treated as passive, and that interfacial heat transfer resistances are negligible. Use the similarity
transformation $\xi = x_1/\sqrt{4\bar{\chi}t}$, where $\bar{\chi} = \bar{\lambda}/(\bar{\rho}\bar{\hat{c}}_p)$, to derive the expression

$$H = \beta \sqrt{4\bar{\chi}t},$$

where

$$\frac{\sqrt{\pi}\beta}{N_{St}} = \frac{\exp(-\beta^2)}{\mathrm{erf}(\beta)} - \sqrt{\alpha} \frac{\lambda}{\bar{\lambda}} \frac{1 - \Theta}{\Theta} \frac{\exp(-\sqrt{\alpha}\beta^2)}{\mathrm{erfc}(\sqrt{\alpha}\beta)},$$

$\Theta = (T_m - T_1)/(T_0 - T_1)$, and $N_{St} = \bar{\hat{c}}_p(T_0 - T_1)/\Delta \hat{h}$ is the Stefan number.

Exercise 14.9   Evaporation of a Liquid into a Stagnant Gas
A pure liquid with density ρ composed of species A partially fills a long cylindrical
container. Above the liquid the cylinder is filled with a stagnant gas that is initially
species B. For $t > 0$, the liquid is allowed to evaporate into the gas, forming a
mixture of species A and B that behaves as an ideal gas. Liquid is added to the
cylinder so that the gas–liquid interface is stationary and located at $x_1 = 0$ and the

gas mixture occupies the region $x_1 > 0$. Species B is insoluble in liquid A, and the equilibrium mole fraction of species A at the gas–liquid interface is x_{Aeq}. The system can be treated as isothermal and there are no chemical reactions. Use the similarity transformation $\xi = x_1/\sqrt{4D_{AB}t}$ to derive an expression for the evaporation rate. Assume the interface can be treated as passive and that interfacial mass transfer resistances are negligible. Also derive an expression to estimate the error introduced by neglecting convective transport.

Exercise 14.10   *Evaporation of a Liquid into a Gas with a Moving Interface*
A stationary liquid with density ρ composed of species A partially fills a container with height H. Above the liquid is a stagnant gas that is initially species B. The height of the liquid is $h(t) < H$ and has an initial value h_0. For $t > 0$, the liquid is allowed to evaporate into the gas forming an ideal gas mixture of A and B. The mole fraction of species A at the top of the container is maintained at a constant value x_{AH}. Species B is not soluble in liquid A, and the equilibrium mole fraction of species A at the gas–liquid interface is x_{Aeq}. The gas mixture has constant molar density c, and constant mass diffusivity D_{AB}. The system isothermal and there are no chemical reactions. Consider the case $c \ll \rho/\tilde{M}_A$ and use the quasi-steady-state approximation (see Section 7.5) to obtain the following expression for the height of the liquid:

$$\left(\frac{h}{H}\right)^2 - \left(\frac{h_0}{H}\right)^2 - 2\left[\left(\frac{h}{H}\right) - \left(\frac{h_0}{H}\right)\right] = 2\frac{c\tilde{M}_A}{\rho} \ln\left(\frac{1 - x_{AH}}{1 - x_{Aeq}}\right)\frac{D_{AB}t}{H^2}.$$

Exercise 14.11   *General Analysis of Sorption Experiments*
The analysis of sorption experiments presented in Section 7.4 was greatly simplified by taking the density of the liquid mixture as constant and by limiting the amount of solute that entered the liquid film. Show that, if these restrictions are removed, the sorption process is governed by the following nonlinear moving boundary problem:

$$\frac{\partial w_A}{\partial t} + v_3\frac{\partial w_A}{\partial x_3} = D_{AB}\left[\frac{\partial^2 w_A}{\partial x_3^2} - \frac{\hat{v}_A - \hat{v}_B}{w_A(\hat{v}_A - \hat{v}_B) + \hat{v}_B}\left(\frac{\partial w_A}{\partial x_3}\right)^2\right],$$

which is solved subject to initial and boundary conditions in (7.39), and

$$\frac{\partial v_3}{\partial x_3} = D_{AB}\frac{\hat{v}_A - \hat{v}_B}{w_A(\hat{v}_A - \hat{v}_B) + \hat{v}_B}\left[\frac{\partial^2 w_A}{\partial x_3^2} - \frac{\hat{v}_A - \hat{v}_B}{w_A(\hat{v}_A - \hat{v}_B) + \hat{v}_B}\left(\frac{\partial w_A}{\partial x_3}\right)^2\right],$$

which is solved subject to (7.37), and

$$\frac{dH}{dt} = v_3(H,t) + \frac{D_{AB}}{1 - w_{Aeq}}\frac{\partial w_A}{\partial x_3}(H,t),$$

which is integrated subject to $H(0) = H_0$. Note that, if $\hat{v}_A - \hat{v}_B = 0$ for all w_A, then $v_3 = 0$ and the first equation becomes linear. However, the problem still involves a moving boundary because of the last equation.

- A unique interfacial velocity vector can be introduced via a Clapeyron-type equation for the ratio of the jumps in momentum density and mass density.
- For every extensive quantity with an ambiguous excess density at the interface, we can introduce a normal velocity of the interface for which this excess density vanishes at all times; differences between the various normal velocities provide the essential information for obtaining interfacial balance equations.
- For mass and momentum, we obtain jump balances relating jumps between bulk phases; for species masses and energy, we obtain balance equations involving time-derivatives of the corresponding excess densities in the particular gauge of a vanishing excess density for the total mass.
- We distinguish between active and passive interfaces; for passive interfaces, all evolution equations for excess densities simplify to jump balances.

15

Interfacial Force–Flux Relations

We have just formulated jump balances for mass and momentum and evolution equations for the species mass and energy excess densities that characterize the physics of the interface. As with the balance equations developed in Chapter 5, balance equations for interfaces require constitutive equations, or force–flux relations, to reconcile the number of equations and unknowns. We found in Chapter 6 that consideration of entropy production and local thermodynamic equilibrium are key ingredients in the formulation of force–flux relations in the bulk. In the present chapter, we use the same approach to formulate force–flux relations for transport within the interface and between the interface and bulk phases.

15.1 Local Equilibrium and Entropy Balance

Up to this point, our discussion has been based entirely on balancing and gauge transformations. We now introduce the idea of local equilibrium thermodynamics to relate the excess energy and entropy densities to temperature, interfacial tension, and chemical potential.

The concept of local equilibrium plays a key role in the description of transport processes, as should be obvious from the detailed discussion in Section 6.1. In bulk phases, we found it quite intuitive to go from global to local equilibrium by considering small boxes. For gases, these boxes may have to be at least of micron size; for a liquid even nanometer-sized boxes may make sense. For interfaces idealized as two-dimensional systems, the concept of local equilibrium becomes much more subtle. Can one assign a local temperature to the dividing interface? How independent is its definition of the states of the surrounding bulk phases, which are in direct contact with the interface?

A nonequilibrium thermodynamic description of interfaces based on Gibbs' approach reviewed in the preceding sections has been initiated by

Waldmann[1] and further developed by Bedeaux and coworkers.[2] The theory of local equilibrium for interfaces has been reviewed and further clarified from the perspective of gauge invariance by Savin *et al.*[3] The essential idea is that, when in contact with nonequilibrium bulk phases, interfaces quickly relax to states that can locally be described by the same set of variables as equilibrium interfaces, with the same thermodynamic relations between these variables.

The shape of the interfacial region or any characteristic property derived from it may be taken as a thermometer for the interfacial temperature of a one-component system. Most conveniently, we can choose the interfacial tension as an interfacial thermometer, that is, we can build a look-up table from the equilibrium results for $\gamma(T)$ to assign a temperature T^s to the dividing interface and to fix its local equilibrium state. The function $\hat{\mu}(T)$ encoding the condition for phase coexistence can then be used to define an unambiguous interfacial chemical potential, $\hat{\mu}^s = \hat{\mu}(T^s)$. The introduction of T^s and $\hat{\mu}^s$ should not be mistaken for a mere notational issue – rather the introduction of local equilibrium for interfaces is a deep physical problem.

Even out of global equilibrium, the excess densities still have a gauge degree of freedom, where the gauge transformations (13.1)–(13.3) now involve jumps of the bulk densities extrapolated to the dividing interface. Note that, under nonequilibrium conditions, the thermodynamic properties of the bulk phases are no longer constant in space, but the slopes in plots like in Figure 13.3 must be very small, so that the concept of extrapolated "bulk densities at the interface" is unproblematic.

When passing from a one-component to a k-component system, according to the Gibbs phase rule (4.37), we need $k - 1$ further independent variables to characterize the local equilibrium state. In view of the conditions for the coexistence of two k-component bulk phases, it is convenient to use $k - 1$ chemical potentials as independent variables, say $\hat{\mu}^s_1, \ldots, \hat{\mu}^s_{k-1}$. In this case, the structure of an interfacial region is characterized by $\gamma = \gamma(T^s, \hat{\mu}^s_1, \ldots, \hat{\mu}^s_{k-1})$, and the condition for phase coexistence is given by $\hat{\mu}^s_k = \hat{\mu}^s_k(T^s, \hat{\mu}^s_1, \ldots, \hat{\mu}^s_{k-1})$.

More intuitive and directly accessible additional variables, for example in a computer simulation, are the gauge-invariant adsorptions $\Upsilon_{\alpha k}$ of components $\alpha = 1, \ldots, k - 1$ relative to component k,

$$\Upsilon_{\alpha k} = \rho^s_\alpha - \rho^s_k \frac{\rho^I_\alpha - \rho^{II}_\alpha}{\rho^I_k - \rho^{II}_k} = -\left(\frac{\partial \gamma}{\partial \hat{\mu}_\alpha}\right)_{T, \hat{\mu}_{\beta \neq \alpha, k}}, \qquad (15.1)$$

[1] Waldmann, *Z. Naturforsch. A* **22** (1967) 1269.
[2] Bedeaux *et al.*, *Physica A* **82** (1976) 438; Bedeaux, *Adv. Chem. Phys.* **64** (1986) 47; Albano & Bedeaux, *Physica A* **147** (1987) 407.
[3] Savin *et al.*, *Europhys. Lett.* **97** (2012) 40002; Schweizer *et al.*, *Phys. Rev. E* **93** (2016) 052803.

where the species bulk densities are taken at the dividing interface. The second equality is obtained using (13.19) and (13.22). The gauge invariance of these adsorptions $\Upsilon_{\alpha k}$ follows from the gauge transformations in (13.4). It may also be convenient to consider the mole number densities $c_\alpha^s = \rho_\alpha^s / \tilde{M}_\alpha$, where $\tilde{M}_\alpha$ is the molecular weight of species α, so that (13.4) implies analogous gauge transformations for c_α^s. We can then define gauge-invariant adsorptions $\tilde{\Upsilon}_{\alpha k}$ by replacing ρ by c everywhere in (15.1). This definition implies $\tilde{\Upsilon}_{\alpha k} = \Upsilon_{\alpha k} / \tilde{M}_\alpha = -(\partial \gamma / \partial \tilde{\mu}_\alpha)_{T, \tilde{\mu}_{\beta \neq \alpha, k}}$, which is known as a *Gibbs adsorption isotherm*. An attractive alternative of defining gauge-invariant adsorptions as single-index quantities is considered in Exercise 15.1.

Now, according to the Gibbs phase rule, the species chemical potentials are constrained by one additional condition for two-phase coexistence. To formulate this condition, we assume that the two bulk phases brought into coexistence can be characterized by the thermodynamic potentials $p^{\mathrm{I}}(T, \hat{\mu}_1, \ldots, \hat{\mu}_k)$ and $p^{\mathrm{II}}(T, \hat{\mu}_1, \ldots, \hat{\mu}_k)$. For planar interfaces, the constraint for the species chemical potentials can be expressed in the elegant form

$$p^{\mathrm{I}}(T, \hat{\mu}_1, \ldots, \hat{\mu}_k) = p^{\mathrm{II}}(T, \hat{\mu}_1, \ldots, \hat{\mu}_k). \tag{15.2}$$

For the further analysis of local equilibrium for moving interfaces, we consider the Euler equation

$$e^s = T^s s^s + \gamma + \sum_{\alpha=1}^{k} \left(\hat{\mu}_\alpha^s + \frac{1}{2} v^{s2} \right) \rho_\alpha^s, \tag{15.3}$$

which we obtain by adding the kinetic energy density $(1/2)\rho^s v^{s2}$ to the local-equilibrium version of (13.16). The gauge invariance of the Euler equation for the general transformation law (13.27) can then be expressed in the form

$$e^{\mathrm{II}} - e^{\mathrm{I}} - T^s(s^{\mathrm{II}} - s^{\mathrm{I}}) - \sum_{\alpha=1}^{k} \left(\hat{\mu}_\alpha^s + \frac{1}{2} v^{s2} \right) (\rho_\alpha^{\mathrm{II}} - \rho_\alpha^{\mathrm{I}}) = -\gamma \, \boldsymbol{\nabla}_\| \cdot \boldsymbol{n}, \tag{15.4}$$

which one can recognize as a nonequilibrium generalization of the Young–Laplace equation. At equilibrium, $v^{\mathrm{I}} = v^{\mathrm{II}} = v^s = \mathbf{0}$, $T^{\mathrm{I}} = T^{\mathrm{II}} = T^s$, and $\hat{\mu}_\alpha^{\mathrm{I}} = \hat{\mu}_\alpha^{\mathrm{II}} = \hat{\mu}_\alpha^s$, so that (15.4) indeed reduces to $p^{\mathrm{II}} - p^{\mathrm{I}} = \gamma \, \boldsymbol{\nabla}_\| \cdot \boldsymbol{n}$. This result is also obtained from the jump momentum balance (14.6) for a system at equilibrium. More generally, the approximate consistency of (14.6) and (15.4) can hence be regarded as an indicator for the validity of the local-equilibrium assumption.

By differentiating the Euler equation (15.3) and using the Gibbs–Duhem equation (13.17), we obtain the fundamental form

$$de^s = T^s \, ds^s + \sum_{\alpha=1}^{k} \left(\hat{\mu}_\alpha^s - \frac{1}{2} v^{s2} \right) d\rho_\alpha^s + v^s \cdot d\boldsymbol{m}^s, \tag{15.5}$$

where $\boldsymbol{v}^{\mathrm{s}} \cdot \boldsymbol{v}^{\mathrm{s}} \, d\rho_\alpha^{\mathrm{s}} + \rho_\alpha^{\mathrm{s}} \boldsymbol{v}^{\mathrm{s}} \cdot d\boldsymbol{v}^{\mathrm{s}} = \boldsymbol{v}^{\mathrm{s}} \cdot d(\rho_\alpha^{\mathrm{s}} \boldsymbol{v}^{\mathrm{s}})$ has been used. A term of the form $\boldsymbol{v}^{\mathrm{s}} \cdot d\boldsymbol{m}^{\mathrm{s}}$ associated with kinetic energy has previously been discussed in (4.6).

By applying the general evolution equation obtained by combining (14.2) and (14.19) for $a = s$, we obtain the balance equation for excess entropy density,

$$\frac{\partial^{\mathrm{s}} s^{\mathrm{s}}}{\partial t} = -\boldsymbol{\nabla}_{\|} \cdot (\boldsymbol{v}_{\mathrm{def}}^{\mathrm{s}} s^{\mathrm{s}} + \boldsymbol{j}_s^{\mathrm{s}}) - s^{\mathrm{s}} \, \boldsymbol{\nabla}_{\|} \cdot \boldsymbol{v}_{\mathrm{tr}}^{\mathrm{s}} + \sigma^{\mathrm{s}}$$

$$+ \, \boldsymbol{n} \cdot \left[(\boldsymbol{v}^{\mathrm{II}} - \boldsymbol{v}^{\mathrm{s}}) s^{\mathrm{II}} + \boldsymbol{j}_s^{\mathrm{II}} - (\boldsymbol{v}^{\mathrm{I}} - \boldsymbol{v}^{\mathrm{s}}) s^{\mathrm{I}} - \boldsymbol{j}_s^{\mathrm{I}} \right] , \qquad (15.6)$$

in the gauge fixed by the condition $\rho^{\mathrm{s}} = 0$. Note that we have added the entropy production rate σ^{s} as a source term in the same way as we had added a source of species mass from chemical reaction in (14.21).

Exercise 15.1   Alternative Definition of Adsorptions

The adsorptions $\Upsilon_{\alpha k}$ correspond to the excess densities ρ_α^{s} in the gauge given by $\rho_k^{\mathrm{s}} = 0$. Define the gauge-invariant adsorptions $\tilde{\Upsilon}_\alpha$ given by ρ_α^{s} in the gauge for which the excess of the total mass density vanishes, that is $\rho^{\mathrm{s}} = 0$.

15.2 One-Component Systems

Following the development in Chapter 6, we first consider one-component systems. As we have already established the gauge invariance of the local-equilibrium assumption, we now apply (15.5) in the gauge $\rho^{\mathrm{s}} = 0$, which for one-component systems gives

$$\frac{\partial^{\mathrm{s}} e^{\mathrm{s}}}{\partial t} = T^{\mathrm{s}} \frac{\partial^{\mathrm{s}} s^{\mathrm{s}}}{\partial t} . \qquad (15.7)$$

Substituting (14.23) and (15.6) into (15.7), using (6.9) to replace the entropy fluxes in the bulk phases, and assuming $\boldsymbol{j}_s^{\mathrm{s}} = \boldsymbol{j}_q^{\mathrm{s}}/T^{\mathrm{s}}$ for the entropy flux within the dividing interface gives

$$T^{\mathrm{s}} \sigma^{\mathrm{s}} = -\boldsymbol{\nabla}_{\|} \cdot (\boldsymbol{v}_{\mathrm{def}}^{\mathrm{s}} e^{\mathrm{s}}) + T^{\mathrm{s}} \, \boldsymbol{\nabla}_{\|} \cdot (\boldsymbol{v}_{\mathrm{def}}^{\mathrm{s}} s^{\mathrm{s}}) - (e^{\mathrm{s}} - T^{\mathrm{s}} s^{\mathrm{s}}) \boldsymbol{\nabla}_{\|} \cdot \boldsymbol{v}_{\mathrm{tr}}^{\mathrm{s}} + \gamma \, \boldsymbol{\nabla}_{\|} \cdot \boldsymbol{v}^{\mathrm{s}}$$

$$+ \, T^{\mathrm{s}} \boldsymbol{j}_q^{\mathrm{s}} \cdot \boldsymbol{\nabla}_{\|} \frac{1}{T^{\mathrm{s}}} - \boldsymbol{\tau}^{\mathrm{s}} : \boldsymbol{\nabla}_{\|} \boldsymbol{v}^{\mathrm{s}}$$

$$+ \, \boldsymbol{n} \cdot \left[(\boldsymbol{v}^{\mathrm{II}} - \boldsymbol{v}^{\mathrm{s}})(e^{\mathrm{II}} - T^{\mathrm{s}} s^{\mathrm{II}}) - (\boldsymbol{v}^{\mathrm{I}} - \boldsymbol{v}^{\mathrm{s}})(e^{\mathrm{I}} - T^{\mathrm{s}} s^{\mathrm{I}}) \right]$$

$$+ \, T^{\mathrm{s}} \boldsymbol{n} \cdot \boldsymbol{j}_q^{\mathrm{II}} \left(\frac{1}{T^{\mathrm{s}}} - \frac{1}{T^{\mathrm{II}}} \right) - T^{\mathrm{s}} \boldsymbol{n} \cdot \boldsymbol{j}_q^{\mathrm{I}} \left(\frac{1}{T^{\mathrm{s}}} - \frac{1}{T^{\mathrm{I}}} \right)$$

$$+ \, \boldsymbol{n} \cdot \left[\boldsymbol{\pi}^{\mathrm{II}} \cdot (\boldsymbol{v}^{\mathrm{II}} - \boldsymbol{v}^{\mathrm{s}}) - \boldsymbol{\pi}^{\mathrm{I}} \cdot (\boldsymbol{v}^{\mathrm{I}} - \boldsymbol{v}^{\mathrm{s}}) \right]$$

$$+ \, \boldsymbol{n} \cdot (\boldsymbol{\pi}^{\mathrm{II}} - \boldsymbol{\pi}^{\mathrm{I}}) \cdot \boldsymbol{v}^{\mathrm{s}} - \boldsymbol{v}^{\mathrm{s}} \cdot (\boldsymbol{\nabla}_{\|} \cdot \boldsymbol{\pi}^{\mathrm{s}}). \qquad (15.8)$$

To obtain (15.8) we have, analogously to the decomposition of the bulk pressure tensor in (6.3), introduced the following decomposition of the surface pressure tensor,

$$\boldsymbol{\pi}^{\mathrm{s}} = -\gamma\boldsymbol{\delta}_{\parallel} + \boldsymbol{\tau}^{\mathrm{s}}, \tag{15.9}$$

where γ is the interfacial tension, and $\boldsymbol{\tau}^{\mathrm{s}}$ is the transverse excess surface stress tensor. Using (13.16) and (15.5), the first line of (15.8) vanishes. Further simplification of (15.8) is based on the desire to identify forces for transport between the bulk phases and the interface. In particular, we combine terms in the third and fourth lines so that forces for energy transport are given by temperature jumps between the interface and each bulk phase. Also, terms for momentum transport in the fifth and sixth lines are expressed in terms of forces involving velocity jumps between the bulk phases, which is achieved using (14.8)–(14.10). After some tedious but straightforward rearrangements, this allows (15.8) to be written as

$$\begin{aligned}
\sigma^{\mathrm{s}} = {}& \boldsymbol{j}_q^{\mathrm{s}} \cdot \boldsymbol{\nabla}_{\parallel}\frac{1}{T^{\mathrm{s}}} - \frac{1}{T^{\mathrm{s}}}\boldsymbol{\tau}^{\mathrm{s}} : \boldsymbol{\nabla}_{\parallel}\boldsymbol{v}^{\mathrm{s}} \\
&+ \boldsymbol{n} \cdot \left[(\boldsymbol{v}^{\mathrm{II}} - \boldsymbol{v}^{\mathrm{s}})h^{\mathrm{II}} + \boldsymbol{j}_q^{\mathrm{II}}\right]\left(\frac{1}{T^{\mathrm{s}}} - \frac{1}{T^{\mathrm{II}}}\right) \\
&- \boldsymbol{n} \cdot \left[(\boldsymbol{v}^{\mathrm{I}} - \boldsymbol{v}^{\mathrm{s}})h^{\mathrm{I}} + \boldsymbol{j}_q^{\mathrm{I}}\right]\left(\frac{1}{T^{\mathrm{s}}} - \frac{1}{T^{\mathrm{I}}}\right) \\
&+ \frac{1}{T^{\mathrm{s}}}\left(\frac{1}{\rho^{\mathrm{II}}} - \frac{1}{\rho^{\mathrm{I}}}\right)^{-1}\left\{T^{\mathrm{s}}\left(\frac{\hat{\mu}^{\mathrm{II}}}{T^{\mathrm{II}}} - \frac{\hat{\mu}^{\mathrm{I}}}{T^{\mathrm{I}}}\right) + \frac{1}{2}\frac{\rho^{\mathrm{I}} + \rho^{\mathrm{II}}}{\rho^{\mathrm{I}} - \rho^{\mathrm{II}}}(\boldsymbol{v}^{\mathrm{II}} - \boldsymbol{v}^{\mathrm{I}})^2\right. \\
&\left.+ \boldsymbol{n} \cdot \left(\frac{\boldsymbol{\tau}^{\mathrm{II}}}{\rho^{\mathrm{II}}} - \frac{\boldsymbol{\tau}^{\mathrm{I}}}{\rho^{\mathrm{I}}}\right) \cdot \boldsymbol{n}\right\}\boldsymbol{n} \cdot (\boldsymbol{v}^{\mathrm{II}} - \boldsymbol{v}^{\mathrm{I}}) \\
&+ \frac{1}{T^{\mathrm{s}}}\left(\frac{1}{\rho^{\mathrm{II}}} - \frac{1}{\rho^{\mathrm{I}}}\right)^{-1}\boldsymbol{n} \cdot \left(\frac{\boldsymbol{\tau}^{\mathrm{II}}}{\rho^{\mathrm{II}}} - \frac{\boldsymbol{\tau}^{\mathrm{I}}}{\rho^{\mathrm{I}}}\right) \cdot (\boldsymbol{v}^{\mathrm{II}} - \boldsymbol{v}^{\mathrm{I}})_{\parallel}.
\end{aligned} \tag{15.10}$$

Notice in the second through sixth lines of (15.10) that entropy production from transport between the bulk phases and the interface involves force–flux pair products, where the forces are expressed in terms of jumps. The forces for energy appear as differences of reciprocal temperature between the interface and the bulk phases. For mass and momentum transport, the force appears in terms of normal and tangential velocity jumps, respectively, between the bulk phases.

In Chapter 6, we learned how useful an analysis of the entropy production is in formulating constitutive assumptions. The same idea is used here for the entropy production within and at the interface. It is important to note that the Curie principle (see Section 6.3), which restricted the coupling between force–flux relations of different tensorial orders because the bulk phase was assumed to be isotropic, does not apply to interfaces, since a dividing interface has a preferred direction given by the unit normal $\boldsymbol{n}$. The simplest way to guarantee that the entropy production is nonnegative is to make each of the contributions in (15.23) nonnegative. More general constitutive equations that allow for coupling are possible.[4] Constitutive equations for multi-component systems are addressed in the following section.

The terms in the first line of the right-hand side of (15.10), which are analogs of the force–flux products in (6.2), are entropy production terms resulting from transport within the interface. To ensure that the first two terms are nonnegative, we assume linear expressions for $\boldsymbol{j}_q^{\mathrm{s}}$ and $\boldsymbol{\tau}^{\mathrm{s}}$,

$$\boldsymbol{j}_q^{\mathrm{s}} = -\lambda^{\mathrm{s}}\,\boldsymbol{\nabla}_{\parallel}T^{\mathrm{s}}, \tag{15.11}$$

where λ^{s} is the thermal conductivity of the interface, and

$$\boldsymbol{\tau}^{\mathrm{s}} = -\eta^{\mathrm{s}}\left[(\boldsymbol{\nabla}_{\parallel}\boldsymbol{v}^{\mathrm{s}})\cdot\boldsymbol{\delta}_{\parallel} + \boldsymbol{\delta}_{\parallel}\cdot(\boldsymbol{\nabla}_{\parallel}\boldsymbol{v}^{\mathrm{s}})^{\mathrm{T}} - (\boldsymbol{\nabla}_{\parallel}\cdot\boldsymbol{v}^{\mathrm{s}})\boldsymbol{\delta}_{\parallel}\right] - \eta_{\mathrm{d}}^{\mathrm{s}}(\boldsymbol{\nabla}_{\parallel}\cdot\boldsymbol{v}^{\mathrm{s}})\boldsymbol{\delta}_{\parallel}, \tag{15.12}$$

where η^{s} and $\eta_{\mathrm{d}}^{\mathrm{s}}$ are the shear and dilatational viscosities, respectively, of the interface. The transport coefficients λ^{s}, η^{s}, and $\eta_{\mathrm{d}}^{\mathrm{s}}$ for the interface must, of course, be nonnegative. The expression in (15.12) is the Boussinesq interfacial fluid model; as with the bulk, alternative interfacial rheological constitutive equations are possible.[5] The diffusive transport (conduction) of energy within an interface described by (15.11) does not appear to have been investigated and is usually neglected.

The entropy source terms in (15.10) having jumps of temperature and normal velocity involve scalars, while the term involving the jump in tangential velocity involves vectors. The most general approach to ensure nonnegative entropy production is to allow for cross effects between force–flux pairs. Here, we choose to omit couplings between mass, momentum, and energy in formulating constitutive equations (for simplicity of notation; if needed, cross effects can be introduced readily[6]).

For the energy flux-temperature difference terms in (15.10), we ensure nonnegativity by assuming the constitutive equations

$$T^{\mathrm{I}} - T^{\mathrm{s}} = -R_{\mathrm{K}}^{\mathrm{Is}}\left[(\boldsymbol{v}^{\mathrm{I}} - \boldsymbol{v}^{\mathrm{s}})h^{\mathrm{I}} + j_q^{\mathrm{I}}\right]\cdot\boldsymbol{n}\,, \tag{15.13}$$

[4] Bedeaux, *Adv. Chem. Phys.* **64** (1986) 47; Albano & Bedeaux, *Physica A* **147** (1987) 407.
[5] Edwards *et al.*, *Interfacial Transport Processes* (Butterworth, 1991); Sagis *et al.*, *Rev. Mod. Phys.*, **83** (2011) 1367.
[6] The coupled transport of mass, energy and charge is described in Chapters 8–10 of Kjelstrup & Bedeaux, *Non-Equilibrium Thermodynamics of Heterogeneous Systems* (World Scientific, 2008); important examples illustrating the physical meaning and relevance of cross effects are discussed in the applications part (Part B) of that book. See also Exercise 15.7.

$$T^{\mathrm{II}} - T^{\mathrm{s}} = R_{\mathrm{K}}^{\mathrm{IIs}} \left[(\boldsymbol{v}^{\mathrm{II}} - \boldsymbol{v}^{\mathrm{s}}) h^{\mathrm{II}} + \boldsymbol{j}_q^{\mathrm{II}} \right] \cdot \boldsymbol{n} \, , \tag{15.14}$$

where the Kapitza resistances $R_{\mathrm{K}}^{\mathrm{Is}}$ and $R_{\mathrm{K}}^{\mathrm{IIs}}$ characterize heat flow through the interface.[7] Note that the Kapitza resistance, which may result in temperature jumps occurring over a distance $\sim 10^{-9}$ m, is entirely different from the heat transfer coefficient discussed in Section 9.3 that describes temperature differences across a thermal boundary layer, which can have a thickness in the range 10^{-3}–10^{1} m. If there is no resistance to heat flow $R_{\mathrm{K}}^{\mathrm{Is}} = R_{\mathrm{K}}^{\mathrm{IIs}} = 0$, then we have

$$T^{\mathrm{I}} = T^{\mathrm{II}} = T^{\mathrm{s}} \, , \tag{15.15}$$

or continuity of temperature at the dividing interface.

We now consider the entropy production terms in (15.10) involving jumps in velocity. For the normal component of velocity, we write the constitutive equation

$$(\boldsymbol{v}^{\mathrm{II}} - \boldsymbol{v}^{\mathrm{I}}) \cdot \boldsymbol{n} = \xi_n \left[T^{\mathrm{s}} \left(\frac{\hat{\mu}^{\mathrm{II}}}{T^{\mathrm{II}}} - \frac{\hat{\mu}^{\mathrm{I}}}{T^{\mathrm{I}}} \right) + \frac{1}{2} \frac{\rho^{\mathrm{I}} + \rho^{\mathrm{II}}}{\rho^{\mathrm{I}} - \rho^{\mathrm{II}}} (\boldsymbol{v}^{\mathrm{II}} - \boldsymbol{v}^{\mathrm{I}})^2 \right.$$
$$\left. + \boldsymbol{n} \cdot \left(\frac{\boldsymbol{\tau}^{\mathrm{II}}}{\rho^{\mathrm{II}}} - \frac{\boldsymbol{\tau}^{\mathrm{I}}}{\rho^{\mathrm{I}}} \right) \cdot \boldsymbol{n} \right] \left(\frac{1}{\rho^{\mathrm{II}}} - \frac{1}{\rho^{\mathrm{I}}} \right)^{-1} , \tag{15.16}$$

where ξ_n is a coefficient characterizing the phase conversion process. It is natural that the difference in chemical potential divided by temperature, with mechanical modifications that are typically negligible (see Exercise 14.7), between the bulk phases occurs in the phase conversion rate in (15.16). If $\xi_n \to \infty$, then mass transfer across the interface occurs with the two phases being in chemical equilibrium. The term having a jump in tangential components of velocity can be made nonnegative using the expression

$$(\boldsymbol{v}^{\mathrm{II}} - \boldsymbol{v}^{\mathrm{I}})_{\parallel} = \boldsymbol{\xi}_{\parallel} \cdot \left(\frac{\boldsymbol{\tau}^{\mathrm{II}}}{\rho^{\mathrm{II}}} - \frac{\boldsymbol{\tau}^{\mathrm{I}}}{\rho^{\mathrm{I}}} \right) \cdot \boldsymbol{n} \left(\frac{1}{\rho^{\mathrm{II}}} - \frac{1}{\rho^{\mathrm{I}}} \right)^{-1} , \tag{15.17}$$

where $\boldsymbol{\xi}_{\parallel}$ is a transverse mobility tensor characterizing. The slip law (15.17) may be considered as purely mechanical, and it involves bulk properties only. If $\boldsymbol{\xi}_{\parallel} = \mathbf{0}$, then we have the *no-slip condition*

$$\boldsymbol{v}_{\parallel}^{\mathrm{II}} = \boldsymbol{v}_{\parallel}^{\mathrm{I}} = \boldsymbol{v}_{\parallel}^{\mathrm{s}} \, , \tag{15.18}$$

or continuity of the tangential components of velocity across the interface, which is commonly used to formulate boundary conditions in fluid mechanics (see Section 7.2 and Chapter 8). The second equality in (15.18) follows using (14.8) or (14.9), and shows that the tangential part of the interface velocity

---

[7] The Kapitza resistance is usually introduced via the temperature difference rather than the difference of inverse temperatures.

is determined by the tangential velocity in the bulk phases when the no-slip condition holds.

Although our derivation of all constitutive boundary conditions relies heavily on the local equilibrium assumption, one does not notice that assumption in (15.17), which is independent of the state of the interface. The interfacial temperature T^s appears in (15.13), (15.14), and (15.16), however, and the local equilibrium assumption obviously plays a crucial role.

As the interfacial temperature is such an important quantity, we would like to have a time-evolution equation for it. Hence, we here produce a temperature equation in view of the facts that the heat flux in the interface is given by the temperature gradient and that the temperature is gauge-invariant, whereas the internal energy is not. For a one-component system with $\gamma = \gamma(T^s)$, we can use (13.13) in the gauge $\rho^s = 0$, which implies

$$\gamma - u^s = T^s \frac{d\gamma}{dT^s}, \qquad -T^s \frac{d^2\gamma}{dT^{s2}} = \frac{du^s}{dT^s}, \tag{15.19}$$

to turn the evolution equation for the excess internal energy density (14.24) into a temperature equation. The result is

$$-T^s \frac{d^2\gamma}{dT^{s2}} \left(\frac{\partial^s T^s}{\partial t} + \boldsymbol{v}_{\mathrm{def}}^s \cdot \boldsymbol{\nabla}_\| T^s \right)$$

$$= T^s \frac{d\gamma}{dT^s} \boldsymbol{\nabla}_\| \cdot \boldsymbol{v}^s - \boldsymbol{\nabla}_\| \cdot \boldsymbol{j}_q^s - \boldsymbol{\tau}^s : \boldsymbol{\nabla}_\| \boldsymbol{v}^s$$

$$+ \boldsymbol{n} \cdot \left[(\boldsymbol{v}^{\mathrm{II}} - \boldsymbol{v}^s) h^{\mathrm{II}} + \boldsymbol{j}_q^{\mathrm{II}} - (\boldsymbol{v}^{\mathrm{I}} - \boldsymbol{v}^s) h^{\mathrm{I}} - \boldsymbol{j}_q^{\mathrm{I}} \right]$$

$$+ \left(\frac{1}{\rho^{\mathrm{II}}} - \frac{1}{\rho^{\mathrm{I}}} \right)^{-1} \left\{ \left[\frac{1}{2} \frac{\rho^{\mathrm{I}} + \rho^{\mathrm{II}}}{\rho^{\mathrm{I}} - \rho^{\mathrm{II}}} (\boldsymbol{v}^{\mathrm{II}} - \boldsymbol{v}^{\mathrm{I}})^2 \right. \right.$$

$$\left. + \boldsymbol{n} \cdot \left(\frac{\boldsymbol{\tau}^{\mathrm{II}}}{\rho^{\mathrm{II}}} - \frac{\boldsymbol{\tau}^{\mathrm{I}}}{\rho^{\mathrm{I}}} \right) \cdot \boldsymbol{n} \right] \boldsymbol{n} \cdot (\boldsymbol{v}^{\mathrm{II}} - \boldsymbol{v}^{\mathrm{I}})$$

$$\left. + \boldsymbol{n} \cdot \left(\frac{\boldsymbol{\tau}^{\mathrm{II}}}{\rho^{\mathrm{II}}} - \frac{\boldsymbol{\tau}^{\mathrm{I}}}{\rho^{\mathrm{I}}} \right) \cdot (\boldsymbol{v}^{\mathrm{II}} - \boldsymbol{v}^{\mathrm{I}})_\| \right\}. \tag{15.20}$$

Notice the similarity between the first line of (6.7) and the first and second lines of (15.20), where one might identify the factor $-T^s \, d^2\gamma/dT^{s2} = T^s \, ds^s/dT^s$ as the heat capacity per unit area of the interface.[8]

---

[8] Alternative and in some cases less general forms of (15.20) can be found elsewhere: Prosperetti, *Meccanica* **14** (1979) 34; Slattery *et al.*, *Interfacial Transport Phenomena* (Springer, 2007).

Exercise 15.2   Alternative Form of Jump Balance for Momentum
Combine (15.9) with the jump balance for momentum in (14.10) to obtain

$$-\left(\frac{1}{\rho^{II}} - \frac{1}{\rho^{I}}\right)^{-1} \boldsymbol{n} \cdot (\boldsymbol{v}^{II} - \boldsymbol{v}^{I})(\boldsymbol{v}^{II} - \boldsymbol{v}^{I}) = \boldsymbol{n} \cdot (\boldsymbol{\tau}^{II} - \boldsymbol{\tau}^{I}) + (p^{II} - p^{I})\boldsymbol{n}$$
$$- \boldsymbol{\nabla}_{\parallel} \cdot \boldsymbol{\tau}^{s} + \boldsymbol{\nabla}_{\parallel}\gamma - \gamma(\boldsymbol{\nabla}_{\parallel} \cdot \boldsymbol{n})\boldsymbol{n}. \quad (15.21)$$

The $\boldsymbol{\nabla}_{\parallel}\gamma$ term is a tangential force due to variation of the interfacial tension, which is known as the Marangoni stress.

Exercise 15.3   Planar Interface with Marangoni and Boussinesq Stresses
Consider a planar interface at $z = h$ between an inviscid gas and a liquid that is an incompressible Newtonian fluid with viscosity η. The velocity field in the liquid has the form $v_r = v_r(r, z), v_\theta = 0, v_z = v_z(r, z)$ and there is no slip at the gas–liquid interface. Use the jump balance for momentum (15.21) and the Boussinesq fluid model (15.12) to obtain

$$\frac{\partial\gamma}{\partial r} + (\eta^{s} + \eta_{d}^{s})\frac{\partial}{\partial r}\left(\frac{1}{r}\frac{\partial}{\partial r}(rv_r^{s})\right) = \eta\left[\frac{\partial v_r}{\partial z}(r, h) + \frac{\partial v_z}{\partial r}(r, h)\right],$$

where the interfacial viscosities η^{s} and η_{d}^{s} are constant.

Exercise 15.4   Velocity Boundary Conditions at a Spherical Surface
Consider a stationary spherical interface having unit normal $\boldsymbol{n} = \boldsymbol{\delta}_r$ and location $r = R$ (see Exercise 14.2) across which there is no mass transfer. For this problem, assume symmetry about the x_3-coordinate so that velocity and temperature fields are independent of ϕ. Further assume γ is constant and $\eta^{s} = \eta_{d}^{s} = 0$. From the jump momentum balance (15.21), derive the following boundary condition for the tangential velocity,

$$\eta\left[\frac{\partial v_\theta}{\partial r}(R, \theta) - \frac{v_\theta(R, \theta)}{R}\right] = \bar{\eta}\left[\frac{\partial \bar{v}_\theta}{\partial r}(R, \theta) - \frac{\bar{v}_\theta(R, \theta)}{R}\right].$$

From the constitutive equation in (15.17) with an isotropic mobility coefficient given by $\boldsymbol{\xi}_{\parallel} = \xi_{\text{slip}}\boldsymbol{\delta}_{\parallel}$, formulate a second boundary condition for the tangential velocity,

$$v_\theta(R, \theta) - \bar{v}_\theta(R, \theta) = \xi_{\text{slip}}\eta\left[\frac{\partial v_\theta}{\partial r}(R, \theta) - \frac{v_\theta(R, \theta)}{R}\right].$$

Exercise 15.5   Temperature Boundary Conditions at a Spherical Surface
For this problem we examine thermal transport at an interface under the same conditions as those in Exercise 15.4. Assume $\lambda^{s} = 0$ and use the temperature equation (15.20) to obtain the temperature boundary condition

$$\lambda\frac{\partial T}{\partial r}(R, \theta) = \bar{\lambda}\frac{\partial \bar{T}}{\partial r}(R, \theta) - \frac{1}{\xi_{\text{slip}}}\left[v_\theta(R, \theta) - \bar{v}_\theta(R, \theta)\right]^2,$$

where λ and $\bar{\lambda}$ are the thermal conductivities of phases I and II, respectively. From the constitutive equations in (15.13) and (15.14), derive the following additional temperature boundary condition,

$$T(R,\theta) - \bar{T}(R,\theta) = (R_{\mathrm{K}}^{\mathrm{Is}} + R_{\mathrm{K}}^{\mathrm{IIs}})\lambda\frac{\partial T}{\partial r}(R,\theta) + \frac{R_{\mathrm{K}}^{\mathrm{IIs}}}{\xi_{\mathrm{slip}}}\left[v_\theta(R,\theta) - \bar{v}_\theta(R,\theta)\right]^2 .$$

15.3 Multi-Component Systems

Our starting point is again the fundamental form (15.5), which we apply in the gauge $\rho^{\mathrm{s}} = 0$. Hence, for multi-component systems we have

$$\frac{\partial^{\mathrm{s}} e^{\mathrm{s}}}{\partial t} = T^{\mathrm{s}}\frac{\partial^{\mathrm{s}} s^{\mathrm{s}}}{\partial t} + \sum_{\alpha=1}^{k}\hat{\mu}_\alpha^{\mathrm{s}}\frac{\partial^{\mathrm{s}}\rho_\alpha^{\mathrm{s}}}{\partial t} . \tag{15.22}$$

Here, our strategy for analyzing the entropy production term is similar to that used in Section 15.2. The first step is substitution of (14.21), (14.23), and (15.6) into (15.22) and using $\boldsymbol{j}_s^{\mathrm{s}} = (\boldsymbol{j}_q^{\mathrm{s}} - \sum_{\alpha=1}^{k}\hat{\mu}_\alpha^{\mathrm{s}}\boldsymbol{j}_\alpha^{\mathrm{s}})/T^{\mathrm{s}}$. After some tedious substitutions and algebra, we obtain (see Exercise 15.6)

$$\begin{aligned}
\sigma^{\mathrm{s}} = {}& \boldsymbol{j}_q^{\mathrm{s}}\cdot\nabla_\parallel\frac{1}{T^{\mathrm{s}}} - \frac{1}{T^{\mathrm{s}}}\boldsymbol{\tau}^{\mathrm{s}}:\nabla_\parallel\boldsymbol{v}^{\mathrm{s}} - \sum_{\alpha=1}^{k}\boldsymbol{j}_\alpha^{\mathrm{s}}\cdot\nabla_\parallel\frac{\hat{\mu}_\alpha^{\mathrm{s}} - \hat{\mu}^{\mathrm{s}}}{T^{\mathrm{s}}} - \frac{1}{T^{\mathrm{s}}}\sum_{\alpha=1}^{k}\nu_\alpha^{\mathrm{s}}\hat{\mu}_\alpha^{\mathrm{s}}\Gamma^{\mathrm{s}} \\
& + \boldsymbol{n}\cdot\left[(\boldsymbol{v}^{\mathrm{II}} - \boldsymbol{v}^{\mathrm{s}})h^{\mathrm{II}} + \boldsymbol{j}_q^{\mathrm{II}}\right]\left(\frac{1}{T^{\mathrm{s}}} - \frac{1}{T^{\mathrm{II}}}\right) \\
& - \boldsymbol{n}\cdot\left[(\boldsymbol{v}^{\mathrm{I}} - \boldsymbol{v}^{\mathrm{s}})h^{\mathrm{I}} + \boldsymbol{j}_q^{\mathrm{I}}\right]\left(\frac{1}{T^{\mathrm{s}}} - \frac{1}{T^{\mathrm{I}}}\right) \\
& + \frac{1}{T^{\mathrm{s}}}\left(\frac{1}{\rho^{\mathrm{II}}} - \frac{1}{\rho^{\mathrm{I}}}\right)^{-1}\left\{T^{\mathrm{s}}\left(\frac{\hat{\mu}^{\mathrm{II}}}{T^{\mathrm{II}}} - \frac{\hat{\mu}^{\mathrm{I}}}{T^{\mathrm{I}}}\right) + \frac{1}{2}\frac{\rho^{\mathrm{I}} + \rho^{\mathrm{II}}}{\rho^{\mathrm{I}} - \rho^{\mathrm{II}}}(\boldsymbol{v}^{\mathrm{II}} - \boldsymbol{v}^{\mathrm{I}})^2\right. \\
& \qquad\qquad\left. + \boldsymbol{n}\cdot\left(\frac{\boldsymbol{\tau}^{\mathrm{II}}}{\rho^{\mathrm{II}}} - \frac{\boldsymbol{\tau}^{\mathrm{I}}}{\rho^{\mathrm{I}}}\right)\cdot\boldsymbol{n}\right\}\boldsymbol{n}\cdot(\boldsymbol{v}^{\mathrm{II}} - \boldsymbol{v}^{\mathrm{I}}) \\
& + \frac{1}{T^{\mathrm{s}}}\left(\frac{1}{\rho^{\mathrm{II}}} - \frac{1}{\rho^{\mathrm{I}}}\right)^{-1}\boldsymbol{n}\cdot\left(\frac{\boldsymbol{\tau}^{\mathrm{II}}}{\rho^{\mathrm{II}}} - \frac{\boldsymbol{\tau}^{\mathrm{I}}}{\rho^{\mathrm{I}}}\right)\cdot(\boldsymbol{v}^{\mathrm{II}} - \boldsymbol{v}^{\mathrm{I}})_\parallel \\
& - \sum_{\alpha=1}^{k}\boldsymbol{n}\cdot\left[(\boldsymbol{v}^{\mathrm{II}} - \boldsymbol{v}^{\mathrm{s}})\rho_\alpha^{\mathrm{II}} - \boldsymbol{m} + \boldsymbol{j}_\alpha^{\mathrm{II}}\right]\left(\frac{\hat{\mu}_\alpha^{\mathrm{s}} - \hat{\mu}^{\mathrm{s}}}{T^{\mathrm{s}}} - \frac{\hat{\mu}_\alpha^{\mathrm{II}} - \hat{\mu}^{\mathrm{II}}}{T^{\mathrm{II}}}\right) \\
& + \sum_{\alpha=1}^{k}\boldsymbol{n}\cdot\left[(\boldsymbol{v}^{\mathrm{I}} - \boldsymbol{v}^{\mathrm{s}})\rho_\alpha^{\mathrm{I}} - \boldsymbol{m} + \boldsymbol{j}_\alpha^{\mathrm{I}}\right]\left(\frac{\hat{\mu}_\alpha^{\mathrm{s}} - \hat{\mu}^{\mathrm{s}}}{T^{\mathrm{s}}} - \frac{\hat{\mu}_\alpha^{\mathrm{I}} - \hat{\mu}^{\mathrm{I}}}{T^{\mathrm{I}}}\right) , \quad (15.23)
\end{aligned}$$

where we have introduced the *flat-average* chemical potentials

$$\hat{\mu}^{\mathrm{I}} = \frac{1}{k}\sum_{\alpha=1}^{k}\hat{\mu}_\alpha^{\mathrm{I}}, \qquad \hat{\mu}^{\mathrm{II}} = \frac{1}{k}\sum_{\alpha=1}^{k}\hat{\mu}_\alpha^{\mathrm{II}}, \qquad \hat{\mu}^{\mathrm{s}} = \frac{1}{k}\sum_{\alpha=1}^{k}\hat{\mu}_\alpha^{\mathrm{s}}, \tag{15.24}$$

and

$$k\,\boldsymbol{m} = \left(\boldsymbol{v}^{\mathrm{I}} - \boldsymbol{v}^{\mathrm{s}}\right)\rho^{\mathrm{I}} = \left(\boldsymbol{v}^{\mathrm{II}} - \boldsymbol{v}^{\mathrm{s}}\right)\rho^{\mathrm{II}}. \tag{15.25}$$

The way of writing the entropy production rate in (15.23) is far from unique. Even in the formulation of Fourier's law, one needs to find the proper splitting into divergence and source terms, based on physical insight. Or, if one wishes to omit cross effects for simplicity (see footnote on p. 274), one should try to arrange terms in such a way that they correspond to uncoupled modes of transport or relaxation; independence of the choice of variables is guaranteed only by allowing for all possible cross effects. So, *why* have we arranged the entropy production in the particular form (15.23)? (The much simpler question of *how* we have obtained that equation, is addressed in Exercise 15.6.) The averaged quantities defined in (15.24) and (15.25) are introduced to facilitate the later formulation of constitutive equations by satisfying certain constraints without breaking the symmetry between the different species.[9] For example, the average quantity $\boldsymbol{m}$ is introduced such that the relative species mass fluxes $(\boldsymbol{v}^{\mathrm{I}} - \boldsymbol{v}^{\mathrm{s}})\rho_{\alpha}^{\mathrm{I}} - \boldsymbol{m}$ in the last line of (15.23) add up to zero, just like the accompanying diffusive mass fluxes $\boldsymbol{j}_{\alpha}^{\mathrm{I}}$. These $\boldsymbol{m}$ terms actually drop out from the last two lines of (15.23) as we have introduced the flat averages of chemical potentials. The systematic occurrence of chemical potentials as deviations from averages is convenient because we can then easily compare this with the one-component case (15.10), which provides very useful guidance for the multi-component case. Flat averages might be surprising because the chemical potential of a large amount of a material in one phase gets the same weight as for a tiny amount of another phase. However, for phase coexistence, the amount of material in the different phases does not matter; only the equality of chemical potentials matters.

Note that the entropy production rate depends only on gauge-invariant surface properties. We see in the last two lines of (15.23) that entropy production from mass transport between the bulk phases and the interface involves force–flux pair products. The forces appear as differences of species chemical potential, relative to the flat-average chemical potential, divided by the difference in temperature between the interface and the bulk phases. For a one-component system, the last two terms in the first line and the seventh and eighth lines of (15.23) vanish, and (15.10) is recovered. For multi-component interfaces, we have the same alternatives for the choice of

---

[9] Indeed, one often chooses $\hat{\mu}_{\alpha} - \hat{\mu}_{k}$ instead of $\hat{\mu}_{\alpha} - \hat{\mu}$ in the various phases, thus breaking the symmetry.

variables as in the bulk (see Section 5.1). Also note that the constitutive equations given in (15.11)–(15.17) are applicable to mixtures.

Exercise 15.6   Interfacial Entropy Production Rate
Starting with the expression in (15.22), derive the interfacial entropy production rate given in (15.23).

To ensure that the third term in the first line of (15.23) is nonnegative, we assume a linear expression for $\dot{j}_\alpha^s$

$$\dot{j}_\alpha^s = -\sum_{\beta=1}^{k} L_{\alpha\beta}^s \nabla_\parallel \frac{\hat{\mu}_\beta^s - \hat{\mu}^s}{T^s}, \tag{15.26}$$

where the $L_{\alpha\beta}^s$ are phenomenological coefficients of a symmetric, positive-semidefinite matrix. Coupling between the fluxes $\dot{j}_\alpha^s$ and $\dot{j}_q^s$ can be handled in the same way as in the bulk (see Exercise 15.7). The last two lines in (15.23) further suggest that the bulk species fluxes lead to chemical potential differences that can be characterized in terms of positive-semidefinite symmetric matrices,

$$\frac{\hat{\mu}_\alpha^I - \hat{\mu}^I}{T^I} - \frac{\hat{\mu}_\alpha^s - \hat{\mu}^s}{T^s} = -\sum_{\beta=1}^{k} R_{\alpha\beta}^{Is} \, \boldsymbol{n} \cdot [(\boldsymbol{v}^I - \boldsymbol{v}^s)\rho_\beta^I - \boldsymbol{m} + \boldsymbol{j}_\beta^I], \tag{15.27}$$

$$\frac{\hat{\mu}_\alpha^{II} - \hat{\mu}^{II}}{T^{II}} - \frac{\hat{\mu}_\alpha^s - \hat{\mu}^s}{T^s} = \sum_{\beta=1}^{k} R_{\alpha\beta}^{IIs} \, \boldsymbol{n} \cdot [(\boldsymbol{v}^{II} - \boldsymbol{v}^s)\rho_\beta^{II} - \boldsymbol{m} + \boldsymbol{j}_\beta^{II}], \tag{15.28}$$

where the matrices $R_{\alpha\beta}^{Is}$ and $R_{\alpha\beta}^{IIs}$ describe the interfacial resistance to mass transfer. The constraint that the sums of all diffusive fluxes for both the interface and the bulk vanish must be taken into account in formulating the matrices $L_{\alpha\beta}^s$, $R_{\alpha\beta}^{Is}$, and $R_{\alpha\beta}^{IIs}$. This can be done elegantly by requiring that all the eigenvectors of these symmetric matrices need to be perpendicular to any vector with k equal components. Hence, the maximum rank of $L_{\alpha\beta}^s$, $R_{\alpha\beta}^{Is}$, or $R_{\alpha\beta}^{IIs}$ is $k-1$, which equals the maximum number of boundary conditions required for the species mass densities in phase I or II (when the continuity equation for each phase is used). More generally, the rank of $R_{\alpha\beta}^{Is}$ or $R_{\alpha\beta}^{IIs}$ must match the rank of the diffusion matrix in the corresponding bulk phase. Analogously to the discussion in the previous section on Kapitza resistances, chemical potential jumps due to interfacial mass transfer resistances $R_{\alpha\beta}^{I,IIs}$ occur over much smaller length scales than the concentration differences described by the mass transfer coefficient introduced in Exercise 10.8.

If there is no resistance to mass transfer $R_{\alpha\beta}^{\mathrm{Is}} = R_{\alpha\beta}^{\mathrm{IIs}} = 0$, from (15.27) and (15.28), we have

$$\frac{\hat{\mu}_\alpha^{\mathrm{I}} - \hat{\mu}^{\mathrm{I}}}{T^{\mathrm{I}}} = \frac{\hat{\mu}_\alpha^{\mathrm{s}} - \hat{\mu}^{\mathrm{s}}}{T^{\mathrm{s}}} = \frac{\hat{\mu}_\alpha^{\mathrm{II}} - \hat{\mu}^{\mathrm{II}}}{T^{\mathrm{II}}}. \tag{15.29}$$

If (15.15) also holds, then we have

$$\hat{\mu}_\alpha^{\mathrm{I}} - \hat{\mu}^{\mathrm{I}} = \hat{\mu}_\alpha^{\mathrm{s}} - \hat{\mu}^{\mathrm{s}} = \hat{\mu}_\alpha^{\mathrm{II}} - \hat{\mu}^{\mathrm{II}}, \tag{15.30}$$

which are $2(k-1)$ independent equations. At equilibrium, we have $\hat{\mu}_\alpha^{\mathrm{I}} = \hat{\mu}_\alpha^{\mathrm{s}} = \hat{\mu}_\alpha^{\mathrm{II}}$ along with the condition in (15.2). In this case, (15.30) can only be satisfied by the flat averages defined in (15.24).

The entropy production due to chemical reaction in the last term in the first line of (15.23) suggests that in the interface we can treat chemical reactions very much as in the bulk (see Section 6.6). As in (6.55), we express this term using molar quantities,

$$\sigma_{\mathrm{react}}^{\mathrm{s}} = -\frac{\Gamma^{\mathrm{s}}}{T^{\mathrm{s}}} \sum_{\alpha=1}^{k} \nu_\alpha^{\mathrm{s}} \hat{\mu}_\alpha^{\mathrm{s}} = -\frac{\tilde{\Gamma}^{\mathrm{s}}}{T^{\mathrm{s}}} \sum_{\alpha=1}^{k} \tilde{\nu}_\alpha^{\mathrm{s}} \tilde{\mu}_\alpha^{\mathrm{s}}, \tag{15.31}$$

where $\tilde{\nu}_\alpha^{\mathrm{s}} = \nu_\alpha^{\mathrm{s}} M_q / M_\alpha$ is the stoichiometric coefficient of species α, and $\tilde{\Gamma}^{\mathrm{s}} = \Gamma^{\mathrm{s}}/M_q$ is the molar rate of reaction per unit area. As in (6.57), it is convenient to write the sum in (15.31) as a difference $x - y$, where x and y are the affinities for product and reactant species, respectively. Hence, writing the constitutive equation for interfacial chemical reactions as

$$\tilde{\Gamma}^{\mathrm{s}} = -L_\Gamma^{\mathrm{s}} \left(e^x - e^y \right) \tag{15.32}$$

leads to a nonnegative entropy production, where L_Γ^{s} is a nonnegative phenomenological coefficient related to the reaction rate constant. The process of active transport in biological systems (see Section 24.2) is an important example where a coupling between chemical reaction and mass transfer at an interface is not only possible, but essential.

Note that the rate of reaction Γ^{s} appears in the balance equation for ρ_α^{s} (14.21), which has been formulated in the gauge $\rho^{\mathrm{s}} = 0$. Also, the rates of reaction Γ^{s} and $\tilde{\Gamma}^{\mathrm{s}}$ in the mass (14.25) and molar (14.27) forms, respectively, of the jump species mass balances further assume $\rho_\alpha^{\mathrm{s}} = 0$. Hence, in general, writing expressions for Γ^{s} or $\tilde{\Gamma}^{\mathrm{s}}$ in terms of excess species densities is not straightforward. For the further investigation of chemical reactions taking place in interfaces, it would be nice to develop a theory of chemical reactions entirely in terms of the gauge-invariant chemical potentials rather than gauge-dependent excess concentrations.

Exercise 15.7   *Force–Flux Relations for Heat and Mass Transfer within an Interface*

Consider a nonisothermal two-component system with cross effects between the diffusive fluxes of mass and energy within an interface. Paralleling the analysis for the bulk phases presented in Section 6.3, allow for a coupling between terms in the first line of (15.23) for entropy production resulting from energy and mass transfer. From this analysis, obtain the expressions

$$j_q^{s\prime} = -\lambda^{s\prime}\, \boldsymbol{\nabla}_\| T^s - \rho_1^s \left[1 - \left(\frac{\partial \hat{\mu}_2^s}{\partial \hat{\mu}_1^s} \right)_{T^s} \right] \left(\frac{\partial \hat{\mu}_1^s}{\partial \rho_1^s} \right)_{T^s} D_{q1}^s\, \boldsymbol{\nabla}_\| \rho_1^s, \qquad (15.33)$$

$$j_1^s = -\rho_1^s \frac{D_{q1}^s}{T^s}\, \boldsymbol{\nabla}_\| T^s - D_{12}^s\, \boldsymbol{\nabla}_\| \rho_1^s, \qquad (15.34)$$

where

$$j_q^{s\prime} = j_q^s - \left[\hat{\mu}_1^s - \hat{\mu}_2^s + T^s \left(\frac{\partial \hat{\mu}_2^s}{\partial T^s} \right)_{\hat{\mu}_1^s} \right] j_1^s. \qquad (15.35)$$

Find expressions analogous to (6.29) and (6.30) for the interfacial transport coefficients $\lambda^{s\prime}$, D_{12}^s, and D_{q1}^s appearing in (15.33) and (15.34).

Exercise 15.8   *Evolution Equations for a Two-Component Interface*

Consider a nonisothermal two-component interface involving species mass and energy transport. From (14.21) and (14.24), obtain the evolution equations for the excess species mass density

$$\frac{\partial^s \rho_1^s}{\partial t} + \boldsymbol{v}_{\mathrm{def}}^s \cdot \boldsymbol{\nabla}_\| \rho_1^s = -\rho_1^s\, \boldsymbol{\nabla}_\| \cdot \boldsymbol{v}^s - \boldsymbol{\nabla}_\| \cdot \boldsymbol{j}_1^s + \nu_1^s \Gamma^s$$

$$+\, \boldsymbol{n} \cdot \left[(\boldsymbol{v}^{II} - \boldsymbol{v}^s)\rho_1^{II} + \boldsymbol{j}_1^{II} - (\boldsymbol{v}^I - \boldsymbol{v}^s)\rho_1^I - \boldsymbol{j}_1^I \right] \qquad (15.36)$$

and interfacial temperature

$$-T^s \left[\left(\frac{\partial^2 \gamma}{\partial T^{s2}} \right)_{\hat{\mu}_1^s} - \rho_1^s \left(\frac{\partial^2 \hat{\mu}_2^s}{\partial T^{s2}} \right)_{\hat{\mu}_1^s} \right] \left(\frac{\partial^s T^s}{\partial t} + \boldsymbol{v}_{\mathrm{def}}^s \cdot \boldsymbol{\nabla}_\| T^s \right)$$

$$= T^s \left(\frac{\partial \gamma}{\partial T} \right)_{\hat{\mu}_1^s} \boldsymbol{\nabla}_\| \cdot \boldsymbol{v}^s$$

$$-\, \boldsymbol{\nabla}_\| \cdot \boldsymbol{j}_q^s - \boldsymbol{\tau}^s : \boldsymbol{\nabla}_\| \boldsymbol{v}^s + \left[\hat{\mu}_1^s - \hat{\mu}_2^s + T^s \left(\frac{\partial \hat{\mu}_2^s}{\partial T^s} \right)_{\hat{\mu}_1^s} \right] (\boldsymbol{\nabla}_\| \cdot \boldsymbol{j}_1^s - \nu_1^s \Gamma^s)$$

$$+\, \boldsymbol{n} \cdot \left[(\boldsymbol{v}^{II} - \boldsymbol{v}^s)h^{II} + \boldsymbol{j}_q^{II} - (\boldsymbol{v}^I - \boldsymbol{v}^s)h^I - \boldsymbol{j}_q^I \right]$$

$$-\, \left[\hat{\mu}_1^s - \hat{\mu}_2^s + T^s \left(\frac{\partial \hat{\mu}_2^s}{\partial T^s} \right)_{\hat{\mu}_1^s} \right] \boldsymbol{n} \cdot \left[(\boldsymbol{v}^{II} - \boldsymbol{v}^s)\rho_1^{II} + \boldsymbol{j}_1^{II} - (\boldsymbol{v}^I - \boldsymbol{v}^s)\rho_1^I - \boldsymbol{j}_1^I \right]$$

$$+\, \left(\frac{1}{\rho^{II}} - \frac{1}{\rho^I} \right)^{-1} \left\{ \left[\frac{1}{2} \frac{\rho^I + \rho^{II}}{\rho^I - \rho^{II}} (\boldsymbol{v}^{II} - \boldsymbol{v}^I)^2 + \boldsymbol{n} \cdot \left(\frac{\boldsymbol{\tau}^{II}}{\rho^{II}} - \frac{\boldsymbol{\tau}^I}{\rho^I} \right) \cdot \boldsymbol{n} \right] \boldsymbol{n} \cdot (\boldsymbol{v}^{II} - \boldsymbol{v}^I) \right.$$

$$\left. +\, \boldsymbol{n} \cdot \left(\frac{\boldsymbol{\tau}^{II}}{\rho^{II}} - \frac{\boldsymbol{\tau}^I}{\rho^I} \right) \cdot (\boldsymbol{v}^{II} - \boldsymbol{v}^I)_\| \right\}. \qquad (15.37)$$

Note the similarity between (15.37) and (6.7). Also note that the term in square brackets appearing both in the third and in the fifth line of (15.37) is the excess energy of mixing within the interface.

- Analogously to the procedure for bulk phases, the entropy production term appearing in the evolution equation for the excess entropy density at an interface can, using the assumption of local thermodynamic equilibrium, be formulated in terms of force–flux pairs.
- Force–flux pairs for transport within the interface lead to analogs of Fourier's, Fick's, and Newton's laws for the diffusive fluxes of energy, species mass, and momentum in terms of gradients of the interfacial temperature, chemical potential divided by temperature, and velocity, respectively.
- Force–flux pairs for transport between the bulk phases and interface lead to expressions for jumps in velocity between the bulk phases, and to jumps between the bulk phases and the interface for temperature and chemical potential.
- The reaction flux for interfacial chemical reactions can be expressed in terms of exponentials of the chemical affinities or, equivalently, in terms of excess densities.

16

Polymer Processing

Polymer materials are ubiquitous in modern society and have diverse applications ranging from food packaging to textiles to microelectronic devices. Polymeric, or plastic, components are commonly found in the vehicles that transport us from place to place and in biomedical devices such as artificial heart valves and knees. Polymer processing involves the transformation of polymer resin, which is usually in powder or granular form, into solid objects having the desired shape and physical properties. The first step in this transformation is usually carried out by *extrusion*, which involves the creation of a continuous flow of molten polymer. In many cases, the molten polymer flows through a die, which may function to shape the material. In other cases, flow from the die is directed, or injected, into a mold with the desired shape. If the molten polymer exiting the die is pulled or drawn, the process is known as *fiber spinning*, and polymeric fibers are produced (see Figure 16.1). The most common polymer processing operations include profile (e.g., rods, tubes) extrusion, injection molding, film blowing, fiber spinning, and calendaring. Several useful texts with more complete descriptions of polymer processing are available.[1]

More recently, *additive manufacturing* has become an alternative to the more conventional polymer processes just mentioned. Additive manufacturing processes, sometimes called 3D printing, involve the precise addition of small amounts of material using computer-controlled positioning based on a digital image of the component to be fabricated. The process known as fused deposition modeling[2] adds molten polymer that flows through a small die. In the selective laser sintering process,[3] a laser is used to selectively

---

[1] Denn, *Polymer Melt Processing* (Cambridge, 2008); Tadmor & Gogos, *Principles of Polymer Processing* (Wiley, 2006).

[2] Turner *et al.*, *Rapid Proto. J.* **20** (2014) 192.

[3] Goodridge *et al.*, *Prog. Mat. Sci.* **57** (2012) 229.

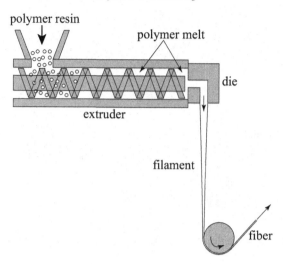

Figure 16.1 Schematic diagram of process for the production of polymer fibers.

melt regions of a polymer powder bed that fuse with already solid portions. It is clear that transport phenomena, at interfaces in particular, play an important role in additive manufacturing processes for polymeric materials.

We saw in Chapter 12 that polymeric materials display nonlinear viscoelastic flow behavior. It is also clear that thermal effects play an important role – both in the melting and solidification of the polymer material and in the dissipation of mechanical energy. These factors, combined with the complex geometries encountered in processing flows, mean that realistic transport analysis of polymer processing operations involves a nonlinear and coupled set of governing equations. Hence, realistic modeling of polymer processes relies heavily on numerical solution of the governing equations using finite difference and finite element methods.[4]

In this chapter, we focus on only three aspects of polymer processing. We begin with an analysis of flow through a circular die. This is followed by an analysis of the extrusion process used to pump molten polymer. The final topic is an analysis of the fiber spinning process. So that we may avoid the need to use sophisticated numerical methods, a large number of assumptions, many of which have limited validity, will be used in order to allow tractable models to be developed.

---

[4] Tucker, *Computer Modeling for Polymer Processing* (Hanser, 1989); Agassant *et al.*, *Polymer Processing, Principles and Modeling* (Hanser, 1989); Osswald & Hernández-Ortiz, *Polymer Processing* (Hanser, 2006).

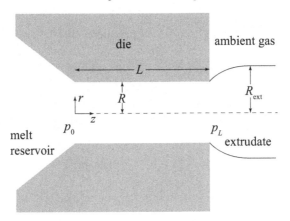

Figure 16.2 Schematic diagram of flow through a circular die.

16.1 Die Flow

Flows of molten polymers through dies of various cross-sectional shapes are common in polymer processing. Here, we consider the simple case of isothermal flow through a circular die of length L and radius R as shown schematically in Figure 16.2. At the entrance to the die, the polymer in the reservoir has pressure p_0, and at the die exit the pressure is p_L. Notice that the extrudate, the liquid polymer exiting the die into a gas, has a radius R_{ext} that is different (usually larger) than the die radius R. Our goal here is to develop a relationship between the mass flow rate $\mathcal{W}$ and pressure drop $p_0 - p_L$. If the polymer were an incompressible Newtonian fluid, this relationship would be the Hagen–Poiseuille equation (8.5).

We consider the steady, isothermal flow of an incompressible fluid. There is symmetry about the axial coordinate and no circumferential flow. The length of the die is much larger than its radius ($R/L \ll 1$), so we assume that entrance and exit effects can be neglected. Based on these assumptions, we postulate the velocity field to have the form $v_r = v_\theta = 0, v_z = v_z(r, z)$. The incompressibility constraint (5.36) further restricts the velocity to $v_z = v_z(r)$. The symmetrized velocity gradient tensor (12.14) takes the form

$$\dot{\gamma} = \begin{pmatrix} 0 & 0 & -\dot{\gamma} \\ 0 & 0 & 0 \\ -\dot{\gamma} & 0 & 0 \end{pmatrix}, \tag{16.1}$$

where $-\dot{\gamma}(r) = dv_z/dr$ is the shear rate. The form of $\dot{\boldsymbol{\gamma}}$ for this flow field, for symmetry reasons (see Section 12.3), implies

$$\boldsymbol{\tau} = \begin{pmatrix} \tau_{rr} & 0 & \tau_{rz} \\ 0 & \tau_{\theta\theta} & 0 \\ \tau_{rz} & 0 & \tau_{zz} \end{pmatrix}, \qquad (16.2)$$

where $\boldsymbol{\tau} = \boldsymbol{\tau}(r)$. Given the high viscosity of polymer liquids, we expect creeping flow. Hence, the equations of motion (5.33), with $\boldsymbol{\pi} = p^L \boldsymbol{\delta} + \boldsymbol{\tau}$, take the form [see (B.13)]

$$\frac{\partial p^L}{\partial r} = -\left[\frac{1}{r}\frac{\partial}{\partial r}(r\tau_{rr}) - \frac{\tau_{\theta\theta}}{r}\right], \qquad \frac{\partial p^L}{\partial \theta} = 0, \qquad \frac{\partial p^L}{\partial z} = -\frac{1}{r}\frac{\partial}{\partial r}(r\tau_{rz}), \quad (16.3)$$

where we have neglected the gravitational force. Assuming that p^L is a sufficiently smooth function requires the axial pressure gradient to be constant, $\partial p^L / \partial z = \Delta p_{\text{die}}/L$, where $\Delta p_{\text{die}} = p_L - p_0$. Hence, integration of the third equation in (16.3) with respect to r gives

$$\tau_{rz}(r) = \tau_R \frac{r}{R}, \qquad (16.4)$$

where $\tau_R = -\Delta p_{\text{die}} R/(2L)$ is the shear stress at the die wall. It is important to note that the linear dependence of shear stress on radial position given in (16.4) requires only that the pressure gradient be constant, and this result therefore is independent of the rheological behavior of the fluid.

At this point, we are more specific about the dependence of $\boldsymbol{\tau}$ on $\dot{\boldsymbol{\gamma}}$. As discussed in Section 12.3, the viscosity of polymeric liquids is a function of the shear rate. Here, we use the generalized Newtonian fluid (12.38) discussed in Exercise 12.9. For the viscosity we take the power-law model $\eta(\dot{\gamma}) = K\dot{\gamma}^{n-1}$, with parameters K and n. For flow of a power-law fluid through a circular die, we have from (16.1) and (12.38)

$$\tau_{rz} = K\left(-\frac{dv_z}{dr}\right)^n. \qquad (16.5)$$

Combination of (16.4) and (16.5) and integration with the no-slip boundary condition $v_z(R) = 0$ gives the velocity field

$$v_z = \left[\frac{-\Delta p_{\text{die}} R}{2KL}\right]^{1/n} \frac{R}{(1/n)+1}\left[1 - \left(\frac{r}{R}\right)^{(1/n)+1}\right]. \qquad (16.6)$$

The velocity distribution for flow of a power-law fluid is shown in Figure 16.3 for several values of the power-law index. From this figure, we see that decreasing n, which corresponds to more pronounced shear thinning (decreasing viscosity), leads to a flattening of the velocity distribution.

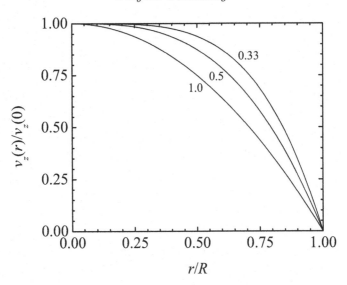

Figure 16.3 Velocity profile for flow through a circular die given by (16.6) for different values of the power law index $n = 1.0, 0.5, 0.33$ from bottom to top.

The mass flow rate through the die is the mass flux integrated over the cross section; for flow of a power-law fluid through a circular die we obtain

$$
\mathcal{W} = \frac{\rho \pi R^3}{1/n + 3} \left[\frac{-\Delta p_{\text{die}} R}{2KL} \right]^{1/n}.
\tag{16.7}
$$

The expression in (16.7) is the analog of the Hagen–Poiseuille equation (8.5) for a power-law fluid. From (16.7) we see that for a shear-thinning polymer ($n < 1$) the pressure drop has a weaker than linear dependence on the flow rate.

As noted earlier, the molten polymer exiting the die has a radius R_{ext} that is larger than the die radius. This phenomenon, known as *extrudate swell*, is related to the normal stresses present in the flow through the die. The presence of a first normal-stress difference N_1 exerts a force on the wall of the die, which is relieved when the polymer exits the die. Tanner[5] carried out an approximate analysis of this complex flow based on the convected Maxwell model (see Section 12.4) and found the expression

$$
\frac{R_{\text{ext}}}{R} \simeq 0.13 + \left\{ 1 + \frac{1}{8} \left(\frac{(N_1)_R}{\tau_R} \right)^2 \right\}^{1/6},
\tag{16.8}
$$

[5] Tanner, *J. Polym. Sci. A* **8** (1970) 2067.

where $(N_1)_R$ and τ_R are the first normal-stress difference and shear stress evaluated at the wall shear rate. The constant 0.13 is included to account for the expansion of a purely viscous Newtonian fluid in the absence of inertial effects, which results from the rearrangement of the velocity field from parabolic to uniform.

There are a number of factors not considered in the analysis presented above that can complicate the analysis of polymer flows through dies. For example, the viscosity of a typical polymer melt is ten million (10^7) times larger than that of water and is also a strong function of temperature, which means viscous dissipation can have a significant effect on the flow.[6] Also, it is well established that wall slip, or violation of the no-slip boundary condition (see Exercise 16.3), occurs in flows of polymer melts at high shear rates.[7] Finally, it is common for dies to have variable cross sections, which means the flow can no longer be treated as one-dimensional.

Exercise 16.1   Extrudate Swell
Use the expression in (16.8) and the rheological data in Figure 12.5 to estimate the extrudate swell if the shear rate at the wall is $\dot{\gamma} = 10 \text{ s}^{-1}$.

Exercise 16.2   Capillary Viscometry
A common method of obtaining the viscosity function $\eta(\dot{\gamma})$ for non-Newtonian fluids is from measurements of pressure drop and flow rate for flow through a circular die or capillary. This method is based on (16.4) and therefore does not require the rheological behavior of the fluid to be known a priori. Derive the following expressions for the mass flow rate,

$$\mathcal{W} = \rho\pi \int_0^R \dot{\gamma}(r) r^2 \, dr = \frac{\pi\rho R^3}{\tau_R^3} \int_0^{\tau_R} \dot{\gamma}(\tau_{rz})\tau_{rz}^2 \, d\tau_{rz}.$$

The first result is obtained from integration by parts using $-\dot{\gamma}(r) = dv_z/dr$, and the second from a change of variable using (16.4). Differentiate this expression to obtain

$$\dot{\gamma}(\tau_R) = \left(\frac{4\mathcal{W}}{\pi\rho R^3}\right)\left[\frac{3}{4} + \frac{d\log[4\mathcal{W}/(\rho\pi R^3)]}{4d\log(\tau_R)}\right].$$

The above expression, which is known as the Rabinowitsch correction, shows that from a plot of flow rate $4\mathcal{W}/(\rho\pi R^3)$ versus pressure drop $\tau_R = (p_0 - p_L)R/(2L)$ the viscosity can be obtained as $\eta = \tau_R/\dot{\gamma}(\tau_R)$.

---

[6] Winter, *Adv. Heat Trans.* **13** (1977) 205.

[7] Brochard & de Gennes, *Langmuir* **8** (1992) 3033; Denn, *Annu. Rev. Fluid Mech.* **33** (2001) 265.

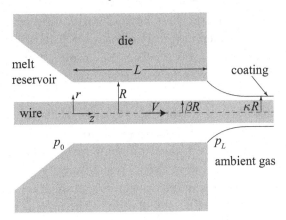

Figure 16.4 Schematic diagram of wire coating flow through a circular die.

Exercise 16.3   Wall Slip in Die Flow

Use the constitutive equation given in (15.17) with an isotropic mobility coefficient given by $\boldsymbol{\xi}_\parallel = \xi_{\text{slip}}\boldsymbol{\delta}_\parallel$ to find the velocity distribution for pressure-driven flow of a power-law fluid through the die shown in Figure 16.2. Also, derive the following relationship between pressure drop and flow rate:

$$\mathcal{W} = \rho\pi\left[\frac{nR^3}{1+3n}\left(\frac{-\Delta p_{\text{die}}\,R}{2KL}\right)^{1/n} - \frac{\xi_{\text{slip}}\,\Delta p_{\text{die}}\,R^3}{2L}\right].$$

Exercise 16.4   Wire Coating Flow

Consider the wire coating process shown schematically in Figure 16.4, where a wire of radius βR moves through a die of radius R with constant velocity V. The movement of the wire drags molten polymer through the die, resulting in a coated wire having radius κR. The molten polymer is modeled as a power-law fluid with constant density ρ. Assume gravitational forces can be neglected and there is no pressure drop across the die, $p_L = p_0$. The velocity field has the form $v_r = v_\theta = 0$, $v_z = v_z(r, z)$. Show that the velocity distribution within the die is given by

$$\frac{v_z}{V} = \frac{(r/R)^{1-1/n} - 1}{\beta^{1-1/n} - 1}.$$

Assuming the ambient gas surrounding the coated wire can be treated as an inviscid fluid, find the following expression for the coating thickness:

$$\kappa^2 = \frac{n-1}{3n-1}\frac{1-\beta^{3-1/n}}{1-\beta^{1-1/n}}.$$

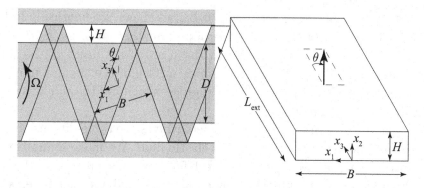

Figure 16.5 Schematics of extruder with screw rotating with angular velocity Ω inside stationary barrel (left) and unwound channel formed between screw and barrel (right).

16.2 Extrusion

Extrusion is the workhorse of polymer processing. An extruder is basically a screw that rotates within a cylindrical cavity. If the screw were being driven into a piece of wood, it would rotate in the opposite direction inside an extruder (for a right-handed screw, the rotation is counter-clockwise). Solid polymer particles are fed into one end of the extruder; melting and the formation of a continuous molten polymer take place in the transition zone, and the pressure in the polymer melt is increased as it passes through the metering zone. At the end of the metering zone, the polymer melt flows into a die, which may have multiple channels with different cross-sectional shapes. Hence, the flow through a die discussed in the previous section occurs because the extruder creates a flow of molten polymer at a pressure sufficient to overcome the pressure drop in the die. Extruders come in a variety of configurations (single- and twin-screw), with various screw types and a wide range of sizes. Here, we focus on the metering zone in which the polymer exists as a continuous melt.

In a single-screw extruder, molten polymer flows within a channel with an approximately rectangular cross section that follows a helical path wrapped around the inner diameter D (or root) of the screw (see Figure 16.5). The channel depth H is the distance between the cylindrical cavity (or barrel) and the screw root, and the channel width B is the distance between successive threads (or flights) on the screw. Here, we consider the simplest screw having constant B and H and a constant number of flights per unit screw length. Typically, $H/D \ll 1$, so it is possible to neglect the curvature along the helical path, which means it can be "unwound" and treated like a straight rectangular channel of length L_{ext} as shown schematically in Figure 16.5.[8] The length of the channel is approximately given by $L_{ext} \approx \pi DN$, where N is the number of flights in the screw. As a result of the rotation of the screw inside the barrel, the upper surface of the channel moves with velocity $V = \Omega D/2$, where Ω is the angular velocity of the screw, at angle θ relative to the channel axis (x_3).

In general, flow in the helical channel involves the three-dimensional, non-isothermal flow of a nonlinear, viscoelastic fluid. We make a number of assumptions so that we may get a feel for how this process works. Given the high viscosity of molten polymers, gravitational and inertial effects can safely be neglected. We assume isothermal flow and that the molten polymer behaves like an incompressible Newtonian fluid with constant viscosity η. Since the channel is narrow $H/B \ll 1$ and long $H/L_{ext} \ll 1$, we postulate the velocity field to have the form $v_1 = v_1(x_2), v_2 = 0, v_3 = v_3(x_2)$. We further assume that the lubrication approximation (see Section 8.4) is valid so that the pressure is independent of x_2.

The assumed velocity field is consistent with the incompressibility constraint (5.36), and the Navier–Stokes equation simplifies to

$$\frac{\partial p^L}{\partial x_1} = \eta \frac{\partial^2 v_1}{\partial x_2^2}, \qquad \frac{\partial p^L}{\partial x_2} = 0, \qquad \frac{\partial p^L}{\partial x_3} = \eta \frac{\partial^2 v_3}{\partial x_2^2}. \qquad (16.9)$$

Integration of the first equation in (16.9) with the boundary conditions $v_1(0) = 0$ and $v_1(H) = -V \sin \theta$ gives

$$v_1 = -\frac{H^2}{2\eta} \frac{\partial p^L}{\partial x_1} \left[\frac{x_2}{H} - \left(\frac{x_2}{H} \right)^2 \right] - \frac{x_2}{H} V \sin \theta. \qquad (16.10)$$

Since there is no net flow in the x_1-direction, we integrate (16.10) over x_2 to obtain an expression for $\partial p^L/\partial x_1$, which leads to

$$v_1 = -V \sin \theta \left[3 \left(\frac{x_2}{H} \right)^2 - 2 \frac{x_2}{H} \right]. \qquad (16.11)$$

The expression in (16.11) is not valid near the flights ($x_1 = \pm B/2$), where there exists a two-dimensional, recirculating flow (see Figure 16.6). The velocity given in (16.11), which is plotted in Figure 16.7, shows the balance between pressure-driven flow near the bottom of the channel and drag flow near the top of the channel.

---

[8] See Chapter 6 of Middleman, *Fundamentals of Polymer Processing* (McGraw-Hill, 1977).

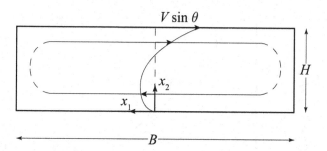

Figure 16.6 Representative streamline showing the recirculating nature of transverse flow within the extruder channel. Solid lines indicate flow described by (16.11).

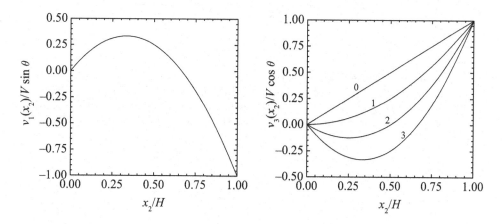

Figure 16.7 Velocity distributions in single-screw extrusion. Left plot shows transverse velocity v_1 given by (16.11); right plot shows the velocity along channel v_3 given by (16.12) for several values of the quantity $H^2 \Delta p_{\mathrm{ext}}/(2\eta L_{\mathrm{ext}} V \cos \theta)$.

Similarly, integration of the third equation in (16.9) with the boundary conditions $v_3(0) = 0$ and $v_3(H) = V \cos \theta$ gives

$$v_3 = -\frac{H^2 \Delta p_{\mathrm{ext}}}{2\eta L_{\mathrm{ext}}} \left[\frac{x_2}{H} - \left(\frac{x_2}{H}\right)^2 \right] + \frac{x_2}{H} V \cos \theta, \qquad (16.12)$$

where we have set $\partial p^{\mathrm{L}}/\partial x_3 \approx \Delta p_{\mathrm{ext}}/L_{\mathrm{ext}}$, where Δp_{ext} is the pressure rise over the channel length L_{ext}. The velocity distribution for several cases is shown in Figure 16.7, ranging from $\Delta p_{\mathrm{ext}} = 0$ to the Δp_{ext} for which there is no net flow along the channel. Multiplying the result in (16.12) by ρ and integrating over the channel cross section gives the following expression for

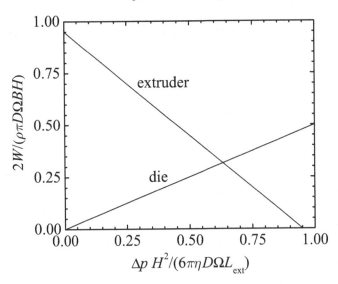

Figure 16.8 Mass flow rate versus pressure difference for single-screw extrusion ($\theta = \pi/10$) of a Newtonian fluid. Line marked "die" is from (8.5) with $3\pi R^4 L_{\text{ext}}/(2BH^3L) = 0.5$, and line marked "extruder" is from (16.13). The intersection of the two lines gives the operating point of the extruder.

the mass flow rate,

$$\mathcal{W} = \frac{\pi}{2}\rho D\Omega\, BH\, \cos\theta - \frac{\rho BH^3\, \Delta p_{\text{ext}}}{12\eta L_{\text{ext}}}, \tag{16.13}$$

where we have used $V = D\Omega/2$. The first term in (16.13) is the drag flow induced by the rotating screw and the second is the pressure-driven flow in the opposite direction. Hence, the net mass flow rate through the extruder involves the competition between drag flow towards the die and pressure-driven flow away from the die.

If the die analyzed in Section 16.1 were connected to the extruder considered here, the mass flow rate $\mathcal{W}$ through each would of course be the same. Also, if both the feed to the extruder and the extrudate are at the same pressure, then the pressure drop across the die equals the pressure increase induced by the extruder, $-\Delta p_{\text{die}} = \Delta p_{\text{ext}}$. Figure 16.8 shows the mass flow rate versus pressure change for the die (8.5) and for the extruder (16.13). The intersection of these lines gives the operating point for the extruder–die configuration, which is "found" by the process. Note that the normalization used in Figure 16.8 means that the slope of the die line, and hence the operating point, depend only on geometric factors for the die and extruder. Although it is oversimplified, this type of analysis can be used to understand the basic design and operation of an extrusion process. More rigorous modeling of the complex flow within extruders can be handled only using numerical methods.[9] It is worth noting that the analysis of flow in an extruder for a generalized Newtonian (i.e., power-law) fluid is considerably more complicated, and the governing equations require numerical solution. For a power-law fluid with $n < 1$, flow in the die is described by (16.7), which, if plotted in Figure 16.8, is a curve that turns upwards, while the extruder is described by a curve that turns downwards.[10] Since shear rates in the die are much larger than in the extruder, the operating point moves to a lower pressure.

Exercise 16.5   Extruder Analysis and Design
Evaluate the rate of work term $\dot{W}$ appearing in (9.34) using the extruder model developed in this section. For the case of an adiabatic process, derive an expression for the temperature rise of the polymer as it passes through the extruder–die system.

16.3 Fiber Spinning

Fiber spinning is a commonly used process for the production of fibers with diameters ranging from a few millimeters down to that of human hair ($\sim 100\,\mu$m). There are several types of fiber spinning processes (melt, dry, and wet), but all involve the pulling, or drawing, of a polymer liquid as it exits a die. The polymer liquid between the die and the device that pulls it is known as the filament. In this section, we develop a model for the melt spinning process focusing on fluid mechanics. Thermal effects are considered separately in an exercise.

We consider the steady flow of a liquid filament with constant density ρ. As shown schematically in Figure 16.9, the filament has radius $R(z)$ and is surrounded by a gas with uniform temperature T_g and pressure p_g. At the reference surface ($z = 0$), the filament has radius R_0 and uniform velocity V_0. The filament has length L, which is the distance between the reference surface and the point of solidification, where the filament has velocity $V_L = D_R V_0$. The ratio of velocities D_R is known as the draw-down ratio, and the force required to pull the solidified filament F_L is known as the take-up force. Analysis of the fiber spinning process has been a subject of great interest since the 1960s.[11]

[9] Sarhangi *et al.*, *Int. J. Num. Meth. Fluids* **70** (2012) 775.
[10] See Chapter 6 of Middleman, *Fundamentals of Polymer Processing* (McGraw-Hill, 1977) and Chapter 12 of Tadmor & Gogos, *Principles of Polymer Processing* (Wiley, 2006).
[11] Denn, *Annu. Rev. Fluid Mech.* **12** (1980) 365.

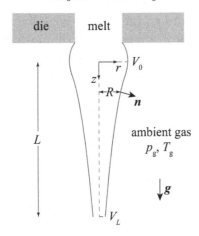

Figure 16.9 Schematic diagram of the fiber spinning process, showing a filament of length L and local radius R. Note that in practice $R \ll L$.

Our goal is to develop a transport model that provides a relationship between D_{R}, F_L, and the final filament radius $R(L)$. There is symmetry about the axial coordinate and no circumferential flow so that the velocity field has the form $v_r = v_r(r, z)$, $v_\theta = 0$, $v_z = v_z(r, z)$. The equations governing this flow are the incompressibility constraint (5.36) and equation of motion (5.33) with $\boldsymbol{\pi} = p^{\mathrm{L}}\boldsymbol{\delta} + \boldsymbol{\tau}$. In a cylindrical coordinate system, these take the form [see (B.8), (B.10), and (B.13)]

$$\frac{1}{r}\frac{\partial}{\partial r}(rv_r) + \frac{\partial v_z}{\partial z} = 0, \tag{16.14}$$

$$\rho\left(v_r\frac{\partial v_r}{\partial r} + v_z\frac{\partial v_r}{\partial z}\right) = -\frac{\partial p^{\mathrm{L}}}{\partial r} - \frac{1}{r}\frac{\partial}{\partial r}(r\tau_{rr}) + \frac{\tau_{\theta\theta}}{r} - \frac{\partial \tau_{rz}}{\partial z}, \tag{16.15}$$

$$\rho\left(v_r\frac{\partial v_z}{\partial r} + v_z\frac{\partial v_z}{\partial z}\right) = -\frac{\partial p^{\mathrm{L}}}{\partial z} - \frac{1}{r}\frac{\partial}{\partial r}(r\tau_{rz}) - \frac{\partial \tau_{zz}}{\partial z}, \tag{16.16}$$

where we have neglected gravity. From the symmetry assumption, $\partial p^{\mathrm{L}}/\partial \theta = 0$.

The surface of the liquid filament is described by $\varphi^{\mathrm{s}}(r, z) = r - R(z) = 0$. According to (13.26), the interface has unit normal $\boldsymbol{n}$ given by

$$\boldsymbol{n} = \frac{\boldsymbol{\delta}_r}{\sqrt{1 + R'^2}} - \frac{R'\boldsymbol{\delta}_z}{\sqrt{1 + R'^2}}, \tag{16.17}$$

where $R' = dR/dz$, which points into the gas phase (I) and away from the liquid phase (II). There is no mass transfer across the stationary interface so that from (14.7) we have $\boldsymbol{n} \cdot \boldsymbol{v}^{\mathrm{II}} = v_n^{\mathrm{s}} = 0$. This leads to the following relation between the velocity components at the surface of the filament,

$$v_r(R, z) = R' v_z(R, z).$$ (16.18)

Note that $R' < 0$, and $|R'| \approx R_0/L \ll 1$.

In the absence of mass transfer and neglecting interfacial tension, the jump balance for momentum (14.6) reduces to $\boldsymbol{n} \cdot \boldsymbol{\pi}^{\mathrm{I}} = \boldsymbol{n} \cdot \boldsymbol{\pi}^{\mathrm{II}}$. Hence, the r- and z-components of the jump balance for momentum take the form

$$\pi_{rr}^{\mathrm{I}} - R' \pi_{rz}^{\mathrm{I}} = \pi_{rr}^{\mathrm{II}} - R' \pi_{rz}^{\mathrm{II}},$$ (16.19)

$$\pi_{rz}^{\mathrm{I}} - R' \pi_{zz}^{\mathrm{I}} = \pi_{rz}^{\mathrm{II}} - R' \pi_{zz}^{\mathrm{II}}.$$ (16.20)

In practice, stresses in the gas can be important and are typically modeled using a drag coefficient. Here, we assume the gas is an inviscid fluid with uniform pressure so that $\boldsymbol{\pi}^{\mathrm{I}} = p_{\mathrm{g}}\boldsymbol{\delta}$. Hence, since p_{g} can be combined with $p^{\mathrm{L}}(R, z)$, we write (16.19) and (16.20) as

$$\tau_{rz}(R, z) = R'[p^{\mathrm{L}}(R, z) + \tau_{zz}(R, z)],$$ (16.21)

$$\tau_{rr}(R, z) = -p^{\mathrm{L}}(R, z),$$ (16.22)

where we have neglected terms of order R'^2.

We now return to the equations governing the velocity and stress fields in the filament (16.14)–(16.16). We saw in Exercise 7.8 the benefit of considering radially averaged quantities in cylindrical geometries with a small aspect ratio. As noted above, the aspect ratio of the filament $|R'| \approx R_0/L$ is small, which can be used to simplify our analysis.[12] Hence, we introduce the following definition for a radially averaged field,

$$\langle(.)\rangle = \frac{2}{R^2} \int_0^R (.) r \, dr,$$ (16.23)

where it is important to keep in mind the axial dependence of the filament radius $R(z)$. If we multiply the continuity equation (16.14) by r and integrate from $r = 0$ to $r = R$, we obtain

$$\frac{d}{dz}\left(R^2 \langle v_z \rangle\right) = 0,$$ (16.24)

where we have used (16.18).

Next, we formulate radially averaged forms of the momentum balances in (16.15) and (16.16). Since $|R'| \ll 1$, we can neglect the inertial terms in

[12] Matovich & Pearson, *Ind. Eng. Chem. Fund.* **8** (1969) 512.

(16.15). Setting the left-hand side of (16.15) to zero, multiplying by r^2, and integrating gives

$$\langle p^{\mathrm{L}} \rangle = -\frac{1}{2} \left(\langle \tau_{rr} \rangle + \langle \tau_{\theta\theta} \rangle \right) + \frac{1}{R^2} \frac{d}{dz} \int_0^R r^2 \tau_{rz} \, dr, \qquad (16.25)$$

where we have used (16.21) and (16.22). Multiplication of (16.16) by r and integrating gives

$$\rho \frac{d}{dz} \left(\pi R^2 \langle v_z^2 \rangle \right) = -\frac{d}{dz} \left[\pi R^2 \left(\langle p^{\mathrm{L}} \rangle + \langle \tau_{zz} \rangle \right) \right], \qquad (16.26)$$

where we have used (16.18) and (16.21).

To proceed, it will be necessary for us to make the assumption that the dependence of v_z on r can be neglected, in which case the continuity equation (16.14) can be used to show that v_r is proportional to r. This is equivalent to treating the flow at each z as simple elongation (see Exercise 12.3) so that $\tau_{rz} = 0$ and $\tau_{\theta\theta} = \tau_{rr}$. This further allows us to set $v_z(R, z) \approx \langle v_z \rangle$ and $\langle v_z^2 \rangle \approx \langle v_z \rangle^2$. Substitution of (16.25) into (16.26) with these approximations gives

$$\mathcal{W} \frac{d \langle v_z \rangle}{dz} = -\frac{d}{dz} \left[\pi R^2 \left(\langle \tau_{zz} \rangle - \langle \tau_{rr} \rangle \right) \right], \qquad (16.27)$$

where we have introduced the mass flow rate $\mathcal{W} = \rho \pi R^2 \langle v_z \rangle$. The averaging procedure and assumptions introduced thus far, which have allowed us to reduce a two-dimensional problem to an effectively one-dimensional problem, are sometimes called the fiber-spinning approximation. We hence must solve the mass and momentum balances in (16.24) and (16.27), which are subject to boundary conditions at the reference surface given by

$$R(0) = R_0, \qquad \langle v_z \rangle(0) = V_0. \qquad (16.28)$$

Exercise 16.6   *Radially Averaged Fiber Spinning Equations*
Use the definition in (16.23) to obtain the radially averaged forms of the mass balance (16.24) and of the momentum balances (16.25) and (16.26).

The tensile force F at any point in the filament is determined as

$$F(z) = -2\pi \int_0^R \pi_{zz}(r, z) r \, dr = -\pi R^2 \langle \pi_{zz} \rangle(z)$$

$$= -\pi R^2 [\langle \tau_{zz} \rangle(z) - \langle \tau_{rr} \rangle(z)], \qquad (16.29)$$

where the second line follows using the simplified form of (16.25), that is

$\langle p^L \rangle = -\langle \tau_{rr} \rangle$. Note that the take-up force at the end of the filament is given by $F_L = F(L)$.

At this point, we make an additional assumption to further simplify our governing equations. For a fiber spinning process involving a high-viscosity polymer liquid, inertial forces can be neglected. In this case, the momentum balance (16.27) simplifies to

$$\frac{d}{dz} \left[\pi R^2 \left(\langle \tau_{zz} \rangle - \langle \tau_{rr} \rangle \right) \right] = 0, \tag{16.30}$$

which, according to (16.29), requires the tensile force to be constant along the filament.

Integration of (16.24) subject to the initial conditions in (16.28) gives

$$R^2 \langle v_z \rangle = R_0^2 V_0 = \frac{W}{\rho \pi}. \tag{16.31}$$

Integration of (16.30), using (16.29) and (16.31), gives

$$\langle \tau_{zz} \rangle - \langle \tau_{rr} \rangle = -\frac{\rho F_L}{W} \langle v_z \rangle. \tag{16.32}$$

To proceed, we must specify a rheological constitutive equation for the polymer liquid. Here, we use the convected Maxwell model given in (12.45). Using the approximations described above, the radially averaged expressions for stress in the liquid filament are given by

$$\langle \tau_{rr} \rangle + \lambda \left[\langle v_z \rangle \frac{d \langle \tau_{rr} \rangle}{dz} + \langle \tau_{rr} \rangle \frac{d \langle v_z \rangle}{dz} \right] = \eta \frac{d \langle v_z \rangle}{dz}, \tag{16.33}$$

$$\langle \tau_{zz} \rangle + \lambda \left[\langle v_z \rangle \frac{d \langle \tau_{zz} \rangle}{dz} - 2 \langle \tau_{zz} \rangle \frac{d \langle v_z \rangle}{dz} \right] = -2\eta \frac{d \langle v_z \rangle}{dz}, \tag{16.34}$$

where λ and η are the relaxation time and viscosity, respectively, of the liquid.

Exercise 16.7   Convected Maxwell Model Stress Equations for Fiber Spinning
Starting with (12.45), derive the radially averaged forms of the stress evolution equations given in (16.33) and (16.34).

We are left with the problem of solving the nonlinear system of equations in (16.32)–(16.34) for $\langle v_z \rangle$, $\langle \tau_{rr} \rangle$, and $\langle \tau_{zz} \rangle$. Substitution of (16.32) into (16.33) and subtracting the result from (16.34) gives

$$\frac{\lambda}{\eta} \langle \tau_{zz} \rangle = 1 - \frac{2}{3} \lambda \frac{\rho F_L}{\eta W} \langle v_z \rangle - \frac{1}{3} \frac{\rho F_L}{\eta W} \langle v_z \rangle \left(\frac{d \langle v_z \rangle}{dz} \right)^{-1}. \tag{16.35}$$

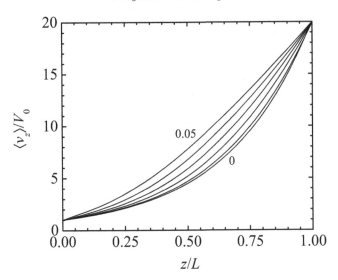

Figure 16.10 Axial velocity along filament for fiber spinning of an upper-convected Maxwell fluid for $N_{\mathrm{De}} = \lambda V_0/L = 0, 0.01, 0.02, 0.03, 0.04, 0.05$ (from bottom to top) obtained by solution of (16.36) for $D_{\mathrm{R}} = 20$. The Newtonian case ($N_{\mathrm{De}} = 0$) is given by (16.39).

Substitution of (16.35) into (16.34) results in a single equation governing $\langle v_z \rangle$ that can be written as[13]

$$\lambda \langle v_z \rangle^2 \frac{d^2 \langle v_z \rangle}{dz^2} - \lambda \langle v_z \rangle \left(1 - \lambda \frac{d\langle v_z \rangle}{dz}\right)\left(\frac{d\langle v_z \rangle}{dz}\right)^2 - \left(\langle v_z \rangle - 3\frac{\eta \mathcal{W}}{\rho F_L}\frac{d\langle v_z \rangle}{dz}\right)\frac{d\langle v_z \rangle}{dz} = 0.$$
(16.36)

This equation can be solved numerically subject to the boundary condition at the end of the filament,

$$\langle v_z \rangle(L) = D_{\mathrm{R}} V_0.$$
(16.37)

It is more convenient (see Exercise 16.9) to use (16.35) to obtain an alternative boundary condition given by,

$$\frac{d\langle v_z \rangle}{dz}(0) = \frac{1}{3}\frac{\rho F_L}{\eta \mathcal{W}}\left[1 - \frac{2}{3}\lambda V_0 \frac{\rho F_L}{\eta \mathcal{W}} - \frac{\lambda}{\eta}\langle \tau_{zz} \rangle(0)\right]^{-1},$$
(16.38)

which is subject to the condition $\langle \tau_{zz} \rangle(0) < \eta/\lambda - (2/3)V_0 \rho F_L/\mathcal{W}$, so that $d\langle v_z \rangle/dz(0) > 0$. Results for a fiber spinning process with $D_{\mathrm{R}} = 20$ are shown in Figure 16.10 for several values of the Deborah number $N_{\mathrm{De}} = \lambda V_0/L$ (see Chapter 12).

[13] Denn *et al.*, *AIChE J.* **21** (1975) 791.

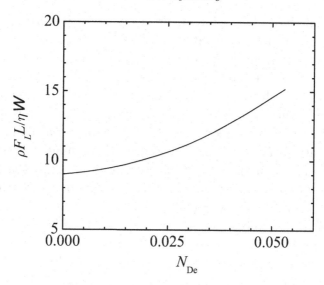

Figure 16.11 Dependence of F_L on N_{De} for the fiber spinning of an upper-convected Maxwell fluid for $D_{\mathrm{R}} = 20$. Note that the maximum possible Deborah number is given by $N_{\mathrm{De}} = (D_{\mathrm{R}} - 1)^{-1}$.

The case of a Newtonian fluid is recovered in the limit $\lambda \to 0$. Setting $\lambda = 0$ in (16.35) or (16.36) and integrating gives

$$\langle v_z \rangle = V_0 \exp\left(\frac{z}{3[\eta \mathcal{W}/(\rho F_L)]} \right) = V_0 D_{\mathrm{R}}^{\frac{z}{L}}, \tag{16.39}$$

where the second equality follows from (16.37). Hence, for the fiber spinning of a Newtonian fluid, neglecting inertia, interfacial tension, and gravity, the take-up force and draw-down ratio are not independent,

$$F_L = 3\frac{\eta \mathcal{W}}{\rho L} \ln D_{\mathrm{R}}. \tag{16.40}$$

The velocity along the filament described by (16.39) is shown in Figure 16.10 for $D_{\mathrm{R}} = 20$. Substitution of (16.39) into (16.31) gives the following expression for the filament radius,

$$R = R_0 D_{\mathrm{R}}^{-\frac{z}{2L}}. \tag{16.41}$$

In practice, the production rate V_0 (or $\mathcal{W}$) and draw-down ratio D_R are prescribed, which sets the radius of the drawn fiber. In general, D_R depends on the take-up force F_L, which determines the stress, and therefore mechanical properties, of the drawn fiber. The dependence of F_L on N_{De} for the fiber spinning of an upper-convected Maxwell fluid is shown in Figure 16.11.

The stability of fiber spinning processes is an important issue.[14] Draw resonance is a phenomenon that is observed during melt spinning when the extrusion rate (V_0) and take-up velocity ($D_R V_0$) are constant. When draw resonance occurs, the radius oscillates along the filament, which generally is undesirable. The phenomenon is a result of the resonant interaction between a disturbance and the residence time of the fluid in the drawing zone. Using linear stability analysis (see Exercise 14.4), it has been shown that the critical draw ratio is $D_R \approx 20$ for the isothermal spinning of a Newtonian fluid in the absence of inertia, interfacial tension, drag, and gravity. Including inertia and gravity tends to stabilize the flow, while interfacial tension and elasticity have a destabilizing effect.[15]

It is worthwhile to briefly discuss some aspects of fiber spinning with viscoelastic fluids. As noted earlier, our simplified model using the rheological constitutive equation for a Maxwell fluid requires the solution of a nonlinear differential equation (16.36). Approximate analytical solution for small Deborah number N_{De}, because the order of (16.36) is reduced for the case $N_{\text{De}} = 0$, results in a singular perturbation problem. Obtaining solutions to singular perturbation problems is considerably more difficult than the perturbation problem considered in Section 8.3. An analytical solution for large N_{De} has also been found. In this limit, the velocity along the filament approaches a linear function, and the maximum possible Deborah number is given by $N_{\text{De}} = (D_R - 1)^{-1}$ (see Figure 16.11).

Exercise 16.8   Fiber Spinning Equations for a Convected Maxwell Fluid
Starting with the momentum balance in (16.32) and stress equations for an upper convected Maxwell fluid in (16.33) and (16.34), derive (16.36).

Exercise 16.9   Fiber Spinning Simulation of a Convected Maxwell Fluid
Normalizing velocity by V_0 and position by L allows (16.36) to be written as

$$N_{\text{De}} \langle v_z \rangle^2 \frac{d^2 \langle v_z \rangle}{dz^2} - N_{\text{De}} \langle v_z \rangle \left(1 - N_{\text{De}} \frac{d \langle v_z \rangle}{dz} \right) \left(\frac{d \langle v_z \rangle}{dz} \right)^2 - \left(\langle v_z \rangle - 3\varepsilon \frac{d \langle v_z \rangle}{dz} \right) \frac{d \langle v_z \rangle}{dz} = 0,$$

where $N_{\text{De}} = \lambda V_0 / L$ and $\varepsilon = \eta W / (\rho F_L L)$. The boundary conditions become $\langle v_z \rangle(0) = 1$, and $\langle v_z \rangle(1) = D_R$, with the alternative condition in (16.35) taking the form

$$\frac{d \langle v_z \rangle}{dz}(0) = \frac{1}{3\varepsilon} \left[1 - \frac{2}{3} \frac{N_{\text{De}}}{\varepsilon} - \frac{\lambda}{\eta} \langle \tau_{zz} \rangle(0) \right]^{-1}.$$

[14] van der Walt *et al.*, *J. Non-Newtonian Fluid Mech.* **208** (2014) 72.
[15] Denn, *Annu. Rev. Fluid Mech.* **12** (1980) 365.

Develop a MATLAB® code to solve this problem and implement it for $D_R = 20$ and several values of $N_{De} \leq 0.05$. Hint: guess a value for ε, use it to determine $d\langle v_z \rangle / d\bar{z}(0)$ with $(\lambda/\eta)\langle \tau_{zz} \rangle(0) \ll 1$, and iterate until $\langle v_z \rangle(1) = D_R$.

Exercise 16.10   *Fiber Spinning with Inertia*

Using (16.24), (16.27), (16.33), and (16.34) with $\lambda = 0$, show for fiber spinning of a Newtonian fluid with inertia that $\langle v_z \rangle$ is governed by

$$\frac{N_{Re}}{3}\langle v_z \rangle \frac{d\langle v_z \rangle}{dz} = \frac{d^2\langle v_z \rangle}{dz^2} - \frac{1}{\langle v_z \rangle}\left(\frac{d\langle v_z \rangle}{dz}\right)^2,$$

where the normalizations $z/L \to z$ and $\langle v_z \rangle/V \to \langle v_z \rangle$ have been used and $N_{Re} = \rho V L/\eta$. Develop a MATLAB® code to solve this problem and implement it for $D_R = 20$ and $N_{Re} = 0.1, 0.3, 1$.

Exercise 16.11   *Thermal Transport in Fiber Spinning*

Thermal transport is critical in the fiber spinning process and is controlled by manipulating heat transfer between the liquid filament and the surrounding gas. To describe thermal transport in the gas phase (I), use Newton's law of cooling (see Section 9.3) to obtain

$$-\lambda\left[\frac{\partial T}{\partial r}(R, z) - R'\frac{\partial T}{\partial z}(R, z)\right] = h[T(R, z) - T_g],$$

where h is the heat transfer coefficient. Show, by neglecting terms of order R'^2 and using the approximation $\langle v_z T \rangle \approx \langle v_z \rangle \langle T \rangle$, that the radially averaged temperature equation, neglecting conduction along the filament, can be written as

$$\hat{c}_p W \frac{d\langle T \rangle}{dz} = -2\pi R h(\langle T \rangle - T_g).$$

- Several factors complicate the analysis of polymer processes, including the coupling of fluid flow and heat transfer, the nonlinear rheological behavior of polymeric liquids, and the complex geometries encountered in processing devices.

- Solutions of the evolution equations for velocity and stress fields for polymer process models can only be found numerically; if analytic solutions are desired, oversimplified models must be considered.

- Pressure-driven flow of polymers through dies is strongly influenced by nonlinear rheological behavior; in particular, this leads to a nonlinear dependence of pressure drop on flow rate.

- Polymer extrusion is a process in which polymer is melted and flows through a die as a result of a superposition of pressure-driven and drag flows, which is caused by the rotation of a screw within a cylindrical cavity where the polymer fills the channel between the screw and the cavity.

- Analysis of the fiber spinning process is facilitated by using radially averaged forms of the mass and momentum balances, which allow one to predict the velocity and stress in terms of controllable process parameters.

17
Transport around a Sphere

There are numerous situations where the transport of mass, energy, or momentum takes place in heterogeneous systems comprised of one material dispersed in a second material. The second material is the continuous phase and surrounds the many particles making up the dispersed phase. In some cases, the heterogeneous system may occur naturally, while in other cases particles are added to the continuous phase to modify its properties. The number of particles, as well as their size, shape, and properties, determine the properties of the composite material. Heterogeneous systems are ubiquitous and make our lives both possible and more comfortable. For example, blood consists of cells suspended in a liquid plasma, and foamed materials used for insulation are often synthetic polymers containing a large number of gas bubbles.

In this chapter, we examine heat transfer and flow in systems containing spherical particles. A good starting place is to consider a single sphere surrounded by a continuous medium of infinite extent. It is interesting to note that the transport problems we consider in this chapter were first solved in the nineteenth and early twentieth centuries. A more general approach to describing transport in heterogeneous systems is presented in Chapter 23.

As shown in Figure 17.1, the center of the sphere is chosen to be the origin of a spherical coordinate system. To simplify things, we consider steady-state problems where a uniform temperature gradient or velocity in the x_3-direction is imposed in the continuous phase far from the sphere. This means there is symmetry about the x_3-direction so that all fields are independent of the ϕ coordinate. After finding temperature and velocity fields around a single sphere, we use these to find expressions for effective transport coefficients in systems containing many spherical particles.

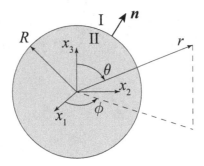

Figure 17.1 Sphere of radius R with its center at the origin of rectangular and spherical coordinate systems surrounded by an infinite medium.

17.1 Heat Transfer around a Sphere

Let us consider a sphere (phase II) with radius R and thermal conductivity $\bar{\lambda}$ surrounded by an infinite medium (phase I) with thermal conductivity λ. Both the sphere and the infinite medium are stationary. Hence, at steady state, the temperature equation (7.3) for both the sphere and surrounding medium simplifies to Laplace's equation, $\nabla^2 T = 0$. Far from the sphere, a constant temperature gradient with magnitude $|\nabla T|_\infty$ in the x_3-direction is maintained in the infinite medium.[1] As noted above, there is symmetry about the x_3-coordinate so that the temperature fields in the medium and in the sphere are not dependent on the ϕ-coordinate. In such cases, the Laplacian can be simplified by the change of variable $\xi = \cos\theta$; the temperature $T(r,\xi)$ is governed by [see (B.16) and note that $d\xi = -\sin\theta\,d\theta$]

$$\frac{\partial}{\partial r}\left(r^2\frac{\partial T}{\partial r}\right) + \frac{\partial}{\partial \xi}\left[(1-\xi^2)\frac{\partial T}{\partial \xi}\right] = 0. \tag{17.1}$$

Here, we use the method of separation of variables and assume a solution of the form $T(r,\xi) = \Psi(r)\Phi(\xi)$. Substitution into (17.1) gives

$$-\frac{1}{\Phi}\frac{d}{d\xi}\left[(1-\xi^2)\frac{d\Phi}{d\xi}\right] = \alpha^2 = \frac{1}{\Psi}\frac{d}{dr}\left(r^2\frac{d\Psi}{dr}\right), \tag{17.2}$$

which follows since a function of ξ and a function of r are equal and therefore must equal the same constant. Bound solutions to the left-hand side of

---

[1] The temperature gradient needs to be very small and the system size large but finite so that the absolute temperature does not become negative.

(17.2), which is Legendre's equation, are obtained only for $\alpha^2 \to \alpha_n^2 = n(n+1)$ for $n = 0, 1, 2, \ldots$, and can be expressed as

$$\Phi_n(\xi) = P_n(\xi), \tag{17.3}$$

where $P_n(\xi)$ is the Legendre polynomial[2] of order n. The right-hand side of (17.2) is an "equidimensional" differential equation, which has a solution for each α_n of the form

$$\Psi_n(r) = A_n r^n + B_n r^{-(n+1)}. \tag{17.4}$$

The A_n and B_n are constants to be determined from boundary conditions. The solution to (17.1) is given by the superposition of (17.3) and (17.4) so that the temperature field in the medium is given by

$$T(r, \xi) = \sum_{n=0}^{\infty} \left[A_n r^n + B_n r^{-(n+1)} \right] P_n(\xi). \tag{17.5}$$

The temperature within the sphere $\overline{T}(r, \xi)$ is governed by the same equation as that for the temperature in the medium (17.1). Hence, for $\overline{T}$ we can write

$$\overline{T}(r, \xi) = \sum_{n=0}^{\infty} \overline{A}_n r^n P_n(\xi), \tag{17.6}$$

where, in order for $\overline{T}(0, \xi)$ to be finite, we have set $\overline{B}_n = 0$.

Three boundary conditions are required to find the constants in (17.5) and (17.6). Far from the sphere $(r \to \infty)$, the boundary condition involving the prescribed temperature gradient is expressed as

$$\frac{\partial T}{\partial r}(\infty, \xi) = |\boldsymbol{\nabla} T|_\infty \xi. \tag{17.7}$$

Combining (17.5) with the boundary condition in (17.7), we find that $A_1 = |\boldsymbol{\nabla} T|_\infty$ and $A_n = 0$ for $n \geq 2$, by using the orthogonality property of $P_n(\xi)$ (cf. footnote on p. 307).

The remaining two boundary conditions are prescribed at the interface between the sphere and the medium $(r = R)$, where we choose the normal

---

[2] Legendre polynomials $P_n(x)$ are defined on the interval $x = [-1, 1]$. When normalized such that $P_n(1) = 1$, for $n = 0, 1, 2, 3$, they are given by

$$P_0(x) = 1, \quad P_1(x) = x, \quad P_2(x) = \frac{1}{2}(3x^2 - 1), \quad P_3(x) = \frac{1}{2}(5x^3 - 3x).$$

The orthogonality condition for $P_n(x)$ means that

$$\int_{-1}^{1} P_n(x) P_m(x) dx = 0 \quad \text{for } n \neq m, \qquad \int_{-1}^{1} P_m^2(x) dx = \frac{2}{2m+1}.$$

vector to point into the medium $(n = \delta_r)$. Both the sphere and the medium are at rest, and there is no mass transfer across the interface so that from (14.7) we have $v^s \cdot n = 0$. If we further neglect transport within the interface, the jump balance for energy (14.26) reduces to the continuity of energy flux across the interface, $n \cdot j_q^I = n \cdot j_q^{II}$. Using Fourier's law (6.4) for the energy flux in both phases leads to the boundary condition

$$\frac{\partial T}{\partial r}(R, \xi) = \beta \frac{\partial \overline{T}}{\partial r}(R, \xi), \qquad (17.8)$$

where $\beta = \overline{\lambda}/\lambda$. The final boundary condition is obtained from the constitutive equations (15.13) and (15.14) for heat flow from the bulk phases to the interface. In the absence of mass transfer across the interface (see Exercise 15.5), these can be combined to obtain the boundary condition

$$T(R, \xi) = \overline{T}(R, \xi) + R_K^s \lambda \frac{\partial T}{\partial r}(R, \xi), \qquad (17.9)$$

where $R_K^s = R_K^{Is} + R_K^{IIs}$ is the sum of the Kapitza resistances, which we assume are constant. Hence, the presence of interfacial resistances to heat transfer leads to a jump in temperature across the interface that is proportional to the heat flux in the bulk phases.

Substitution of (17.5) and (17.6) into (17.8) and (17.9) leads to a pair of algebraic equations for each n. When solved (see Exercise 17.1), the following expressions for the temperature fields in the sphere and medium are obtained,

$$\frac{\overline{T}(r, \xi) - T_0}{R|\nabla T|_\infty} = \frac{3}{2 + \beta(1 + 2N_{Ka})}\left(\frac{r}{R}\right)\xi, \qquad (17.10)$$

$$\frac{T(r, \xi) - T_0}{R|\nabla T|_\infty} = \left(\frac{r}{R}\right)\xi + \frac{1 - \beta(1 - N_{Ka})}{2 + \beta(1 + 2N_{Ka})}\left(\frac{R}{r}\right)^2\xi, \qquad (17.11)$$

where $N_{Ka} = R_K^s \lambda / R$ is the dimensionless Kapitza resistance. It is worth noting that the relative importance of interfacial resistance increases with decreasing particle size so that its effect can be quite important in nanoscale devices.[3] Note that $r\xi = r\cos\theta = x_3$, which means the temperature gradient within the sphere is uniform and is parallel to the temperature gradient imposed in the medium far from the sphere. The second term in (17.11) is the disturbance of temperature in the medium due to the presence of the sphere, which we see decays as r^{-2}. Figure 17.2 shows the temperature field predicted by (17.10) and (17.11) for cases with $N_{Ka} = 0$.

---

[3] Cahill *et al.*, *J. Appl. Phys.* **93** (2003) 793; Cahill *et al.*, *Appl. Phys. Rev.* **1** (2014) 011305.

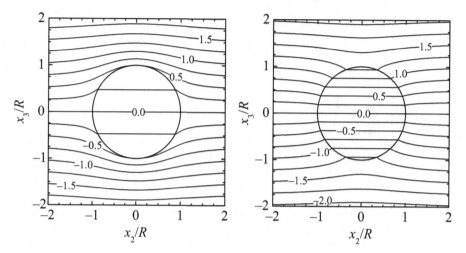

Figure 17.2 Normalized temperature $(T - T_0)/(R|\nabla T|_\infty)$ isotherms in the x_2–x_3 plane around a sphere imbedded in an infinite medium predicted by (17.10) and (17.11) with $N_{\text{Ka}} = 0$ and $\beta = 4$ (left) and $\beta = 1/4$ (right).

Exercise 17.1   Temperature Fields about a Spherical Particle
Show that substitution of (17.5) and (17.6) into (17.8) and (17.9) leads to the temperature fields given in (17.10) and (17.11).

Exercise 17.2   Temperature Jump due to Kapitza Resistance
Substitute (17.10) and (17.11) into (17.9) to obtain the following expression for the temperature jump across the interface at $r = R$,

$$\frac{T(R,\theta) - \overline{T}(R,\theta)}{R|\nabla T|_\infty} = \frac{3\beta N_{\text{Ka}}}{2 + \beta(1 + 2N_{\text{Ka}})} \cos\theta,$$

and explain why the jump is not uniform over the surface of the sphere.

Exercise 17.3   Boundary Condition with Interfacial Thermal Conductivity
Use the interfacial energy balance in (15.20) with (15.11) to derive the boundary condition

$$\lambda \frac{\partial T}{\partial r}(R,\theta) = \overline{\lambda} \frac{\partial \overline{T}}{\partial r}(R,\theta) - \frac{\lambda^{\text{s}}}{R^2 \sin\theta} \frac{\partial}{\partial \theta}\left[\frac{\partial T^{\text{s}}}{\partial \theta}(R,\theta)\sin\theta\right].$$

17.2 Effective Medium Theory

We now consider the effect that a large number of particles has on the transport of energy in a material. The transport equations governing the temperature field in the heterogeneous system shown in the left side of

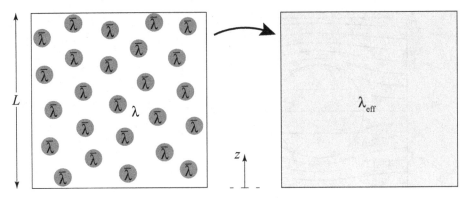

Figure 17.3 Schematic diagram of heterogeneous system composed of multiple spheres with thermal conductivity $\bar{\lambda}$ imbedded in a medium with thermal conductivity λ, and its representation as an effectively continuous medium with effective thermal conductivity λ_{eff}.

Figure 17.3 would require numerical methods. Instead, we develop a method to estimate the effective thermal conductivity λ_{eff} of an effectively continuous medium, as shown conceptually in the right side of Figure 17.3. Let V_{eff} be the total volume of a system comprised of a continuous phase with thermal conductivity λ and volume V_{con} and a dispersed phase with thermal conductivity $\bar{\lambda}$ and volume V_{dis}. If there are k particles of the same size, then each particle has volume $V_0 = V_{\text{dis}}/k$. The volume V_{eff} should be large enough to contain a large number of particles, but small enough that $(V_{\text{eff}})^{1/3}$ is much smaller than the characteristic length of the geometry to be described. In other words, we assume the two-phase system is effectively a continuous, single-phase system. The approach just described is known as *effective medium theory* and is based on the volume average of a quantity,

$$\langle (.) \rangle = \frac{1}{V_{\text{eff}}} \int_{V_{\text{eff}}} (.)dV. \tag{17.12}$$

In this section, we use an intuitive application of effective medium theory. A more formal treatment of volume averaging is presented in Chapter 23.

Our next step is to express Fourier's law in a form applicable to effective medium theory. Let T' be the temperature at a point within V_{eff} which has area A and thickness L over which there exists temperature difference $\Delta T = \int_L |\nabla T'|dz$. Integrating this equation over A, we obtain $\Delta T/L = \langle |\nabla T'| \rangle$. The rate of energy transfer over A is $\dot{Q} = \int_A |j_q|dA$, which, if integrated over L, leads to $\dot{Q}/A = \langle |j_q| \rangle$. By analogy with Fourier's law (6.4), we write

$\dot{Q}/A \propto \Delta T/L$, or, in terms of volume-averaged quantities,

$$\langle \boldsymbol{j}_q \rangle = -\lambda_{\mathrm{eff}} \langle \boldsymbol{\nabla} T' \rangle, \tag{17.13}$$

where λ_{eff} is the effective thermal conductivity within V_{eff}. Using (17.12) and noting that $V_{\mathrm{eff}} = V_{\mathrm{con}} + V_{\mathrm{dis}}$, we can write (17.13) as

$$\langle \boldsymbol{j}_q \rangle = -\lambda \langle \boldsymbol{\nabla} T' \rangle - \frac{\overline{\lambda} - \lambda}{V_{\mathrm{eff}}} \int_{V_{\mathrm{dis}}} \boldsymbol{\nabla} \overline{T} \, dV, \tag{17.14}$$

where the second term isolates the effect of the dispersed phase. In general, the presence of each particle influences the temperature field in all other particles, making the integral in (17.14) difficult to evaluate.

To simplify things, we consider the case where the distance between particles is large, or the system is dilute. As shown in Figure 17.2, the disturbance of the temperature field in the medium decays rapidly with distance from the sphere. Hence, it is reasonable to assume for a dilute system that the temperature field for each particle is unaffected by the presence of other particles. This means we need only evaluate the integral over the volume of each particle V_0; for spherical particles of the same size, the contribution from each particle is the same. Now, from (17.10), we see the temperature gradient is uniform throughout the sphere, which greatly simplifies evaluation of the integral over the particle volume. Hence, using (17.10), the integral in (17.14) can be written as

$$\frac{1}{V_{\mathrm{eff}}} \int_{V_{\mathrm{dis}}} \boldsymbol{\nabla} \overline{T} \, dV = \frac{k}{V_{\mathrm{eff}}} \int_{V_0} \boldsymbol{\nabla} \overline{T} \, dV = \phi \frac{3}{2 + \beta(1 + 2N_{\mathrm{Ka}})} |\boldsymbol{\nabla} T|_{\infty} \delta_3, \tag{17.15}$$

where $\phi = kV_0/V_{\mathrm{eff}} \ll 1$ is the volume fraction of the dispersed phase. Now, since $\langle \boldsymbol{\nabla} T' \rangle = |\boldsymbol{\nabla} T|_{\infty} \delta_3 + O(\phi)$, the result in (17.15) to first order in ϕ is linear in $\langle \boldsymbol{\nabla} T \rangle$. By substitution of (17.15) into (17.14) and comparing this result with (17.13), we obtain

$$\frac{\lambda_{\mathrm{eff}}}{\lambda} = 1 + \frac{3(\beta - 1)}{2 + \beta(1 + 2N_{\mathrm{Ka}})} \phi. \tag{17.16}$$

The expression in (17.16) with interfacial resistances neglected ($N_{\mathrm{Ka}} = 0$) is equivalent to that found by Maxwell[4] in 1873 for the effective dielectric constant of a composite. It is worthwhile to look at the prediction of (17.16)

---

[4] Maxwell, *A Treatise on Electricity and Magnetism* (Oxford, 1904), used a different approach for $N_{\mathrm{Ka}} = 0$ and obtained the expression

$$\frac{\lambda_{\mathrm{eff}}}{\lambda} = 1 + \frac{3\phi(\beta - 1)}{2 + \beta - \phi(\beta - 1)} = 1 + \frac{3(\beta - 1)}{2 + \beta} \phi + O(\phi^2).$$

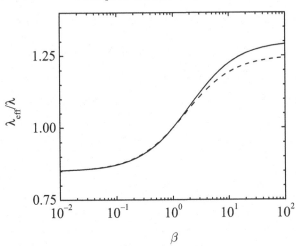

Figure 17.4 Normalized effective thermal conductivity as a function of thermal conductivity ratio β predicted by (17.16) for $\phi = 0.1$ with $N_{\text{Ka}} = 0$ (solid line) and $N_{\text{Ka}} = 0.1$ (dashed line).

for limiting values of the thermal conductivity ratio β, which, in the absence of interfacial resistances ($N_{\text{Ka}} = 0$), are given by

$$\frac{\lambda_{\text{eff}}}{\lambda} = \begin{cases} 1 + 3\phi & \text{for} \quad \beta \to \infty, \\ 1 - \frac{3}{2}\phi & \text{for} \quad \beta \to 0. \end{cases} \tag{17.17}$$

Figure 17.4 shows the relative effective thermal conductivity as a function of β. From this figure, we see for $\phi = 0.1$ that most of the maximum possible change in λ_{eff} takes place when the particle has a thermal conductivity that is 10 times larger/smaller than that of the medium. Expressions for the effective thermal conductivity of composites that take into account particle–particle interactions and non-spherical particles can be found elsewhere.[5]

Exercise 17.4   Effective Thermal Conductivity
Elaborate the steps involved in going from (17.15) to (17.16).

17.3 Creeping Flow around a Sphere

We now consider steady flow around a sphere and refer again to Figure 17.1. In this case, the sphere (phase II) is a rigid solid, and the infinite medium (phase I) surrounding the sphere is a Newtonian fluid with constant density

---

[5] Nan *et al.*, *J. Appl. Phys.* **81** (1997) 6692.

ρ and constant viscosity η. The flow is incompressible, so the equations governing the velocity and pressure fields are the constraint on the velocity in (5.36), that is $\nabla \cdot \boldsymbol{v} = 0$, and the Navier–Stokes equation (7.8). The velocity far from the sphere is given by $\boldsymbol{V} = V\boldsymbol{\delta}_3$. Our goal is to find the velocity and pressure fields and use these to calculate the force exerted by the fluid on the sphere.

There is symmetry about the x_3-coordinate so that the velocity field has the form $v_r = v_r(r,\theta)$, $v_\theta = v_\theta(r,\theta)$, $v_\phi = 0$. Applying the $n - m$ procedure (see Section 7.5) to the steady-state form of (7.8), we have the quantities $r, \theta, v_r, v_\theta, \mathcal{P}, \eta, \rho, R$, and V, which have units of length, time, and mass, so that $n - m = 6$. As we are interested in velocity and pressure as functions of position, there remains only one dimensionless parameter. If we use R, V, and η to set the basic units, we obtain the dimensionless form of the Navier–Stokes equation in (7.50) with $N_{\mathrm{Re}} = \rho V R / \eta$. If ρ is expressed in the chosen units of mass and length, we have $N_{\mathrm{Re}} = \rho$. For $N_{\mathrm{Re}} \ll 1$, the inertial term can hence be neglected, which is the creeping flow limit. We consider this case here, so that (7.8) simplifies to

$$\nabla p^{\mathrm{L}} = \eta \nabla^2 \boldsymbol{v}, \qquad (17.18)$$

where we have chosen to neglect gravity.[6] The linearity of (17.18), which is sometimes called the Stokes equation, has lead to the development of numerous solution methods for a wide range of problems for creeping flow.[7] The pseudo-pressure p^{L} is governed by $\nabla^2 p^{\mathrm{L}} = 0$, as one realizes by taking the divergence of (17.18) for constant viscosity.

As discussed in Section 5.1, it is sometimes convenient to express the components of the velocity in terms of a stream function ψ. For this flow, these relations are derived from $\nabla \cdot \boldsymbol{v} = 0$ expressed in spherical coordinates [see (B.15)] and are given by

$$v_r = \frac{1}{r^2 \sin\theta} \frac{\partial \psi}{\partial \theta}, \qquad v_\theta = -\frac{1}{r \sin\theta} \frac{\partial \psi}{\partial r}. \qquad (17.19)$$

The equation that governs the stream function is derived by substitution of (17.19) into (17.18) and eliminating pressure p^{L}. This straightforward, but rather tedious, procedure (see Exercise 17.5) gives

$$\left[\frac{\partial^2}{\partial r^2} + \frac{\sin\theta}{r^2} \frac{\partial}{\partial \theta} \left(\frac{1}{\sin\theta} \frac{\partial}{\partial \theta} \right) \right]^2 \psi = E^4 \psi = 0, \qquad (17.20)$$

where the operator E^2 is contained in the square brackets (see Exercise 7.6).

---

[6] If gravity is included, $p^{\mathrm{L}} \to \mathcal{P}$ (see Exercise 17.7).
[7] Happel & Brenner, *Low Reynolds Number Hydrodynamics* (Martinus Nijhoff, 1973).

By introducing the stream function, our fluid mechanics problem has been transformed from a system of a first-order and two second-order differential equations that govern $v_r(r, \theta)$, $v_\theta(r, \theta)$, and $p^L(r, \theta)$ into the single fourth-order differential equation (17.20) that governs $\psi(r, \theta)$.

Exercise 17.5   Derivation of the Stream Function Equation
The r- and θ-components of (17.18) are given by

$$\frac{\partial p^L}{\partial r} = \eta \left[\frac{1}{r^2} \frac{\partial^2}{\partial r^2} (r^2 v_r) + \frac{1}{r^2 \sin \theta} \frac{\partial}{\partial \theta} \left(\sin \theta \frac{\partial v_r}{\partial \theta} \right) \right], \tag{17.21}$$

and

$$\frac{1}{r} \frac{\partial p^L}{\partial \theta} = \eta \left[\frac{1}{r^2} \frac{\partial}{\partial r} \left(r^2 \frac{\partial v_\theta}{\partial r} \right) + \frac{1}{r^2} \frac{\partial}{\partial \theta} \left(\frac{1}{\sin \theta} \frac{\partial}{\partial \theta} (\sin \theta \, v_\theta) \right) + \frac{2}{r^2} \frac{\partial v_r}{\partial \theta} \right]. \tag{17.22}$$

Show that substitution of (17.19) into (17.21) and (17.22) and assuming p^L is a sufficiently smooth function leads to (17.20).

The solution of (17.20) requires four boundary conditions. Two of these are obtained from the velocity far from the sphere $(r \to \infty)$,

$$v_r(\infty, \theta) \to V \cos \theta, \qquad \frac{\partial \psi}{\partial \theta}(\infty, \theta) \to V r^2 \sin \theta \cos \theta, \tag{17.23}$$

$$v_\theta(\infty, \theta) \to -V \sin \theta, \qquad \frac{\partial \psi}{\partial r}(\infty, \theta) \to V r \sin^2 \theta, \tag{17.24}$$

where the second equations in (17.23) and (17.24) have been obtained from the first equations and the relations in (17.19). The remaining two boundary conditions are prescribed at the phase interface between the sphere and fluid $(r = R)$, which is stationary $(v^s = 0)$ with normal vector $n = \delta_r$. There is no mass transfer across the interface, and the sphere is stationary, so that from (14.7) we can write

$$v_r(R, \theta) = 0, \qquad \frac{\partial \psi}{\partial \theta}(R, \theta) = 0. \tag{17.25}$$

To obtain a second boundary condition at the solid–fluid interface, we use the constitutive equation for the jump in the tangential component of velocity across an interface given in (15.17). The jump balance for momentum (14.6) for an interface without mass transfer and interfacial tension simplifies to $n \cdot \pi^I = n \cdot \pi^{II}$. For τ^I, we use Newton's law of viscosity (6.5) for an incompressible fluid, and we use $\xi_\parallel = \xi_{slip} \delta_\parallel$ for the transverse mobility tensor.[8] Combining these, we obtain (see Exercise 15.4)

$$v_\theta(R,\theta) = \xi_{\text{slip}}\eta \left[r\frac{\partial}{\partial r}\left(\frac{v_\theta}{r}\right)\right]_{r=R}, \qquad \frac{\partial\psi}{\partial r}(R,\theta) = \xi_{\text{slip}}\eta \left[r^2\frac{\partial}{\partial r}\left(\frac{1}{r^2}\frac{\partial\psi}{\partial r}\right)\right]_{r=R}.$$
(17.26)

A boundary condition of the type given in (17.26) is sometimes referred to as the Navier slip law. A no-slip boundary condition is recovered when the slip coefficient is zero, $\xi_{\text{slip}} = 0$.

The solution of (17.20) for $\psi(r,\theta)$ can be obtained by means of the separation of variables method used in the previous section. However, instead of introducing an arbitrary function for the θ-dependence, we can use boundary conditions to find a more specific form. Integration of the boundary conditions for ψ in (17.23) and (17.24) suggests

$$\psi(r,\theta) = f(r)\sin^2\theta,$$
(17.27)

where $f(r)$ is an unknown function. Substitution of (17.27) into (17.20) produces an equidimensional equation that governs f. Hence, the form given in (17.27) is not only consistent with the boundary conditions in (17.23) and (17.24), but also preserves the angular dependence in the E^2 operator.[9] The solution of the resulting equidimensional equation for f is given by

$$f(r) = C_1 r^{-1} + C_2 r + C_3 r^2 + C_4 r^4.$$
(17.28)

The constants C_n are determined from the boundary conditions in (17.23)–(17.26) that, when combined with (17.27), give the following expression for the stream function,

$$\psi(r,\theta) = V\left[\frac{1}{2}\left(\frac{r}{R}\right)^2 - \frac{3}{4}\frac{1+2\Lambda_\xi}{1+3\Lambda_\xi}\left(\frac{r}{R}\right) + \frac{1}{4}\frac{1}{1+3\Lambda_\xi}\left(\frac{R}{r}\right)\right]R^2\sin^2\theta,$$
(17.29)

where $\Lambda_\xi = \xi_{\text{slip}}\eta/R$ is a dimensionless slip coefficient. Stream functions predicted by (17.29) are shown in Figure 17.5 for no-slip, $\Lambda_\xi = 0$, and perfect slip, $\Lambda_\xi \to \infty$, cases. The expression in (17.29) with $\Lambda_\xi = 0$ was found by Stokes in 1851 and, besides being one of the first, is one of the most important solutions to a viscous flow problem.

---

[8] Note that this introduces an additional dimensional quantity so that the solution will depend also on $\xi_{\text{slip}}\eta/R$.
[9] The operator E^2 defined in (17.20) acting on the stream function in (17.27) gives

$$[E^2]^2\psi = E^4(f\sin^2\theta) = \sin^2\theta\left(\frac{d^2}{dr^2} - \frac{2}{r^2}\right)^2 f.$$

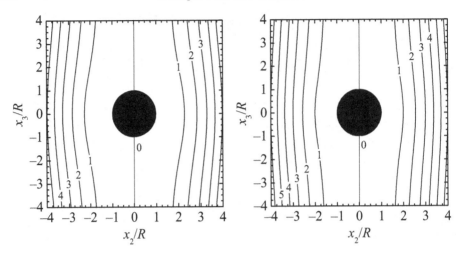

Figure 17.5 Normalized stream function $\psi/(VR^2)$ in the x_2–x_3 plane for creeping flow around a sphere predicted by (17.29) with $\Lambda_\xi = 0$ (left) and $\Lambda_\xi \to \infty$ (right).

The velocity can be found by substituting (17.29) into (17.19), which gives

$$\frac{v_r}{V} = \left[1 - \frac{3}{2}\frac{1 + 2\Lambda_\xi}{1 + 3\Lambda_\xi}\left(\frac{R}{r}\right) + \frac{1}{2}\frac{1}{1 + 3\Lambda_\xi}\left(\frac{R}{r}\right)^3\right]\cos\theta, \qquad (17.30)$$

$$\frac{v_\theta}{V} = -\left[1 - \frac{3}{4}\frac{1 + 2\Lambda_\xi}{1 + 3\Lambda_\xi}\left(\frac{R}{r}\right) - \frac{1}{4}\frac{1}{1 + 3\Lambda_\xi}\left(\frac{R}{r}\right)^3\right]\sin\theta. \qquad (17.31)$$

The pressure field can be found by substituting (17.30) and (17.31) into (17.21) and (17.22) and integrating with the boundary condition $p^L(\infty, \theta) \to 0$. Alternatively, the pressure can be found by solving $\nabla^2 p^L = 0$, which will have a solution of the form given in (17.5) with $A_n = 0$ to satisfy the boundary condition at infinity. From (17.21) we obtain a boundary condition for $\partial p^L/\partial r(R, \theta)$, which is proportional to $\cos\theta = \xi$. Hence, the orthogonality condition (cf. footnote on p. 307) means that only the $n = 1$ term survives. Either method gives

$$\frac{p^L}{\eta V/R} = -\frac{3}{2}\frac{1 + 2\Lambda_\xi}{1 + 3\Lambda_\xi}\left(\frac{R}{r}\right)^2\cos\theta, \qquad (17.32)$$

which completes the solution for creeping flow around a sphere. Note that, for perfect slip at the interface $\Lambda_\xi \to \infty$, this solution corresponds to that for flow around an inviscid gas bubble. Pressure contours from (17.32) are shown in Figure 17.6, where we see the presence of large pressure gradients with opposite signs near the upstream and downstream surfaces of the sphere.

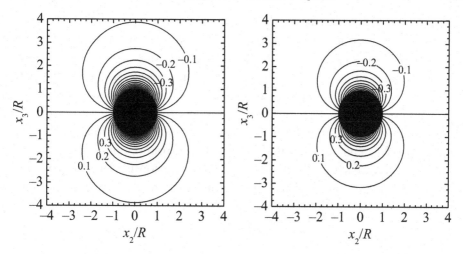

Figure 17.6 Normalized pressure contours $p^L/(\eta V/R)$ in the x_2-x_3 plane for creeping flow around a sphere from (17.32) with $\Lambda_\xi = 0$ (left) and $\Lambda_\xi \to \infty$ (right).

We now use these results to find the force exerted on the sphere. The general expression for the force exerted by a fluid on a solid is given in (8.39). In this expression, $\boldsymbol{n}$ is the unit normal vector to A pointing away from the fluid. Hence, for flow around a sphere we have $\boldsymbol{n} = -\boldsymbol{\delta}_r$, and $\boldsymbol{\pi}$ on the sphere surface can be determined from expressions we just found for the velocity and pressure fields. The result of this calculation (see Exercise 17.6) for the no-slip ($\Lambda_\xi = 0$) case leads to

$$\mathcal{F}_{\rm s} = 6\pi\eta RV, \tag{17.33}$$

which is known as Stokes' law. Two-thirds of the force in (17.33) are due to viscous forces exerted on the sphere surface (skin drag), and the remaining one-third comes from pressure forces that result from the perturbation of the velocity field as it passes the sphere (form drag). The friction, or drag, coefficient for creeping flow around a sphere is $\zeta = 6\pi\eta R$, which is the inverse of the mobility coefficient (see Exercise 6.6).

Stokes' law (17.33), which shows that $\mathcal{F}_{\rm s}$ scales with ηRV, has been found to be valid for $N_{\rm Re} \lesssim 0.1$. Interestingly, the relative importance of inertial forces becomes larger as the distance from the sphere increases. In other words, (17.18) is not uniformly valid for flow around a sphere, which makes finding corrections for small but finite $N_{\rm Re}$ a significant challenge. It is also interesting to note that a solution of (17.18) does not exist for flow normal to a long cylinder in an infinite fluid, which is known as Stokes' paradox.

To understand why this is the case, let us use dimensional analysis. For flow past a cylinder with radius $R \ll L$, we are interested in finding the force per unit length $\mathcal{F}'_s = \mathcal{F}_s/L$ on the cylinder. For creeping flow, the dimensionless quantity $\mathcal{F}'_s/(\eta V)$ must be constant so that $\mathcal{F}'_s$ is independent of R, which is clearly unphysical. If inertia is brought back in, then $\mathcal{F}'_s/(\eta V)$ is a function of N_{Re}, which is clearly more physical. The book by Happel and Brenner[10] contains extensive discussions on corrections to Stokes' law for finite N_{Re}, as well as analysis for flow around non-spherical particles and flow in multi-particle systems. Also, a solution for time-dependent flow around a sphere, along with a discussion of the history of this problem, can be found elsewhere.[11]

Exercise 17.6   Stokes' Law
Evaluation of the integral in (8.39) requires an expression for the pressure tensor, which for incompressible Newtonian fluids is given by $\boldsymbol{\pi} = p^L \boldsymbol{\delta} - \eta \left[\boldsymbol{\nabla v} + (\boldsymbol{\nabla v})^T \right]$. Use (17.30)–(17.32) to show that (8.39) gives

$$\mathcal{F}_s = -\int_0^{2\pi} \int_0^\pi \boldsymbol{n} \cdot \boldsymbol{\pi}(R) R^2 \sin\theta \, d\theta \, d\phi = 6 \frac{1 + 2\Lambda_\xi}{1 + 3\Lambda_\xi} \pi \eta R V \boldsymbol{\delta}_3.$$

Note that the force is in the direction of the imposed velocity in the fluid $\boldsymbol{V} = V\boldsymbol{\delta}_3$, so the above result is independent of coordinate system. Consider a sphere moving with constant velocity $\boldsymbol{v}_b$ through a fluid that is otherwise at rest. Use the results in the subsection on transport in non-inertial coordinate systems (see Section 11.1) to show this expression gives the force on the sphere by setting $\boldsymbol{V} \to -\boldsymbol{v}_b$.

Exercise 17.7   Terminal Velocity of a Sphere
Consider a sphere that moves with constant velocity $\boldsymbol{v}_b = v_b \boldsymbol{\delta}_3$ through a fluid that is otherwise at rest due to the gravitational force $\boldsymbol{g} = -g\boldsymbol{\delta}_3$. The velocity is determined by a balance between the gravitational force on the sphere and the force exerted by the fluid on the sphere. Use the results from Exercise 17.6 to show that this velocity, known as the terminal velocity, is given by

$$v_b = \frac{2}{9} \frac{1 + 3\Lambda_\xi}{1 + 2\Lambda_\xi} \frac{R^2(\rho - \bar{\rho})g}{\eta},$$

where $\bar{\rho}$ is the density of the spherical particle. This expression has been used to obtain the viscosity of liquids in a "falling ball viscometer."

Exercise 17.8   Flow past a Sphere in Vector Form
For $\Lambda_\xi = 0$, show that the velocity and pressure fields for flow past a sphere given

[10] Happel & Brenner, *Low Reynolds Number Hydrodynamics* (Martinus Nijhoff, 1973).
[11] Schieber *et al.*, *J. Non-Newtonian Fluid Mech.* **200** (2013) 3.

in (17.30), (17.31), and (17.32) can be expressed in the vector form

$$\boldsymbol{v} = \left[\left(1 - \frac{3}{4}\frac{R}{r} - \frac{1}{4}\frac{R^3}{r^3}\right)\boldsymbol{\delta} - \frac{3}{4}\left(\frac{R}{r^3} - \frac{R^3}{r^5}\right)\boldsymbol{rr}\right]\cdot\boldsymbol{V},$$

and

$$p^{\mathrm{L}} = -\frac{3}{2}\eta R\frac{\boldsymbol{r}\cdot\boldsymbol{V}}{r^3}.$$

Verify that these expressions do indeed provide the solution for flow past a sphere. Hint: transform the solution to rectangular coordinates.

Exercise 17.9   Motion of a Bubble Driven by Marangoni Stress
Consider the motion of a spherical bubble of radius R in an otherwise stationary liquid driven by a constant temperature gradient with magnitude $|\boldsymbol{\nabla}T|_\infty$ in the x_3-direction that is maintained far from the bubble (see Figure 17.1). The non-uniform interfacial temperature $T^{\mathrm{s}} = T^{\mathrm{s}}(\theta)$ leads to a Marangoni stress $d\gamma = \gamma_T\,dT^{\mathrm{s}}$, where $\gamma_T = d\gamma/dT^{\mathrm{s}}$ is constant. The liquid is an incompressible Newtonian fluid with constant η and λ, and the bubble is an inviscid fluid ($\bar\eta \ll \eta$) with small thermal conductivity ($\bar\lambda \ll \lambda$). At the gas–liquid interface $r = R$, there is no mass transfer and the no-slip assumption holds. Use the interfacial energy balance in (15.20) with $\boldsymbol{j}_q^{\mathrm{s}} = \boldsymbol{0}$ and jump balance for momentum in (15.21) with $\boldsymbol{\tau}^{\mathrm{s}} = \boldsymbol{0}$ to derive the boundary conditions

$$\lambda\frac{\partial T}{\partial r}(R,\theta) = -\frac{\gamma_T}{R\sin\theta}T(R,\theta)\frac{\partial}{\partial\theta}[v_\theta(R,\theta)\sin\theta],$$

$$\eta\left[\frac{\partial v_\theta}{\partial r}(R,\theta) - \frac{v_\theta(R,\theta)}{R}\right] = -\frac{\gamma_T}{R}\frac{\partial T}{\partial\theta}(R,\theta).$$

17.4 Einstein's Transport Theories

We have already mentioned Albert Einstein several times, but it still may be surprising to see that an entire section bears his name in a book on transport phenomena. Almost everyone knows Einstein's formula for the equivalency between mass and energy $E = mc^2$, and many know his profound 1905 contributions to our understanding of the photoelectric effect, Brownian motion, and special relativity (see Section 12.6). What is less known is that during the same year Einstein completed his doctoral thesis entitled "Eine neue Bestimmung der Moleküldimensionen" (A New Determination of Molecular Dimensions). The focus of Einstein's thesis was to develop theories that would allow the determination of the Avogadro number $\tilde{N}_{\mathrm{A}}$. However, it contains two results that have profound implications in the field of transport phenomena. Both results apply to a dilute suspension of spheres: one

Figure 17.7 Albert Einstein (1879–1955) in 1910.

is an expression for the effective viscosity of a suspension, and the other is for the particle diffusivity.

First, we see how Einstein found an expression for the effective viscosity η_{eff} of a suspension.[12] The volume V_{eff} is comprised of a fluid phase with viscosity η and volume V_{con} and a large number of spherical particles with total volume V_{dis} in the dilute (volume fraction $\phi \ll 1$) limit so that particle–particle interactions can be neglected. Moreover, so that we may take advantage of the nicely linear Stokes' flow equation (17.18), our analysis is restricted to creeping flows.

The approach followed by Einstein is based on finding the increased rate of energy dissipation due to the presence of the particles in the fluid. As we saw in Exercise 8.13, the rate of work (per unit area) done on a fluid at a moving solid surface with unit normal $\boldsymbol{n}$ (directed into the solid) is given by $-\boldsymbol{v} \cdot \boldsymbol{\pi} \cdot \boldsymbol{n}$. Here, we consider a stationary solid, and are interested in the rate of work done by the fluid, which, since the flow is steady, is equal to the rate at which energy is dissipated. The rate of work done by the fluid in V_{eff} on a solid bound by surface A_{eff} with unit normal $\boldsymbol{n}$ (directed into the

[12] Straumann, arXiv:physics (2005) 0504201.

fluid) can be written as

$$\dot{W} = -\int_{A_{\text{eff}}} v_i \pi_{ij} n_j \, dA. \qquad (17.34)$$

We are using the Einstein summation convention,[13] which has already been introduced (cf. footnote on p. 233), but we recognize that not everyone is interested in relativistic hydrodynamics. Interestingly, Einstein himself did not benefit from this convention in the present analysis. He introduced it only in the context of general relativity – more than a decade after publishing his dissertation and his work on special relativity.

For a fluid with a single suspended spherical particle, we write for the velocity $\boldsymbol{v} = \boldsymbol{v}^{(0)} + \boldsymbol{v}^{(1)}$, where $\boldsymbol{v}^{(0)}$ is the velocity in the absence of a particle, and $\boldsymbol{v}^{(1)}$ is the perturbation caused by the particle. For the unperturbed flow, we write

$$v_i^{(0)} = \kappa_{ij}^{(0)} x_j, \qquad (17.35)$$

where $\boldsymbol{\kappa} = (\boldsymbol{\nabla v})^{\mathrm{T}}$ is the transpose of the velocity gradient tensor. The base flow given in (17.35) is defined such that $\kappa_{ij}^{(0)}$ is constant, symmetric, and traceless. In other words, we consider a uniform, irrotational, and in-compressible base flow which, without loss of generality, may be thought of as an extensional flow with diagonal $\kappa_{ij}^{(0)}$ (see Exercise 12.3). Similarly, for the pressure tensor, we write $\boldsymbol{\pi} = \boldsymbol{\pi}^{(0)} + \boldsymbol{\pi}^{(1)}$. For the assumed base flow, we have $\pi_{ij}^{(0)} = -2\eta \kappa_{ij}^{(0)}$. Hence, retaining only terms to first order in the perturbation, we can write (17.34) as

$$\dot{W} = 2\eta \kappa_{ij}^{(0)} \kappa_{ij}^{(0)} V_{\text{eff}} - \int_{A_{\text{eff}}} v_i^{(0)} \pi_{ij}^{(1)} n_j \, dA - \int_{A_{\text{eff}}} v_i^{(1)} \pi_{ij}^{(0)} n_j \, dA. \qquad (17.36)$$

The first term, which has been obtained using the divergence theorem (cf. footnote on p. 60), is the dissipation from the homogeneous base flow in V_{eff}.

Our task is now to solve Stokes' equation (17.18) for flow around a spherical particle. On the surface of the particle, we have $v_i = 0$, and far from the particle we have $v_i = v_i^{(0)}$. As verifying a solution may be much simpler than constructing it (see Exercise 17.8), here we give only the solution, which can

---

[13] In this context, the Einstein summation convention can be stated as follows: repeated indices in a term imply summation over that index with values $1, 2, 3$; an unrepeated index can take the values 1, 2, or 3. For example, the integrand in (17.34) represents a nine-term sum,

$$\boldsymbol{v} \cdot \boldsymbol{\pi} \cdot \boldsymbol{n} = \sum_{i=1}^{3} v_i \boldsymbol{\delta}_i \cdot \sum_{k,j=1}^{3} \pi_{kj} \boldsymbol{\delta}_k \boldsymbol{\delta}_j \cdot \sum_{l=1}^{3} n_l \boldsymbol{\delta}_l = \sum_{i,j=1}^{3} v_i \pi_{ij} n_j = v_i \pi_{ij} n_j,$$

where the summation convention is used to obtain the last equality.

be expressed as

$$v_i^{(1)} = -\frac{5}{2}R^3\kappa_{jk}^{(0)}\frac{x_ix_jx_k}{r^5} - R^5\left(\kappa_{ij}^{(0)}\frac{x_j}{r^5} - \frac{5}{2}\kappa_{jk}^{(0)}\frac{x_ix_jx_k}{r^7}\right), \tag{17.37}$$

$$p^{L(1)} = -5\eta R^3\kappa_{ij}^{(0)}\frac{x_ix_j}{r^5}. \tag{17.38}$$

Notice the similarity between these equations and those for the flow around a sphere discussed in Exercise 17.8.

Exercise 17.10   *Elongational Flow past a Sphere in Rectangular Coordinates*
Verify that (17.37) and (17.38) provide a divergence-free solution of (17.18) with $v_i^{(1)} = -\kappa_{ij}^{(0)}x_j$ for $r = R$ and $v_i^{(1)} \to 0$ for $r \to \infty$.

Exercise 17.11   *Alternative Expression for the Velocity Field*
Rewrite (17.37) in the third-order-derivative form

$$v_i^{(1)} = \frac{1}{6}R^4\left(\kappa_{ik}^{(0)}\frac{\partial^2}{\partial x_j\,\partial x_j} - \kappa_{jk}^{(0)}\frac{\partial^2}{\partial x_i\,\partial x_j}\right)\frac{\partial}{\partial x_k}\left(5\frac{r}{R} + \frac{R}{r}\right). \tag{17.39}$$

We now take A_{eff} to be the surface of a very large sphere around the origin with radial normal vector $n_i = x_i/r$. We use (17.37) keeping only the first term decaying as $1/r^2$, and (17.38) to evaluate the two integrals in (17.36). The result is (see Exercise 17.12)

$$\dot{W} = 2\eta\kappa_{ij}^{(0)}\kappa_{ij}^{(0)}\left(V_{\text{eff}} + \frac{2}{3}\pi R^3\right) = 2\eta\kappa_{ij}^{(0)}\kappa_{ij}^{(0)}V_{\text{eff}}\left(1 + \frac{1}{2}\phi\right). \tag{17.40}$$

Next, we must compute the rate of dissipation for a fluid with a large number of particles. The velocity perturbation (17.39) results from a sphere at the origin. If the sphere is at the position $\boldsymbol{x}'$, we need only replace $r = |\boldsymbol{x}|$ by $|\boldsymbol{x} - \boldsymbol{x}'|$ in (17.39), which demonstrates the usefulness of this alternative representation. For a sphere positioned randomly inside a large sphere with volume V_{eff} and radius R_{eff}, we need to calculate the average (see Exercise 17.13)

$$\frac{1}{V_{\text{eff}}}\int_{V_{\text{eff}}}\left(5\frac{|\boldsymbol{x} - \boldsymbol{x}'|}{R} + \frac{R}{|\boldsymbol{x} - \boldsymbol{x}'|}\right)d^3x'$$
$$= \frac{4\pi R^3}{3V_{\text{eff}}}\left(\frac{3R_{\text{eff}}^2 - r^2}{2R^2} + \frac{15R_{\text{eff}}^4 + 10R_{\text{eff}}^2r^2 - r^4}{4R^4}\right). \tag{17.41}$$

Only the last term survives after the three differentiations in (17.39), and the velocity perturbation becomes linear in position,

$$v_i^{(1)} = -\frac{4\pi R^3}{3V_{\text{eff}}} \kappa_{ij}^{(0)} x_j. \tag{17.42}$$

The averaging over random positions corresponds to the presence of many spheres in a large volume, so that the total velocity field is given by the base flow and the properly weighted contribution of the spheres according to their volume fraction,

$$v_i = \kappa_{ij}^{(0)} x_j (1 - \phi). \tag{17.43}$$

The simplicity of this result may be surprising. However, if there is no preferred position of the spheres, the vectorial structure of the result can only be introduced by the base flow field. When the modified homogeneous flow field (17.43) is used to evaluate (17.34), the result for the energy dissipation of the effective fluid is obtained in exactly the same way as the first term in (17.36) for the pure base flow,

$$\dot{W} = 2\eta_{\text{eff}} \kappa_{ij}^{(0)} \kappa_{ij}^{(0)} V_{\text{eff}} (1 - 2\phi). \tag{17.44}$$

Comparing (17.40) with (17.44) and keeping only terms to first order in ϕ, we obtain Einstein's famous result[14]

$$\frac{\eta_{\text{eff}}}{\eta} = 1 + \frac{5}{2}\phi. \tag{17.45}$$

It is of interest to note that, due to a mathematical error made by Einstein, the originally published version of (17.45) was missing the factor of 5/2, which was eventually discovered as the result of discrepancies between the theory and experimental observations.

The more well-known result from Einstein's dissertation is his theory for the diffusivity D of a particle in a suspension. This elegant derivation is based on equating the motion of suspended particles due to diffusion and the presence of an external force (typically, the gravitational field). The presence of a concentration gradient of particles is the driving force for a diffusive flux which, in a steady state, must be balanced by the flow resulting from the external force. This is analogous to the argument made in Exercise 6.6 to derive the Nernst–Einstein equation (6.31). The friction coefficient, which can be found from Stokes' law (17.33) as in Exercise 17.6, is $\zeta = 6\pi\eta R$. Combining these, we obtain the Stokes–Einstein relation

$$D = \frac{k_{\text{B}} T}{6\pi\eta R}. \tag{17.46}$$

---

[14] Einstein, *Ann. der Phys.* **324** (1906) 289.

The simple expression in (17.46) has deep and profound implications. It shows the connection between thermal fluctuations on the microscopic level and their manifestation on the macroscopic level as friction. On the practical side, the Stokes–Einstein relation provides reasonable estimates for diffusivities in simple systems and, as we will see later, is the cornerstone for interpreting experiments for measuring the properties of complex fluids (see Chapter 25).

Exercise 17.12   Evaluation of the Rate of Work
Evaluate the integrals in (17.36) over the surface of a very large sphere around a suspended sphere of radius R to arrive at (17.40).

Exercise 17.13   Averaging for Flow around a Randomly Positioned Sphere
Calculate the average in (17.41).

- Systems containing spherical particles provide an important example of heterogeneous systems comprised of one material dispersed in a second material; detailed solutions of transport problems around spheres can be used to develop effective medium theories for heterogeneous systems.
- By solving the steady-state temperature equation around a single sphere with an imposed temperature gradient far from the sphere, we obtain an expression for the effective thermal conductivity of a heterogeneous material containing dispersed spheres at small concentration.
- By solving the steady-state Navier–Stokes equation without inertia for flow around a single sphere, we derive Stokes' law for the friction coefficient of a sphere in terms of the fluid viscosity and radius of the sphere.
- By evaluating the increase in the rate of energy dissipation resulting from flow around a sphere and neglecting interactions between multiple spheres, we obtain Einstein's expression for the effective viscosity of a suspension at small concentration.

18

Bubble Growth and Dissolution

The growth and collapse of gas or vapor bubbles in liquids driven by diffusive transport occurs in natural, biological, and technological systems. A few examples are bubbles growing in volcanic magmas,[1] gas bubble collapse in biological fluids,[2] and the growth of bubbles in polymeric foam production.[3] More familiar examples include the bubbles formed when water is boiled to cook pasta and when a bottle of champagne is opened. Not surprisingly, many studies have been conducted in order to develop transport models for the prediction of bubble growth and collapse rates.

Studies on the dynamics of bubble growth and collapse appear to begin with Rayleigh,[4] who excluded mass transfer (phase change) between the bubble and the surrounding liquid so that the dynamics are controlled by hydrodynamic effects only. For systems with a phase change, the transport of mass or energy in the liquid surrounding the bubble becomes important. The dynamics of diffusion-induced bubble growth and collapse, both with and without hydrodynamic effects,[5] have been studied extensively.

In this chapter, we formulate the equations that govern the diffusion-induced growth and dissolution of a spherical bubble surrounded by a liquid of infinite extent (see Figure 18.1). Our goal is to formulate and solve the equations that govern the time evolution of the bubble radius. We focus on the growth and collapse of a bubble induced by mass diffusion in the liquid surrounding the bubble. Much of this development, however, can also be applied to bubble growth and collapse induced by thermal transport in one-component systems. Not surprisingly, interfacial transport plays an important role in the dynamics of bubble growth and collapse.

[1] Proussevitch & Sahagian, *J. Geophys. Res.* **101** (1996) 793.
[2] Srinivasan *et al.*, *J. Appl. Physiol.* **86** (1999) 732.
[3] Koopmans *et al.*, *Adv. Mater.* **12** (2000) 1873.
[4] Rayleigh, *Phil. Mag.* **34** (1917) 94.
[5] Barlow & Langlois, *IBM J.* **6** (1962) 329; Epstein & Plesset, *J. Chem. Phys.* **18** (1950) 1505.

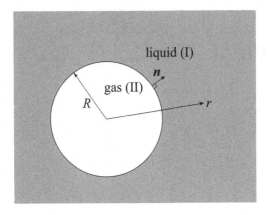

Figure 18.1 Schematic diagram of a gas bubble with radius R surrounded by an infinite liquid.

18.1 Balance Equations in the Bulk Phases

We begin by formulating balance equations for the bubble and the surrounding liquid. We assume spherical symmetry so that all fields depend only on r, the radial position from the center of the bubble, and time t. The system is isothermal with uniform temperature T_0; there are no chemical reactions and the gravitational force is neglected.

The bubble is stationary and suspended in a two-component liquid that is otherwise at rest so that the velocity is purely radial, $\boldsymbol{v} = v\boldsymbol{\delta}_r$. Hence, the liquid surrounding the bubble is described by the scalar fields ρ, p, v, and the solute mass fraction w_1. This fluid, which we take to be Newtonian, has constant viscosity $\eta \; (= \rho\nu)$ and constant diffusivity D.

Initially, the bubble has radius R_0 and is surrounded by a fluid with uniform pressure $p_0 + \Delta p$, and the initial solute mass fraction is w_{10}. At $t = 0^-$ the pressure in the liquid far from the bubble is changed to p_0. If the fluid comprising the bubble and the liquid were allowed to reach equilibrium, the solute mass fraction in the liquid would be $w_{1\mathrm{eq}}$, which is determined by the solubility of the gas at p_0. The characteristic liquid density is defined as $\rho_0 = \rho(p_0, w_{10})$.

Transport within the liquid is governed by (7.1), (7.2), and (7.4), along with an equation of state for density $\rho = \rho(p, w_1)$. As we did in Chapter 7, here we identify the conditions necessary for the velocity field to be divergence free, that is $\boldsymbol{\nabla} \cdot \boldsymbol{v} = 0$. Writing (7.5) [using (B.15)], we have

$$\frac{1}{r^2}\frac{\partial}{\partial r}(r^2 v) = -\frac{1}{\rho c_T^2}\frac{Dp}{Dt} + \rho(\hat{v}_1 - \hat{v}_2)\frac{Dw_1}{Dt}. \qquad (18.1)$$

Since the velocity is purely radial, the flow is irrotational so that the vorticity vanishes $w = 0$ (see Exercise 5.9). In this case, it is convenient to use the equation of motion in (7.7). The r-component is [using (B.15) and (B.17)] given by

$$\rho \left(\frac{\partial v}{\partial t} + v \frac{\partial v}{\partial r} \right) = -\frac{\partial p}{\partial r} + \frac{4}{3} \eta \frac{\partial}{\partial r} \left[\frac{1}{r^2} \frac{\partial}{\partial r} (r^2 v) \right], \tag{18.2}$$

where we have (for convenience) neglected the bulk viscosity η_d.

Bubble dynamics are governed by the diffusive transport of species mass in the liquid so we choose as our characteristic time R_0^2/D, with R_0 as the characteristic length.[6] Since the velocity is induced by diffusive transport, the characteristic velocity is D/R_0.

For a typical bubble growth (collapse) process, the dependence of density on composition is small, and the second term on the right-hand side of (18.1) can be neglected. The first term can also be neglected since it is reasonable to expect $N_{Ma} = (D/R_0)/c_T \ll 1$, where N_{Ma} is the Mach number (see Section 7.5). In this case, (18.1) can easily be integrated, which leads to

$$v = \frac{f}{r^2}, \tag{18.3}$$

where $f(t)$ is an unknown function to be determined later. If (18.3) is substituted into the right-hand side of (18.2), we see that viscous effects do not affect the pressure field in the liquid. We will see in the following section, however, that viscous stresses at the interface can play an important role.

We have chosen R_0^2/D to be the characteristic time for diffusion-induced bubble growth (collapse), while the characteristic time for momentum transport in (18.2) is R_0^2/ν. The ratio of these time scales is $D/\nu = 1/N_{Sc}$, where N_{Sc} is the Schmidt number. For ordinary liquids $N_{Sc} \gg 1$ (see Table 7.1), which means that, on the time scale of diffusive transport in the liquid, we can safely neglect the left-hand side of (18.2). Hence, the pressure gradient in the liquid vanishes, and the pressure and density are uniform having values $p = p_0$ and $\rho = \rho_0$, respectively.

For the solute mass balance in the liquid we use the evolution equation in (7.4) for the solute mass fraction. Using (18.3), the diffusion equation governing the solute mass fraction is [using (B.16)] given by

$$\frac{\partial w_1}{\partial t} + \frac{f}{r^2} \frac{\partial w_1}{\partial r} = \frac{D}{r^2} \frac{\partial}{\partial r} \left(r^2 \frac{\partial w_1}{\partial r} \right), \tag{18.4}$$

which is subject to the initial condition $w_1(r,0) = w_{10}$. Representative

---

[6] For a CO_2 bubble with $R_0 = 1$ mm in water $R_0^2/D \approx 10^3$ s.

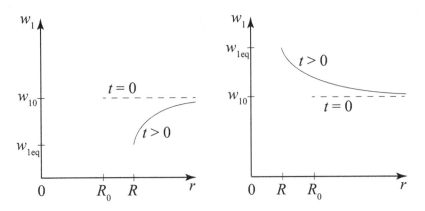

Figure 18.2 Representative concentration field w_1 in liquid $(r > R)$ versus radial position r for bubble growth $(R > R_0)$ and collapse $(R < R_0)$.

concentration profiles for the cases of bubble growth $(w_{1eq} - w_{10} < 0)$ and collapse $(w_{1eq} - w_{10} > 0)$ are shown in Figure 18.2.

The one-component bubble is assumed to be an ideal gas with equation of state given by $\bar{\rho} = A\bar{p}$, where $A = \tilde{M}_1/(\tilde{R}T_0)$ is a constant ($\tilde{M}_1$ is the molecular weight of the solute). The gas behaves as a Newtonian fluid with constant viscosity $\bar{\eta} \ (= \bar{\nu}\bar{\rho}_0)$, where $\bar{\rho}_0 = Ap_0$. The velocity within the bubble is radial $\bar{\boldsymbol{v}} = \bar{v}\boldsymbol{\delta}_r$, so that the bubble is described by the fields $\bar{\rho}$, $\bar{p}$, and $\bar{v}$.

The continuity equation (7.1) for the bubble is given by

$$\frac{\partial \bar{\rho}}{\partial t} + \bar{v}\frac{\partial \bar{\rho}}{\partial r} = -\frac{\bar{\rho}}{r^2}\frac{\partial}{\partial r}(r^2\bar{v}). \tag{18.5}$$

The radial component of the equation of motion (7.7) for the bubble can be written as

$$\bar{\rho}\left(\frac{\partial \bar{v}}{\partial t} + \bar{v}\frac{\partial \bar{v}}{\partial r}\right) = -\frac{\partial \bar{p}}{\partial r} + \frac{4}{3}\bar{\eta}\frac{\partial}{\partial r}\left[\frac{1}{r^2}\frac{\partial}{\partial r}\left(r^2\bar{v}\right)\right]. \tag{18.6}$$

The time scale for sound wave propagation within the bubble is $R_0/\bar{c}_T$, where $\bar{c}_T$ is the isothermal speed of sound for the gas within the bubble. Clearly, this time scale is much smaller than the time scale for mass diffusion R_0^2/D, so (18.6) implies that the pressure within the bubble is uniform, $\bar{p} = \bar{p}(t)$. Since the pressure $\bar{p}$ is uniform within the bubble, the density is also uniform $\bar{\rho} = \bar{\rho}(t)$. In this case, we can integrate (18.5) to obtain the velocity within the bubble

$$\bar{v} = -\frac{r}{3\bar{\rho}}\frac{d\bar{\rho}}{dt}, \tag{18.7}$$

where we have enforced regularity at the origin.

Let us summarize our progress so far. We have shown that the pressure within the bubble is uniform, and we found an expression for the velocity in terms of the rate of change of the bubble density. We have argued that the pressure in the surrounding liquid is uniform and found an expression for the velocity in terms of an unknown function. Finally, we have formulated an evolution equation for the solute mass fraction in the liquid that is of the convection–diffusion form. All of these results have been obtained by treating the two phases (bubble and liquid) in isolation.

18.2 Balance Equations at the Interface

The interface between the bubble and liquid is a spherical surface $r = R$ with unit normal pointing into the liquid $\boldsymbol{n} = \boldsymbol{\delta}_r$ making the liquid phase I and the bubble phase II (see Figure 18.1). The notation used in the previous section already identifies the phases by the absence (phase I) or presence (phase II) of an overbar. From (14.12), the velocity of this surface can be written as $\boldsymbol{n} \cdot \boldsymbol{v}^s_{\mathrm{tr}} = \boldsymbol{n} \cdot \boldsymbol{v}^s = dR/dt$.

The jump balances for mass and momentum at a spherical interface have been considered in Exercise 14.2. The jump balance for total mass (14.17) can be written as

$$\rho_0 \left[v(R) - \frac{dR}{dt} \right] = \bar{\rho} \left[\bar{v}(R) - \frac{dR}{dt} \right]. \tag{18.8}$$

Substitution of (18.3) and (18.7) into (18.8) leads to the result

$$\frac{f}{R^2} = v(R) = \frac{dR}{dt} - \frac{1}{3\rho_0 R^2} \frac{d}{dt} \left(\bar{\rho} R^3 \right). \tag{18.9}$$

We see that a convective velocity is induced in the liquid when mass is exchanged between the bubble and liquid due to the density difference between the two phases. The radial component of the jump balance for momentum given in (14.18) takes the form

$$\rho_0 [v(R) - \bar{v}(R)] \left[v(R) - \frac{dR}{dt} \right] + p - 2\eta \frac{\partial v}{\partial r}(R) = \bar{p} - 2\bar{\eta} \frac{\partial \bar{v}}{\partial r}(R) - \gamma \frac{2}{R}, \tag{18.10}$$

where γ is the interfacial tension which is assumed to be constant.[7] Again, since $N_{\mathrm{Sc}} \gg 1$ and $\bar{\eta}/\eta \ll 1$, by substituting (18.3) and (18.9) into (18.10), we obtain

$$\bar{p} = p_0 + \gamma \frac{2}{R} + \eta \frac{4}{R} \left[\frac{dR}{dt} - \frac{1}{3\rho_0 R^2} \frac{d}{dt} \left(\bar{\rho} R^3 \right) \right], \tag{18.11}$$

---

[7] As discussed in Section 15.1, in general, $\gamma = \gamma(\hat{\mu}^s_1)$.

where we have set $p = p_0$. Hence, the pressure within the bubble is different from that in the liquid because of surface tension and viscous stresses. Note that, if the third term on the right-hand side of (18.11) is neglected, we obtain the Young–Laplace equation (13.25). By applying jump balances for mass and momentum we have obtained expressions for the velocity in the liquid and pressure within the bubble in terms of the bubble radius and density.

Since we have already considered the jump balance for the total mass, we need only consider a jump balance for the solute mass. Here, we take the interface to be passive and set the excess mass density of the solute to zero. In this case, the jump balance for the solute mass (14.25), using (18.9), takes the form

$$\frac{1}{3R^2}\frac{d}{dt}\left(\bar{\rho}R^3\right) = \frac{\rho_0 D}{1 - w_1(R)}\frac{\partial w_1}{\partial r}(R), \tag{18.12}$$

which expresses the balance between the rate of change of mass within the bubble and the diffusive flux of species mass at the surface of the bubble. The result in (18.12) provides an evolution equation for the bubble radius $R(t)$ and through (18.9) an expression for the velocity in the liquid.

Exercise 18.1   Jump Balance for Solute Mass for Bubble Growth (Collapse)
Starting with the jump balance for solute mass (14.25), derive (18.12).

Exercise 18.2   Hydrodynamically Controlled Bubble Growth (Collapse)
In the absence of diffusive transport in the liquid, the right-hand side of (18.12) vanishes, and the characteristic time is R_0^2/ν. Show for hydrodynamically controlled bubble growth (collapse) in an incompressible liquid, using (18.2), that the evolution equation for the bubble radius is given by

$$\rho_0\left[R\frac{d^2R}{dt^2} + \frac{3}{2}\left(\frac{dR}{dt}\right)^2\right] + \eta\frac{4}{R}\frac{dR}{dt} + \gamma\frac{2}{R} = \bar{p} - p(\infty). \tag{18.13}$$

To integrate (18.13), which is known as the *Rayleigh–Plesset* equation, an expression relating bubble pressure to radius is required.

18.3 Boundary Conditions

To solve the diffusion equation in the liquid (18.4), two boundary conditions are required. Far from the bubble, the concentration field remains at its initial value. Hence, we can write the boundary condition $w_1(\infty, t) = w_{10}$. Our final task is the formulation of a boundary condition for the solute mass fraction at the interface $w_1(R, t)$. To find this, we use the constitutive boundary conditions from Chapter 15 along with thermodynamic relations presented in Chapter 4.

We first consider the constitutive equations for chemical potential jumps in (15.27) and (15.28) for multi-component interfaces. For an isothermal system, we have $T = \bar{T} = T^s = T_0$. We have assumed the bubble is a one-component fluid although thermodynamically this is not allowed. Hence, we write $\bar{\rho}_1 \approx \bar{\rho}$, $\bar{\rho}_2 \approx 0$, and $\bar{j}_1 \approx 0$, which allows (15.27) and (15.28) to be written as (see Exercise 18.3)

$$\hat{\mu}_1(R) - \hat{\mu}_1^s - [\hat{\mu}_2(R) - \hat{\mu}_2^s] = -R^s \rho_0 \left\{ [2w_1(R) - 1]\left(v(R) - \frac{dR}{dt}\right) - 2D\frac{\partial w_1}{\partial r}(R)\right\},$$
(18.14)

$$\bar{\hat{\mu}}_1(R) - \hat{\mu}_1^s - [\bar{\hat{\mu}}_2(R) - \hat{\mu}_2^s] = -\bar{R}^s \rho_0 \left(v(R) - \frac{dR}{dt}\right),$$
(18.15)

where $R^s = 2T_0 R_{11}^{Is}$ and $\bar{R}^s = 2T_0 R_{11}^{IIs}$. Note that, while we treat the bubble as a one-component fluid, the flat average chemical potentials in (15.27) and (15.28) mean that $\bar{\hat{\mu}}_2$ must exist, even if the bubble contains only a few molecules of species 2. Since species 2 is not transferred across the interface, this implies $\hat{\mu}_2(R) = \hat{\mu}_2^s = \bar{\hat{\mu}}_2(R)$. Hence, using (18.9) and (18.12), we can write (18.14) and (18.15) as

$$\hat{\mu}_1(R) - \hat{\mu}_1^s = R^s \frac{\rho_0 D}{1 - w_1(R)}\frac{\partial w_1}{\partial r}(R),$$
(18.16)

$$\bar{\hat{\mu}}_1(R) - \hat{\mu}_1^s = -\bar{R}^s \frac{\rho_0 D}{1 - w_1(R)}\frac{\partial w_1}{\partial r}(R).$$
(18.17)

The expressions in (18.16) and (18.17) can be used to determine the jump in chemical potential across the interface, and to determine $\hat{\mu}_1^s$, which would be needed if the interfacial tension were allowed to be a function of concentration $\gamma = \gamma(\hat{\mu}_1^s)$. Subtracting (18.16) from (18.17), using (18.12) and recalling $A = \tilde{M}_1/(\tilde{R}T_0)$, we can write

$$A\Lambda = \bar{\hat{\mu}}_1(R) - \hat{\mu}_1(R) = -(\bar{R}^s + R^s)\frac{1}{3\rho_0 R^2}\frac{d}{dt}\left(\bar{\rho}R^3\right),$$
(18.18)

which defines the dimensionless parameter Λ. From (18.16) and (18.17), we see that, in the absence of interfacial resistances for mass transfer ($R^s = \bar{R}^s = 0$), the chemical potential of the solute is continuous across the interface, $\bar{\hat{\mu}}_1(R) = \hat{\mu}_1^s = \hat{\mu}_1^s(R)$, and phase change occurs between bulk phases that are in equilibrium. If phase change does not take place at equilibrium, the expression in (18.18) predicts a jump in the solute chemical

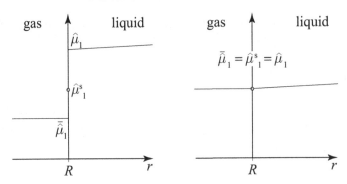

Figure 18.3 Representative solute chemical potentials for the gas phase $\bar{\hat{\mu}}_1$, interface $\hat{\mu}_1^s$, and liquid $\tilde{\hat{\mu}}_1$ near the gas–liquid interface for bubble growth showing nonequilibrium (left) and equilibrium (right) cases.

potential between the gas and liquid phases (see Figure 18.3) with a sign that is negative for bubble growth and positive for bubble collapse.

The formulation of a boundary condition for $w(R, t)$ requires expressions for the chemical potential of the solute in the bubble and liquid phases. For the bubble, which we have assumed is an ideal gas, we use the familiar (see Exercise 4.1) expression $\bar{\hat{\mu}}_1 = \hat{\mu}_1^{0,\mathrm{ig}}(T_0, p) + A \ln(\bar{p})$, where $\hat{\mu}_1^{0,\mathrm{ig}}(T_0, p)$ is the chemical potential of the solute at an ideal gas reference state. For a gas that has a small solubility in a liquid we use Henry's law (see Exercise 4.9). In this case, we write the chemical potential of the solute in the liquid as $\hat{\mu}_1 = \hat{\mu}_1^{0,\mathrm{ig}}(T_0, p) + A \ln(k_H x_1)$. Taking the difference between the solute chemical potentials and using Λ defined in (18.18), we obtain

$$w_1(R, t) = \frac{\bar{p}}{k_w} e^{-\Lambda}, \qquad (18.19)$$

where $k_w = \tilde{M}/(\tilde{M}_1 k_H)$ is taken to be constant. If the two phases are in equilibrium ($\Lambda = 0$), and if hydrodynamic effects are negligible ($\bar{p} = p_0$), then the boundary condition in (18.19) simplifies to $w_1(R, t) = w_{1\mathrm{eq}}$, where $w_{1\mathrm{eq}} = p_0/k_w$. We now have the required boundary conditions to solve (18.4).

Exercise 18.3   Constitutive Equations for Chemical Potential Jumps
Starting with the constitutive equations in (15.27) and (15.28), derive the expressions in (18.16) and (18.17).

18.4 Bubble Growth and Collapse Dynamics

We have just formulated rather general forms of the interface balance equations that govern bubble growth or collapse induced by the diffusion of species mass. Solution of the convection–diffusion equations in the liquid with the simplified (passive) forms of the interface balance equations still represents a significant challenge. The presence of a convective term and a moving boundary, along with the coupling of the diffusive transport in the liquid to momentum transport at the interface, make the governing equations nonlinear.

We see from (18.18) that nonequilibrium effects ($\Lambda \neq 0$) at the interface are likely to occur when the rate of bubble growth or collapse is large.[8] However, the importance of these effects is at present not well understood. Bubble growth and collapse dynamics driven by mass diffusion that include nonequilibrium effects at the interface can be found elsewhere.[9] To simplify things somewhat, in this section we consider the case when the two phases are in equilibrium ($\Lambda = 0$).

At this point, it is useful to collect the relevant equations from the three previous sections. Using (18.9), we can write (18.4) as

$$\frac{\partial w_1}{\partial t} + \left[\frac{dR}{dt} - \epsilon \left(\frac{\bar{p}}{p_0} \frac{dR}{dt} + \frac{R}{3p_0} \frac{d\bar{p}}{dt} \right) \right] \left(\frac{R}{r} \right)^2 \frac{\partial w_1}{\partial r} = \frac{D}{r^2} \frac{\partial}{\partial r} \left(r^2 \frac{\partial w_1}{\partial r} \right),$$
$$(18.20)$$

where we have introduced the parameter $\epsilon = \bar{\rho}_0/\rho_0$. The bubble pressure is obtained from (18.11), which we write as

$$\bar{p} = p_0 + \frac{2\gamma}{R} + \frac{4\eta}{R} \left[\frac{dR}{dt} - \epsilon \left(\frac{\bar{p}}{p_0} \frac{dR}{dt} + \frac{R}{3p_0} \frac{d\bar{p}}{dt} \right) \right]. \tag{18.21}$$

The initial and boundary conditions for (18.20) take the form

$$w_1(r, 0) = w_{10}, \qquad w_1(\infty, t) = w_{10}, \qquad w_1(R, t) = w_{1\text{eq}} \frac{\bar{p}}{p_0}, \tag{18.22}$$

where we have used $w_{1\text{eq}} = p_0/k_w$. The evolution equation for the bubble radius comes from (18.12), which can be written as

$$\frac{dR}{dt} = -\frac{R}{3p_0} \frac{d\bar{p}}{dt} + \frac{D}{\epsilon[1 - w_1(R, t)]} \frac{\partial w_1}{\partial r}(R, t), \tag{18.23}$$

and is solved subject to the initial condition

$$R(0) = R_0. \tag{18.24}$$

The system of equations (18.20)–(18.24) can only be solved numerically, for example, using finite difference methods. Since the domain has a moving boundary and is semi-infinite, it is advantageous to use a coordinate

[8] Bornhorst & Hatsopoulos, *J. Appl. Mech.* **34** (1967) 847.
[9] Öttinger, Bedeaux & Venerus, *Phys. Rev. E* **80** (2009) 021606.

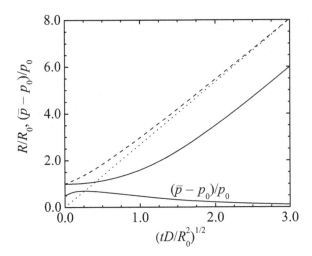

Figure 18.4 Bubble radius and pressure versus time during bubble growth for the following parameter values: $N_w = 1$ and $\epsilon = 0.001$. Solid lines are for the diffusion-induced case where $N_\eta = 0.1$, $N_{\Delta p} = 0.5$, and $N_\gamma = 0.25$. The dashed line is the diffusion-controlled case $N_\eta = N_\gamma = 0$; the dotted line is for (18.30) with $\beta = 1.34$.

transformation before the governing equations are discretized.[10] Solutions of these equations have been obtained for diffusion-induced bubble growth in Newtonian fluids[11] and in viscoelastic and yield-stress fluids.[12]

Given the large number of parameters in (18.20)–(18.24), it is useful to apply dimensional analysis (the $n - m$ procedure, see Section 7.5). We identify the independent quantities $r, t, w_1, R, \bar{p}/p_0$ (variables) and $\bar{\rho}_0/\rho_0, D, \eta/p_0, \gamma/p_0, w_{10}, w_{1eq}, R_0$ (parameters) so that $n = 12$. These quantities involve the dimensions of length and time so that $m = 2$, which leads to $n - m = 10$. Note that the nonlinearity of (18.20)–(18.22) does not allow additive or multiplicative normalization, which would eliminate w_{10} and w_{1eq} from the list of independent quantities. Nevertheless, it is convenient (see Section 18.5) to use the normalized solute mass fraction $(w_1 - w_{10})/(w_{1eq} - w_{10})$, which is a decreasing function of r/R_0 for both bubble growth and collapse (see Figure 18.2). To study the dynamics of bubble growth or collapse, we examine R/R_0 and $\bar{p}/p_0$ as functions of tD/R_0^2. This leaves five dimensionless parameters, which are chosen as $\epsilon = \bar{\rho}_0/\rho_0$, $N_w = (w_{10} - w_{1eq})/[\epsilon(1 - w_{1eq})]$, $N_\eta = \eta D/(R_0^2 p_0)$, $N_\gamma = \gamma/(R_0 p_0)$, and

[10] Duda & Vrentas, *AIChE J.* **15** (1969) 351.
[11] Venerus & Yala, *AIChE J.* **43** (1997) 2948.
[12] Venerus *et al.*, *J. Non-Newtonian Fluid Mech.* **75** (1998) 55; Venerus, *J. Non-Newtonian Fluid Mech.* **215** (2015) 53.

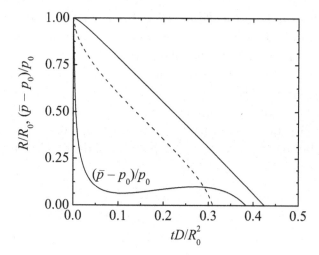

Figure 18.5 Bubble radius and pressure versus time during bubble collapse for the following parameter values: $N_w = -1$ and $\epsilon = 0.001$. Solid lines are for the diffusion-induced case where $N_\eta = 0.1$, $N_{\Delta p} = 2$, and $N_\gamma = 0.5$. The dashed line is for the diffusion-controlled case $N_\eta = N_\gamma = 0$.

$N_{\Delta p} = (1 + \Delta p/p_0)^{-1} = w_{1\text{eq}}/w_{10}$. Note that for bubble growth $N_w > 0$, and for bubble collapse $N_w < 0$.

The evolution of bubble radius and that of pressure during bubble growth are shown in Figure 18.4 for $N_{\Delta p} = 0.5$, which corresponds to a pressure reduction by a factor of two. Initially, the pressure within the bubble increases, which is the result of viscous stress in the liquid. An increase in bubble pressure, according to the boundary conditions in (18.22), leads to a decrease in the driving force for diffusive transport. The pressure increase also leads to a density increase, and together these result in a reduced initial rate of bubble growth. At later stages, both interfacial and viscous stresses decrease, causing the pressure and density to decrease, leading to an increase in the rate of bubble growth. Eventually, the bubble pressure approaches that of the liquid, and the bubble radius increases linearly with the square root of time. The other cases shown in this figure by the dashed and dotted lines are discussed in the following section.

Figure 18.5 shows the evolution of bubble radius and pressure during bubble collapse for $N_{\Delta p} = 2$, which corresponds to a pressure increase by a factor of two. In contrast to bubble growth, during bubble collapse, the contributions of viscous and interfacial forces to bubble pressure have opposite signs and increase in magnitude with time. There is a rapid initial decrease of pressure within the bubble, which is the result of solute transport from the bubble to the liquid while interfacial tension and viscous stresses resist bubble collapse. After the initial decrease, the pressure reaches an approximately constant value, which indicates a balance between interfacial and viscous forces. According to the boundary conditions in (18.22), a decrease in bubble pressure leads to an increase in the driving force for diffusive transport. At later stages, the viscous force overtakes the interfacial force, and there is a sharp decrease in pressure, which drops below ambient pressure, just before bubble dissolution is complete. The dashed line in Figure 18.5 is discussed in the following section.

18.5 Diffusion-Controlled Bubble Dynamics

In this section, we consider a special case of the model discussed in the previous section that has a number of important applications. This special case is realized when both viscous ($N_\eta \ll 1$) and interfacial ($N_\gamma \ll 1$) stresses are negligible. Dropping the interfacial (second) and viscous (third) stress terms in (18.21), we see that the bubble pressure is constant, having value $\bar{p} = p_0$. In this case, the second boundary condition in (18.22) becomes $w_1(R, t) = w_{1\text{eq}}$. Dynamics are hence determined entirely by mass diffusion, and we refer to this as *diffusion-controlled* bubble growth or collapse. For this case, we write the governing equations in dimensionless form using the dimensionless (and normalized) variables introduced in the previous section. The evolution equation for solute concentration (18.20) simplifies to

$$\frac{\partial w_1}{\partial t} + (1 - \epsilon)\frac{dR}{dt}\left(\frac{R}{r}\right)^2\frac{\partial w_1}{\partial r} = \frac{1}{r^2}\frac{\partial}{\partial r}\left(r^2\frac{\partial w_1}{\partial r}\right), \tag{18.25}$$

and the initial and boundary conditions in (18.22) take the form

$$w_1(r, 0) = 0, \qquad w_1(\infty, t) = 0, \qquad w_1(R, t) = 1. \tag{18.26}$$

The evolution equation for bubble radius (18.23) becomes

$$\frac{dR}{dt} = -N_w\frac{\partial w_1}{\partial r}(R, t), \tag{18.27}$$

which is solved subject to

$$R(0) = 1. \tag{18.28}$$

The system of equations in (18.25)–(18.28) also applies to the diffusion-controlled growth or dissolution of a one-component solid particle (see Exercise 18.4). Replacing w_1 by a normalized temperature and redefining N_w in (18.25)–(18.28) leads to a set of equations governing phase change processes (e.g., melting, vaporization) in one-component systems (see Exercise 18.5).

While greatly simplified, the problem defined by (18.25)–(18.28) is still a nonlinear, moving-boundary problem. Numerical solutions for diffusion-controlled bubble growth and collapse are presented in Figures 18.4 and 18.5 (dashed lines). From these figures, we see that, when diffusive transport is coupled to momentum transport, the rate of bubble growth or collapse is retarded. The primary reason for this is the reduction of the driving force for diffusive transport, which occurs when the bubble pressure is affected by viscous stresses and interfacial tension. For the case of bubble growth shown in Figure 18.4, we see that, as the bubble radius becomes large, viscous and interfacial stresses become negligible, and the rate of bubble growth approaches the diffusion-controlled case.

Approximate solutions for the diffusion-controlled case can be found using perturbation methods (see Section 8.3). As noted earlier, the nonlinearity arises because of the convective transport term in (18.25) and the moving boundary governed by (18.27). The nonlinear terms involve the parameters ϵ and N_w, so it is natural to use these as perturbation parameters. To first order in N_w, the following expression for bubble radius has been found,[13]

$$R^2 = 1 + 2N_w(t + 2\sqrt{t/\pi}),\tag{18.29}$$

which we leave as an exercise to derive. It should be noted that higher-order solutions are not uniformly valid and therefore cannot be handled by simple perturbation methods.[14]

For the case of bubble growth from zero initial bubble radius ($R_0 = 0$), equations (18.25)–(18.28) possess a self-similar solution[15] (see Section 7.5), and the transformation $\xi = r/\sqrt{4t}$ is used. The resulting expression for the bubble radius is given by

$$R = 2\beta\sqrt{t},\tag{18.30}$$

where $\beta = \beta(N_w, \epsilon)$ is a constant that is found by solving a nonlinear algebraic equation (see Exercise 18.8). The evolution of bubble radius given by the similarity solution is shown in Figure 18.4.

We have several examples where the bubble radius increases linearly with the square root of time. This behavior is observed in the similarity solution (18.30) and at sufficiently large times for both diffusion-induced and diffusion-controlled bubble growth shown in Figure 18.4. From (18.27), we see that the rate of bubble growth is proportional to the gradient at the moving interface $\partial w_1/\partial r(R, t)$. Let δ be the boundary layer thickness, which is

[13] Duda & Vrentas, *AIChE J.* **15** (1969) 351.
[14] Vrentas *et al.*, *Chem. Eng. Sci.* **38** (1983) 1927.
[15] Scriven, *Chem. Eng. Sci.* **10** (1959) 1.

roughly the distance from the bubble radius at which the concentration field is unaffected by the presence of the bubble, then $\partial w_1/\partial r(R,t) \approx -1/\delta$. If the boundary layer thickness increases at the same rate as the bubble radius ($\delta \propto R$), then according to (18.27) we see that $R\,dR/dt$ is independent of time, or that R^2 increases linearly with time.

Exercise 18.4   Quasi-steady Growth or Dissolution of a Solid Particle
For the growth or dissolution of a solid particle it is reasonable to set $\epsilon = 1$. Solve (18.25)–(18.28) for the case $|N_w| \ll 1$ and derive the following expression for bubble radius,

$$R^2 = 1 + 2N_w t.$$

Exercise 18.5   Growth or Shrinkage of a Spherical Particle Due to Phase Change
Consider a one-component system where the growth or dissolution of a spherical particle (solid or vapor) is induced by phase change with the liquid surrounding it. Initially, the liquid has uniform temperature T_0 and the particle has radius R_0. Throughout the process, the particle has uniform temperature T_{eq}, the temperature at which phase change takes place. Show that the normalized liquid temperature is governed by

$$\frac{\partial T}{\partial t} + (1-\epsilon)\frac{dR}{dt}\left(\frac{R}{r}\right)^2 \frac{\partial T}{\partial r} = \frac{1}{r^2}\frac{\partial}{\partial r}\left(r^2 \frac{\partial T}{\partial r}\right),$$

$$T(r,0) = 0, \qquad T(\infty,t) = 0, \qquad T(R,t) = 1,$$

and that the bubble radius is governed by

$$\frac{dR}{dt} = -\frac{N_{St}}{\epsilon}\frac{\partial T}{\partial r}(R,t),$$

$$R(0) = 1,$$

where $N_{St} = \hat{c}_p(T_0 - T_{eq})/\Delta\hat{h}_{eq}$ is the Stefan number and $\Delta\hat{h}_{eq} = \bar{\hat{h}}(T_{eq}) - \hat{h}(T_{eq})$. Note that the equations above are equivalent to those in (18.25)–(18.28).

Exercise 18.6   Melting of a Spherical Particle Surrounded by a Gas
A solid spherical particle of radius R_0 with uniform initial temperature T_{eq}, which is the melting point of the solid, is surrounded by a gas having uniform temperature T_g except near the surface of the particle. Since $T_g > T_{eq}$, the particle melts, forming a spherical liquid shell $R(t) < r \leq R_0$. There is no mass transfer at the liquid–gas interface ($r = R_0$), and heat transfer is governed by Newton's law of cooling (see Section 9.3) with heat transfer coefficient h. For this analysis, assume the density of the solid $\bar{\rho}$ is approximately equal to that of the liquid ρ. Use the quasi-steady-state

approximation (see Section 7.5) to derive an expression for the dimensionless solid particle radius $R/R_0 \to R$ as a function of the dimensionless time $\chi t/R_0^2 \to t$:

$$\frac{1-N_{Bi}}{3N_{Bi}}(1-R^3) + \frac{1}{2}(1-R^2) = N_{St}t,$$

where $N_{Bi} = hR_0/\lambda$ and $N_{St} = \hat{c}_p(T_g - T_{eq})/(\hat{h} - \bar{\hat{h}})$. Also, derive an expression that can be used to predict the time required for the solid particle to melt $R(t_{melt}) = 0$.

Exercise 18.7   Slow Bubble Growth or Dissolution
To solve (18.25)–(18.28), it is convenient to use the coordinate transformation $x = r/R$ to immobilize the moving boundary and the change of variable $w_1 = u/x$. Apply these to show that the integrated form of (18.27) leads to the following expression for the bubble radius,

$$R^2 = 1 + 2N_wt - 2N_w\int_0^t \frac{\partial u}{\partial x}(1, t')dt'.$$

Using the perturbation expansion $u = u^{(0)} + N_w u^{(1)} + \cdots$, show that the zeroth-order solution is given by

$$u^{(0)}(x,t) = 1 - \operatorname{erf}\left(\frac{x-1}{2\sqrt{t}}\right),$$

and that this leads to the expression given in (18.29), which is valid for $N_w \ll 1$.

Exercise 18.8   Similarity Solution for Particle Growth from Zero Initial Radius
Use the similarity transformation $\xi = r/\sqrt{4t}$ (see Exercises 7.9 and 7.11) on (18.25)–(18.27) to derive the expression for R given in (18.30), and show that the parameter β can be found from

$$N_w = 2\beta^3 \exp[(3-2\epsilon)\beta^2]\int_\beta^\infty \frac{1}{x^2}\exp\left[-x^2 - 2(1-\epsilon)\frac{\beta^3}{x}\right]dx.$$

Hence, the nonlinearity of (18.25)–(18.27) has been reduced to the solution of the above nonlinear equation for β for given values of N_w and ϵ.

Exercise 18.9   Thin Boundary Layer Approximation
An approximate solution to (18.25)–(18.28) can be found using the method of weighted residuals. This method is implemented by assuming a solution for w_1 within the boundary layer. For example, consider

$$w_1 = \begin{cases} \left(1 - \dfrac{r-R}{\delta}\right)^2 & \text{for } R \le r \le R + \delta, \\ 0 & \text{for } r > R + \delta. \end{cases}$$

Show that multiplication of (18.20) by r^2 and integrating over r from R to $R + \delta$ using the above expression gives

$$\frac{\delta}{R} + \frac{1}{2} \left(\frac{\delta}{R} \right)^2 + \frac{1}{10} \left(\frac{\delta}{R} \right)^3 = \frac{1 - \epsilon N_w}{N_w} \frac{R^3 - 1}{R^3}.$$

Use the above for the case of a thin boundary layer $\delta/R \ll 1$ in (18.27) to derive the following expression for the bubble radius,

$$\frac{dR}{dt} = \frac{2N_w^2}{1 - \epsilon N_w} \frac{R^2}{R^3 - 1}.$$

The thin boundary layer approximation is valid for $N_w \gtrsim 10$. Note that for $R \gg 1$, the bubble radius has a linear dependence on $\sqrt{t}$.

- The description of bubble growth and collapse dynamics driven by diffusive transport requires a set of evolution equations based on balance equations for the two bulk phases, interfacial balance equations, and interfacial constitutive equations.
- Spherical symmetry and the irrotational nature of the flow allow the balance equations for mass and momentum to be integrated, yielding simple expressions for radial velocity and pressure.
- The evolution equation for solute concentration in the liquid has the form of a convection–diffusion equation and, under typical conditions, the pressure and thus density within the bubble are uniform.
- Jump balances for mass and momentum can be used to obtain an expression relating pressures in the bubble and liquid and an evolution equation for the bubble radius in terms of the diffusive flux of solute at the bubble–liquid interface.
- Interfacial force–flux relations can be used to obtain expressions for chemical potential jumps applicable to mass transfer processes between phases that are not in equilibrium.
- The coupling of hydrodynamic effects at the bubble–liquid interface with diffusive transport in the liquid gives diffusion-induced bubble growth and collapse phenomena a rich dynamic behavior.

19

Semi-Conductor Processing

It is nearly impossible to go one day, or maybe even one hour, without using a device that contains at least one integrated circuit. Computers, phones, and most appliances all rely on one or more integrated circuits to function. Integrated circuits, or microelectronic devices, can also be found in the automobiles, trains, and jets that allow us to so easily move from place to place and in the medical devices that monitor or maintain our health. The first integrated circuits made in the early 1970s (see Figure 19.1) were used in hand-held calculators that could carry out only the most basic algebraic operations. The remarkable increase in the power of microelectronic devices has resulted from the ability to make the devices smaller and smaller.

How are these amazing devices made? The answer is through a sophisticated, multi-step process that begins with the production of a high-purity, single-crystal semiconductor material. The most commonly used material in microelectronic devices is silicon (Si), which is one of the most abundant elements on earth. Thin disks, or wafers, of single-crystal Si are then subjected to a series of steps that involve selectively modifying, or adding or removing,

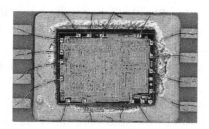

Figure 19.1 Image of the Intel® 4004 microprocessor *circa* 1970 with overall dimensions of ⌣ 10 mm × 10 mm. This device contains 2300 transistors and has device features of size ⌣ 10 μm. Current microprocessors have device features with sizes ⌣ 25 nm and over 500 million transistors.

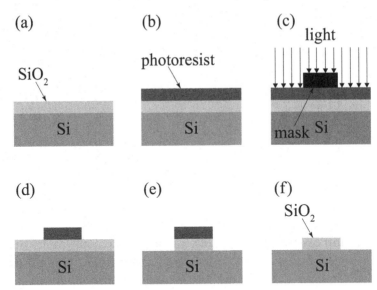

Figure 19.2 Schematic diagram of selected semi-conductor processing steps: (a) Si wafer with SiO$_2$ layer; (b) surface coated with positive photoresist film; (c) exposure of photoresist through mask; (d) photoresist after selective removal; (e) selective removal of SiO$_2$ layer by etching; (f) SiO$_2$ feature left after photoresist removal.

thin layers of material with different properties. These steps include oxidation, deposition, implantation, and etching, which involve chemical reaction and diffusion. Spatial selectivity is achieved by a process known as lithography, which involves coating a surface with a thin film of a photosensitive material and then exposing it to light through a mask having a desired pattern. The photoresist film is applied using a process known as spin coating. Figure 19.2 schematically shows the sequence of steps that might be used to produce a single feature on a wafer. Keep in mind that such features have dimensions as small as 10 nm in modern integrated circuits. This sequence is repeated many times so that layer-by-layer the device is produced. Readers interested in a more complete description of the processes involved in the fabrication of microelectronic devices are referred to the book by Jaeger.[1]

In this chapter we examine three of the processing steps just mentioned: crystal growth, oxidation, and spin coating. Our goal is to develop transport models that will allow us to predict relevant process variables in terms of process parameters. The processes considered here involve moving

---

[1] Jaeger, *Introduction to Microelectronic Fabrication* (Prentice Hall, 2002).

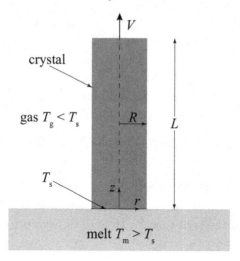

Figure 19.3 Schematic diagram of Czochralski crystal growth process.

boundaries, which means we will rely heavily on the interfacial balance equations developed in Chapter 14. The inherent complexity of these processes also means that dimensional analysis and simplifying assumptions will be necessary to reduce mathematical complexity while still keeping the most essential physical phenomena.

19.1 Crystal Growth

The starting point for modern integrated circuits is a high-purity, single-crystal semiconductor such as silicon, Si. The most commonly used method for producing single-crystal Si is named after the chemist Jan Czochralski (1885–1953). Other methods include Bridgman-Stockbarger and floating zone techniques.[2]

Let us consider the Czochralski crystal growth (CZCG) process shown schematically in Figure 19.3. Crystalline Si is formed by the solidification of molten Si as it is pulled from the melt, forming a cylindrically shaped solid. The rate at which the crystal is formed is dominated by thermal conduction within the solid and by the transport of energy between the solid and both the melt and surrounding gas. For pure Si, the phase change takes place at a fixed temperature $T_s \approx 1700$ K. To keep things simple, we assume the interface to be planar and fixed at the origin of a stationary coordinate system.

[2] Brown, *AIChE J.* **34** (1988) 881.

The cylindrical crystal has constant radius R and time-dependent length L, and we take the density ρ, specific heat $\hat{c}_p$, and thermal conductivity λ to be constants. The solid is surrounded by a well-mixed gas with a uniform temperature T_g, except near the crystal surface. We assume there is symmetry about the z-axis and that the solid has uniform velocity $v = V\delta_z$. From (7.21), the temperature field in the solid $T(r, z, t)$ takes the form [see (B.9)]

$$\rho\hat{c}_p\left(\frac{\partial T}{\partial t} + V\frac{\partial T}{\partial z}\right) = \lambda\left[\frac{1}{r}\frac{\partial}{\partial r}\left(r\frac{\partial T}{\partial r}\right) + \frac{\partial^2 T}{\partial z^2}\right],\qquad(19.1)$$

which indicates that the convective transport of energy may also affect the temperature field. Since the interface position is fixed, the growth rate of the crystal equals the pull rate V. Hence, the evolution equation for crystal length L is simply

$$\frac{dL}{dt} = V,\qquad(19.2)$$

which is solved subject to the initial condition $L(0) = L_0$. To solve (19.1), we will need to specify an initial condition and boundary conditions at $r = 0, R$ and $z = 0, L$.

We have a problem involving the time evolution of coupled dependent variables, namely the temperature field T and the crystal length L. There are two time scales to consider: the time scale for the crystal length to change, R/V, and the time scale for diffusive energy transfer, R^2/χ, where $\chi = \lambda/(\rho\hat{c}_p)$ is the thermal diffusivity. Here, we use R/V and R to make time and length, respectively, dimensionless so that (19.2) becomes $dL/dt = 1$. The dimensionless form of (19.1) then has a left-hand side that is multiplied by the Péclet number $N'_{Pe} = VR/\chi$ (see Section 7.5). In a typical CZCG process, $R \approx 10$ cm, $V \approx 10^{-3}$ cm/s, and $\chi \approx 10^{-1}$ cm^2/s, so that $N'_{Pe} \approx 0.1$. Hence, the time scale for diffusive energy transport is roughly 10 times smaller than the time scale for length change. This seems like a perfect opportunity to invoke the quasi-steady-state approximation discussed in Section 7.5. Based on this argument, the temperature equation (19.1) simplifies to

$$\frac{1}{r}\frac{\partial}{\partial r}\left(r\frac{\partial T}{\partial r}\right) + \frac{\partial^2 T}{\partial z^2} = 0.\qquad(19.3)$$

The elliptic partial differential equation in (19.3) requires four boundary conditions, two of which are straightforward to write. Along the center of the solid ($r = 0$) symmetry leads to

$$\frac{\partial T}{\partial r}(0, z, t) = 0.\qquad(19.4)$$

For the second boundary condition, we assume equilibrium at the solid–liquid interface ($z = 0$), which is justified since the rate of phase change is slow in the CZCG process. This implies that interfacial (Kapitza) resistances are negligible so that (15.13) and (15.14) reduce to continuity of temperature across the interface with a constant value of T_s,

$$T(r, 0, t) = T_s. \tag{19.5}$$

We now consider boundary conditions at the solid–gas interfaces at $r = R$ and $z = L$. The unit normal vector $\boldsymbol{n}$ points into the gas phase (I) and away from the solid phase (II). There is no mass transfer across these interfaces, so (14.7) applies. The jump balance for energy (14.26) reduces to the continuity of energy flux across the interface, $\boldsymbol{j}_q^{\mathrm{I}} \cdot \boldsymbol{n} = \boldsymbol{j}_q^{\mathrm{II}} \cdot \boldsymbol{n}$. We further assume $T^{\mathrm{I}} = T^{\mathrm{II}}$, which implies the absence of interfacial resistances. Since we know the temperature of the gas only away from the surface (T_g), neither the continuity of energy flux nor continuity of temperature across the interface can be used to formulate a useful boundary condition. If we use the jump energy balance with Fourier's law (6.4) for the solid phase $\boldsymbol{j}_q^{\mathrm{II}} = -\lambda\,\boldsymbol{\nabla}T^{\mathrm{II}}$, we still need an expression for $\boldsymbol{j}_q^{\mathrm{I}}$. The temperature field in the gas, where natural convection and radiative energy transport would lead to complex transport equations, can be found only by using numerical methods.[3] Alternatively, we can use the heat transfer coefficient introduced in Section 9.3 and Newton's law of cooling (9.26). Since $\boldsymbol{n}$ points into the gas phase, we can write (9.26) as $\boldsymbol{j}_q^{\mathrm{I}} \cdot \boldsymbol{n} = h_g(T^{\mathrm{II}} - T_g)$, where h_g is the heat transfer coefficient for the gas phase. We hence arrive at the following boundary conditions,

$$-\lambda\frac{\partial T}{\partial r}(R, z, t) = h_g[T(R, z, t) - T_g], \tag{19.6}$$

$$-\lambda\frac{\partial T}{\partial z}(r, L, t) = h_g[T(r, L, t) - T_g]. \tag{19.7}$$

The solution of (19.3)–(19.7), which can be found using the method of separation of variables,[4] is given by

$$\frac{T - T_g}{T_s - T_g} = 2\sum_{n=1}^{\infty} \frac{N_{\mathrm{Bi}}J_0(\alpha_n r/R)}{(N_{\mathrm{Bi}}^2 + \alpha_n^2)J_0(\alpha_n)}$$
$$\times \frac{\alpha_n \cosh[\alpha_n(L - z)/R] + N_{\mathrm{Bi}}\sinh[\alpha_n(L - z)/R]}{\alpha_n \cosh(\alpha_n L/R) + N_{\mathrm{Bi}}\sinh(\alpha_n L/R)}, \tag{19.8}$$

where J_0 is the Bessel functions of the first kind, and the α_n are determined

---

[3] Brown, *AIChE J.* **34** (1988) 881; Prasad *et al.*, *Adv. Heat Trans.* **30** (1997) 313.
[4] Carslaw and Jaeger, *Conduction of Heat in Solids* (Oxford, 1959).

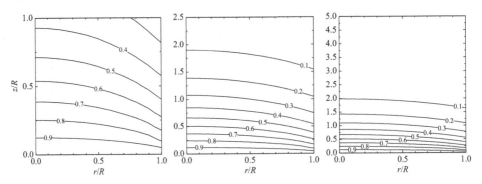

Figure 19.4 Normalized temperature $(T - T_g)/(T_s - T_g)$ isotherms in the Czochralski crystal growth process given by (19.8) with $N_{Bi} = 1.0$ for three crystal lengths $L/R = 1, 2.5, 5$ (from left to right).

by $\alpha_n J_1(\alpha_n) - N_{Bi} J_0(\alpha_n) = 0$. The dimensionless number $N_{Bi} = h_g R/\lambda$ is known as the Biot number, which indicates the rate of convective heat transfer in the fluid relative to conduction heat transfer in the solid. Keep in mind that (19.8) is the quasi-steady-state solution to (19.2)–(19.7) with $L = L_0 + Vt$. Figure 19.4 shows the temperature fields predicted by (19.8) for several crystal lengths. From this figure, we see that, with increasing crystal length, radial temperature variations diminish. Also note that the location and shape of the isotherms do not change significantly when $L/R \gtrsim 5$, which means the crystal length becomes effectively infinite.

Up to this point, we have not discussed the influence of the melt on the CZCG process. It turns out that energy transport in the melt is extremely important and in fact determines the crystal pull rate V. Natural convection (see Exercise 7.10) induces complex flows that strongly influence the melt temperature field and hence energy transport at the crystal–melt interface. In some cases, a magnetic field is used to stabilize the flow, and the melt crucible is rotated to promote symmetry. As with the gas phase, detailed transport models of thermal transport in the melt are extremely complex and can be solved only using numerical methods.[5] So that we may get a feel for the importance of thermal transport in the melt, we invoke Newton's law of cooling (9.26). Hence, we assume a uniform melt temperature T_m away from the crystal–melt interface and that complex phenomena in the melt are captured in a heat transfer coefficient h_m.

The phase change that occurs at the crystal–melt interface ($z = 0$), which earlier we assumed to be at equilibrium, can be described by the balance

---

[5] Brown, *AIChE J.* **34** (1988) 881; Prasad *et al.*, *Adv. Heat Trans.* **30** (1997) 313.

equations introduced in Chapter 14. We have assumed the interface is stationary ($v^s = 0$), and here set the unit normal to the interface n to point into the melt, which now is phase I. If we neglect kinetic energy contributions, the jump balance for energy (14.26) simplifies to

$$h^{\mathrm{I}} v_z^{\mathrm{I}} + (j_q^{\mathrm{I}})_z = h^{\mathrm{II}} v_z^{\mathrm{II}} + (j_q^{\mathrm{II}})_z. \tag{19.9}$$

Hence, there is a jump in the energy flux across the interface due to the difference in enthalpy densities $h^{\mathrm{I}} - h^{\mathrm{II}}$ between the liquid and solid phases. The jump balance for mass (14.5) for this case gives $\rho^{\mathrm{I}} v_z^{\mathrm{I}} = \rho^{\mathrm{II}} v_z^{\mathrm{II}}$. Taking this into account, and using Newton's law of cooling and Fourier's law for the fluxes in phases I and II, respectively, (19.9) can be written as

$$\rho V(\hat{h}^{\mathrm{I}} - \hat{h}^{\mathrm{II}}) = \lambda \left\langle \frac{\partial T}{\partial z}(r, 0, t) \right\rangle - h_{\mathrm{m}}(T_{\mathrm{s}} - T_{\mathrm{m}}). \tag{19.10}$$

The angular brackets in (19.10) indicate the radial average of a quantity $\langle (.) \rangle = 2/R^2 \int_0^R (.) r \, dr$, which is introduced since both T_{s} and T_{m} are uniform so that radial variations at the interface are not allowed. The validity of using a radially averaged temperature is examined in an exercise (see Exercise 19.1).

The energy balance in (19.10) shows how the crystal pull rate V, or rate of solidification, is determined by a balance between the rate of conduction away from the interface and the rate of heat addition from the melt. The term for the temperature gradient at the crystal–melt interface in (19.10) can be evaluated using (19.8), which is given by

$$-\left\langle \frac{\partial T}{\partial z}(r, 0, t) \right\rangle = \frac{T_{\mathrm{s}} - T_{\mathrm{g}}}{R} 4 \sum_{n=1}^{\infty} \frac{N_{\mathrm{Bi}}^2}{\alpha_n (N_{\mathrm{Bi}}^2 + \alpha_n^2)}$$
$$\times \frac{\alpha_n \sinh(\alpha_n L/R) + N_{\mathrm{Bi}} \cosh(\alpha_n L/R)}{\alpha_n \cosh(\alpha_n L/R) + N_{\mathrm{Bi}} \sinh(\alpha_n L/R)}. \tag{19.11}$$

Since this expression depends on the crystal length L, for a constant pull rate V, (19.10) can be satisfied only if the melt temperature T_{m} is allowed to be a function of L. Alternatively, if T_{m} is constant, the crystal pull rate V from (19.10) can be substituted into (19.2), which could be solved (numerically) to find L as a function of time. In practice, both the melt temperature and the pull rate are controlled to optimize the production rate and maintain crystal quality.

Exercise 19.1   The Fin Approximation
At stages of the crystal growth process when $L \gg R$, temperature variations in the radial direction become much smaller than those in the axial direction (see

Figure 19.3). However, the surface area for conduction on the cylindrical surface of the crystal is much larger than the top of the crystal. In situations such as this, where the domain has a large aspect ratio, it is common to approximate the temperature distribution in the direction having the smaller dimension (the radial direction in this case) by its average. A radially averaged quantity was used in the formulation of the jump balance for energy in (19.10). Show that the two-dimensional problem defined by (19.3)–(19.7) can be reduced to the following one-dimensional problem,

$$\frac{d^2\langle T\rangle}{dz^2} - 2\frac{h_g}{\lambda R}(\langle T\rangle - T_g) = 0, \tag{19.12}$$

$$\langle T\rangle(0,t) = T_s, \qquad -\lambda\frac{d\langle T\rangle}{dz}(L,t) = h_g[\langle T\rangle(L,t) - T_g]. \tag{19.13}$$

The averaging procedure used to obtain (19.12) is known as the *fin approximation*, so named because fins are often used to increase the area for heat transfer. Solve (19.12) subject to (19.13) to obtain

$$-\frac{d\langle T\rangle}{dz}(0,t) = \frac{T_s - T_g}{R}\frac{\sqrt{2N_{Bi}}\sinh(\sqrt{2N_{Bi}}L/R) + N_{Bi}\cosh(\sqrt{2N_{Bi}}L/R)}{\cosh(\sqrt{2N_{Bi}}L/R) + \sqrt{N_{Bi}/2}\sinh(\sqrt{2N_{Bi}}L/R)}. \tag{19.14}$$

Evaluate the expressions in (19.11) and (19.14) for $L = 10R$ as a function of N_{Bi} to find the range of validity for the fin approximation.

Exercise 19.2   CZCG Processing
For CZCG with a fixed melt temperature, the crystal pull rate V is determined by the energy balance in (19.10). Use R, R^2/χ, and $T_s - T_g$ to rescale length, time, and temperature, respectively, and show that (19.10) can be expressed as

$$\frac{dL}{dt} = N_{St}\left[\left\langle\frac{\partial T}{\partial z}(r,0,t)\right\rangle - \bar{N}_{Bi}\frac{T_m - T_s}{T_s - T_g}\right],$$

where $N_{St} = \hat{c}_p(T_s - T_g)/(\hat{h}^I - \hat{h}^{II})$ is the Stefan number (see Exercise 14.8) and $\bar{N}_{Bi} = h_m R/\lambda$. Using results from Exercise 19.1, estimate the time in hours required to grow a crystal with $L = 20$ with $N_{Bi} = \bar{N}_{Bi} = 1$, $(T_m - T_s)/(T_s - T_g) = 1.1$, and $N_{St} = 0.1$. Assume the initial crystal length can be neglected, $L_0 \ll L$.

Exercise 19.3   Conduction in an Anisotropic Slab with a Heat Source
Consider a solid rectangular slab having thickness d in the x_3-direction surrounded by a gas having uniform temperature T_0. The slab is effectively infinite in the x_1- and x_2-directions, and is heated at the origin by a line source at $x_3 = 0$ with power per unit length P_0. Heat transfer at the surfaces $x_3 = \pm d/2$ is governed by Newton's law of cooling (9.26) with heat transfer coefficient h. Conduction within the slab is anisotropic and governed by (6.23) with a diagonal thermal conductivity tensor: $\lambda = \lambda_{11}\delta_1\delta_1 + \lambda_{22}\delta_2\delta_2 + \lambda_{33}\delta_3\delta_3$. Assume temperature variations in the x_3-direction

are small so that the fin approximation can be used with the average temperature $\langle T \rangle = \int_{-d/2}^{d/2} T\, dx_3/d$. Show that the steady-state temperature distribution in the slab is given by

$$\frac{\langle T \rangle - T_0}{P_0/\lambda_0} = \frac{1}{2\pi\sqrt{\alpha_1\alpha_2}} K_0\left(\sqrt{2N_{\text{Bi}}\left[\frac{(x_1/d)^2}{\alpha_1} + \frac{(x_2/d)^2}{\alpha_2}\right]}\right).$$

where $K_0(x)$ is the modified Bessel function of the second kind, $\alpha_i = \lambda_{ii}/\lambda_0$, and $N_{\text{Bi}} = hd/\lambda_0$. Hint: use the change in variables $x = x_1/\sqrt{\alpha_1}$ and $y = x_2/\sqrt{\alpha_2}$ and solve the transformed equation using Fourier transforms. Compare isotherms for the anisotropic and isotropic $(\boldsymbol{\lambda} = \lambda_0\boldsymbol{\delta})$ cases.

19.2 Silicon Oxidation

Usually, the first step after producing a Si wafer is the formation of a thin layer of amorphous SiO_2. The most common method for creating this film is through a chemical reaction known as oxidation. Typically, these reactions are carried out at atmospheric pressure and temperatures in the range 1100–1200 K. In the dry oxidation process, the Si surface is exposed to O_2 gas, and the chemical reaction $Si + O_2 \rightarrow SiO_2$ takes place. The process is shown schematically in Figure 19.5 at a stage where the thickness of the SiO_2 layer is h. The reaction takes place only at the Si–SiO_2 interface, so it is treated as a heterogeneous chemical reaction. Our goal is to obtain an expression for h as a function of time.

The oxide layer growth rate is determined by the diffusion of O_2 through the SiO_2 layer and the reaction taking place at the Si–SiO_2 interface. For our analysis, we assume isothermal, one-dimensional diffusion in the x_3-direction and that the gas–SiO_2 interface $(x_3 = 0)$ is stationary. Since we are dealing with a chemical reaction, it is convenient to work with molar concentration variables. The SiO_2 film is taken to be a binary mixture with O_2 such that $c_{O_2} \ll c_{SiO_2}$. At $x_3 = 0$, we assume that the O_2 in the SiO_2 film is in equilibrium with the O_2 in the gas so that the molar density of O_2 is maintained at a constant value $c_{O_2\text{eq}}$. The molar form of the species continuity equation is given in (5.18). For O_2, this equation simplifies to

$$\frac{\partial c_{O_2}}{\partial t} + \frac{\partial}{\partial x_3}[c_{O_2}(v_{O_2})_3] = 0, \tag{19.15}$$

where we have excluded the term for homogeneous chemical reaction. The SiO_2 produced is an immobile solid so that $(v_{SiO_2})_3 = 0$, and its molar density c_{SiO_2} is assumed to be constant. Hence, the species continuity equation for SiO_2 is irrelevant. Using these assumptions and (5.19), (5.20), and (6.27),

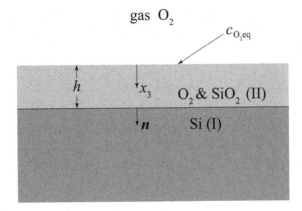

Figure 19.5 Schematic representation of Si oxidation process showing SiO_2 layer having thickness h.

it is possible to derive (see Exercise 19.4) an expression for the molar flux in (19.15) given by

$$c_{O_2}(v_{O_2})_3 = -D\frac{\partial c_{O_2}}{\partial x_3}, \tag{19.16}$$

where D is the diffusivity of O_2 in SiO_2.

Boundary conditions at the SiO_2–Si interface can be formulated using the balance equations in Chapter 14. The planar interface at $x_3 = h$ is uniform and has unit normal vector $\boldsymbol{n} = \boldsymbol{\delta}_3$ making the Si phase I and the SiO_2–O_2 phase II. Here, we treat the interface as passive so that all of the excess species densities are zero simultaneously ($\rho_\alpha^s = 0$). Hence, the molar form of the jump balance for species mass (14.27) can be written as,

$$c_\alpha^I[(v_\alpha^I)_3 - v_n^s] = c_\alpha^{II}[(v_\alpha^{II})_3 - v_n^s] + \tilde{\nu}_\alpha^s\tilde{\Gamma}^s. \tag{19.17}$$

Note that $c_{Si}^{II} = 0$, and we will assume the unreacted Si is a one-component phase so that $c_{O_2}^I = c_{SiO}^I = 0$. We take the reaction to be a pseudo-first-order reaction and "irreversible" so that from (15.32) we can write

$$\tilde{\Gamma}^s = L_\Gamma^s \exp\left(\frac{\tilde{\mu}_{O_2eq}^s}{\tilde{R}T}\right) = L_\Gamma^s \exp\left(\frac{\tilde{\mu}_{O_2eq}^{II}}{\tilde{R}T}\right) = k^s c_{O_2eq}^{II}, \tag{19.18}$$

where the second equality is obtained by assuming continuity of chemical potentials at the interface $\tilde{\mu}_{O_2eq}^s = \tilde{\mu}_{O_2eq}^{II}$ and the third by treating the mixture as ideal. Hence, the reaction is described by the mass action law with rate constant k^s.

The problem we are considering does not have a geometric length that can be used to make spatial position dimensionless. In such cases, it is natural to use the transport coefficients to identify a characteristic length. A possible choice is to use D/k^s as the unit of length, which seems reasonable since it is the interplay of diffusion and reaction that determines the growth of the film thickness h. A proper unit of time is $D/(k^s)^2$, and c_{O_2eq} is used as the unit for molar densities. Assuming that D is constant, substitution of (19.16) into (19.15) leads to the dimensionless equation

$$\frac{\partial c_{O_2}}{\partial t} = \frac{\partial^2 c_{O_2}}{\partial x_3^2}. \tag{19.19}$$

The boundary condition at the O_2–SiO_2 interface is simply

$$c_{O_2}(0, t) = 1. \tag{19.20}$$

Using (19.18) and writing (19.17) for O_2 and SiO_2, we obtain with $v_n^s = dh/dt$ the boundary condition

$$\frac{\partial c_{O_2}}{\partial x_3}(h, t) + c_{O_2}(h, t) = 0, \tag{19.21}$$

and the following evolution equation for the film thickness

$$\frac{dh}{dt} = -\varepsilon \frac{\partial c_{O_2}}{\partial x_3}(h, t), \tag{19.22}$$

where $\varepsilon = c_{O_2eq}/c_{SiO_2}$.[6] The boundary condition in (19.21) shows the relation between the flux to the interface and the rate of reaction, while the expression in (19.22) relates the flux to the rate of film growth.

The problem defined by (19.19)–(19.22), which is known as a Stefan problem, looks relatively simple. However, notice that the boundary condition in (19.21) is prescribed at a position h that evolves in time according to (19.22). This means that we have a moving-boundary problem (see Section 7.4), which makes the system of equations nonlinear. To facilitate finding a solution to this problem, it is useful to apply the coordinate transformation $\bar{x}_3 = x_3/h = [0, 1]$. Applying this transformation, (19.19)–(19.22) become (see Exercise 19.5)

$$\left(\frac{\partial c_{O_2}}{\partial t} - \frac{\bar{x}_3}{h} \frac{dh}{dt} \frac{\partial c_{O_2}}{\partial \bar{x}_3} \right) h^2 = \frac{\partial^2 c_{O_2}}{\partial \bar{x}_3^2}, \tag{19.23}$$

$$c_{O_2}(0, t) = 1, \tag{19.24}$$

---

[6] By writing (19.17) for Si, the velocity in the SiO_2 phase is given by

$$(v_{Si})_3 = \left(1 - \frac{c_{SiO_2}}{c_{Si}} \right) \frac{dh}{dt}.$$

The ratio $c_{SiO_2}/c_{Si} \approx 0.5$ so that roughly one-half of the oxide film thickness is outside the location of the original Si surface.

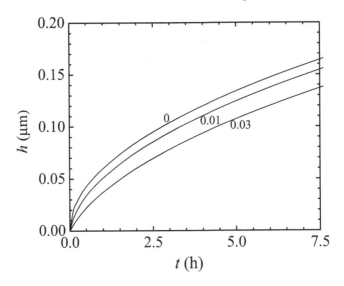

Figure 19.6 Oxide film thickness h in μm versus time t in hours given by (19.28) with $h \to hk^s/D$ and $t \to t(k^s)2/D$, and the following parameter values: $\varepsilon = 10^{-5}$, $D = 5 \times 10^{-14}$ m^2/s. Different curves correspond to different values of $D/k^s = 0.0, 0.01, 0.03$ μm (top to bottom).

$$\frac{\partial c_{O_2}}{\partial \bar{x}_3}(1,t) + h c_{O_2}(1,t) = 0, \tag{19.25}$$

$$h\frac{dh}{dt} = -\varepsilon \frac{\partial c_{O_2}}{\partial \bar{x}_3}(1,t). \tag{19.26}$$

Note that, in the transformed coordinate system, the boundary condition in (19.25) is immobilized, but the diffusion equation (19.23) is now nonlinear.

Since $\varepsilon \ll 1$, we invoke the quasi-steady-state approximation and need only solve the steady-state form of (19.23). Using the boundary conditions in (19.24) and (19.25), we obtain

$$c_{O_2} = \frac{1 + (1 - \bar{x}_3)h}{1 + h}. \tag{19.27}$$

Substitution of (19.27) into (19.26) and integration with $h(0) = 0$ gives

$$h^2 + 2h = 2\varepsilon t. \tag{19.28}$$

The expression for the film thickness in (19.28) is the well-known result found by Deal and Grove, which predicts linear, followed by square-root, growth.[7] The occurrence of $h^2 + 2h$ suggests that we have identified the proper scale for introducing dimensionless lengths. Figure 19.6 shows a plot of the film

[7] Deal & Grove, *J. Appl. Phys.* **36** (1965) 3770.

thickness versus time predicted by (19.28) where we see that it takes 3–5 hours to grow a film having a thickness of roughly 100 nm. This model is known to give reasonable predictions for oxide film thickness growth, except at early stages of growth when the film is thin. A correction to this solution obtained using a perturbation method involving the parameter ε is left as an exercise.

Exercise 19.4   Molar Flux with Immobile Species
Derive the expression in (19.16).

Exercise 19.5   Coordinate Transformation for Moving Boundary
Use the transformation $\bar{x}_3 = x_3/h$ to derive (19.23)–(19.26) from (19.19)–(19.22).

Exercise 19.6   Silicon Oxidation with Rapid Reaction
In the limit that the oxidation reaction is rapid compared with the rate of diffusion $D/k^s \to 0$, the concentration of O_2 at the Si–SiO$_2$ interface approaches zero. In this case, it is not possible to identify characteristic quantities for length and time. Hence, the spatial position and time variables in (19.19)–(19.22) are dimensional, with D appearing in front of the second-order derivative in (19.19), and the boundary condition in (19.21) is replaced by $c_{O_2}(h, t) = 0$. Use the similarity transformation $\xi = x_3/\sqrt{4Dt}$ to find the following expression for the film thickness,[8]

$$h = 2\beta\sqrt{Dt}, \qquad \beta = \varepsilon\frac{\exp(-\beta^2)}{\sqrt{\pi}\,\mathrm{erf}(\beta)}.$$

The nonlinearity is now reduced to the solution of a transcendental equation.

Exercise 19.7   Perturbation Solution for Slow Oxidation
A perturbation solution (see Section 8.3) to (19.23)–(19.26) can be found[9] by substitution of $c_{O_2} = c_{O_2}^{(0)} + \varepsilon c_{O_2}^{(1)} + \cdots$ and $h = h^{(0)} + \varepsilon h^{(1)} + \cdots$. Show that the zeroth-order solutions $c_{O_2}^{(0)}$ and $h^{(0)}$ are given by (19.27) and (19.28), respectively. Find the expression for the first-order correction to the film thickness given by

$$h^{(1)} = -\frac{2}{3\sqrt{1 + 4\varepsilon t}} + \frac{4 + 2(6 - \sqrt{1 + 4\varepsilon t})2\varepsilon t - 3\sqrt{1 + 4\varepsilon t}\ln(1 + 4\varepsilon t)}{6(1 + 4\varepsilon t)}.$$

19.3 Spin Coating

As discussed at the beginning of this chapter, spin coating is the process by which thin liquid films are formed on a wafer for the purpose of creating

---

[8] Peng *et al.*, *J. Vac. Sci. Tech. B* **14** (1996) 3317.
[9] Mhetar & Archer, *J. Vac. Sci. Tech. B* **16** (1999) 2121.

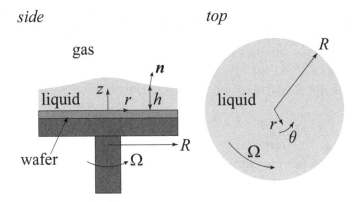

Figure 19.7 Schematic diagram of spin coating process. Note that the film thickness h is exaggerated for clarity.

patterned structures. A typical photoresist is a polymer solution with a viscosity that typically is 100 times larger than that of water. During the spin coating process, the solvent evaporates, leaving a thin ($\sim$ 1–10 μm) polymer film that undergoes a chemical reaction when exposed to light. It is critical that the polymer film has uniform thickness over the surface of the wafer.

We now develop a simple model[10] for the spin coating process shown schematically in Figure 19.7. The formation of a thin film is governed by the balance between centrifugal and viscous forces. We assume that the fluid is a Newtonian liquid with constant density ρ and viscosity η. The liquid film coats a solid disk with radius R that rotates with constant angular velocity Ω, which is typically 1000 revolutions per minute. The gas above the film is assumed to be inviscid, and we assume that there is no mass transfer at the gas–liquid interface at $z = h(r, t)$. The initial thickness of the film, which may be non-uniform, has characteristic value h_0, and we introduce the parameter $\beta = h_0/R \ll 1$. Finally, we assume symmetry about the z-axis and that the flow is incompressible. Based on these assumptions, the velocity field will have the form $v_r = v_r(r, z, t)$, $v_\theta = v_\theta(r, z, t)$, $v_z = v_z(r, z, t)$.

Before considering the mass and momentum balances that govern the velocity and pressure fields, it is useful to first formulate boundary conditions for velocity. The disk is an impermeable solid, and we assume that there is no slip (15.18) at the solid–liquid interface, so that

$$v_r(r, 0, t) = 0, \qquad v_\theta(r, 0, t) = \Omega r, \qquad v_z(r, 0, t) = 0. \tag{19.29}$$

[10] Emslie *et al.*, *J. Appl. Phys.* **29** (1958) 858.

To formulate boundary conditions at the gas–liquid interface, we will need to consider jump balances for mass and momentum. First, we need an expression for the unit normal vector to the surface n, which we choose to point into the gas phase (phase I). If we describe the surface by $\varphi^s(r, z, t) = z - h(r, t) = 0$, then according to (13.26) the normal to the surface is given by

$$n = \frac{-(\partial h/\partial r)\boldsymbol{\delta}_r + \boldsymbol{\delta}_3}{\sqrt{1 + (\partial h/\partial r)^2}}, \tag{19.30}$$

which we see is time-dependent. Using (14.15), we obtain

$$\frac{\partial h}{\partial t} + (\boldsymbol{v}^s \cdot \boldsymbol{n})\sqrt{1 + (\partial h/\partial r)^2} = 0. \tag{19.31}$$

We have assumed that there is no mass transfer across this interface, which from (14.7) implies continuity of the normal velocities across the interface, $\boldsymbol{v}^{\mathrm{II}} \cdot \boldsymbol{n} = \boldsymbol{v}^s \cdot \boldsymbol{n}$. From (19.30) and (19.31), we obtain

$$v_z(r, h, t) - v_r(r, h, t)\frac{\partial h}{\partial r} = \frac{\partial h}{\partial t}. \tag{19.32}$$

In the absence of mass transfer, the jump balance for momentum in (15.21) reduces to,

$$\boldsymbol{n} \cdot \boldsymbol{\tau}^{\mathrm{I}} + p^{\mathrm{I}}\boldsymbol{n} = \boldsymbol{n} \cdot \boldsymbol{\tau}^{\mathrm{II}} + p^{\mathrm{II}}\boldsymbol{n} - \gamma(\boldsymbol{\nabla}_\| \cdot \boldsymbol{n})\boldsymbol{n}, \tag{19.33}$$

where we have taken the interfacial tension γ to be constant and set $\boldsymbol{\tau}^s = \boldsymbol{0}$. For the Newtonian liquid phase, we have $\boldsymbol{\tau}^{\mathrm{II}} = -\eta^{\mathrm{II}}[\boldsymbol{\nabla}\boldsymbol{v} + (\boldsymbol{\nabla}\boldsymbol{v})^{\mathrm{T}}]^{\mathrm{II}}$; since the gas phase is an inviscid fluid, we set $\boldsymbol{\tau}^{\mathrm{I}} = \boldsymbol{0}$. After substitution into (19.33), the r-, θ-, and z-components of the jump balance for momentum are given by

$$\eta\left[\frac{\partial v_r}{\partial z}(r, h, t) + \frac{\partial v_z}{\partial r}(r, h, t)\right] = \left[p^{\mathrm{L}}(r, h, t) + \gamma\frac{\partial^2 h}{\partial r^2} - 2\eta\frac{\partial v_r}{\partial r}(r, h, t)\right]\frac{\partial h}{\partial r}, \tag{19.34}$$

$$\frac{\partial v_\theta}{\partial z}(r, h, t) = 0, \tag{19.35}$$

$$p^{\mathrm{L}}(r, h, t) + \gamma\frac{\partial^2 h}{\partial r^2} = 2\eta\frac{\partial v_z}{\partial z}(r, h, t) - \eta\left[\frac{\partial v_r}{\partial z}(r, h, t) + \frac{\partial v_z}{\partial r}(r, h, t)\right]\frac{\partial h}{\partial r}, \tag{19.36}$$

where we have used $\boldsymbol{\nabla}_\| \cdot \boldsymbol{n} \approx -\partial^2 h/\partial r^2$, which holds for $\partial h/\partial r \sim \beta \ll 1$, and absorbed p^{I} into the pseudo-pressure p^{L}. The boundary conditions in the second equation in (19.29) and (19.35) suggest $v_\theta(r, z, t) = \Omega r$. In other words, the tangential velocity of the liquid is uniform and equal to that of the disk.

In light of this, we consider the mass and momentum balances for the liquid film expressed in a rotating (non-inertial) coordinate system (see Section 11.1). As with the centrifuge analysis, here we have $\boldsymbol{c} = \boldsymbol{0}$ and $\boldsymbol{\omega} = \Omega \boldsymbol{\delta}_z$. Given the complexity of this flow problem, it is useful to write the momentum balance in dimensionless form using scaling. Here, we choose $V = \rho \Omega^2 h_0^3 / \eta$ to scale velocity, which reflects the fact that flow in the liquid film is dictated by a balance between centrifugal and viscous forces. To make position and time dimensionless, we use h_0 and h_0 / V, respectively. Dropping the primes, we have the constraint on velocity $\boldsymbol{\nabla} \cdot \boldsymbol{v} = 0$, and the momentum balance (11.3) can be written as

$$N_{\mathrm{Re}}^2 \left(\frac{\partial \boldsymbol{v}}{\partial t} + \boldsymbol{v} \cdot \boldsymbol{\nabla} \boldsymbol{v} \right) = \nabla^2 \boldsymbol{v} - \boldsymbol{\nabla} \mathcal{P} - [2 N_{\mathrm{Re}}(\boldsymbol{\omega} \times \boldsymbol{v}) + \boldsymbol{\omega} \times (\boldsymbol{\omega} \times \boldsymbol{r})], \quad (19.37)$$

where we have scaled pressure by $\eta V / h_0$ and the angular velocity vector $\boldsymbol{\omega}$ by Ω. Note that the Reynolds number, defined here as $N_{\mathrm{Re}} = \rho \Omega h_0^2 / \eta$, appears to the second power on the left-hand side and to the first power in front of the Coriolis term. Typically, $N_{\mathrm{Re}} \ll 1$, which translates to $\Omega \lesssim 10^4$ rad/s, so we neglect the inertial and Coriolis terms in (19.37), and note that the latter is consistent with the assumption $v_\theta(r, z, t) = \Omega r$.

The constraint on the velocity field can be written as

$$\frac{1}{r} \frac{\partial}{\partial r}(r v_r) + \frac{\partial v_z}{\partial z} = 0. \quad (19.38)$$

Dropping terms based on $N_{\mathrm{Re}} \ll 1$, the dimensional r- and z-components of (19.37) can be written as

$$\frac{\partial^2 v_r}{\partial z^2} = -\frac{\rho \Omega^2}{\eta} r, \quad (19.39)$$

$$\frac{\partial^2 v_z}{\partial z^2} = \frac{1}{\eta} \frac{\partial \mathcal{P}}{\partial z}, \quad (19.40)$$

where we have also neglected terms $O(\beta^2)$. Hence, we have invoked the quasi-steady-state approximation, and the only remaining time-evolution equation is for the film thickness h in (19.32). The boundary conditions in (19.34) and (19.36), again neglecting terms $O(\beta^2)$, become

$$\frac{\partial v_r}{\partial z}(r, h, t) = 0, \quad (19.41)$$

$$p^{\mathrm{L}}(r, h, t) = 2\eta \frac{\partial v_z}{\partial z}(r, h, t) - \gamma \frac{\partial^2 h}{\partial r^2}. \quad (19.42)$$

By identification of appropriate dimensionless parameters, namely N_{Re} and β, to estimate the relative magnitudes of terms in the mass and

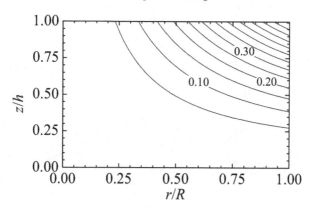

Figure 19.8 Normalized stream function $\psi/(Vh_0^2)$ for velocity within a uniform film given by (19.43) and (19.44) in terms of the stream function given in (19.49) as found in Exercise 19.8. Contours are for constant values of ψ/h^3.

momentum balances, we are able to see that the physics of this flow is captured by the balance of centrifugal and viscous forces in (19.39) and by incompressibility in (19.38).

Integration of (19.39) and using the boundary conditions in the first equation in (19.29) and (19.41) gives

$$v_r = r \left[\frac{z}{h} - \frac{1}{2} \left(\frac{z}{h} \right)^2 \right] h^2, \tag{19.43}$$

where we have normalized r, z, and h by h_0, and velocity by $V = \rho \Omega^2 h_0^3/\eta$. To find v_z, we substitute (19.43) into (19.38) and integrate from $z = 0$ with the third equation in (19.29), which gives

$$v_z = - \left[\left(\frac{z}{h} \right)^2 - \frac{1}{3} \left(\frac{z}{h} \right)^3 \right] h^3 - \frac{1}{2} r z^2 \frac{\partial h}{\partial r}. \tag{19.44}$$

Note the strong dependence of the velocity on film thickness in (19.43) and (19.44). The stream function (see Exercise 19.8) representation of this flow is shown in Figure 19.8. The pseudo-pressure field is found by substitution of (19.44) into (19.40), which is left as an exercise.

Our goal is to obtain an expression for the evolution of the film thickness h. Substitution of (19.43) and (19.44) into (19.32) gives

$$\frac{\partial h}{\partial t} = -\frac{1}{3r} \frac{\partial}{\partial r} \left(r^2 h^3 \right). \tag{19.45}$$

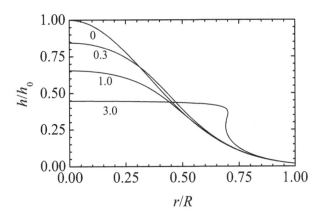

Figure 19.9 Evolution of film profile predicted by (19.48) with an initially Gaussian profile $h' = \exp(-4r'^2)$. Different profiles are for different values of time $t = 0, 0.3, 1.0, 3.0$ (from top to bottom).

The first-order partial differential equation (19.45) can be solved by the method of characteristics (see Exercise 11.2). Along curves $r(t)$ satisfying the ordinary differential equation

$$\frac{dr}{dt} = rh^2 \tag{19.46}$$

the function $h(t, r(t))$ satisfies the ordinary differential equation

$$\frac{dh}{dt} = \frac{\partial h}{\partial t} + \frac{\partial h}{\partial r}\frac{dr}{dt} = -\frac{2}{3}h^3. \tag{19.47}$$

The solution of (19.46) and (19.47) gives the position of a point r, h on the interface, parameterized by t,

$$h = h'\left(1 + \frac{4}{3}h'^2t\right)^{-1/2}, \qquad r = r'\left(1 + \frac{4}{3}h'^2t\right)^{3/4}, \tag{19.48}$$

where r', h' is the location of a point on the surface at $t = 0$. An example of the film profile evolution is shown in Figure 19.9 for a profile that is initially Gaussian. Note that the profile becomes nearly uniform in a relatively short time. It should also be noted that the steep profile at $r \approx 0.7$ for $t = 3.0$ shown in Figure 19.9 is not consistent with the analysis used to obtain (19.48), which assumes a mildly curved free surface.

The model we have just considered neglects a number of important effects, such as the evaporation of the solvent during spinning and the presence of a contact line (see Exercise 13.6) where gas, liquid, and solid phases co-exist. More complicated models of the spin coating process can be found elsewhere.[11]

Exercise 19.8   Stream Function for Spin Coating Flow
Show that the stream function scaled by $V h_0^2$ is, for the velocity field in (19.43) and (19.44) assuming a uniform film thickness, given by

$$\psi = r^2 \left[\left(\frac{z}{h} \right)^2 - \frac{1}{3} \left(\frac{z}{h} \right)^3 \right] h^3. \tag{19.49}$$

Exercise 19.9   Pseudo-pressure Field for Spin Coating Flow
The gravitational vector is given by $\boldsymbol{g} = -\boldsymbol{\nabla}\phi = -g\boldsymbol{\delta}_z$. Derive the following expression for the pseudo-pressure field in the liquid film,

$$p^{\mathrm{L}} = \frac{1}{N_{\mathrm{Fr}}} \left(1 - \frac{z}{h} \right) h - \left[1 + 2\frac{z}{h} - \left(\frac{z}{h} \right)^2 \right] h^2 - \left(1 + \frac{z}{h} \right) hr \frac{\partial h}{\partial r} + \frac{1}{N_{\mathrm{We}}} \frac{\partial^2 h}{\partial r^2}, \tag{19.50}$$

where p^{L} has been normalized by $\eta V/h_0$. The Froude number is given by $N_{\mathrm{Fr}} = \Omega^2 h_0/g$, and the Weber number by $N_{\mathrm{We}} = \rho\Omega^2 h_0^3/\gamma$.

- Processes such as crystal growth, oxidation, and spin coating used to fabricate semi-conductor devices have dynamic interfaces that involve mass, momentum, and energy transport as well as chemical reaction.
- Analysis of thermal transport in the Czochralski crystal growth process allows prediction of the crystal temperature distribution and growth rate, which can be simplified by use of the quasi-steady-state and fin approximations.
- Analysis of the silicon oxidation process, which involves diffusion and heterogeneous chemical reaction, is simplified using the quasi-steady-state approximation, and it leads to an expression for the growth rate of the oxide film thickness.
- Analysis of the hydrodynamics of the spin coating process, which is governed by a balance between centrifugal and viscous forces, leads to expressions for the velocity field and time evolution of the film thickness profile.

---

[11] Flack *et al.*, *J. Appl. Phys.* **56** (1984) 1199.

20

Refreshing Topics in Equilibrium Statistical Mechanics

In Chapter 6, we introduced a number of transport coefficients, such as the thermal conductivity or the shear viscosity. We saw how to measure such phenomenological quantities in Chapter 7, and we have frequently used them in many of the subsequent chapters. Is there any way of predicting these transport properties theoretically, of calculating them from the interactions between atoms or molecules? The molecular approach to transport phenomena is the topic of this and the subsequent chapters. In particular, we will develop some key elements of *kinetic theory* (see Chapters 21 and 22), but also experimental ideas based on the molecular view (see Chapters 25 and 26).

There actually is a similar question for equilibrium systems: how can we predict or calculate thermodynamic material properties, such as the heat capacity or the isothermal compressibility? We know that all thermodynamic material information is contained in any thermodynamic potential, so that we need to calculate a convenient thermodynamic potential. In the present chapter, we refresh the reader's memories of statistical mechanics in a similar spirit to what we did for equilibrium thermodynamics in Chapter 4. As our discussion of equilibrium thermodynamics was largely based on the classical textbook by Callen, it is interesting to note that the original edition (1960) of that book was entitled *Thermodynamics* (with the word "thermostatics" appearing in the subtitle given on the front page of the book, but not on the cover) whereas the second edition (1985) was entitled *Thermodynamics and an Introduction to Thermostatistics* (read carefully to notice the difference between thermostatics and thermostatistics). Whereas thermodynamics is a topic in its own right, there is a clear need to supplement it by statistical mechanics in order to be able to calculate the required thermodynamic material information. The development of kinetic theory in the subsequent chapters relies on knowledge of equilibrium statistical mechanics. Einstein's theory of fluctuations turns out to be particularly useful for our purposes.

20.1 The Fundamental Statistical Ensemble

Thermodynamic systems involve a large number of microscopic degrees of freedom, typically of the order of 10^{23}. In contrast, only a few degrees of freedom are needed to describe the thermodynamic behavior of a system at equilibrium. For example, according to Chapter 4, the thermodynamic state of a gas can be fully described by the energy U, the volume V, and the number of particles $\tilde{N}$ as independent variables. How can we achieve this enormous reduction in the number of degrees of freedom? How does the thermodynamic information arise in this *coarse-graining* procedure?

Averages play an important role in coarse graining. For example, if we want to determine the pressure in a gas, we need to know how many particles with a certain momentum hit a certain area of the wall in a certain period of time. The larger the area and the time interval, the larger the number of particle collisions and the more reliable the estimate of the pressure. The measurement of a thermodynamic quantity typically involves averaging in space and time.

In a theoretical approach, time averaging would require the calculation of the evolution of the entire system of 10^{23} degrees of freedom. For interacting systems, this is an absolutely hopeless task. Even the specification of 10^{23} initial conditions is beyond our possibilities. But hope comes from *ergodic theory*. The essential idea is that, in the course of the time evolution, certain particle configurations occur with a certain frequency. If we have a list of possible configurations and the probabilities of their occurrence, we know how to calculate averages over microstates. This is the idea of *statistical ensembles*.

As the characterization of configurations includes the positions of all particles in a system, spatial averaging is implied. Shown in Figure 20.1 are two randomly chosen configurations of a gas of six particles in boxes of the same size and with equal energy, as indicated by the sameness of the mean-square values of the velocity vectors. The configurational knowledge also includes the momenta of the particles. In the calculation of pressure, the momentum transfer of the particles to the wall can be obtained only if we know the particle momenta. It may be necessary to include further configurational properties, such as spin, to arrive at a complete microscopic characterization of the particle configurations or microstates. Hence, it may sometimes be necessary to describe the microstates by quantum mechanics.

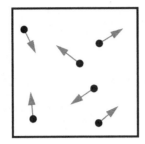 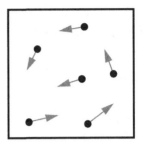

Figure 20.1 Two equivalent particle configurations for a gas of spherical particles.

The most fundamental ensemble for theoretical developments is obtained by considering an isolated system and fixing all the extensive thermodynamic variables, say U, V, and $\tilde{N}$. Figure 20.1 gives an example of two configurations of such an ensemble, which is known as the *microcanonical ensemble*. Which of the two configurations displayed in the figure has a larger probability of occurrence? As there is no reason for preferring one over the other configuration, the microcanonical ensemble assigns the same probability to every microstate compatible with the extensive thermodynamic variables. In short, in a microcanonical ensemble we average over all microstates, all having the same probability. The averaging procedure is fully determined by the number Ω of microstates consistent with given extensive variables, say $\Omega(U, V, \tilde{N})$ for a gas. The assumption of equal probabilities can actually be supported by a symmetry argument. In the words of Callen,[1] "the equal probabilities of permissible states for a closed system in equilibrium is a consequence of time reversal symmetry of the relevant quantum mechanical laws." However, there still is the issue of whether a trajectory starting at one configuration will eventually reach the other one, or come very close to it.

The phrase "number of microstates" should not be taken too literally. In classical mechanics, we deal with a continuous phase space, and Ω is not found by counting but rather by integrating. In quantum mechanics, we actually have discrete states (for a finite volume), and the counting procedure is more well-defined. Moreover, we cannot insist on a precise value of the internal energy but rather need to count the states with energies between U and $U + \Delta U$, where ΔU is large compared with microscopic energies and small compared with macroscopic energies. The large number of degrees of freedom is essential for this procedure. For a meaningful quantum

---

[1] See p. 468 of Callen, *Thermodynamics* (Wiley, 1985).

mechanical counting, many energy eigenvalues should fall within the interval from U to $U + \Delta U$.

20.2 More Practical Ensembles

Although the microcanonical ensemble is most suitable for theoretical considerations and can be taken as a fundamental reference ensemble, it turns out to be rather inconvenient for practical calculations. For the extensive variables U, V, and $\tilde{N}$, according to (4.14)–(4.17), the proper thermodynamic potential is given by the entropy $S(U, V, \tilde{N})$. From equilibrium thermodynamics, we already know that there exist more convenient choices of variables that can be introduced by Legendre transforms. The corresponding operation to simplify calculations in statistical mechanics is to switch to another ensemble. We here sketch how to obtain the *canonical ensemble* associated with the variables T, V, and $\tilde{N}$. The ideas for switching from an extensive to an intensive variable (e.g., from U to T), however, are very general. By a further switch from particle number to chemical potential, we arrive at the *grand canonical ensemble*.

The canonical ensemble can be derived by considering the system of interest in contact with a thermal reservoir of a given temperature rather than an isolated system. The system plus the reservoir constitute an isolated system which can be treated by the microcanonical ensemble. For the system alone, however, microstates do not have the same probability. Rather, the probabilities are proportional to the famous exponential Boltzmann factors $\exp\{-E_j/(k_\mathrm{B}T)\}$ for a microstate j with energy E_j. The normalization factor for the probabilities is the *canonical partition sum*, Z, as a function of the temperature and the extensive mechanical state variables, say $Z(T, V, \tilde{N})$ for a gas. For the example of the gas illustrated in Figure 20.1, the energy E_j associated with a particle configuration j would simply be the sum of all the kinetic energies of the particles.

The occurrence of weight factors of the exponential form is related to the logarithmic form of the Shannon entropy associated with the probability densities in phase space (that is, ensembles) in information theory. If we prescribe the average of the internal energy and otherwise minimize the information content of the ensemble, then an exponential factor containing temperature as a Lagrange multiplier arises. For every extensive variable treated in a similar way, an exponential factor and a Lagrange multiplier serving as the conjugate intensive variable arise. This is the essence of *switching ensembles*, which is the counterpart of Legendre transformations in equilibrium thermodynamics.

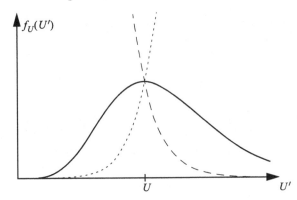

Figure 20.2 Peaked distribution of a macroscopic variable such as the internal energy U' in a canonical ensemble.

How can different ensembles be equivalent? To be equivalent to a microcanonical reference ensemble in calculating averages, the *fluctuations* of the macroscopic variables in any alternative ensemble must be sufficiently small. Sharply peaked distributions typically arise from the product of a decreasing exponential factor, like the Boltzmann factor, and a rapidly increasing number of microstates. An example is shown in Figure 20.2; the example of the figure actually corresponds to the probability for finding the internal energy U' in the canonical ensemble for an ideal gas of only four particles for given average energy U. With increasing number of particles, the peak becomes sharper, and for 10^{23} particles it becomes extremely sharp. It is important to note that different ensembles agree on averages, but not on fluctuations. For example, energy fluctuates in the canonical ensemble but has a fixed value in the microcanonical ensemble.

20.3 Calculation of Thermodynamic Potentials

Knowing the idea of statistical ensembles, and the microcanonical and canonical ensembles in particular, we should now address the issue of predicting thermodynamic information. For complete thermodynamic information, we should calculate a thermodynamic potential. For the microcanonical ensemble, we need to construct the entropy $S(U, V, \tilde{N})$ from the number of microstates $\Omega(U, V, \tilde{N})$, and, for the canonical ensemble, we need to construct the Helmholtz free energy $F(T, V, \tilde{N})$ from the canonical partition sum $Z(T, V, \tilde{N})$.

Let us start with the more fundamental microcanonical ensemble. In Figure 20.3, we show an isolated system consisting of two isolated subsystems.

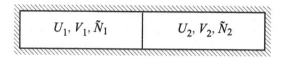

Figure 20.3 Two subsystems brought into contact to identify the statistical expression for the entropy.

If the numbers of microstates for the subsystems are $\Omega_1(U_1, V_1, \tilde{N}_1)$ and $\Omega_2(U_2, V_2, \tilde{N}_2)$, then the combined system has

$$\Omega = \Omega_1(U_1, V_1, \tilde{N}_1)\,\Omega_2(U_2, V_2, \tilde{N}_2) \tag{20.1}$$

microstates because each microstate of subsystem 1 can be combined with any microstate of subsystem 2 to obtain a microstate of the total system. If we remove the wall between the subsystems, additional states become available, for example, by moving a particle from subsystem 1 to subsystem 2, so that

$$\Omega(U_1 + U_2, V_1 + V_2, \tilde{N}_1 + \tilde{N}_2) > \Omega_1(U_1, V_1, \tilde{N}_1)\,\Omega_2(U_2, V_2, \tilde{N}_2). \tag{20.2}$$

The quantity Ω is multiplicative instead of additive, but it increases just like the entropy when the subsystems can exchange particles, energy, or volume. Going from multiplicative to additive is the magic of logarithms. It is hence most natural to make the identification

$$S(U, V, \tilde{N}) = k_B \ln \Omega(U, V, \tilde{N})\,. \tag{20.3}$$

The value of the proportionality constant k_B, which is known as Boltzmann's constant, still needs to be determined. The constant would be different if we were to use a logarithm with a different base rather than the natural logarithm.

If one uses (20.3) to calculate the entropy of an ideal gas, one is faced with a major disappointment: the resulting entropy is not even extensive. It grows faster than $\tilde{N}$, actually as $\tilde{N} \ln \tilde{N}$, which means that we are overestimating the number of microstates. Hence, we define

$$S(U, V, \tilde{N}) = k_B \ln \left[\frac{\Omega(U, V, \tilde{N})}{\tilde{N}!} \right]\,, \tag{20.4}$$

so that $S(U, V, \tilde{N})$ has all the nice properties of an entropy, where Stirling's formula shows why $\tilde{N} \ln \tilde{N}$ is removed by $-\ln \tilde{N}!$. In particular, the entropy of an ideal gas is properly reproduced by (20.4). The factor $\tilde{N}!$ can be interpreted in a very simple way: renumbering of the particles does not lead to a different microstate. In that sense, the particles are *indistinguishable*.

This is a perfectly natural idea in quantum mechanics, but it is known as Gibbs' paradox in classical statistical mechanics. Long before the advent of quantum mechanics, Gibbs had recognized this surprising indistinguishability very clearly, and he tried to present the unavoidable as natural in an ensemble interpretation:[2] "If two phases differ only in that certain entirely similar particles have changed places with one another, are they to be regarded as identical or different phases? If the particles are regarded as indistinguishable, it seems in accordance with the spirit of the statistical method to regard the phases as identical. In fact, it might be urged that in such an ensemble of systems as we are considering no identity is possible between the particles of different systems except that of qualities, and if ν particles of different systems are described as entirely similar to one another and to ν of another system, nothing remains on which to base the identification of any particular particle of the first system with any particular particle of the second."

In the microcanonical ensemble, even the calculation of the entropy of an ideal gas requires some effort. Although the particles are noninteracting, they are coupled by fixing the value of the total internal energy. If one particle happens to have a particularly large energy, the other particles have to compensate for that by having correspondingly smaller energies. In the canonical approach, it turns out that one needs to evaluate only single-particle integrals to obtain the Helmholtz free energy so that calculations for noninteracting systems become much simpler.

In the canonical ensemble, the individual microstates are weighted by Boltzmann factors. Their normalization, given by the canonical partition sum over all microstates, is

$$Z = \sum_j e^{-E_j/(k_{\mathrm{B}}T)} \,, \tag{20.5}$$

which contains the macroscopically relevant information, where $E_j(V, \tilde{N})$ is the energy of microstate j with $\tilde{N}$ particles in a volume V. As a sum over possible states, Z clearly contains entropic information, but it is properly corrected for energetic effects via the Boltzmann factors. It is hence appealing to express the Helmholtz free energy $F = U - TS$ as (see Exercise 20.2)

$$F = -k_{\mathrm{B}}T \ln Z \,. \tag{20.6}$$

By rewriting sums over microstates as sums over energy intervals, for which we already know the number of microstates in any interval, we can actually

---

[2] See pp. 187–188 of Gibbs, *Principles in Statistical Mechanics* (New York, 1902).

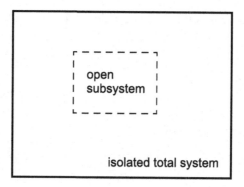

Figure 20.4 Open subsystem with fluctuating internal energy and particle number within a large equilibrium bath.

derive the validity of (20.6) from our knowledge about the microcanonical ensemble.

Exercise 20.1   Third Law from Statistical Mechanics
Explain the third law of thermodynamics from the perspective of statistical mechanics.

Exercise 20.2   Recipe for Calculating Helmholtz Free Energy
Verify that the function $F(T, V, \tilde{N})$ defined in (20.6) is indeed the Helmholtz free energy by falling back on the properties of the microcanonical ensemble.

20.4 Einstein's Fluctuation Theory

With the statistical background presented in the preceding sections, we can now do more than calculate ensemble averages – we can consider fluctuations. However, different ensembles are designed to be equivalent for averages only, not for fluctuations. We hence need to argue carefully to obtain a valid theory of fluctuations, as originally developed by Einstein.[3]

Let us consider a large system obtained by fixing a volume in an even larger equilibrium system that serves as a bath (see Figure 20.4). The internal energy U and the mole number N in the probe volume fluctuate. If the probability of observing a macrostate of the subsystem is proportional to the number of corresponding microstates, then, according to (20.4), the probability of macroscopic fluctuations must be proportional to $\exp\{S/k_B\}$. The equilibrium state has the maximum entropy and hence the maximum

---

[3] See, for example, Section 10.B of Reichl, *Modern Course in Statistical Physics* (University of Texas Press, 1980).

probability for given extensive variables. The second-order Taylor expansion of the entropy around its maximum implies a Gaussian distribution for the fluctuations of the extensive variables around their equilibrium values. The variances and covariances of the fluctuations can hence be calculated by a straightforward procedure from the second-order derivatives of the entropy, which are the well-known thermodynamic material properties. The second-order Taylor expansion

$$\Delta S = \frac{1}{2}\frac{\partial^2 S(U,N)}{\partial U^2}\Delta U^2 + \frac{\partial^2 S(U,N)}{\partial U\,\partial N}\Delta U\,\Delta N + \frac{1}{2}\frac{\partial^2 S(U,N)}{\partial N^2}\Delta N^2 \quad (20.7)$$

characterizes the joint fluctuations of U and N in the fixed volume, where Δ here indicates the deviation from the average. Note that first-order terms are absent because we expand around a maximum.

Fluctuations of the extensive quantities U and N in the subsystem volume are a natural concept. Fluctuations of the temperature T and the chemical potential $\tilde{\mu}$, on the other hand, are a less natural concept. We define the fluctuating temperature as the equilibrium temperature associated with the fluctuating extensive variables,

$$T + \Delta T = T(U + \Delta U, N + \Delta N). \quad (20.8)$$

Assuming small fluctuations, we can use first-order Taylor expansion to obtain

$$\Delta\left(\frac{1}{T}\right) = \frac{\partial(1/T)}{\partial U}\Delta U + \frac{\partial(1/T)}{\partial N}\Delta N = \frac{\partial^2 S(U,N)}{\partial U^2}\Delta U + \frac{\partial^2 S(U,N)}{\partial U\,\partial N}\Delta N,$$
$$(20.9)$$

where the expression (4.15) for $1/T$ has been used. For the chemical potential, we proceed in the same way to obtain

$$-\Delta\left(\frac{\tilde{\mu}}{T}\right) = \frac{\partial^2 S(U,N)}{\partial U\,\partial N}\Delta U + \frac{\partial^2 S(U,N)}{\partial N^2}\Delta N. \quad (20.10)$$

By inserting the last two results into (20.7), we find the alternative formulations

$$\Delta S = \frac{1}{2}\left(\Delta U\,\Delta\left(\frac{1}{T}\right) - \Delta N\,\Delta\left(\frac{\tilde{\mu}}{T}\right)\right)$$
$$= -\frac{1}{2}\left[\frac{\partial U(T,N)}{\partial T}\frac{\Delta T^2}{T^2} + \frac{1}{T}\frac{\partial^2 F(T,N)}{\partial N^2}\Delta N^2\right], \quad (20.11)$$

where a Maxwell relation has been used to realize that the mixed terms proportional to $\Delta T\,\Delta N$ cancel. From (4.34) or (4.35), we obtain the following second moments for fixed V,

$$\langle \Delta U^2 \rangle = k_{\mathrm{B}} T^2 C_V \,, \qquad \langle \Delta T^2 \rangle = \frac{k_{\mathrm{B}} T^2}{C_V} . \qquad (20.12)$$

Note that the mean-square temperature fluctuations are inversely proportional to the heat capacity and hence to the system size. More generally, the fluctuations of all intensive quantities, including the densities of extensive quantities, are inversely proportional to system size. In other words, for the amplitude of fluctuations, system size *does* matter. The last line of (20.11) moreover implies that, in a fixed volume, the temperature fluctuates independently from the mole number (and hence independently from the mole, number, or mass densities).

Exercise 20.3   Temperature Fluctuations
How large are the temperature fluctuations in 18 grams (1 mol) of water at 25 °C and standard pressure?

Exercise 20.4   Fluctuations and System Size
Rewrite the expressions for the fluctuations in (20.12) in terms of the internal energy density u, specific heat $\hat{c}_{\hat{v}}$, and volume V.

Exercise 20.5   Density Fluctuations
Derive the density fluctuations

$$\langle \Delta \rho^2 \rangle = \frac{\kappa_T \rho^2 k_{\mathrm{B}} T}{V} .$$

- The goal of statistical mechanics is to calculate thermodynamic potentials of macroscopic systems from the microscopic, or molecular, properties of these systems.
- Statistical ensembles assigning probabilities to microstates, such as the fundamental microcanonical ensemble for isolated systems and the more practical canonical and grand canonical ensembles for systems in contact with reservoirs, provide the proper setting for calculating various thermodynamic potentials.
- Statistical mechanics enriches thermodynamics by a theory of fluctuations; the strength of fluctuations is intimately related to thermodynamic material properties and decreases with system size.

21

Kinetic Theory of Gases

Hydrodynamics is a core theme of transport phenomena. In the phenomenological approach of Chapters 5 and 6, we developed the hydrodynamic equations (summarized in Section 7.1) from balance equations for conserved quantities combined with thermodynamic arguments leading to constitutive equations for the fluxes of the conserved quantities. In the present chapter, we wish to develop a full kinetic theory that leads to the hydrodynamic equations, including microscopic expressions for the transport coefficients, thus connecting the theme of hydrodynamics from the macroscopic and microscopic perspectives.

We first present an elementary kinetic theory approach so that the reader can develop a feeling for the mechanical model of a monatomic gas, the relevant length scales, and the phenomena that can be explained. All transport coefficients are obtained in terms of the mean free path. We then motivate Boltzmann's kinetic equation in an elementary way and show how it leads to molecular expressions for the pressure tensor and the heat flux vector, from which more rigorous expressions for the corresponding transport coefficients can be derived.

This chapter is a condensed and simplified version of Section 7.1 of the textbook *Beyond Equilibrium Thermodynamics*[1] (presented with permission from Wiley; copyright by John Wiley & Sons). The scope of the present chapter is restricted to the presentation of some basic ideas, equations, predictions, and conclusions. More details on some concepts and derivations, including a number of exercises, can be found in the above-mentioned textbook. Much more about the kinetic theory of gases can be found in a famous

---

[1] Öttinger, *Beyond Equilibrium Thermodynamics* (Wiley, 2005).

handbook article by Grad[2] and in a classic book by Hirschfelder, Curtiss, and Bird.[3]

21.1 Model of a Rarefied Gas

We begin with some elementary remarks that should help to supply or refresh an understanding for the properties of rarefied gases, thus providing a background and reference for some of the more sophisticated developments of the subsequent sections. The mean free path is recognized as a key concept, and it leads directly to the functional forms of the transport coefficients, that is, the self-diffusion coefficient, the viscosity, and the thermal conductivity of gases in terms of temperature and pressure.

As a simple model, we assume that a gas consists of $\tilde{N}$ small structureless spherical particles of diameter d and mass m, which obey the laws of classical mechanics. We make the following assumptions.

- The mean distance between particles is so large compared with their diameter that, to obtain equilibrium properties, they can be treated as point particles.

- There are no long-range interactions; the particles of the gas interact only during short-range collisions.

- The conservation laws for momentum and energy can be used for the elastic collisions between particles and for the reflections of particles at the walls of the container.

More precisely, it is possible to ignore one aspect of the interaction between particles, namely, nonideal gas behavior; the effect of collisions on the temporal evolution of the gas, however, remains finite and is described by Boltzmann's kinetic equation.[4] Such a situation can be realized in the following limit. We assume $d \to 0$ and $\tilde{N} \to \infty$ such that the effect of collisions between the particles of decreasing size remains important, which translates into $\tilde{N}d^2 = $ constant (see the estimate of the mean free path given below). In this limit, we have $\tilde{N}d^3 \to 0$, so that the volume fraction of the spherical particles in a container of fixed volume vanishes, and ideal gas behavior is thus approached. This formal limit, which is useful for justifying the validity of Boltzmann's kinetic equation in a rigorous manner, is known as the *Grad limit*.

[2] Grad, Principles of the Kinetic Theory of Gases (1958).
[3] Hirschfelder, Curtiss & Bird, *Molecular Theory of Gases and Liquids* (Wiley, 1954).
[4] See p. 214–216 of Grad, Principles of the Kinetic Theory of Gases (1958).

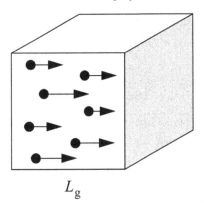

L_{g}

Figure 21.1 Gas particles with a positive velocity component toward the right wall of a cubic container (particles with a negative velocity component are not shown). [Reprinted with permission from Öttinger, *Beyond Equilibrium Thermodynamics* (Wiley, 2005). Copyright by John Wiley & Sons.]

To illustrate the ideal gas behavior of our hard-sphere gas, we calculate the pressure on the walls of a container (see Figure 21.1). If $\tilde{N}_j$ is the number of particles having a positive velocity component v_j toward the right wall, then the fraction $v_j \Delta t/L_{\mathrm{g}}$ of those particles will be reflected at the wall during the time interval Δt, where L_{g} denotes the size of the cubic container. The momentum transfer to the wall by a reflected particle with velocity v_j is $2mv_j$. The total momentum transferred to the wall per unit time, divided by the area of the wall, then gives the pressure

$$p = \frac{1}{L_{\mathrm{g}}^2} \sum_j 2mv_j \frac{v_j}{L_{\mathrm{g}}} \tilde{N}_j = \frac{2m}{L_{\mathrm{g}}^3} \sum_j \tilde{N}_j v_j^2. \qquad (21.1)$$

We note that $\sum_j \tilde{N}_j = \tilde{N}/2$ because only half of the particles have a positive velocity component toward the right wall and that, on average, the square of each velocity component is one-third of the square of the full velocity vector in three-dimensional space. Hence, we obtain

$$p = \frac{\tilde{N}}{V} m \frac{1}{3} \langle \boldsymbol{v}^2 \rangle = \frac{2}{3} \frac{\tilde{N}}{V} \frac{1}{2} m \langle \boldsymbol{v}^2 \rangle \qquad (21.2)$$

or the law of Boyle and Mariotte in the form

$$pV = \frac{2}{3} E. \qquad (21.3)$$

An alternative implication of (21.2) is

$$\langle v^2 \rangle = \frac{3p}{\rho}, \tag{21.4}$$

which gives a root-mean-square velocity of $1840\,\text{m/s}$ for hydrogen, $490\,\text{m/s}$ for nitrogen, and $460\,\text{m/s}$ for oxygen under normal conditions.

Exercise 21.1   Average Speed
Show that, for an isotropic Gaussian velocity distribution around zero, the average speed is given by

$$\langle |v| \rangle = \sqrt{\frac{8}{3\pi} \langle v^2 \rangle} = \sqrt{\frac{8k_\text{B}T}{\pi m}}. \tag{21.5}$$

21.2 Mean Free Path and Transport Coefficients

To calculate the pressure, we considered the particle–wall interactions. To understand transport phenomena in a gas, it is important to find the frequency of collisions between particles. Because we can determine the characteristic particle velocity from (21.4) and (21.5), we can equivalently determine the mean free path between collisions. We here focus on the functional form of the mean free path rather than prefactors.

Consider a spherical shell of radius r and thickness Δr around a particle. The number of particles in the shell is $4\pi r^2 \Delta r\, n$, where $n = \tilde{N}/V$ is the number density of particles. Because each particle blocks an area of πd^2 for the collision-free passage of a particle of equal size, the fraction of the solid angle covered by the particles in the shell, or the probability for a collision, is $\pi d^2 \Delta r\, n$. It is remarkable that this result does not depend on r. By accumulating the equal contributions from a number of shells with thickness Δr, the mean free path l_mfp is determined by the condition $\pi d^2 l_\text{mfp} n \approx 1$ because this implies a probability for a collision of order unity. A more rigorous calculation for moving target particles gives a smaller result by $1/\sqrt{2}$,

$$l_\text{mfp} = \frac{1}{\sqrt{2}\pi d^2 n} = \frac{l_\text{md}^3}{\sqrt{2}\pi d^2}, \tag{21.6}$$

where $l_\text{md} = n^{-1/3}$ is the characteristic distance between particles. The expression (21.6) for the mean free path was obtained by Maxwell and used by Boltzmann.[5]

---

[5] See p. 65 of Boltzmann, *Gastheorie, Teil I* (Barth, 1896).

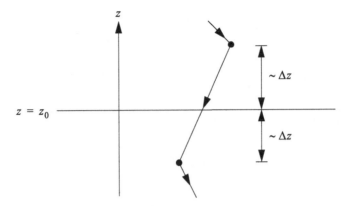

Figure 21.2 A particle has a collision a distance Δz above the plane and transports property a to another particle a distance Δz below the plane. [Reprinted with permission from Öttinger, *Beyond Equilibrium Thermodynamics* (Wiley, 2005). Copyright by John Wiley & Sons.]

For an ideal gas under normal conditions, one knows the number density and finds $l_{\mathrm{md}} = 30\,\text{Å}$. When we approximate the diameter of a diatomic gas molecule by four times the Bohr radius, $d \approx 2\,\text{Å}$, the mean free path turns out to be $0.15\,\mu\text{m}$. According to the first equality in (21.6), the Grad limit $\tilde{N}d^2 = \text{constant}$ in a fixed volume V corresponds to a constant mean free path at a negligible volume fraction of the particles.

It is possible to describe the transport of mass, momentum, and energy in gases in a unified manner, where we here follow the treatment of Reichl.[6] The same arguments can be found already in Boltzmann's classical lectures.[7] Let us assume that $a = a(z)$ is the molecular property to be transported and that it varies in the vertical direction, as indicated by the dependence on the height coordinate z (see Figure 21.2). Let us draw an imaginary plane in the gas at $z = z_0$. When a particle crosses the plane, it transports the value of a it obtained in its last collision and transfers that to another particle in its next collision.

Let $a(z + \Delta z)$ be the value of a transported across the plane $z = z_0$ from $z > z_0$. The distance Δz above the plane where the particle had its last collision is of the order of the mean free path l_{mfp}, and the same is true for the distance Δz below the plane at which the next collision takes place (see Figure 21.2). The average number of particles crossing the plane per unit area and per unit time is of the order of $n \langle |\boldsymbol{v}| \rangle$, where $\langle |\boldsymbol{v}| \rangle$ is the average

[6] See Section 13.B.4 of Reichl, *Modern Course in Statistical Physics* (University of Texas Press, 1980).

[7] See Section 11 of Boltzmann, *Gastheorie, Teil I* (Barth, 1896).

speed of the particles. For the flux of a transported in the positive z-direction per unit area and per unit time, we hence obtain

$$j_a(z) \approx n \langle |\boldsymbol{v}| \rangle [a(z_0 - \Delta z) - a(z_0 + \Delta z)] \approx -n \langle |\boldsymbol{v}| \rangle l_{\text{mfp}} \frac{da(z)}{dz}. \quad (21.7)$$

The proper prefactor of order unity in (21.7) has, for example, been determined by a more detailed geometrical consideration for the diffusion coefficient;[8] the final result is

$$j_a(z) = -\frac{1}{3} n \langle |\boldsymbol{v}| \rangle l_{\text{mfp}} \frac{da(z)}{dz} \quad (21.8)$$

for the flux of any molecular property a. The dimensions of the flux j_a are those of a per unit area and time.

Let us first consider the mole fraction of tracer particles, $a(z) = x_1(z)$. We then obtain for the number of tracer particles crossing the plane per unit area and per unit time

$$j_1 = -\frac{1}{3} n \langle |\boldsymbol{v}| \rangle l_{\text{mfp}} \frac{dx_1}{dz}. \quad (21.9)$$

A comparison with (6.27) for the mass flux caused by diffusion results in the following expression for the diffusion coefficient,[9]

$$D_{12} = \frac{1}{3} \langle |\boldsymbol{v}| \rangle l_{\text{mfp}} = \frac{\langle |\boldsymbol{v}| \rangle}{3\sqrt{2}\pi d^2 n} = \frac{2}{3d^2 \sqrt{m}} \frac{1}{p} \left(\frac{k_{\mathrm{B}} T}{\pi} \right)^{3/2}. \quad (21.10)$$

Quite intuitively, the diffusion coefficient increases with temperature, maybe even more strongly than anticipated, and decreases under pressure.

If the gas in Figure 21.1 is sheared between horizontally moving plates, we choose $a(z) = m \langle v_{\text{horizontal}}(z) \rangle$ as the average momentum per particle. We then obtain the momentum flux

$$j_\tau = -\frac{1}{3} nm \langle |\boldsymbol{v}| \rangle l_{\text{mfp}} \frac{d \langle v_{\text{horizontal}}(z) \rangle}{dz}. \quad (21.11)$$

A comparison with (6.5) results in the following expression for the viscosity,

$$\eta = \frac{1}{3} nm \langle |\boldsymbol{v}| \rangle l_{\text{mfp}} = \frac{m \langle |\boldsymbol{v}| \rangle}{3\sqrt{2}\pi d^2} = \frac{2\sqrt{m}}{3\pi d^2} \left(\frac{k_{\mathrm{B}} T}{\pi} \right)^{1/2}. \quad (21.12)$$

The viscosity of the gas turns out to be independent of density or pressure and increases with temperature. In 1859, when James Clerk Maxwell found this counterintuitive behavior of viscosity by using dimensional analysis and

---

[8] See p. 463 of Reichl, *Modern Course in Statistical Physics* (University of Texas Press, 1980).

[9] With the idea of tracer particles, we have formally introduced a two-component system so that we can make contact with the binary diffusion coefficient introduced in Section 6.3. If the tracer particles are of the same type as the other particles, one speaks of self-diffusion.

simple arguments that revealed the essential physics, he was so surprised
that, together with his wife Katherine, he undertook detailed experiments
in their home laboratory to confirm this "remarkable result."[10] The increase
of the viscosity of gases with increasing temperature has been verified experimentally. One should also note that larger particles have a shorter mean
free path and hence a smaller viscosity.

If the temperature of the gas varies in the vertical direction, then we
choose $a(z) = m \langle v(z)^2 \rangle /2 = (3/2)k_\mathrm{B}T(z)$, and

$$j_q = -\frac{1}{2}n \langle |v| \rangle l_\mathrm{mfp}k_\mathrm{B}\frac{dT(z)}{dz} \tag{21.13}$$

is the heat flux. A comparison with (6.4) results in the following expression
for the thermal conductivity,

$$\lambda = \frac{1}{2}n \langle |v| \rangle l_\mathrm{mfp}k_\mathrm{B}. \tag{21.14}$$

The molecular expressions (21.10), (21.12), and (21.14) for the transport
coefficients have been derived by very simplistic arguments, but they describe fairly well the observed qualitative behavior of transport coefficients
in rarefied gases. If we form ratios of the transport coefficients for gases,
the mean free path drops out and the ratios are given by static averages
available in equilibrium thermodynamics. For example, we obtain

$$\frac{\eta}{D_{12}} = nm = \rho \tag{21.15}$$

and

$$\frac{\lambda}{\eta} = \frac{3}{2}\frac{k_\mathrm{B}}{m}, \tag{21.16}$$

which is the heat capacity per unit mass at constant volume for an ideal gas
(see Table A.2 in Appendix A). The thermal conductivity of gases thus has
the same temperature dependence as the viscosity, and it is also independent
of the pressure or density. However, the prefactors in the expressions (21.10),
(21.12), and (21.14) should not be taken too seriously; additional factors of
order unity should rather be expected in the ratios (21.15) and (21.16).
Experimental results for these ratios and the predictions of a more rigorous
kinetic theory are compiled in Table 21.1, where it should be noted that the
heat capacity per unit mass at constant pressure is $\hat{c}_p = \frac{5}{2}k_\mathrm{B}/m$ for an ideal
gas.[11] Note that the ratios in the last two columns of Table 21.1 are the
Prandtl and Schmidt numbers previously introduced in Table 7.1.

[10] See p. xv and p. 351 of Maxwell, *Theory of Heat* (Dover, 2001).
[11] The data are taken from Table 1.2-3 of Hirschfelder, Curtiss & Bird, *Molecular Theory of Gases
and Liquids* (Wiley, 1954).

Table 21.1 *Prandtl and Schmidt numbers for gases at* $0\,^\circ$C

Gas	$N_{\mathrm{Pr}} = \dfrac{\eta \hat{c}_p}{\lambda}$	$N_{\mathrm{Sc}} = \dfrac{\eta}{\rho D_{12}}$
Ne	0.66	0.73
Ar	0.67	0.75
H_2	0.71	0.73
N_2	0.71	0.74
O_2	0.72	0.74
CO_2	0.75	0.71
CH_4	0.74	0.70
(21.15), (21.16)	5/3	1
Kinetic theory	2/3	5/6

Exercise 21.2   Collision Frequency
Estimate the frequency of collisions between gas particles for hydrogen under normal conditions.

Exercise 21.3   Viscosity of Nitrogen Gas
Estimate the viscosity of nitrogen gas at room temperature.

21.3 Boltzmann's Equation

The elementary kinetic theory considerations of the preceding subsection gave us the correct functional form of the transport coefficients for the dependence on temperature and pressure as well as on size and mass of the molecules. The prefactors, however, are not correct. To obtain quantitatively correct predictions, one needs a more rigorous kinetic theory. For rarefied gases, such a theory can be built on Boltzmann's equation. One can get not only improved results for the hard-sphere gas but also predictions resulting from a more detailed description of the interactions between molecules.

An elementary derivation of Boltzmann's kinetic equation can be based on the evolution equation for the single-particle distribution function

$$\frac{\partial f(\boldsymbol{r}, \boldsymbol{p})}{\partial t} = -\left(\frac{\boldsymbol{p}}{m} \cdot \frac{\partial}{\partial \boldsymbol{r}} - \frac{\partial \phi^{(\mathrm{e})}(\boldsymbol{r})}{\partial \boldsymbol{r}} \cdot \frac{\partial}{\partial \boldsymbol{p}} \right) f(\boldsymbol{r}, \boldsymbol{p}) + \left[\frac{\partial f(\boldsymbol{r}, \boldsymbol{p})}{\partial t} \right]_{\mathrm{coll}}. \quad (21.17)$$

The function $f(\boldsymbol{r}, \boldsymbol{p})$ is the probability density for finding a particle with momentum $\boldsymbol{p}$ at the position $\boldsymbol{r}$. The first term on the right-hand side of this equation corresponds to a deterministic change of position $\boldsymbol{r}$ with the velocity $\boldsymbol{p}/m$ and a deterministic change of momentum $\boldsymbol{p}$ due to an external force $-\partial\phi^{(e)}(\boldsymbol{r})/\partial\boldsymbol{r}$ for a single particle of mass m. The second term describes the change of the single-particle distribution function due to the occurrence of collisions with the other particles in the system.

The collision term in (21.17) consists of a gain and a loss contribution. If a particle with the given momentum $\boldsymbol{p}$ collides with another particle with any momentum $\boldsymbol{p}'$, this leads to a decrease of $f(\boldsymbol{r},\boldsymbol{p})$. If two particles with momenta $\boldsymbol{q}$ and $\boldsymbol{q}'$ collide such that one of the particles acquires the momentum $\boldsymbol{p}$, this leads to an increase of $f(\boldsymbol{r},\boldsymbol{p})$. For a precise formulation, we introduce the transition rate $w(\boldsymbol{q},\boldsymbol{q}'|\boldsymbol{p},\boldsymbol{p}')$ for a pair of particles with initial momenta $\boldsymbol{p}$, $\boldsymbol{p}'$ to the final momenta $\boldsymbol{q}$, $\boldsymbol{q}'$. For an elastic collision, $w(\boldsymbol{q},\boldsymbol{q}'|\boldsymbol{p},\boldsymbol{p}')$ vanishes unless momentum and energy are conserved,

$$\boldsymbol{p} + \boldsymbol{p}' = \boldsymbol{q} + \boldsymbol{q}', \tag{21.18}$$

$$\boldsymbol{p}^2 + \boldsymbol{p}'^2 = \boldsymbol{q}^2 + \boldsymbol{q}'^2. \tag{21.19}$$

The two free parameters in the choice of $\boldsymbol{q}$ and $\boldsymbol{q}'$ [the six vector components minus the four conditions (21.18), (21.19)] correspond to the solid angle into which a colliding particle is scattered. Furthermore, the transition rate $w(\boldsymbol{q},\boldsymbol{q}'|\boldsymbol{p},\boldsymbol{p}')$ has the following symmetry properties corresponding to time reversal and particle exchange in the two-particle collisions,

$$w(\boldsymbol{q},\boldsymbol{q}'|\boldsymbol{p},\boldsymbol{p}') = w(\boldsymbol{p},\boldsymbol{p}'|\boldsymbol{q},\boldsymbol{q}') \tag{21.20}$$

and

$$w(\boldsymbol{q},\boldsymbol{q}'|\boldsymbol{p},\boldsymbol{p}') = w(\boldsymbol{q}',\boldsymbol{q}|\boldsymbol{p}',\boldsymbol{p}). \tag{21.21}$$

We can now write

$$\left[\frac{\partial f(\boldsymbol{r},\boldsymbol{p})}{\partial t}\right]_{\text{coll}} = \int w(\boldsymbol{q},\boldsymbol{q}'|\boldsymbol{p},\boldsymbol{p}')[f(\boldsymbol{r},\boldsymbol{q})f(\boldsymbol{r},\boldsymbol{q}') - f(\boldsymbol{r},\boldsymbol{p})f(\boldsymbol{r},\boldsymbol{p}')]d^3p'\,d^3q\,d^3q',$$
$$\tag{21.22}$$

and (21.17) with (21.22) is Boltzmann's kinetic equation. The assumption that the probability for a two-particle collision is proportional to the product $f(\boldsymbol{r},\boldsymbol{p})f(\boldsymbol{r},\boldsymbol{p}')$ is Boltzmann's celebrated *Stoßzahlansatz*. It has been assumed that the collisions are local so that the same argument $\boldsymbol{r}$ appears in all single-particle distribution functions occurring in Boltzmann's equation. If $f(\boldsymbol{r},\boldsymbol{p})$ is normalized to a number density,

$$\int f(\boldsymbol{r},\boldsymbol{p})d^3r\,d^3p = \tilde{N}, \tag{21.23}$$

then the transition rates for two-particle collisions must be independent of the number of particles in the system; they depend only on the impact conditions and on the interaction potential.

21.4 *H* **Theorem and Local Maxwellian Distribution**

In this section, we consider two important consequences of Boltzmann's equation (21.17) and (21.22): the *H* theorem and the local Maxwellian distribution in velocity space. We first consider the time evolution of the functional

$$H_{\rm B} = \int f(\boldsymbol{r}, \boldsymbol{p}) \ln f(\boldsymbol{r}, \boldsymbol{p}) d^3 r \, d^3 p. \qquad (21.24)$$

By using Boltzmann's equation, we obtain

$$\frac{dH_{\rm B}}{dt} = \int [\ln f(\boldsymbol{r}, \boldsymbol{p}) + 1] \frac{\partial f(\boldsymbol{r}, \boldsymbol{p})}{\partial t} d^3 r \, d^3 p, \qquad (21.25)$$

where the constant term in the square bracket does not contribute to the integral because of the normalization (21.23) of f for all times. In principle, we should have used $\ln[f(\boldsymbol{r}, \boldsymbol{p})/f_0]$ instead of $\ln f(\boldsymbol{r}, \boldsymbol{p})$ in (21.24), where the constant f_0 is required to make the argument of the logarithm dimensionless. However, the normalization condition implies that this would change $H_{\rm B}$ only by an irrelevant additive constant. The non-collision term in (21.17) leads to total derivatives under the integral and hence to surface terms, which can be neglected for an isolated system, so that we arrive at

$$\frac{dH_{\rm B}}{dt} = \int \left[\frac{\partial f(\boldsymbol{r}, \boldsymbol{p})}{\partial t} \right]_{\rm coll} \ln f(\boldsymbol{r}, \boldsymbol{p}) d^3 r \, d^3 p. \qquad (21.26)$$

By using (21.22) and the symmetry properties (21.20), (21.21), we obtain

$$\frac{dH_{\rm B}}{dt} = -\frac{1}{4} \int w(\boldsymbol{q}, \boldsymbol{q}' | \boldsymbol{p}, \boldsymbol{p}')[f(\boldsymbol{r}, \boldsymbol{q})f(\boldsymbol{r}, \boldsymbol{q}') - f(\boldsymbol{r}, \boldsymbol{p})f(\boldsymbol{r}, \boldsymbol{p}')]$$
$$\times \{\ln[f(\boldsymbol{r}, \boldsymbol{q})f(\boldsymbol{r}, \boldsymbol{q}')] - \ln[f(\boldsymbol{r}, \boldsymbol{p})f(\boldsymbol{r}, \boldsymbol{p}')]\} d^3 r \, d^3 p \, d^3 p' \, d^3 q \, d^3 q', \qquad (21.27)$$

and hence

$$\frac{dH_{\rm B}}{dt} \leq 0. \qquad (21.28)$$

The final inequality results from the fact that $w(\boldsymbol{q}, \boldsymbol{q}' | \boldsymbol{p}, \boldsymbol{p}')$ is nonnegative, and, in view of the monotonic behavior of the logarithm, the two remaining factors involving differences must have the same sign.

We have thus found a functional $H_{\rm B}$ that can only decrease or stay constant in the course of the time evolution described by Boltzmann's equation. Equation (21.28) is *Boltzmann's H theorem*, which suggests that, as an increasing function with proper dimensions, $S = -k_{\rm B} H_{\rm B}$ plays the role of an entropy. Furthermore, Boltzmann's equation clearly describes an irreversible time evolution.

A steady state in the evolution of the single-particle distribution function can be reached when $dH_{\mathrm{B}}/dt = 0$; at this condition, collisions establish an equilibration in momentum space at each position r. In such a situation, we obtain

$$f(\boldsymbol{r},\boldsymbol{p})f(\boldsymbol{r},\boldsymbol{p}') = f(\boldsymbol{r},\boldsymbol{q})f(\boldsymbol{r},\boldsymbol{q}'), \tag{21.29}$$

where the momenta $\boldsymbol{p}$ and $\boldsymbol{p}'$ are arbitrary and the choice of $\boldsymbol{q}$ and $\boldsymbol{q}'$ is restricted by the conservation laws (21.18), (21.19). Equation (21.29) can be rewritten as an additive conservation law,

$$\ln f(\boldsymbol{r},\boldsymbol{p}) + \ln f(\boldsymbol{r},\boldsymbol{p}') = \ln f(\boldsymbol{r},\boldsymbol{q}) + \ln f(\boldsymbol{r},\boldsymbol{q}'). \tag{21.30}$$

Because momentum and energy are the only conserved quantities restricting the choice of $\boldsymbol{q}$ and $\boldsymbol{q}'$, we conclude

$$\ln f(\boldsymbol{r},\boldsymbol{p}) = \lambda_0(\boldsymbol{r}) + \boldsymbol{\lambda}_1(\boldsymbol{r}) \cdot \boldsymbol{p} + \lambda_2(\boldsymbol{r})\boldsymbol{p}^2, \tag{21.31}$$

which implies a Gaussian distribution in momentum space. The interpretation of the factors λ_0, $\boldsymbol{\lambda}_1$, λ_2 becomes more obvious if we rewrite (21.31) as

$$f_{\rho,v,T}^{\mathrm{Maxw}}(\boldsymbol{r},\boldsymbol{p}) = \frac{\rho(\boldsymbol{r})}{m\sqrt{2\pi m k_{\mathrm{B}} T(\boldsymbol{r})}^3} \exp\left\{-\frac{[\boldsymbol{p} - m\boldsymbol{v}(\boldsymbol{r})]^2}{2m k_{\mathrm{B}} T(\boldsymbol{r})}\right\}, \tag{21.32}$$

which is known as the *Maxwellian momentum distribution*. With our knowledge about Gaussian distributions from Sections 2.3 and 2.6, we get the interpretation of $\rho(\boldsymbol{r})$, $\boldsymbol{v}(\boldsymbol{r})$, and $T(\boldsymbol{r})$ as the mass density, velocity, and temperature. The H theorem and this Maxwellian distribution are immediate but important consequences of Boltzmann's equation.

21.5 Boltzmann's Equation and Hydrodynamics

The general role of Boltzmann's kinetic equation has been described very clearly by Grad:[12]

The principal use of the Boltzmann equation is when there are significant changes in gas flow properties over a mean-free-path or over a collision time. There is no reason to doubt its validity in shock waves of arbitrary strength or in sound waves of arbitrary frequency; it is, indeed, the only way of studying such problems. From this viewpoint we must consider the use of the Boltzmann equation to compute the values of transport coefficients as a secondary application.

---

[12] See p. 214 of Grad, Principles of the Kinetic Theory of Gases (1958).

These remarks indicate that Boltzmann's equation contains much more than standard hydrodynamics. Historically, major efforts in generalizing hydrodynamics were made to understand the aerodynamics of satellites and space stations in the outer limits of our atmosphere. The same generalized equations are nowadays used also in microfluidics.[13]

Boltzmann's kinetic equation is also the starting point for obtaining kinetic theory expressions for the transport coefficients of rarefied gases. The hydrodynamic fields are given in terms of $f(\boldsymbol{r}, \boldsymbol{p})$ as

$$\int m f(\boldsymbol{r}, \boldsymbol{p}) d^3 p = \rho(\boldsymbol{r}), \tag{21.33}$$

$$\int \boldsymbol{p} f(\boldsymbol{r}, \boldsymbol{p}) d^3 p = \boldsymbol{m}(\boldsymbol{r}) = \rho(\boldsymbol{r}) \boldsymbol{v}(\boldsymbol{r}), \tag{21.34}$$

and

$$\int \frac{\boldsymbol{p}^2}{2m} f(\boldsymbol{r}, \boldsymbol{p}) d^3 p = \frac{1}{2} \rho(\boldsymbol{r}) \boldsymbol{v}(\boldsymbol{r})^2 + \frac{3}{2} \frac{\rho(\boldsymbol{r})}{m} k_{\mathrm{B}} T(\boldsymbol{r}). \tag{21.35}$$

The evolution of these moments in the absence of an external force can be obtained from Boltzmann's equation. We find the following expressions for the time derivatives of the hydrodynamic fields,

$$\frac{\partial \rho(\boldsymbol{r})}{\partial t} = -\frac{\partial}{\partial \boldsymbol{r}} \cdot [\boldsymbol{v}(\boldsymbol{r}) \rho(\boldsymbol{r})], \tag{21.36}$$

$$\frac{\partial}{\partial t} [\boldsymbol{v}(\boldsymbol{r}) \rho(\boldsymbol{r})] = -\frac{\partial}{\partial \boldsymbol{r}} \cdot [\boldsymbol{v}(\boldsymbol{r}) \boldsymbol{v}(\boldsymbol{r}) \rho(\boldsymbol{r})]$$
$$- \frac{\partial}{\partial \boldsymbol{r}} \cdot \int \frac{[\boldsymbol{p} - m\boldsymbol{v}(\boldsymbol{r})][\boldsymbol{p} - m\boldsymbol{v}(\boldsymbol{r})]}{m} f(\boldsymbol{r}, \boldsymbol{p}) d^3 p, \tag{21.37}$$

and

$$\frac{\partial T(\boldsymbol{r})}{\partial t} = -\boldsymbol{v}(\boldsymbol{r}) \cdot \frac{\partial T(\boldsymbol{r})}{\partial \boldsymbol{r}}$$
$$- \frac{1}{\rho(\boldsymbol{r}) \hat{c}_{\hat{v}}} \int \frac{[\boldsymbol{p} - m\boldsymbol{v}(\boldsymbol{r})][\boldsymbol{p} - m\boldsymbol{v}(\boldsymbol{r})]}{m} f(\boldsymbol{r}, \boldsymbol{p}) d^3 p : \frac{\partial}{\partial \boldsymbol{r}} \boldsymbol{v}(\boldsymbol{r})$$
$$- \frac{1}{\rho(\boldsymbol{r}) \hat{c}_{\hat{v}}} \frac{\partial}{\partial \boldsymbol{r}} \cdot \int \left[\frac{\boldsymbol{p}}{m} - \boldsymbol{v}(\boldsymbol{r}) \right] \frac{[\boldsymbol{p} - m\boldsymbol{v}(\boldsymbol{r})]^2}{2m} f(\boldsymbol{r}, \boldsymbol{p}) d^3 p, \tag{21.38}$$

where $\hat{c}_{\hat{v}} = 3k_{\mathrm{B}}/(2m)$ and where the collision term does not contribute directly to these time derivatives in view of the local conservation laws for mass, momentum, and kinetic energy. By comparing these equations for

---

[13] Karniadakis *et al.*, *Microflows and Nanoflows* (Springer, 2005); Struchtrup, *Transport Equations for Rarefied Gas Flows* (Springer, 2005).

a rarefied gas with the general evolution equations in (5.5), (5.32), and in the first line of (6.7), we actually recognize molecular expressions for the momentum and internal energy fluxes in terms of the single-particle distribution function,

$$\pi = \int \frac{[\boldsymbol{p} - m\boldsymbol{v}(\boldsymbol{r})][\boldsymbol{p} - m\boldsymbol{v}(\boldsymbol{r})]}{m} f(\boldsymbol{r}, \boldsymbol{p}) d^3 p \qquad (21.39)$$

and

$$\boldsymbol{j}_q = \int \left[\frac{\boldsymbol{p}}{m} - \boldsymbol{v}(\boldsymbol{r})\right] \frac{[\boldsymbol{p} - m\boldsymbol{v}(\boldsymbol{r})]^2}{2m} f(\boldsymbol{r}, \boldsymbol{p}) d^3 p. \qquad (21.40)$$

In these equations, we can nicely recognize the non-convective flux of momentum and kinetic internal energy.

The expressions (21.39) and (21.40) for the fluxes are useful only if we have a solution $f(\boldsymbol{r}, \boldsymbol{p})$ of Boltzmann's equation. Equation (21.17) is a nonlinear integro-differential equation for the single-particle distribution function, and it is generally impossible to find analytical solutions. It is hence natural to look for approximate solutions, in particular, to employ expansion techniques. A very thorough discussion of such expansions was given by Grad.[14] By means of the so-called first-order Chapman–Enskog expansion, one realizes that the fluxes in (21.39) and (21.40) are proportional to the gradients of velocity and temperature, respectively, as previously realized in Chapter 6. Moreover, the viscosity and the thermal conductivity can be expressed in terms of the transition rate $w(\boldsymbol{q}, \boldsymbol{q}'|\boldsymbol{p}, \boldsymbol{p}')$ introduced in (21.22). The entries in the last line of Table 21.1 are obtained for a particular simple interaction between the gas particles leading to a convenient transition rate.

Exercise 21.4   *Conserved Moments in Collisions*
Show that

$$\int \left[\frac{\partial f(\boldsymbol{r}, \boldsymbol{p})}{\partial t}\right]_{\text{coll}} d^3 p = 0, \quad \int \boldsymbol{p} \left[\frac{\partial f(\boldsymbol{r}, \boldsymbol{p})}{\partial t}\right]_{\text{coll}} d^3 p = \boldsymbol{0},$$

$$\int \boldsymbol{p}^2 \left[\frac{\partial f(\boldsymbol{r}, \boldsymbol{p})}{\partial t}\right]_{\text{coll}} d^3 p = 0. \qquad (21.41)$$

Exercise 21.5   *Molecular Expressions for Fluxes*
Derive the expressions (21.36)–(21.38) for the time derivatives of the hydrodynamic fields.

---

[14] Grad, *Phys. Fluids* **6** (1963) 147.

- Mechanical modeling of a gas as a collection of hard spherical particles allows us to calculate the ideal gas pressure and the mean free path between particle collisions.
- Within numerical prefactors of order unity, the transport coefficients can be expressed in terms of the mean free path, thus obtaining their temperature and pressure dependences; for example, within this elementary kinetic theory, the viscosity of a gas is found to be independent of pressure and to increase with temperature.
- Boltzmann's kinetic equation is the basis for a more detailed kinetic theory of gases; it allows us to identify an entropy function implying irreversibility (H theorem) and to derive the Maxwellian momentum distribution.
- Boltzmann's kinetic equation can be used to derive molecular expressions for the fluxes of momentum and energy; when these are evaluated within the first-order Chapman–Enskog expansion, one obtains molecular expressions for the viscosity and the thermal conductivity.

22

Kinetic Theory of Polymeric Liquids

In Chapter 12, we recognized the need to go beyond balance equations for describing the flow behavior of complex fluids. We introduced a relaxation equation for the pressure tensor, known as the (upper) convected Maxwell model. Can we strengthen our phenomenological arguments in favor of that model by kinetic theory? Can we obtain microscopic expressions for the parameters of that model? Can we use molecular insight to develop more realistic models? These questions are addressed in the present chapter in the context of polymeric liquids, which certainly are an important class of complex fluids.

Here, we consider both dilute solutions and melts of linear polymer molecules. In both situations, we end up with the upper convected Maxwell model, but with different expressions for the model parameters. This observation gives us back a bit of the universality of the equations governing transport that we lost by going beyond balance equations.

Polymer kinetic theory employs diffusion equations in polymer conformation space. We study a transport problem in an abstract conformation space to find an expression for the pressure tensor, which then governs momentum transport in real space. Transport of probability in an abstract space is used to understand transport of momentum in real space. We thus develop the transport motif to a higher level, where our initial consideration of diffusion equations in abstract spaces in Chapter 2 nicely supports the implementation of the idea. We are not interested in all details of the probability density in the high-dimensional polymer conformation space which, for example, consists of all the monomer position vectors. We are interested only in the pressure tensor, which is obtained as a particular average performed with that probability density. Brownian dynamics is the ideal tool for solving such problems; we have already developed that tool in Chapter 3. Moreover,

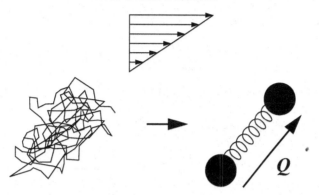

Figure 22.1 Under flow conditions, such as the indicated shear flow, polymer molecules become stretched and oriented. To develop a simple kinetic theory, we model the polymer conformations by a dumbbell.

the theory of stochastic differential equations developed in Section 3.2 provides a convenient tool for formulating the kinetic theory of polymers in dilute solutions. A number of transport themes are coming together in the present chapter.

22.1 Dilute Solutions

In the Hookean dumbbell model,[1] polymer molecules in dilute solution are modeled by two spherical beads connected by a Hookean spring (see Figure 22.1). The solvent is modeled as an isothermal, incompressible Newtonian fluid which is completely characterized by its temperature T and its viscosity η_s. As we consider dilute polymer solutions, no interactions between different dumbbells need to be considered. The beads represent large chain segments so that the continuum description of the solvent is justified. The springs represent entropic effects due to the elimination of most of the molecular degrees of freedom of a polymer chain. A linear spring force reflects Gaussian statistics of the polymer conformations. This is so because the logarithm of a Gaussian probability density leads to a quadratic entropic potential and hence to a linear force law.

Our first goal is to formulate equations of motion for the interacting beads of a dumbbell, which we treat as Brownian particles. For a one-dimensional Brownian particle moving in a dimensionless potential $\Phi(x) = Hx^2/(2k_BT)$, where H is a Hookean spring constant, k_B is Boltzmann's constant, and T is temperature, the diffusion equation (2.24) with constant diffusion coefficient

---

[1] Hermans, *Physica* **10** (1943) 777.

D implies the equivalent stochastic differential equation (see Section 3.1)

$$dX_t = -\frac{D}{2k_{\mathrm{B}}T} H X_t \, dt + \sqrt{D} \, dW_t \,, \tag{22.1}$$

where the increments dW_t are independent Gaussian random variables with zero mean and variance dt. For a more obvious interpretation of (22.1), we introduce the friction coefficient $\zeta = 2k_{\mathrm{B}}T/D$ and obtain[2]

$$dX_t = -\frac{1}{\zeta} H X_t \, dt + \sqrt{\frac{2k_{\mathrm{B}}T}{\zeta}} \, dW_t \,. \tag{22.2}$$

After multiplying by ζ, (22.2) can be recognized as a force balance. The sum of the Hookean spring force and the random force on the right-hand side compensates for the frictional force proportional to the velocity of the Brownian particle. Neglect of the inertial force (mass times acceleration) is consistent with the irregularity of Brownian trajectories, which would be smoothed by inertial forces (making the trajectories differentiable).

We are now ready to write down the stochastic differential equations describing the motion of the two beads of a dumbbell with positions $\boldsymbol{R}_1, \boldsymbol{R}_2$ in three dimensions,

$$d\boldsymbol{R}_1 = \boldsymbol{v}(\boldsymbol{R}_1) \, dt - \frac{1}{\zeta} H(\boldsymbol{R}_1 - \boldsymbol{R}_2) \, dt + \sqrt{\frac{2k_{\mathrm{B}}T}{\zeta}} \, d\boldsymbol{W}_1 \tag{22.3}$$

and

$$d\boldsymbol{R}_2 = \boldsymbol{v}(\boldsymbol{R}_2) \, dt - \frac{1}{\zeta} H(\boldsymbol{R}_2 - \boldsymbol{R}_1) \, dt + \sqrt{\frac{2k_{\mathrm{B}}T}{\zeta}} \, d\boldsymbol{W}_2 \,, \tag{22.4}$$

where we have assumed that the frictional force arises from the deviation of the bead velocity from a given solvent velocity field $\boldsymbol{v}(\boldsymbol{R})$ and suppressed the time argument. If we introduce the connector vector $\boldsymbol{Q} = \boldsymbol{R}_2 - \boldsymbol{R}_1$, we arrive at the single stochastic differential equation

$$d\boldsymbol{Q} = \boldsymbol{\kappa} \cdot \boldsymbol{Q} \, dt - \frac{2}{\zeta} H\boldsymbol{Q} \, dt + \sqrt{\frac{4k_{\mathrm{B}}T}{\zeta}} \, d\boldsymbol{W} \,. \tag{22.5}$$

To arrive at (22.5), we have assumed that the velocity variations over the dimensions of a polymer molecule are so small that we can safely use the first-order Taylor expansion $\boldsymbol{v}(\boldsymbol{R}_2) - \boldsymbol{v}(\boldsymbol{R}_1) = \boldsymbol{\kappa} \cdot (\boldsymbol{R}_2 - \boldsymbol{R}_1)$, where the velocity gradient $\boldsymbol{\kappa} = (\boldsymbol{\nabla}\boldsymbol{v})^{\mathrm{T}}$ is independent of position but may depend on time

---

[2] The more usual form $\zeta = k_{\mathrm{B}}T/D$ of the Nernst–Einstein equation (6.31) would result if we had not introduced a factor of $1/2$ in the second-order derivative term of the diffusion equation (3.5).

(time-dependent homogeneous flow). Moreover, we have replaced $d\mathbf{W}_2 - d\mathbf{W}_1$ by $\sqrt{2}\,d\mathbf{W}$ because the sum or difference of two independent Gaussian random variables with the same variance is another Gaussian random variable with doubled variance.

Equation (22.5) can be interpreted very easily. The flow tends to stretch and orient the dumbbells according to the first term on the right-hand side. The second term describes the retraction of the dumbbells under the influence of the Hookean force. The dumbbells do not shrink to zero size because the beads are kicked around by Brownian random forces according to the third term, which also randomizes the dumbbell orientation.

In polymer kinetic theory, it is more common to model in terms of diffusion equations in polymer conformation space rather than in terms of stochastic differential equations.[3] We have established (see Chapter 3) the general equivalence between the stochastic differential equation (3.7) and the diffusion equation (2.34), provided that the relationship (3.6) holds. Hence, we write the diffusion equation for the probability density $p = p(t, \mathbf{Q})$ in the dumbbell configuration space that is equivalent to (22.5) as

$$\frac{\partial p}{\partial t} = -\frac{\partial}{\partial \mathbf{Q}} \cdot \left[\left(\boldsymbol{\kappa} \cdot \mathbf{Q} - \frac{2H}{\zeta} \mathbf{Q} \right) p \right] + \frac{2k_{\mathrm{B}}T}{\zeta} \frac{\partial}{\partial \mathbf{Q}} \cdot \frac{\partial}{\partial \mathbf{Q}} p. \qquad (22.6)$$

According to the straightforward multivariate generalization of Section 2.3, the diffusion equation (22.6), as a linear equation in $\mathbf{Q}$, can be solved in terms of a time-dependent Gaussian solution, provided that the initial condition is Gaussian. As the two beads of a dumbbell are on an equal footing, that is, the direction of $\mathbf{Q}$ does not matter, the average of $\mathbf{Q}$ must vanish. All the information is hence contained in the second-moment tensor $\boldsymbol{\Theta} = \langle \mathbf{QQ} \rangle_t$, which satisfies the evolution equation (see Exercise 22.1)

$$\frac{\partial \boldsymbol{\Theta}}{\partial t} = \boldsymbol{\kappa} \cdot \boldsymbol{\Theta} + \boldsymbol{\Theta} \cdot \boldsymbol{\kappa}^{\mathrm{T}} - \frac{4H}{\zeta} \boldsymbol{\Theta} + \frac{4k_{\mathrm{B}}T}{\zeta} \boldsymbol{\delta}. \qquad (22.7)$$

The steady solution in the absence of flow is

$$\boldsymbol{\Theta}_{\mathrm{eq}} = \frac{k_{\mathrm{B}}T}{H} \boldsymbol{\delta}, \qquad (22.8)$$

which means the second-moment tensor is isotropic at equilibrium.

Once we have found the equation of motion and its solution, the next step is to obtain an expression for the pressure tensor. In principle, nonequilibrium thermodynamics imposes a relationship between reversible dynamics

---

[3] A standard reference for polymer kinetic theory is Bird, Curtiss, Armstrong & Hassager, *Dynamics of Polymeric Liquids. Volume 2. Kinetic Theory* (Wiley, 1987); for a systematic reformulation of that standard approach in terms of stochastic differential equations, see Öttinger, *Stochastic Processes in Polymeric Fluids* (Springer, 1996).

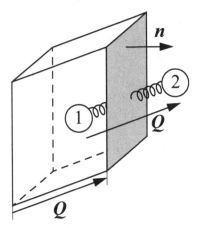

Figure 22.2 The volume that must be occupied by bead 1 when the connector vector $\boldsymbol{Q}$ intersects the shaded rectangle with unit normal vector $\boldsymbol{n}$.

and the pressure tensor. We here prefer to rely on an intuitive argument based on Figure 22.2. If the area of the shaded rectangle is A and $\boldsymbol{n} \cdot \boldsymbol{Q} > 0$, the total number of dumbbells with connector vectors around $\boldsymbol{Q}$ and contributing to the exchange of momentum through the shaded rectangle is $n_{\mathrm{p}} \boldsymbol{n} \cdot \boldsymbol{Q} \, A \, p \, d^3 Q$, where n_{p} is the number density of dumbbells, that is, the number density of polymers. The force exerted from the negative side of the shaded rectangle on the positive side is $-H\boldsymbol{Q}$. For $\boldsymbol{n} \cdot \boldsymbol{Q} < 0$, both factors change sign. By integrating over all possible $\boldsymbol{Q}$, we thus obtain the pressure tensor

$$\boldsymbol{\tau} = n_{\mathrm{p}} k_{\mathrm{B}} T \boldsymbol{\delta} - n_{\mathrm{p}} H \langle \boldsymbol{Q}\boldsymbol{Q} \rangle \,, \tag{22.9}$$

where the isotropic contribution has been added so that $\boldsymbol{\tau}$ vanishes at equilibrium.[4] From (22.7), we hence obtain an evolution equation for the pressure tensor,

$$\frac{\partial \boldsymbol{\tau}}{\partial t} = \boldsymbol{\kappa} \cdot \boldsymbol{\tau} + \boldsymbol{\tau} \cdot \boldsymbol{\kappa}^{\mathrm{T}} - n_{\mathrm{p}} k_{\mathrm{B}} T \left(\boldsymbol{\kappa} + \boldsymbol{\kappa}^{\mathrm{T}} \right) - \frac{4H}{\zeta} \boldsymbol{\tau} \,. \tag{22.10}$$

Starting from a kinetic theory for Hookean dumbbells, we have recovered the convected Maxwell model given in (12.45). In addition to corroborating our phenomenological arguments developed in Chapter 12, we have found molecular expressions for the modulus $G = n_{\mathrm{p}} k_{\mathrm{B}} T$ and the relaxation time $\lambda = \zeta/(4H)$ appearing in the phenomenological equation. The modulus is predicted to be proportional to temperature and the number density of

[4] In view of our incompressibility assumption, the isotropic pressure contribution must actually be determined from the constraint condition.

polymers. Such a result for the modulus has previously been motivated by an analysis of the entropy balance (see solution to Exercise 12.11). To appreciate the result for the relaxation time, we connect ζ and H to the diffusion coefficient and the size of the polymer molecules,

$$D\lambda = \frac{2k_BT}{\zeta}\frac{\zeta}{4H} = \frac{1}{2}\frac{k_BT}{H}. \tag{22.11}$$

For the dumbbell model, the diffusion coefficient for the polymer is of the order of the diffusion coefficient for a bead, and, according to (22.8), k_BT/H is of the order of the equilibrium size of the polymer molecules. Equation (22.11) hence states that, within a relaxation time, the polymer diffuses by a distance of the order of its own size. This detailed insight into the connection between relaxation and diffusion is one of the results of our kinetic theory.

By adding the Newtonian solvent contribution $-\eta_s(\kappa+\kappa^T)$ to the pressure tensor, the upper convected Maxwell model is turned into the Oldroyd B model, as discussed in Exercise 12.12. We have here provided a clear motivation for adding a solvent contribution.

Our kinetic theory for dilute polymer solutions can be generalized in various ways. We here discuss chain models, nonlinear springs, and hydrodynamic interactions.

- For long linear molecules, it is natural to pass from a dumbbell model to a bead–spring–chain model. For Hookean springs, one arrives at the Rouse model,[5] for which one can still derive a closed-form rheological constitutive equation. By transforming to the eigenmodes of the bead–spring chain, one obtains a collection of independent dumbbells with a spectrum of relaxation times, thus arriving at a multimode generalization of the upper-convected Maxwell model. In particular, one obtains an additional prediction for the dependence of the relaxation times on molecular weight.
- In contrast to Hookean springs, real polymer chains are only finitely extensible. A popular modification is the FENE (finitely extensible nonlinear elastic) spring force law, $-HQ/(1 - Q^2/Q_0^2)$. One can additionally introduce short-range repulsive forces between any pair of beads, known as excluded-volume forces, to model the expansion of polymer molecules in good solvents. Except for approximate versions, it is impossible to derive closed-form constitutive equations for nonlinear interactions.
- When a bead moves through the solvent, it perturbs the solvent flow field. The perturbations propagate quickly through the solvent and affect the motion of the other beads. Such effects are known as hydrodynamic

---

[5] Rouse, *J. Chem. Phys.* **21** (1953) 1272.

interactions. They affect the collective motions in bead–spring models in a crucial way. In particular, the molecular-weight dependence of the relaxation times becomes significantly weaker than in the Rouse model.

As most of the above generalizations make it impossible to derive closed-form constitutive equations for the pressure tensor, how can we then use them for actual flow calculations? An interesting possibility is to combine Brownian dynamics simulations for the kinetic theory with numerical methods for solving the momentum balance equation (5.33) subject to (5.36).[6] More concretely, one distributes a large number of model polymers (e.g., dumbbells) over the flow domain and obtains the pressure tensor to be used in the momentum balance from local ensemble averages. Solving the momentum balance gives the velocity field, which describes the motion of the polymers, from which the velocity gradient is obtained and used in the force balance equation for the model polymers.

Exercise 22.1   Derivation of Hookean Dumbbell Moment Equation
Derive the second-moment equation (22.7) from the diffusion equation (22.6).

Exercise 22.2   Simulation of a Hookean Dumbbell in Steady Shear Flow
Simulate a Hookean dumbbell in steady shear flow, and observe the tumbling motion of the dumbbell.

Exercise 22.3   A Finite Deformation Tensor
The finite deformation tensor $E(t, t')$ is defined as the time-ordered exponential of the time integral of the velocity gradient tensor, $\int_{t'}^{t} \kappa(t'')dt''$, that is,

$$E(t, t') = \delta + \int_{t'}^{t} dt_1\, \kappa(t_1) + \int_{t'}^{t} dt_1 \int_{t'}^{t_1} dt_2\, \kappa(t_1) \cdot \kappa(t_2)$$
$$+ \int_{t'}^{t} dt_1 \int_{t'}^{t_1} dt_2 \int_{t'}^{t_2} dt_3\, \kappa(t_1) \cdot \kappa(t_2) \cdot \kappa(t_3) + \cdots. \qquad (22.12)$$

Explain why $E(t, t')$ is referred to as a time-ordered exponential. Show that $E(t, t')$ satisfies the differential equation

$$\frac{\partial E(t, t')}{\partial t} = \kappa(t) \cdot E(t, t'). \qquad (22.13)$$

Exercise 22.4   Finite Deformation Tensor for Elongational Flows
Why is the evaluation of time-ordered exponentials for the elongational flows introduced in (12.20) and (12.22) particularly simple?

---

[6] Laso & Öttinger, *J. Non-Newtonian Fluid Mech.* **47** (1993) 1; Hulsen, van Heel & van den Brule, *J. Non-Newtonian Fluid Mech.* **70** (1997) 79.

Exercise 22.5   Solution of the Equation of Motion for a Hookean Dumbbell
Solve the linear stochastic differential equation (22.5) in terms of the finite deformation tensor $E(t, t')$ defined in (22.12). Use the solution to evaluate the time dependence of the second-moment tensor.

Exercise 22.6   Stress Growth in Start-Up of Steady Shear Flow
Use the explicit expression for the second-moment tensor found in Exercise 22.5 to determine the stress growth in start-up of steady shear flow.

22.2 Melts

In dilute solutions, it is possible to consider the dynamics of a polymer molecule independently of what all the other polymer molecules do. In concentrated polymer solutions and melts, one expects massive interactions between the polymer molecules. Nevertheless, we here try to model the dynamics of individual polymer chains or chain segments to keep the calculations for melts tractable. How can such an enormously helpful simplification be justified?

As long as the molecular weight of the polymer molecules in a melt is sufficiently small, we can assume that they possess the same type of dynamic modes as in a dilute solution, but now in an effective high-viscosity environment provided by the surrounding polymer molecules. Dilute-solution models are still applicable, at least approximately. If the molecular weight of the polymer molecules in the melt becomes larger, we expect that the molecules will begin to entangle, where the most famous mechanical analog is a plate of spaghetti (without sauce). The reader can alternatively imagine selling shoelaces, keeping the entire stock in a big pile, and allowing customers to pull out their favorite ones (to guarantee a massively entangled state). One way to look at the situation is to identify a network of strands between entanglements where, by definition, the strands between entanglement points are basically unconstrained (otherwise we would introduce additional entanglements). Another way to handle the complicated situation is to assume that a single polymer chain in a melt can move only through snake-like (reptational) motions in a cage formed by entanglements with other chains. These two options lead to the so-called temporary network[7] and reptation[8] models. We here focus on the simpler temporary network model.

---

[7] Green & Tobolsky, *J. Chem. Phys.* **14** (1946) 80; Lodge, *Trans. Faraday Soc.* **52** (1956) 120.
[8] See Doi & Edwards, *Theory of Polymer Dynamics* (Oxford, 1986) for an introduction to the basic versions of the reptation model.

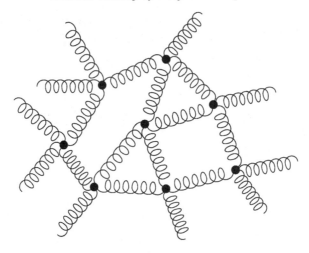

Figure 22.3 Temporary network conformation.

In the temporary network model, the conformation of a melt is repre-
sented by a network of springs, where the network structure changes in time
(see Figure 22.3). The small black circles in Figure 22.3 represent two-chain
entanglements, and the springs represent chain segments between entanglements. If the entanglement density is not too large, we expect Gaussian
statistics of the chain segments between entanglements, and we hence use
Hookean springs. These Hookean springs remind us of the dumbbells used
for modeling polymers in dilute solution. As Hookean networks transform
affinely, each network strand feels the externally imposed deformation, just
as we assumed in the dumbbell model. However, due to the network connec-
tivity, the individual segments cannot retract, and their ends cannot diffuse
freely. Instead, we have the destruction and creation of network strands
by the destruction and creation of entanglements through chain motion.
Elongated segments disappear by disentanglement; a characteristic size and
random orientation is given to the newly created strands by choosing their
conformations according to the equilibrium distribution, p_{eq}. If the time
scale for the destruction and creation of network strands is λ, we can write
the following evolution equation for the conformational probability density
$p = p(t, \mathbf{Q})$,

$$\frac{\partial p}{\partial t} = -\frac{\partial}{\partial \mathbf{Q}} \cdot (\boldsymbol{\kappa} \cdot \mathbf{Q}\, p) + \frac{1}{\lambda} \left(p_{eq} - p \right). \tag{22.14}$$

The simplest temporary network model, which is known as the Lodge model,
is obtained for constant λ. Note that we have chosen the rates $1/\lambda$ for
destruction and creation to be equal. As a consequence, the probability
density p remains normalized in the course of time. The steady state of

(22.14) in the absence of flow is p_{eq}, which we assume to be of the Gaussian form

$$p_{eq}(\boldsymbol{Q}) = (2\pi L^2)^{-3/2} \exp\left\{-\frac{1}{2}\frac{\boldsymbol{Q}^2}{L^2}\right\}, \qquad (22.15)$$

with a characteristic equilibrium segment length scale L. The second-moment equation following from (22.14) is (see Exercise 22.7)

$$\frac{\partial \boldsymbol{\Theta}}{\partial t} = \boldsymbol{\kappa} \cdot \boldsymbol{\Theta} + \boldsymbol{\Theta} \cdot \boldsymbol{\kappa}^{\mathrm{T}} + \frac{1}{\lambda}\left(L^2 \boldsymbol{\delta} - \boldsymbol{\Theta}\right). \qquad (22.16)$$

This equation coincides with the second-moment equation for Hookean dumbbells with $\lambda = \zeta/(4H)$ and $L^2 = k_{\mathrm{B}}T/H$, where H is now the spring constant for network segments. As we are dealing with a system of Hookean springs, we again use the expression (22.9) for the pressure tensor. The only difference is that n_{p} is no longer the density of polymer molecules but rather the density of network strands. Therefore, the modulus G is given by $n_{\mathrm{seg}}k_{\mathrm{B}}T$, where the segment number density is much larger than n_{p}, at least for highly entangled polymers. Experimental measurements for the modulus G, such as the ones shown in Figure 12.3, provide a convenient way of estimating the number density of entanglements in a melt (see Exercise 22.9).

We have formulated two significantly different kinetic theories for dilute polymer solutions and polymer melts, (22.6) and (22.14), and ended up with the same upper-convected Maxwell model for the pressure tensor in both theories. This suggests that there is a certain universality in the basic modeling of the viscoelasticity of polymeric fluids. For dilute solutions, an additional solvent contribution arises naturally. For polymer melts, the fast degrees of freedom associated with length scales shorter than the segment size can also lead to a Newtonian contribution to the pressure tensor.

There are a number of natural generalizations for temporary network models. Not all network strands can be expected to be of equal length. We hence deal with a distribution of Hookean spring constants and a spectrum of relaxation times. Moreover, the creation and destruction rates for network strands can be allowed to depend on the segment conformation vector $\boldsymbol{Q}$ (see Exercise 22.10). If one moreover allows for nonlinear springs, the network strands are no longer deformed affinely and cannot be considered as independent strands. All these generalizations require simulation techniques which go beyond Brownian dynamics (because of the creation and destruction of strands).[9]

Within the temporary network model, we do not obtain a molecular prediction for the relaxation time λ, as it is assumed as one of the input

[9] Biller & Petruccione, *J. Chem. Phys.* **92** (1990) 6322.

parameters. We hence offer a separate argument to learn more about the relaxation time of entangled polymers. In the reptation picture, we consider the diffusion of a polymer molecule along its own contour. Once the molecule has diffused out of its own contour of length L_c, the original conformation has relaxed entirely. The relaxation time is hence given by L_c^2/D_c, where D_c is the diffusion coefficient for diffusion along the contour. As the total friction that the entire molecule experiences in motion along its contour is proportional to the molecular weight $\tilde{M}$, the Nernst–Einstein equation (6.31) tells us that $D_c \propto \tilde{M}^{-1}$. As the contour length is proportional to the molecular weight, $L_c \propto \tilde{M}$, we find that the relaxation time λ is proportional to $\tilde{M}^3$. Also, since $G = n_{\mathrm{seg}} k_B T$, which is independent of $\tilde{M}$, the polymer contribution to viscosity is given by $\eta_p = G\lambda \propto \tilde{M}^3$. This strong dependence on the molecular weight (see Figure 12.3) is the origin of the enormous relaxation times and viscosities encountered for polymer melts.

Exercise 22.7   Derivation of Temporary Network Moment Equation
Derive the second-moment equation (22.16) from the evolution equation (22.14) for the probability density of the temporary network model.

Exercise 22.8   Neo-Hookean Limit
In the limit $\lambda \to \infty$, we expect the temporary network model to become a fixed network model, that is, a model for an elastic solid admitting large deformations. Find the corresponding expression for the pressure tensor. How does such a neo-Hookean material differ from a Hookean solid given by (12.17)? [Hint: use the results of Exercise 22.6.]

Exercise 22.9   Network Strand Density from Storage Modulus
Elaborate how experimental data for the storage modulus G' and the loss modulus G'' (see Figure 12.3) can be used to extract the modulus G and to estimate the network strand density of the temporary network model.

Exercise 22.10   Conformation-Dependent Rate Parameter
Discuss the changes resulting from a conformation-dependent rate parameter $1/\lambda(\boldsymbol{Q})$ in the basic kinetic-theory equation (22.14) for the temporary network model.

- Polymer kinetic theories are usually based on evolution equations for probability densities in polymer conformation space. For models of dilute solutions, the evolution equations are equivalent to stochastic differential equations of motion. For the temporary network model of melts, creation and destruction of network strands are involved. Separate arguments are required for obtaining the stress tensor.

- For the simplest kinetic-theory models, one can derive closed-form rheological constitutive equations; both for dilute solutions and for melts, one arrives at the upper-convected Maxwell and Oldroyd B models.

- Polymer kinetic theory typically provides us with molecular expressions for the phenomenological parameters in constitutive equations for the stress tensor, and they can be readily generalized in various ways.

- Stochastic simulation techniques for kinetic-theory models can be used to obtain rheological material information in given flows or to solve complex flow problems in combination with numerical integrators for the momentum balance equation.

23

Transport in Porous Media

Porous materials, both synthetic and naturally occurring, are everywhere. A material is classified as porous if it consists of fluid-filled voids, or pores, within a solid matrix. Substances such as rock, soil, and biological tissues (e.g., bones), and man-made materials such as cements, ceramics, and membranes, are all examples of *porous media*. The unique properties of these materials can be rationalized only by taking into consideration the structure of the pores, which can range in size from nanometers to millimeters. Flow in porous media is important in geology, environmental science, and petroleum engineering. Porous materials with gas-filled pores have smaller average densities and effective thermal conductivities than their solid counterparts, making them lightweight insulators. In some cases, porous materials are used because of their high surface area per unit volume, which can be as large as 10^7 m^2/m^3.[1] For example, catalytic reactions, which typically occur at solid surfaces coated with expensive transition metals such as Ni, Au, or Pt, are often carried out in porous materials.

It should be evident that transport phenomena in porous media are significant in numerous applications. However, a direct application of the equations of transport phenomena developed in preceding chapters is, because of the extremely complex geometry of the porous structure, entirely impractical. It should also be clear that interfacial transport between the fluid and solid phases plays a critical role. The use of words such as "average" and "effective" suggests that smoothed (reduced) forms of the balance equations are important and that transport processes are described by multiscale models. Several different methods have been used to analyze mass, energy, and

---

[1] A porous material with surface area per unit volume 10^7 m^2/m^3 and a volume of 8 cm^3, or two ordinary cubes of sugar, would have a surface area of 8000 m^2, or roughly the area of a football field.

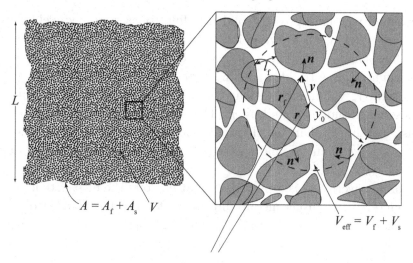

Figure 23.1 Schematic representation of a porous medium with a magnified view of averaging volume V_{eff} at position r.

momentum transport in porous media.[2] In this chapter, we follow the approach presented in Whitaker's educational book and apply this to the analysis of diffusion and reaction in a porous catalyst. The well-known equation describing this system, which typically is presented without derivation in transport phenomena textbooks, is given by

$$\frac{\partial \langle c_A \rangle_i}{\partial t} = D_{\text{eff}} \nabla^2 \langle c_A \rangle_i - k_{\text{eff}} \langle c_A \rangle_i , \qquad (23.1)$$

where $\langle c_A \rangle_i$ is an average concentration within the porous medium and D_{eff} and k_{eff} are the effective diffusion coefficient and reaction rate constant. The goals of the present chapter are to demonstrate how transport equations for porous media, such as the one in (23.1), are formulated and to show their usefulness in modeling this important class of materials.

23.1 Volume Averaging

The formulation of transport equations applicable to porous media is based on the technique of *volume averaging*. The left image in Figure 23.1 shows a macroscopic view of a porous object with volume V bound by a surface with area A in contact with both fluid A_f and solid A_s phases. The right image shows a magnified view of the porous medium where shaded and

---

[2] Bear & Bachmat, *Introduction to Modeling of Transport Phenomena in Porous Media* (Kluwer, 1990); Slattery, *Advanced Transport Phenomena* (Cambridge, 1999); Whitaker, *The Method of Volume Averaging* (Kluwer, 1999).

unshaded regions represent solid and fluid phases, respectively. Note that the unit normal vector $\boldsymbol{n}$ points into the solid phase (I) and away from the fluid phase (II). The spherical volume V_{eff} at position $\boldsymbol{r}$ denotes the averaging volume (right image in Figure 23.1), which is the sum of fluid V_{f} and solid V_{s} volumes. Position relative to the center of V_{eff} is given by $\boldsymbol{y}$, whose magnitude is bound by y_0. The averaging volume is chosen such that it is much larger than the characteristic length of the porous structure and much smaller than the characteristic length of the macroscopic system, that is $l_{\text{f}} \ll y_0 \ll L$.

Volume averaging is a procedure that exploits the disparate length scales encountered in porous media and allows for the formulation of transport equations having spatial resolution $y_0 \sim (V_{\text{eff}})^{1/3}$. We have already used volume averaging in our application of effective medium theory (see Section 17.2) to predict the effective thermal conductivity of a composite comprised of spheres. That analysis, which used the average in (17.12), was restricted to dilute systems and relied on the availability of an exact solution for the temperature field in the two-phase system. This is not the situation for porous materials, which typically have complex geometries and cannot be treated as dilute.

The formulation of transport equations for porous media involves several different averaged quantities. Here, we are interested in describing transport within the fluid phase and introduce the *superficial average* of quantity a (e.g., concentration, temperature) in the fluid phase, which is defined as

$$\langle a(\boldsymbol{r}) \rangle_{\text{s}} = \frac{1}{V_{\text{eff}}} \int_{V_{\text{f}}} a(\boldsymbol{r}_{\text{f}}) d^3 \boldsymbol{r}_{\text{f}}, \tag{23.2}$$

where $\boldsymbol{r}_{\text{f}} = \boldsymbol{r} + \boldsymbol{y}$ (see right image in Figure 23.1). Similarly, the *intrinsic average* in the fluid phase is given by,

$$\langle a(\boldsymbol{r}) \rangle_{\text{i}} = \frac{1}{V_{\text{f}}} \int_{V_{\text{f}}} a(\boldsymbol{r}_{\text{f}}) d^3 \boldsymbol{r}_{\text{f}} = \frac{\langle a(\boldsymbol{r}) \rangle_{\text{s}}}{\varepsilon}, \tag{23.3}$$

where $\varepsilon = V_{\text{f}}/V_{\text{eff}}$ is the porosity. It is important to recognize the different meanings of $V_{\text{f}} = V_{\text{f}}(\boldsymbol{r})$ as it appears in (23.2) and (23.3): as $\int_{V_{\text{f}}}$ it indicates the domain over which the integration is performed, and otherwise V_{f} indicates the (local) fluid phase volume within V_{eff}. We will also need the *area average* of a, which is defined by

$$\langle a(\boldsymbol{r}) \rangle_{\text{a}} = \frac{1}{A_{\text{fs}}} \int_{A_{\text{fs}}} a(\boldsymbol{r}_{\text{f}}) d^2 \boldsymbol{r}_{\text{f}}, \tag{23.4}$$

where A_{fs} is (the area of) the fluid–solid interface within V_{eff}. Note that the surface area per unit volume is $a_{\text{fs}} = A_{\text{fs}}/V_{\text{eff}} \approx l_{\text{f}}^{-1}$.

To handle spatial derivatives appearing in transport equations, we introduce the *spatial average theorem* for the gradient of the field a

$$\langle \boldsymbol{\nabla} a(\boldsymbol{r}) \rangle_{\mathrm{s}} = \boldsymbol{\nabla} \langle a(\boldsymbol{r}) \rangle_{\mathrm{s}} + \frac{1}{V_{\mathrm{eff}}} \int_{A_{\mathrm{fs}}} \boldsymbol{n} a(\boldsymbol{r}_{\mathrm{f}}) d^2 \boldsymbol{r}_{\mathrm{f}}, \qquad (23.5)$$

and for the divergence of vector field $\boldsymbol{a}$

$$\langle \boldsymbol{\nabla} \cdot \boldsymbol{a}(\boldsymbol{r}) \rangle_{\mathrm{s}} = \boldsymbol{\nabla} \cdot \langle \boldsymbol{a}(\boldsymbol{r}) \rangle_{\mathrm{s}} + \frac{1}{V_{\mathrm{eff}}} \int_{A_{\mathrm{fs}}} \boldsymbol{n} \cdot \boldsymbol{a}(\boldsymbol{r}_{\mathrm{f}}) d^2 \boldsymbol{r}_{\mathrm{f}}. \qquad (23.6)$$

Note that the same $\boldsymbol{\nabla}$ operator is used on the left- and right-hand sides of
(23.5) and (23.6); however, differentiation on the left-hand sides is meaningful only within V_{f}.[3]
A second ingredient of volume averaging involves the spatial decompo-
sition of fields. While $\langle a \rangle_{\mathrm{s}}$ is the natural quantity for applying the spatial
average theorems, our interest is in transport occurring within the fluid
phase so that $\langle a \rangle_{\mathrm{i}}$ is the more relevant quantity. Hence, for the field a in the
fluid we write

$$a = \langle a \rangle_{\mathrm{i}} + a', \qquad (23.7)$$

where a' is the deviation of a from $\langle a \rangle_{\mathrm{i}}$. Spatial variations of $\langle a \rangle_{\mathrm{i}}$ occur over
the length scale L, while those for a' occur on the much smaller length scale
l_{f}. Hence, the decomposition in (23.7) allows the separation of phenomena
occurring on different length scales within porous media. Using (23.7), we
can write

$$\int_{A_{\mathrm{fs}}} \boldsymbol{n} a(\boldsymbol{r}_{\mathrm{f}}) d^2 \boldsymbol{r}_{\mathrm{f}} = \int_{A_{\mathrm{fs}}} \boldsymbol{n} \langle a(\boldsymbol{r}_{\mathrm{f}}) \rangle_{\mathrm{i}} d^2 \boldsymbol{r}_{\mathrm{f}} + \int_{A_{\mathrm{fs}}} \boldsymbol{n} a'(\boldsymbol{r}_{\mathrm{f}}) d^2 \boldsymbol{r}_{\mathrm{f}}$$

$$= \int_{A_{\mathrm{fs}}} \boldsymbol{n} \langle a(\boldsymbol{r}) \rangle_{\mathrm{i}} d^2 \boldsymbol{r}_{\mathrm{f}} + \int_{A_{\mathrm{fs}}} \boldsymbol{n} \boldsymbol{y} \cdot \boldsymbol{\nabla} \langle a(\boldsymbol{r}) \rangle_{\mathrm{i}} d^2 \boldsymbol{r}_{\mathrm{f}} + \cdots$$

$$+ \int_{A_{\mathrm{fs}}} \boldsymbol{n} a'(\boldsymbol{r}_{\mathrm{f}}) d^2 \boldsymbol{r}_{\mathrm{f}}$$

$$\approx -V_{\mathrm{eff}} (\boldsymbol{\nabla} \varepsilon) \langle a(\boldsymbol{r}) \rangle_{\mathrm{i}} - V_{\mathrm{eff}} (\boldsymbol{\nabla} \langle \boldsymbol{y} \rangle_{\mathrm{s}}) \cdot \boldsymbol{\nabla} \langle a(\boldsymbol{r}) \rangle_{\mathrm{i}}$$

$$+ \int_{A_{\mathrm{fs}}} \boldsymbol{n} a'(\boldsymbol{r}_{\mathrm{f}}) d^2 \boldsymbol{r}_{\mathrm{f}}, \qquad (23.8)$$

where a Taylor series expansion for $\langle a(\boldsymbol{r}_{\mathrm{f}}) \rangle_{\mathrm{i}}$ has been used to obtain the
second equality, and the third equality is obtained by pulling the fields at $\boldsymbol{r}$

---

[3] The spatial average theorem can be derived using several approaches and has been the subject
of extensive discussion. For example, see Anderson & Jackson, *Ind. Eng. Chem. Fund.* **6** (1967)
527; Slattery, *AIChE J.* **13** (1967) 1066; Whitaker, *AIChE J.* **13** (1967) 420; Howes & Whitaker
Chem. Eng. Sci. **40** (1985) 1387.

outside the integral and evaluating the spatial moments of the porous media (see Exercise 23.1). Combining (23.5) and (23.8), we obtain

$$\langle \nabla a(\mathbf{r}) \rangle_s \approx \varepsilon \, \nabla \langle a(\mathbf{r}) \rangle_i + \frac{1}{V_{\text{eff}}} \int_{A_{\text{fs}}} n a'(\mathbf{r}_{\text{f}}) d^2 \mathbf{r}_{\text{f}} , \qquad (23.9)$$

where we have assumed for disordered porous media that $\nabla \langle y \rangle_s$ is diagonal and has elements that are small compared with unity. The integral term in (23.9) describes the interaction of the fluid with the solid surfaces and is sometimes referred to as a spatial filter. Using a similar procedure, we can write

$$\int_{A_{\text{fs}}} a(\mathbf{r}_{\text{f}}) d^2 \mathbf{r}_{\text{f}} = \int_{A_{\text{fs}}} \langle a(\mathbf{r}_{\text{f}}) \rangle_i \, d^2 \mathbf{r}_{\text{f}} + \int_{A_{\text{fs}}} a'(\mathbf{r}_{\text{f}}) d^2 \mathbf{r}_{\text{f}}$$

$$= \int_{A_{\text{fs}}} \langle a(\mathbf{r}) \rangle_i \, d^2 \mathbf{r}_{\text{f}} + \int_{A_{\text{fs}}} \mathbf{y} \cdot \nabla \langle a(\mathbf{r}) \rangle_i \, d^2 \mathbf{r}_{\text{f}} + \cdots$$

$$+ \int_{A_{\text{fs}}} a'(\mathbf{r}_{\text{f}}) d^2 \mathbf{r}_{\text{f}}$$

$$\approx A_{\text{fs}} \langle a(\mathbf{r}) \rangle_i + A_{\text{fs}} \langle \mathbf{y} \rangle_a \cdot \nabla \langle a(\mathbf{r}) \rangle_i + \int_{A_{\text{fs}}} a'(\mathbf{r}_{\text{f}}) d^2 \mathbf{r}_{\text{f}} .$$

$$(23.10)$$

In the problem considered in this chapter, the deviation a' from $\langle a \rangle_i$ results from the interplay between diffusion and heterogeneous reaction occurring over the length scale l_{f}. In such cases, we expect $a' \ll \langle a \rangle_i$ and that a' is strongly correlated with $\mathbf{n}$ within V_{eff}. For this reason, we neglect the last term in the third equality of (23.10) while retaining the last term in the third equality of (23.8). Further assuming $|\langle \mathbf{y} \rangle_a| \ll y_0$, we can combine (23.4) and (23.10) to obtain the useful approximation,

$$\langle a(\mathbf{r}) \rangle_a \approx \langle a(\mathbf{r}) \rangle_i . \qquad (23.11)$$

Exercise 23.1   *Volume-Average Identities*
Using (23.2) and (23.5), show that

$$\frac{1}{V_{\text{eff}}} \int_{A_{\text{fs}}} \mathbf{n} \, d^2 \mathbf{r}_{\text{f}} = -\nabla \varepsilon, \qquad \frac{1}{V_{\text{eff}}} \int_{A_{\text{fs}}} \mathbf{n} y \, d^2 \mathbf{r}_{\text{f}} = -\nabla \langle y \rangle_s .$$

Exercise 23.2   *Volume-Averaged Form of Fourier's Law*
Consider a heterogeneous system comprised of k spheres having radius R and thermal conductivity $\bar{\lambda}$ dispersed in a medium having thermal conductivity λ contained within a volume V_{eff} (see Section 17.2). The volume-averaged form of Fourier's law

(6.4) using (23.5) can be written as

$$\langle \boldsymbol{j}_q \rangle_s = -\langle \lambda \, \boldsymbol{\nabla} T \rangle_s = -\lambda \langle \boldsymbol{\nabla} T \rangle_s - \frac{\lambda}{V_{\text{eff}}} \int_{A_{\text{fs}}} n T(\boldsymbol{r}_{\text{f}}) d^2 \boldsymbol{r}_{\text{f}} \,,$$

taking λ to be uniform within V_{eff}. Use (17.11) with $\bar{\lambda}/\lambda \ll 1$ and $N_{\text{Ka}} = 0$ to derive an expression for λ_{eff}, and compare this with the result in (17.16).

23.2 Diffusion and Reaction in Porous Catalysts

We consider a first-order, heterogeneous reaction $A \to C$ in the presence of inert species B, which takes place at the catalytic surfaces within a porous particle. In practice, particles with sizes $L \sim 10$ mm and pore sizes $l_{\text{f}} \sim 1$ μm or smaller are used. These particles are exposed to a gas containing the reactant in packed or fluidized beds (see Section 23.3).

The balance equation for species mass (5.21) can be written as

$$\frac{\partial c_\alpha}{\partial t} = -\boldsymbol{\nabla} \cdot (\boldsymbol{v}^* c_\alpha + \boldsymbol{J}_\alpha^*) \,, \tag{23.12}$$

where we have excluded the term for homogeneous chemical reaction. We assume the fluid mixture is ideal and the process is isothermal and isobaric. As in Section 10.2, we consider a mixture that is dilute in species A ($x_A \ll 1$) and therefore is also dilute in species C ($x_C \ll 1$). This means that $x_B \approx 1$, and the molar density of the mixture c is approximately constant. In the absence of homogeneous reaction, for constant molar density c, (5.22) simplifies to $\boldsymbol{\nabla} \cdot \boldsymbol{v}^* = 0$. Since the velocity is zero on all solid surfaces, we have $\boldsymbol{v}^* = \boldsymbol{0}$ everywhere in the fluid phase. The balance equation for species A is obtained from (23.12), which simplifies to

$$\frac{\partial c_A}{\partial t} = -\boldsymbol{\nabla} \cdot \boldsymbol{J}_A^* \,. \tag{23.13}$$

Given the small pore sizes $l_{\text{f}} \approx 10^2$ nm that may be encountered in porous media, for gases where $l_{\text{mfp}} \approx 10^2$ nm (see Section 21.2), it is likely that $N_{\text{Kn}} = l_{\text{mfp}}/l_{\text{f}} \approx 1$. In such situations, diffusion is no longer Fickian (see Section 6.3) but rather occurs by *Knudsen diffusion*,[4] named after the physicist Martin Hans Christian Knudsen (1871–1949), where molecular collisions with the confining walls (pore surfaces) are as frequent as collisions between molecules. To avoid this complication, we restrict our analysis to situations where $N_{\text{Kn}} = l_{\text{mfp}}/l_{\text{f}} \ll 1$ so that Fick's law is applicable. For gases, this means our analysis is valid only for systems having pores with $l_{\text{f}} \approx 1$ μm or larger.

[4] Knudsen, *Ann. der Phys.* **331** (1909) 75; Pollard & Present, *Phys. Rev.* **73** (1948) 762.

Assuming Fickian diffusion occurs within the fluid-filled pores, and again taking advantage of the dilute mixture approximation (see Section 10.2), we have from (10.10) the expression $J_A^* = -D_{AB} \nabla c_A$. Substitution into (23.13) gives

$$\frac{\partial c_A}{\partial t} = \nabla \cdot (D_{AB} \nabla c_A), \qquad (23.14)$$

which applies at all points within the fluid phase. To formulate boundary conditions for (23.14), we treat the interfaces at A_{fs} as passive so that the molar form of the jump balance for species mass (see Exercise 14.6) is applicable. The interface is stationary $v^s = 0$, and if we further neglect diffusive transport within the interface $J_\alpha^s = 0$, (14.27) simplifies to

$$n \cdot (v^{*\mathrm{II}} c_A^{\mathrm{II}} + J_A^{*\mathrm{II}}) + \tilde{\nu}_A^s \tilde{\Gamma}^s = 0, \qquad (23.15)$$

which reflects the fact that the solid phase (I) is impermeable. The surface reaction is of first order and "irreversible," so from (15.32) we can write

$$\tilde{\Gamma}^s = L_\Gamma^s \exp\left(\frac{\tilde{\mu}_A^s}{\tilde{R}T}\right) = L_\Gamma^s \exp\left(\frac{\tilde{\mu}_A^{\mathrm{II}}}{\tilde{R}T}\right) = k^s c_A^{\mathrm{II}}, \qquad (23.16)$$

where the second equality is obtained by assuming $\tilde{\mu}_A^s = \tilde{\mu}_A^{\mathrm{II}}$ and the third because of the ideality of the gas mixture. Hence, the reaction is described by the mass action law with rate constant k^s. Substitution of (23.16) into (23.15) leads to the following boundary condition at fluid–solid interfaces:

$$-n \cdot D_{AB} \nabla c_A = k^s c_A \qquad \text{at } A_{\mathrm{fs}}, \qquad (23.17)$$

where we have also used $v^{*\mathrm{II}} = 0$. In principle, (23.14) can be solved at all points within the fluid subject to the boundary condition in (23.17). However, the complex geometry of the surface A_{fs} makes this impractical.

Our goal then is to formulate a more tractable form of the problem given by (23.14) and (23.17) where details of the porous structure are smoothed out using the volume-averaging equations presented in the preceding section. Integrating (23.14) over V_{f}, dividing by V_{eff} and applying (23.6) gives

$$\frac{\partial \langle c_A \rangle_s}{\partial t} = \nabla \cdot \langle D_{AB} \nabla c_A \rangle_s + \frac{1}{V_{\mathrm{eff}}} \int_{A_{\mathrm{fs}}} n \cdot D_{AB} \nabla c_A(r_{\mathrm{f}}) d^2 r_{\mathrm{f}}, \qquad (23.18)$$

where we have taken V_{f} to be independent of time. Using the boundary condition (23.17) in the surface integral, we can write (23.18) as

$$\frac{\partial \langle c_A \rangle_s}{\partial t} = \nabla \cdot (D_{AB} \langle \nabla c_A \rangle_s) - a_{\mathrm{fs}} k^s \langle c_A \rangle_a, \qquad (23.19)$$

where we have assumed D_{AB} and k^s are constant within V_{eff} and used (23.4).

Using the spatial average theorem in (23.9) and the result in (23.11), (23.19) can be written as

$$\varepsilon \frac{\partial \langle c_A \rangle_i}{\partial t} = \nabla \cdot (\varepsilon D_{AB} \nabla \langle c_A \rangle_i) + \nabla \cdot \left(\frac{D_{AB}}{V_{eff}} \int_{A_{fs}} n c'_A(r_f) d^2 r_f \right) - a_{fs} k^s \langle c_A \rangle_i ,$$
(23.20)

which follows since ε is independent of time.

The equation governing the intrinsic average molar density $\langle c_A \rangle_i$ in (23.20) contains a term involving the surface integral of the concentration deviation $c'_A = c_A - \langle c_A \rangle_i$. Rewriting this in terms of $\langle c_A \rangle_i$ is referred to as solving the "closure problem." To obtain an equation that governs c'_A, we divide (23.20) by ε and subtract this result from (23.14), which gives

$$\frac{\partial c'_A}{\partial t} = \nabla \cdot (D_{AB} \nabla c'_A) - \frac{\nabla \varepsilon}{\varepsilon} \cdot (D_{AB} \nabla \langle c_A \rangle_i) + \frac{a_{fs} k^s}{\varepsilon} \langle c_A \rangle_i$$
$$- \frac{1}{\varepsilon} \nabla \cdot \left(\frac{D_{AB}}{V_{eff}} \int_{A_{fs}} n c'_A(r_f) d^2 r_f \right) .$$
(23.21)

Similarly, from (23.17), we obtain

$$-n \cdot D_{AB} \nabla c'_A - k^s c'_A = n \cdot D_{AB} \nabla \langle c_A \rangle_i + k^s \langle c_A \rangle_i \qquad \text{at } A_{fs}. \quad (23.22)$$

From (23.21) and (23.22), we see that $\langle c_A \rangle_i$ acts as a source for c'_A.

We now have a coupled system of equations governing $\langle c_A \rangle_i$ and c'_A. At this point, it may appear that we have failed (miserably) to achieve our stated goal of formulating a more tractable form of (23.14) and (23.17). To reach this goal, we must make several assumptions to simplify (23.21) and (23.22). First, since c'_A varies over the length scale l_f and $\langle c_A \rangle_i$ varies over the length $L \gg l_f$, we can assume the quasi-steady-state approximation (see Section 7.5) holds for (23.21). The difference in these length scales also allows us to neglect the non-local diffusion (surface integral) term in (23.21). Finally, we can neglect the diffusive source term in both (23.21) and (23.22) if $c'_A \ll \langle c_A \rangle_i$ and $N_{Th}^2 \lesssim L/l_f$, where $N_{Th} = L\sqrt{a_{fs}k^s/D_{AB}}$ is known as the Thiele number (modulus).[5] Hence, we write (23.21) and (23.22) as

$$\nabla^2 c'_A = -\frac{a_{fs} k^s}{\varepsilon D_{AB}} \langle c_A \rangle_i ,$$
(23.23)

$$-n \cdot \nabla c'_A = n \cdot \nabla \langle c_A \rangle_i + \frac{k^s}{D_{AB}} \langle c_A \rangle_i \qquad \text{at } A_{fs}, \quad (23.24)$$

where we have also assumed D_{AB} is constant over the length scale l_f.

[5] Thiele, *Ind. Eng. Chem.* **31** (1939) 916.

Transport in Porous Media

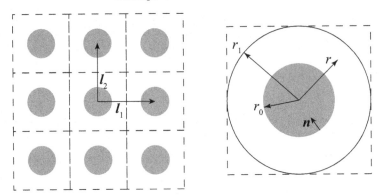

Figure 23.2 Schematic diagram of spatially periodic unit cells (left) and a single unit consisting of a spherical solid surrounded by a spherical fluid (right).

Rather than solving (23.23) and (23.24) over the entire domain, we instead consider a local problem on a spatially periodic unit cell such that $n(r+l_i) = n(r)$ with $i = 1, 2, 3$, where the l_i are lattice vectors relating the positions of adjacent unit cells as shown in Figure 23.2 (left). For the local problem, we assume that (23.23) and (23.24) can be replaced by

$$\nabla^2 c'_A = -\frac{a_{fs} k^s}{\varepsilon D_{AB}} \langle c_A \rangle_i (r), \qquad (23.25)$$

$$-n \cdot \nabla c'_A = n \cdot \nabla \langle c_A \rangle_i (r) + \frac{k^s}{D_{AB}} \langle c_A \rangle_i (r) \qquad \text{at } A_{fs}. \qquad (23.26)$$

The formulation of the local problem is completed by specifying a periodic boundary condition consistent with the assumed periodic array of unit cells,

$$c'_A(r + l_i) = c'_A(r). \qquad (23.27)$$

By considering this much simpler local problem, which requires $l_i \ll L$, $\langle c_A \rangle_i (r)$ and $\nabla \langle c_A \rangle_i (r)$ can be treated as constant over the unit cell. The local deviations c'_A within the unit cell can be assumed to depend only on these constant values and on the position and orientation of the interface A_{fs} in the unit cell. We hence write

$$c'_A = c' + s' \langle c_A \rangle_i (r) + b' \cdot \nabla \langle c_A \rangle_i (r), \qquad (23.28)$$

where the functions c', s', and b' are closure variables, which act as low-order polynomial basis functions accounting for the influence of the local arrangement of the interface A_{fs}.

We are (finally) in a position to achieve our stated goal. Substitution of (23.28) into (23.20) leads to

$$\varepsilon \frac{\partial \langle c_A \rangle_i}{\partial t} = \nabla \cdot (\varepsilon \mathbf{D}_{\text{eff}} \cdot \nabla \langle c_A \rangle_i) + \nabla \cdot (\mathbf{u} \langle c_A \rangle_i) - a_{\text{fs}} k^s \langle c_A \rangle_i, \qquad (23.29)$$

where the effective diffusivity tensor $\mathbf{D}_{\text{eff}}$ is given by

$$\mathbf{D}_{\text{eff}} = D_{AB} \left(\boldsymbol{\delta} + \frac{1}{V_f} \int_{A_{\text{fs}}} \mathbf{n} \mathbf{b}'(\mathbf{r}_f) d^2 \mathbf{r}_f \right), \qquad (23.30)$$

and the velocity $\mathbf{u}$ induced by chemical reaction is given by

$$\mathbf{u} = \frac{\varepsilon D_{AB}}{V_f} \int_{A_{\text{fs}}} \mathbf{n} s'(\mathbf{r}_f) d^2 \mathbf{r}_f, \qquad (23.31)$$

where we have taken into account that the concentration deviation c'_A in (23.28) exists only in V_f. Equation (23.29) is a diffusion equation with both drift and source terms that governs the intrinsic average reactant concentration $\langle c_A \rangle_i$, for which (23.1) is a special case. Typically, $\nabla \cdot (\mathbf{u} \langle c_A \rangle_i) \ll a_{\text{fs}} k \langle c_A \rangle_i$, and the drift term can be neglected.

The equations governing the closure variables c', s', and $\mathbf{b}'$ can be obtained by substitution of (23.28) into (23.25)–(23.27). Since c' is arbitrary (and without loss of generality can be set equal to zero), and since we are neglecting the contribution associated with s' (note that $\mathbf{u} = \mathbf{0}$ for a symmetric unit cell), we need only consider $\mathbf{b}'$, which is governed by

$$\nabla^2 \mathbf{b}' = \mathbf{0}, \qquad (23.32)$$

$$-\mathbf{n} \cdot \nabla \mathbf{b}' = \mathbf{n} \qquad \text{at } A_{\text{fs}}, \qquad (23.33)$$

$$\mathbf{b}'(\mathbf{r} + \mathbf{l}_i) = \mathbf{b}'(\mathbf{r}). \qquad (23.34)$$

In general, the solution of (23.32)–(23.34) and evaluation of the integral in (23.30), which is a measure of the local tortuosity of the porous medium, must be found numerically.

A simpler, but reasonably accurate, model[6] considers a spherically symmetric unit cell consisting of a solid sphere of radius r_0 with $\mathbf{n} = -\boldsymbol{\delta}_r$ surrounded by a fluid sphere having radius r_1 (see Figure 23.2, right) so that $\mathbf{b}' = b'_r \boldsymbol{\delta}_r$. The boundary condition in (23.34) is replaced with $b'_r(r_1) = 0$, which leads to the solution (see Exercise 23.3)

$$b'_r = -r \frac{1 - (r_1/r)^3}{1 + 2/(1 - \varepsilon)}, \qquad (23.35)$$

[6] Chang, *AIChE J.* **29** (1982) 846.

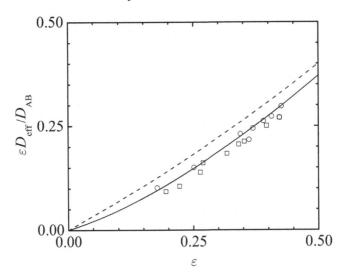

Figure 23.3 Comparison of experimental effective diffusivities in porous media comprised of spheres (circles) and sand (squares) and model predictions from (23.32)–(23.34) for a periodic array of spheres (solid curve) and for (23.37) (dashed curve): $\varepsilon D_{\text{eff}}/D_{\text{AB}} = 2\varepsilon/(3 - \varepsilon)$. [Adapted from Whitaker, *The Method of Volume Averaging* (Kluwer, 1999).]

where $\varepsilon = 1 - (r_0/r_1)^3$. Using (23.35) in (23.30), we obtain

$$\boldsymbol{D}_{\text{eff}} = D_{\text{AB}} \left[\boldsymbol{\delta} - \frac{3}{4\pi(r_1^3 - r_0^3)} \int_0^{2\pi} \int_0^{\pi} \boldsymbol{nn} b_r'(r_0) r_0^2 \sin\theta \, d\theta \, d\phi \right] = D_{\text{eff}} \boldsymbol{\delta} \,,$$

(23.36)

where

$$D_{\text{eff}} = \frac{2}{3 - \varepsilon} D_{\text{AB}} \,.$$

(23.37)

Note that, even though the periodic cell model for the porous medium is not isotropic (see Figure 23.2, left), $\boldsymbol{D}_{\text{eff}}$ is isotropic.

As shown in Figure 23.3, the prediction from the model given by (23.32)–(23.34), which is based on a large number of approximations, is in reasonably good agreement with experimental data.[7] This can be understood by noting that in (23.30) the closure variable $\boldsymbol{b}'$ affects $\boldsymbol{D}_{\text{eff}}$ only through an integral over the fluid–solid surface. It is also worth noting that (23.37) is equivalent to Maxwell's result (cf. footnote on p. 311) for the effective thermal conductivity in (17.16) by setting $\beta = 0$ and $\phi = 1 - \varepsilon$.

---

[7] Ryan *et al.*, *Chem. Eng. Sci.* **35** (1980) 10; Quintard & Whitaker, *Chem. Eng. Sci.* **48** (1993) 2537.

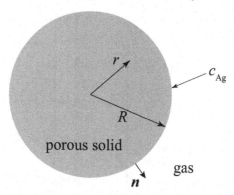

Figure 23.4 Schematic diagram of a spherical porous catalyst particle with radius R surrounded by a gas that maintains concentration c_{Ag} at the particle surface.

In spite of its apparent success, we feel compelled to mention an important issue concerning the volume-averaging method and the equations it produces. The volume-averaging method just elaborated is *not* an example of coarse graining in the thermodynamic sense. While dissipative processes are altered by the modification of transport coefficients, no new dissipative processes are introduced. For example, the effective diffusivity $\boldsymbol{D}_{\mathrm{eff}}$ describes the "hindered" diffusion that results from the tortuous paths within porous materials. Hence, volume averaging should be characterized as a reduction strategy as opposed to coarse graining.[8]

Exercise 23.3   Closure Problem Solution on a Sphere
The closure problem for the periodic unit cell shown in Figure 23.2 is given by

$$\frac{d}{dr}\left(\frac{1}{r^2}\frac{d}{dr}(r^2 b'_r)\right) = 0,$$

$$\frac{db'_r}{dr}(r_0) = -1, \qquad b'_r(r_1) = 0.$$

Obtain the solution in (23.35), and make a contour plot of b'_r/r_1 in the x_1–x_2 plane for $\varepsilon = 0.875, 0.657$. Use this solution to obtain the result in (23.36).

23.3 Catalytic Reactor Analysis

We now consider diffusion and reaction in a spherical catalyst particle of radius R surrounded by a gas (see Figure 23.4). Initially, the particle is free

[8] Gorban *et al.*, *Model Reduction and Coarse Graining* (Springer, 2006).

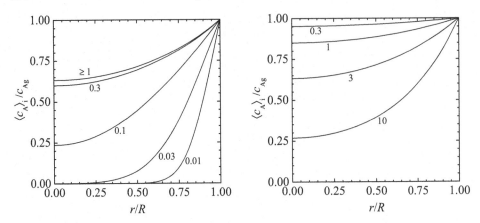

Figure 23.5 Reactant concentration profile within the catalyst particle from (23.40) for $N_{\mathrm{Da}} = 3$ at different times $D_{\mathrm{eff}}t/R^2 = 0.01, 0.03, 0.1, 0.3, 1$ (left); steady-state reactant concentration profile from (23.41) for different values of $N_{\mathrm{Da}} = 0.3, 1, 3, 10$ (right).

of reactant, and, for $t > 0$, the concentration of the reactant at the particle surface has the value c_{Ag}. To keep things simple, we assume the catalyst particle is an isotropic porous medium with uniform porosity ε and that D_{AB} is constant. In this case, (23.29) can be written as

$$\frac{\partial \langle c_{\mathrm{A}} \rangle_{\mathrm{i}}}{\partial t} = D_{\mathrm{eff}} \frac{1}{r^2} \frac{\partial}{\partial r} \left(r^2 \frac{\partial \langle c_{\mathrm{A}} \rangle_{\mathrm{i}}}{\partial r} \right) - k_{\mathrm{eff}} \langle c_{\mathrm{A}} \rangle_{\mathrm{i}}, \qquad (23.38)$$

where $k_{\mathrm{eff}} = a_{\mathrm{fs}} k^{\mathrm{s}} / \varepsilon$ is the effective reaction rate constant. The initial and boundary conditions for (23.38) are given by

$$\langle c_{\mathrm{A}} \rangle_{\mathrm{i}}(r, 0) = 0, \qquad \frac{\partial \langle c_{\mathrm{A}} \rangle_{\mathrm{i}}}{\partial r}(0, t) = 0, \qquad \langle c_{\mathrm{A}} \rangle_{\mathrm{i}}(R, t) = c_{\mathrm{Ag}}. \qquad (23.39)$$

Alternatively, the boundary condition at R could be formulated using a mass transfer coefficient (see Exercise 10.8) to take into account a resistance to mass transfer in the gas phase. The solution of (23.38) and (23.39), which is left as an exercise, is given by

$$\frac{\langle c_{\mathrm{A}} \rangle_{\mathrm{i}}}{c_{\mathrm{Ag}}} = 2\pi \frac{R}{r} \sum_{n=1}^{\infty} \frac{n(-1)^n}{n^2 \pi^2 + N_{\mathrm{Da}}} \sin\left(\frac{n\pi r}{R}\right) \left\{ \exp\left[-(n^2 \pi^2 + N_{\mathrm{Da}}) \frac{D_{\mathrm{eff}} t}{R^2} \right] - 1 \right\},$$
$$(23.40)$$

where $N_{\mathrm{Da}} = k_{\mathrm{eff}} R^2 / D_{\mathrm{eff}}$ is the effective Damköhler number (cf. footnote on p. 178). From Figure 23.5 (left), where the evolution of reactant concentration within the catalyst particle is plotted, we see that a steady state is reached when the rate of diffusion is balanced by the rate of reaction.

It is of interest to examine the steady-state solution of (23.38) and (23.39), which is given by

$$\frac{\langle c_A \rangle_i}{c_{Ag}} = \frac{\sinh\left(\sqrt{N_{Da}}\, r/R\right)}{(r/R)\sinh\left(\sqrt{N_{Da}}\right)}. \tag{23.41}$$

In Figure 23.5 (right), we see that the average of the steady-state concentration of reactant decreases with increasing N_{Da}. Hence, increasing the reaction rate constant k_{eff}, by increasing the temperature for example, also leads to a decrease in reactant concentration $\langle c_A \rangle_i$. Since the reaction rate is proportional to $k_{eff}\langle c_A \rangle_i$, optimal design of catalyst particles must take the interaction of these effects into account.

A commonly used design parameter known as the effectiveness factor[9] η is the ratio of the total rate of reactant consumption for a catalyst particle to that which would be observed if the reactant concentration were at its maximum value throughout the particle. Hence, we can define η as

$$\eta = \frac{\int_V k_{eff}\langle c_A \rangle_i \, dV}{\int_V k_{eff} c_{Ag} \, dV}. \tag{23.42}$$

For a sphere, using (23.41), we obtain

$$\eta = \frac{\int_0^R k_{eff}\langle c_A \rangle_i r^2 \, dr}{\int_0^R k_{eff} c_{Ag} r^2 \, dr} = \frac{3N'_{Th}\coth(3N'_{Th}) - 1}{3N'_{Th}{}^2}, \tag{23.43}$$

where have introduced a modified Thiele number $N'_{Th} = N_{Th}/\sqrt{\varepsilon f(\varepsilon)} = \sqrt{N_{Da}}/3$, with $L = V/A$ being the characteristic length for an arbitrarily shaped particle having volume V and surface area A. Figure 23.6 shows the dependence of η on N'_{Th} for spherical and thin disk catalyst particles (the curve for cylindrical particles lies between the two curves). For small values of N'_{Th} where $\eta \to 1$, the process is reaction-limited, while for large values of N'_{Th} the process is diffusion-limited, leading to a decrease in η. The dependence of η on the Thiele number is not strongly affected by reaction order. Thermal transport, which has not been considered here, can have a significant effect; for exothermic reactions, it can lead to $\eta > 1$.

As mentioned earlier, catalyst particles are often used in a packed bed reactor. In this configuration, shown schematically in Figure 23.7, gas containing the reactant flows in the z-direction. Reactant from the gas surrounding the particles diffuses and reacts within the particles according to the model just developed for a single particle. We can treat the packed bed as a porous medium and identify the averaging volume $V_{eff,b}$ (right image in

[9] Aris, *Chem. Eng. Sci.* **262** (1957) 6.

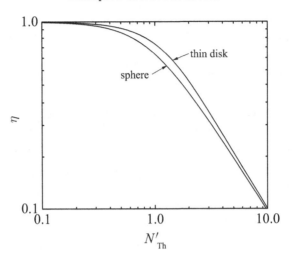

Figure 23.6 Effectiveness factor η for catalyst particles with spherical (23.43) and thin disk (23.52) shapes as a function of the modified Thiele modulus N'_{Th}.

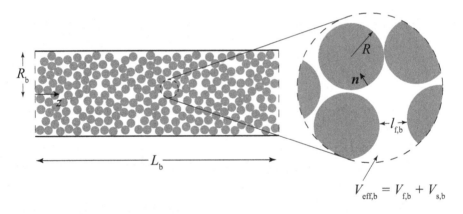

Figure 23.7 Schematic diagram of tubular packed bed reactor having length L_b and radius R_b containing spherical catalyst particles with magnified view of averaging volume $V_{\text{eff},b}$.

Figure 23.7). For randomly packed spheres where $R \ll R_b$, the bed porosity is $\varepsilon_b = V_{f,b}/V_{\text{eff},b} \gtrsim 0.36$.[10]

We now apply the volume-averaging method presented in Section 23.1 to (23.12) with $\alpha = \text{A}$. Integrating over $V_{f,b}$, dividing by $V_{\text{eff},b}$, and applying

---

[10] Scott & Kilgour *J. Phys. D: Appl. Phys.* **2** (1969) 863

(23.6), we obtain

$$\frac{\partial \langle c_A \rangle_{s,b}}{\partial t} = -\boldsymbol{\nabla} \cdot \langle \boldsymbol{v}^* c_A + \boldsymbol{J}_A^* \rangle_{s,b} - \frac{1}{V_{eff,b}} \int_{A_{fs,b}} \boldsymbol{n} \cdot (\boldsymbol{v}^* c_A + \boldsymbol{J}_A^*)(\boldsymbol{r}_{f,b}) d^2 \boldsymbol{r}_{f,b},$$

(23.44)

where $\langle c_A \rangle_{s,b}$ is the superficial-average reactant concentration in the packed bed and $A_{fs,b}$ is the area of fluid–solid surfaces within $V_{eff,b}$. Within $V_{eff,b}$, the flux at $A_{fs,b}$ is uniform, so we can write (23.44) as

$$\frac{\partial \langle c_A \rangle_{s,b}}{\partial t} + \boldsymbol{\nabla} \cdot \langle \boldsymbol{v}^* c_A \rangle_{s,b} = \boldsymbol{\nabla} \cdot (D_{AB} \, \boldsymbol{\nabla} \langle c_A \rangle_{s,b}) - a_{fs,b} \boldsymbol{n} \cdot \boldsymbol{N}_A, \quad (23.45)$$

where $\boldsymbol{N}_A = \boldsymbol{v}^* c_A + \boldsymbol{J}_A^*$ is the local reactant flux due to the reaction taking place within the catalyst particles. Note that the surface area per unit volume of the packed bed $a_{fs,b} = A_{fs,b}/V_{eff,b} \approx l_{f,b}^{-1}$ does not include the surface area within the catalyst particles; for a packed bed containing spherical particles $a_{fs,b} = 3(1 - \varepsilon_b)/R$. On comparing (23.19) and (23.45), we see that they have the same form with the exception of a convective transport term in the latter. Using procedures[11] similar to those followed in the previous section, (23.45) can be written as

$$\varepsilon_b \frac{\partial \langle c_A \rangle_{i,b}}{\partial t} + \boldsymbol{\nabla} \cdot (\varepsilon_b \langle \boldsymbol{v}^* \rangle_{i,b} \langle c_A \rangle_{i,b}) = \boldsymbol{\nabla} \cdot (\varepsilon_b \boldsymbol{D}_{eff,b} \, \boldsymbol{\nabla} \langle c_A \rangle_{i,b}) - a_{fs,b} \boldsymbol{n} \cdot \boldsymbol{N}_A,$$

(23.46)

where $\langle c_A \rangle_{i,b}$ is the intrinsic-average reactant concentration. The effective diffusivity tensor $\boldsymbol{D}_{eff,b}$ is given by (23.30) plus a term accounting for dispersion (see Section 11.2) that results from convective transport. If the molar density of the gas within the packed bed reactor is constant, it is straightforward to show that the volume-averaged form of $\boldsymbol{\nabla} \cdot \boldsymbol{v}^* = 0$ is given by

$$\boldsymbol{\nabla} \cdot (\varepsilon_b \langle \boldsymbol{v}^* \rangle_{i,b}) = 0, \quad (23.47)$$

where we have used $\boldsymbol{n} \cdot \boldsymbol{v}^* = 0$ at $A_{fs,b}$.

We apply these equations to a packed bed reactor at steady state and, for convenience, neglect diffusive transport and assume all fields are uniform in the radial direction. In this case, (23.46) and (23.47) lead to

$$\langle v_z^* \rangle_{i,b} \frac{d \langle c_A \rangle_{i,b}}{dz} = -\frac{3}{R} \frac{1 - \varepsilon_b}{\varepsilon_b} \boldsymbol{n} \cdot \boldsymbol{N}_A, \quad (23.48)$$

where $\langle v_z^* \rangle_{i,b}$ is constant. This mass balance for the catalytic reactor bed is linked to our previous result for a single catalyst particle by the flux

---

[11] See Chapter 3 of Whitaker, *The Method of Volume Averaging* (Kluwer, 1999).

appearing on the right-hand side of (23.48). We hence write for the local reactant flux

$$\boldsymbol{n} \cdot \boldsymbol{N}_A = D_{\text{eff}} \frac{d\langle c_A \rangle_i}{dr}(R) = \frac{D_{\text{eff}}}{R} \left[\sqrt{N_{\text{Da}}} \coth(\sqrt{N_{\text{Da}}}) - 1 \right] \langle c_A \rangle_{i,b}, \quad (23.49)$$

where we have used (23.41) with $c_{Ag} \to \langle c_A \rangle_{i,b}$. Using (23.37) and substituting (23.49) into (23.48), we obtain

$$\langle v_z^* \rangle_{i,b} \frac{d\langle c_A \rangle_{i,b}}{dz} = -3 \frac{1 - \varepsilon_b}{\varepsilon_b} \frac{k^s}{R} \left[\frac{6 D_{AB}}{(3 - \varepsilon) R k^s} \eta N_{\text{Th}}'^2 \right] \langle c_A \rangle_{i,b}, \quad (23.50)$$

where η is given by (23.42). From (23.50), we see that the reactant concentration decreases exponentially along the packed bed reactor. The quantity in square brackets contains all of the information (i.e., $R, \varepsilon, a_{\text{fs}}$) about the structure of the catalyst particles and accounts for the interplay of reaction and diffusion within the particles. If non-porous particles were used and the reaction took place only at the particle surfaces, then $\boldsymbol{n} \cdot \boldsymbol{N}_A(A_{\text{fs,b}}) = k^s \langle c_A \rangle_{i,b}$, and the quantity in square brackets would be unity.

Typically, the goal of chemical reactor design is to achieve a desired reactant conversion (to product) at the smallest residence time, which is proportional to the ratio of the reactor size to throughput (see Exercises 5.7 and 10.10). Integration of (23.50) over the length of the packed bed reactor gives the following expression for the dimensionless residence time,

$$\frac{L_b}{\langle v_z^* \rangle_{i,b}} \frac{k^s}{R} = \frac{\varepsilon_b}{3(1 - \varepsilon_b)} \left[\frac{6 D_{AB}}{(3 - \varepsilon) R k^s} \eta N_{\text{Th}}'^2 \right]^{-1} \ln \left(\frac{\langle c_A \rangle_{i,b}(0)}{\langle c_A \rangle_{i,b}(L_b)} \right). \quad (23.51)$$

Hence, the residence time is proportional to the reactant conversion (the logarithm of the inlet and outlet reactant concentrations) and has a rather weak dependence on the bed porosity ε_b. The residence time is reduced by increasing the quantity in square brackets, which accounts for reaction and diffusion within the porous catalysts. From Figure 23.6 we see for $N_{\text{Th}}' \gg 1$ that $\eta N_{\text{Th}}'^2 \propto N_{\text{Th}}'$, which suggests that the catalyst particles be designed so that the process operates in the diffusion-limited regime.

Exercise 23.4   *Solution of Diffusion–Reaction Equation for a Sphere*
Use the results from Exercises 7.12 and 10.6 to solve (23.38) and (23.39) and obtain the expression in (23.40).

Exercise 23.5   *Effectiveness Factor for a Thin Disk*
Consider a disk-shaped catalytic particle having radius R and thickness $2H$ where

$H/R \ll 1$. Using the definition for the effectiveness factor in (23.42), obtain

$$\eta = \frac{\tanh(N'_{\mathrm{Th}})}{N'_{\mathrm{Th}}}. \tag{23.52}$$

Exercise 23.6   Effectiveness Factor with External Mass Transfer Resistance
Consider a spherical catalytic particle having radius R surrounded by a gas where there is a resistance to mass transfer between the gas and particle that can be described using a mass transfer coefficient (see Exercise 10.8). The reactant concentration far from the particle surface is c_{Ag}. Formulate a boundary condition at the particle surface and solve for the steady-state reactant concentration within the particle. Use this result in (23.42) to derive

$$\eta = \frac{N_{\mathrm{Sh}}[3N'_{\mathrm{Th}} \coth(3N'_{\mathrm{Th}}) - 1]}{3{N'_{\mathrm{Th}}}^2[N_{\mathrm{Sh}} + \sqrt{N_{\mathrm{Da}}} \coth(3N'_{\mathrm{Th}})]}, \tag{23.53}$$

where $N_{\mathrm{Sh}} = \tilde{k}_{\mathrm{m}} R/(c D_{\mathrm{eff}})$ is the Sherwood number.

Exercise 23.7   Flow through Porous Media
The volume-averaging approach can also be applied to flow in porous media.[12] For steady creeping (Stokes) flow through a rigid porous medium, the volume-averaged form of (7.8) can be written as

$$\langle \boldsymbol{v} \rangle_{\mathrm{s}} = -\frac{1}{\eta} \boldsymbol{\kappa} \cdot \boldsymbol{\nabla} \langle \mathcal{P} \rangle_{\mathrm{i}} + \boldsymbol{\kappa} \cdot \nabla^2 \langle \boldsymbol{v} \rangle_{\mathrm{i}}, \tag{23.54}$$

where $\boldsymbol{\kappa}$ is the permeability tensor. The superficial-average velocity $\langle \boldsymbol{v} \rangle_{\mathrm{s}} = \varepsilon \langle \boldsymbol{v} \rangle_{\mathrm{i}}$ is constrained by $\boldsymbol{\nabla} \cdot \langle \boldsymbol{v} \rangle_{\mathrm{s}} = 0$. If the second term (known as the Brinkman correction) is not included, this is a form of *Darcy's law*,[13] which was found empirically by the hydraulic engineer Henry Darcy (1803–1858). Consider flow through the packed bed reactor shown schematically in Figure 23.7 with porosity ε_{b} and isotropic permeability tensor $\boldsymbol{\kappa} = \kappa_{\mathrm{b}} \boldsymbol{\delta}$. Derive an expression relating the mass flow rate $\mathcal{W}$ and pressure drop $\Delta \langle \mathcal{P} \rangle_{\mathrm{i}}$, and simplify for the case $\varepsilon_{\mathrm{b}} R_{\mathrm{b}}^2 / \kappa \gg 1$.

Exercise 23.8   Squeezing Flow in a Liquid Film Bound by a Porous Disk
The flow of liquid films between porous solids occurs in many applications such as load bearings and in the joints between bones. Consider the flow of a Newtonian fluid with constant ρ and η in a film with initial thickness H_0 between two disks of radius R (see Figure 23.8). The stationary lower disk is impermeable, while the upper disk has a permeable layer with thickness b and permeability κ. A load $\mathcal{F}_{\mathrm{s}}$ is

---

[12] See Chapter 4 of Whitaker, *The Method of Volume Averaging* (Kluwer, 1999).
[13] Darcy, *Les fontaines publiques de la ville de Dijon* (Dalmont, 1856).

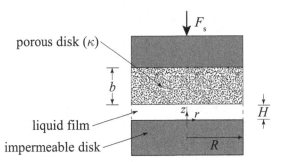

Figure 23.8 Schematic diagram of liquid film between porous and impermeable disks.

applied to the upper disk, inducing a squeeze flow in the liquid film ($\dot{H} = dH/dt < 0$) that, in turn, causes the liquid to flow in the porous disk. Flow in the porous disk is governed by Darcy's law (see Exercise 23.7) and, since $R \gg H$, the lubrication approximation holds in the liquid film. Apply the quasi-steady-state approximation and derive an expression for the force $\mathcal{F}_s$ as a function of the film thickness H.

- The technique of volume averaging couples transport within bulk phases to transport between bulk phases and interfaces.
- Volume-averaged forms of transport equations are expressed in terms of volume-averaged fields and involve effective transport coefficients.
- For diffusion in porous catalyst particles with heterogeneous chemical reaction, the volume-averaged form of the species mass balances have the form of a diffusion equation with a source term.
- The effectiveness factor, which is the actual rate of chemical reaction within a catalyst particle normalized by the ideal rate of reaction, is a useful quantity to understand the relative rates of reaction and diffusion in porous catalyst particles.
- Volume-averaged equations can be applied to create multiscale models that involve transport phenomena occurring over a wide range of length scales.

24

Transport in Biological Systems

Life depends on transport. In this chapter, we discuss molecular motors and ion pumps, which perform key functions in living organisms. Both examples of molecular machines are based on activated transport, that is, transport enabled by the energy provided by a chemical reaction, typically the hydrolysis of adenosine triphosphate (ATP). Molecular motors convert chemical energy into mechanical work (such as locomotion, carrying around stuff in cells); ion pumps transport ions against concentration gradients. In both cases, our goal is to develop kinetic theories to predict phenomenological transport coefficients from more microscopic descriptions. Both kinetic theories are based on stochastic equations. For ion pumps, we rely on a diffusion equation introduced in Chapter 2 and repeatedly occurring ever since. For molecular motors, we rely on a master equation as a useful new tool.

24.1 Molecular Motors

Purposeful movement is one of the characteristics of living organisms. This ability is based on molecular motors that convert chemical into mechanical energy. Indeed, molecular motors are the key to *active transport*, which keeps living organisms away from deadly equilibrium. Our description and modeling of such tiny machines is strongly influenced by Chapter 16 of the comprehensive textbook entitled *Physical Biology of the Cell* by Phillips, Kondev, and Theriot.[1]

Transport of ions through cell membranes against ion concentration gradients, already mentioned for motivation in Section 1.1, provides an example of active transport. Even large molecules, such as proteins, nucleic acids, sugars, or lipids, can be pushed or pulled through holes in membranes, which is typically achieved by *translocation motors*. In the present chapter, however,

---

[1] Phillips *et al.*, *Physical Biology of the Cell* (Garland Science, 2009).

we rather focus on *translational motors*. Transport in nerve cells and muscle contraction are our motivating examples for looking at translational motors. *Rotational motors* and *polymerization motors*, where the latter exert force by the assembly or disassembly of filaments, are not discussed here; however, the modeling of translational molecular motors in terms of master equations as developed in this chapter can readily be adapted to rotational and polymerization motors.

The basic principle of translational molecular motors is walking on lines, which are also referred to as tracks. Two visionary lines from Johnny Cash's famous song *I Walk the Line* (1956) address two key issues in understanding molecular motors, namely "I keep the ends out for the tie that binds" and "You've got a way to keep me on your side," as we elaborate in the following.

Translational molecular motors consist of motor proteins, such as kinesins and myosins, moving in a particular direction along asymmetric tracks, such as microtubules and actin filaments in the cytoskeleton (the cellular scaffolding contained in a cell's cytoplasm). A motor protein typically possesses a head domain that binds to the track and a tail domain that binds to the cargo ("the tie that binds"). It moves along the track by taking a series of discrete steps of fixed size given by the distance between proper functional groups on the track for binding to the head domain. The motor head domain catalyzes the hydrolysis of ATP and, with the help of the released energy, undergoes a large-scale conformational change that, accompanied by binding and debinding, allows it to take a step along the track. The number of steps that a motor protein can take on a single track before falling off is known as processivity ("a way to keep me on your side").

For further motivation and illustration, we describe two important examples of such translational molecular motors in some detail.

1 Transport of cargo, such as proteins or vesicles, in a neuron (nerve cell) from the site of synthesis in the cell body to the synapses at the end of the axon (nerve fibre). This may not sound impressive, but note that the length of an axon can be several meters, for example, in a giraffe, so that this intracellular transport phenomenon, which takes place at a speed of 20–40 cm/day, is truly macroscopic. The molecular motor protein responsible for this cargo transport process in neurons is from the family of kinesins, and the tracks it walks on are microtubules.

2 Muscle contraction. The molecular motor protein behind muscle contraction is from the family of myosins, and it moves along actin filaments. For example, the muscle myosin II is a double-stranded molecule of length

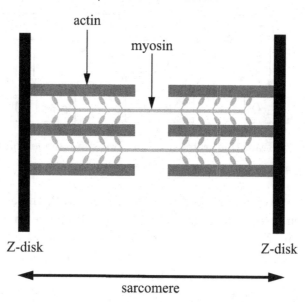

Figure 24.1 A sarcomere as the basic unit of a muscle. The ends of the myosin filaments are actually attached to the Z-disk by means of highly elastic molecules which, for simplicity, are not shown in the figure. [For a more realistic sketch see, for example, Kim *et al.*, *Trends Cell Biol.* **18** (2008) 264.]

150 nm and a molecular weight of order 500 kg/mol. Several hundred individual myosin II molecules self-assemble to form a *thick filament* of almost 2 μm length and 60 nm thickness, from which the head domains stick out (see Figure 24.1). The myosin molecules in the two halves of the thick filament point into opposite directions. The basic unit of a muscle, known as a sarcomere, consists of two parallel Z-disks (from the German word "Zwischenscheibe") to which the ends of actin filaments (*thin filaments*) are attached. Through an alternating arrangement of myosin filaments walking on actin filaments, muscle contraction is achieved. Typical velocities are of the order of 0.5–1 μm/s, and typical step sizes range from 8 nm (kinesin) to 74 nm (myosin V), so that some 10 steps per second need to be taken.

The simplest model of molecular motors accounts for (i) the position za of a motor protein along the filament track, where z is an integer and a is the step size corresponding to the distance between successive binding sites on the track, and (ii) the internal state $i = 0, 1$ associated with ATP hydrolysis and the corresponding conformational change of the motor head domain. By allowing for the different types of transition rates $\Gamma_A^\pm$ and $\Gamma_B^\pm$

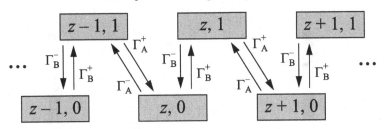

Figure 24.2 Transition rates for the two-state model of translational motors.

between the states z, i as illustrated in Figure 24.2, the asymmetry of the track is taken into account in a simple way. Of course, these rates depend on the magnitude F of the applied force, which we assume to be applied in the backward direction. The energy to move and work against this force, or load, is provided by the ATP hydrolysis. If the ratio of transition rates Γ_A^+/Γ_A^- is large because the transition from the internal state 1 to 0 releases energy, then, in the presence of an applied backward force F, the ratio

$$\frac{\Gamma_A^+(F)}{\Gamma_A^-(F)} = \frac{\Gamma_A^+(0)}{\Gamma_A^-(0)} \exp\left(-\frac{aF}{k_B T}\right) \tag{24.1}$$

may still be larger than unity (k_B is Boltzmann's constant, T is the absolute temperature) Hence, the motor protein can move against the force in the forward direction. For notational simplicity, we here assume that the full displacement a takes place in the ATP hydrolysis step. Of course, we should expect the transition rates to depend also on the concentration of ATP.

The simplest way of satisfying (24.1) is by assuming that only one of the two rates Γ_A^+ and Γ_A^- depends on F. We thus arrive at the two basic models,

$$\Gamma_A^+ = c_+ e^{f_0 - f}, \quad \Gamma_A^- = c_-, \quad \Gamma_B^+ = \frac{1}{c_+}, \quad \Gamma_B^- = \frac{1}{c_-}, \tag{24.2}$$

$$\Gamma_A^+ = c_+ e^{f_0}, \quad \Gamma_A^- = c_- e^{f}, \quad \Gamma_B^+ = \frac{1}{c_+}, \quad \Gamma_B^- = \frac{1}{c_-}, \tag{24.3}$$

where the coefficients c_+ and c_- may depend on the ATP concentration and $f = aF/(k_B T)$ is the dimensionless applied force appearing in (24.1). Note that we have chosen units of time such that $\Gamma_A^- \Gamma_B^- = 1$ (for $f = 0$). The parameter f_0 is the energy provided by the hydrolysis of an ATP molecule divided by the thermal energy $k_B T$. Realistic values of f_0 are around 20.

If $p(t, z, i)$ represents the probability of finding a molecular motor protein at time t at position z and in the internal state i, the transition rates indicated in Figure 24.2 imply the following time-evolution equations, also known as master equations,[2]

$$\frac{\partial p(t, z, 0)}{\partial t} = \Gamma_A^+ p(t, z-1, 1) + \Gamma_B^- p(t, z, 1) - (\Gamma_A^- + \Gamma_B^+) p(t, z, 0) \quad (24.4)$$

and

$$\frac{\partial p(t, z, 1)}{\partial t} = \Gamma_A^- p(t, z+1, 0) + \Gamma_B^+ p(t, z, 0) - (\Gamma_A^+ + \Gamma_B^-) p(t, z, 1). \quad (24.5)$$

Note that we have neglected the possibility that the motor protein might fall off the track. The natural way to solve these coupled linear ordinary differential equations for $p(t, z, i)$ would be to introduce Fourier series in z so that two coupled linear equations for each Fourier mode are obtained (see Exercise 24.3). We here follow a more direct strategy for obtaining the long-time behavior of the solutions in an intuitive way.

Assume that the motor protein initially is at position z and in the internal state 0. We consider a cycle of two successive steps, at the end of which the protein is at $z-1$, z, or $z+1$, and back to the internal state 0. The respective probabilities can be calculated easily because the probabilities of each step are proportional to the rates,

$$P_{z-1} = P^- = \frac{\Gamma_A^- \Gamma_B^-}{\Gamma}, \qquad P_z = \frac{\Gamma_A^+ \Gamma_A^- + \Gamma_B^+ \Gamma_B^-}{\Gamma}, \qquad P_{z+1} = P^+ = \frac{\Gamma_A^+ \Gamma_B^+}{\Gamma}, \tag{24.6}$$

with

$$\Gamma = (\Gamma_A^- + \Gamma_B^+)(\Gamma_A^+ + \Gamma_B^-). \tag{24.7}$$

The average first and second moments of the displacement in a single cycle are given by

$$M_{\text{cycle}}^{(1)} = (P^+ - P^-) a \tag{24.8}$$

and

$$M_{\text{cycle}}^{(2)} = (P^+ + P^-) a^2. \tag{24.9}$$

If we look at N successive cycles, which are stochastically independent, the corresponding moments are

$$M_N^{(1)} = N M_{\text{cycle}}^{(1)}, \tag{24.10}$$

and

$$M_N^{(2)} = N M_{\text{cycle}}^{(2)} + N(N-1) \left(M_{\text{cycle}}^{(1)} \right)^2. \tag{24.11}$$

[2] van Kampen, *Stochastic Processes in Physics and Chemistry* (North Holland, 1992).

We next wish to understand the stochastic process of motor protein walking in the course of time. We hence need to know the number of cycles N occurring up to time t. For a one-step process with a single rate Γ', the distribution of the corresponding stochastic variable N_t would be a Poisson distribution with the moments $\langle N_t \rangle = \Gamma' t$ and $\langle N_t(N_t - 1) \rangle = (\Gamma' t)^2$. For our two-step model, the corresponding result is slightly more complicated (see Exercise 24.1),

$$\langle N_t \rangle = \frac{\Gamma}{\Gamma_{\text{tot}}^2}(\Gamma_{\text{tot}} t - 2) \tag{24.12}$$

and

$$\langle N_t(N_t - 1) \rangle = \frac{\Gamma^2}{\Gamma_{\text{tot}}^4}(\Gamma_{\text{tot}}^2 t^2 - 6\Gamma_{\text{tot}} t + 12), \tag{24.13}$$

where we have introduced the sum of all four transition rates as

$$\Gamma_{\text{tot}} = \Gamma_A^+ + \Gamma_A^- + \Gamma_B^+ + \Gamma_B^-. \tag{24.14}$$

In the expressions (24.12) and (24.13), we have neglected exponentially decaying terms of the form $\exp(-\Gamma_{\text{tot}} t)$. By combining (24.10) and (24.11) with (24.12) and (24.13), we obtain the following expressions for the rate of change of the average and variance of the displacement,

$$A_0 = \frac{\Gamma}{\Gamma_{\text{tot}}} M_{\text{cycle}}^{(1)} = \frac{\Gamma_A^+ \Gamma_B^+ - \Gamma_A^- \Gamma_B^-}{\Gamma_{\text{tot}}} a, \tag{24.15}$$

$$
\begin{aligned}
D_0 &= \frac{\Gamma}{\Gamma_{\text{tot}}} M_{\text{cycle}}^{(2)} - 2\frac{\Gamma^2}{\Gamma_{\text{tot}}^3}\left(M_{\text{cycle}}^{(1)}\right)^2 \\
&= \frac{\Gamma_A^+ \Gamma_B^+ + \Gamma_A^- \Gamma_B^-}{\Gamma_{\text{tot}}} a^2 - 2\frac{(\Gamma_A^+ \Gamma_B^+ - \Gamma_A^- \Gamma_B^-)^2}{\Gamma_{\text{tot}}^3} a^2.
\end{aligned} \tag{24.16}
$$

Note that the time-dependent variance of the displacement of the motor protein obtained by averaging (24.10) and (24.11) can also be written in the illuminating form

$$\langle N_t \rangle \left[M_{\text{cycle}}^{(2)} - \left(M_{\text{cycle}}^{(1)}\right)^2 \right] + \left(\langle N_t^2 \rangle - \langle N_t \rangle^2 \right) \left(M_{\text{cycle}}^{(1)} \right)^2. \tag{24.17}$$

One contribution stems from the variance of a cycle, and a second contribution comes from the variation in the number of cycles. The diffusion coefficient D_0 is the growth rate of the variance (24.17).

If one is interested in a time-resolved view of a single cycle, one can look at the cycle or dwell time distribution for two successive rate processes with

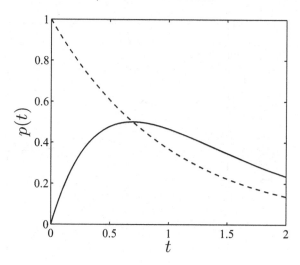

Figure 24.3 Dwell time distribution (24.18) for $\Gamma_1 = 1$, $\Gamma_2 = 2$ (continuous line); for comparison, the single exponential decay e^{-t} is also shown (dashed line).

rates Γ_1 and Γ_2, such as $\Gamma_1 = \Gamma_B^+$ and $\Gamma_2 = \Gamma_A^+$ for an overall step to the right. The result for the dwell time distribution is

$$p(t) = \Gamma_1 \Gamma_2 \int_0^\infty dt_1\, e^{-\Gamma_1 t_1} \int_0^\infty dt_2\, e^{-\Gamma_2 t_2}\, \delta(t_1 + t_2 - t) = \frac{e^{-\Gamma_1 t} - e^{-\Gamma_2 t}}{(1/\Gamma_1) - (1/\Gamma_2)}.$$
$$(24.18)$$

Figure 24.3 illustrates that this result is very different from the exponentially decaying dwell time distribution of a single rate process, which would result as the limit if one of the rate constants became very large so that only the slow step remains visible. Therefore, the number of relevant steps in a cycle may be inferred from the shape of the cycle time distribution.

An important characteristic of a molecular motor is the average drift velocity as a function of the backward force against which the motor has to work. Figure 24.4 shows the behavior near the stall force, at which the motor comes to rest and is forced to move in the backward direction, for the rate parameters introduced in (24.2) and (24.3) with coefficients $c_+ = 1$, $c_- = 0.2$, and $f_0 = 20$. The dashed curve corresponding to (24.3) has the nice feature that the motor velocity for $f = 0$ is maintained almost up to the stall force f_0. For the continuous curve corresponding to (24.2), the velocity is reduced to less than half of its initial value already at $0.9 f_0$, whereas the much slower velocity in the backward direction beyond the stall force does not seem to be much of an advantage.

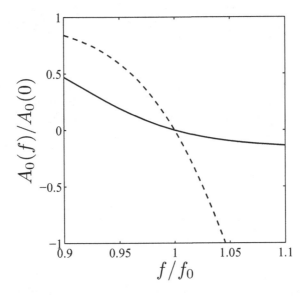

Figure 24.4 Normalized forward drift velocity as a function of the backward pulling force f divided by the stall force f_0; the continuous and the dashed lines correspond to the rate parameters (24.2) and (24.3), respectively, where the coefficients are chosen as $c_+ = 1$, $c_- = 0.2$, and $f_0 = 20$.

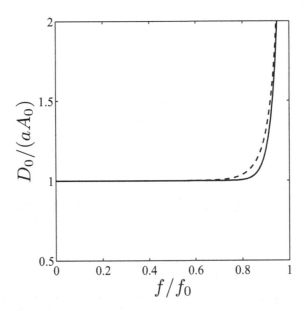

Figure 24.5 Randomness as a function of the backward pulling force f divided by the stall force f_0; the continuous and the dashed lines correspond to the rate parameters (24.2) and (24.3), respectively, where the coefficients are chosen as $c_+ = 1$, $c_- = 1$, and $f_0 = 20$.

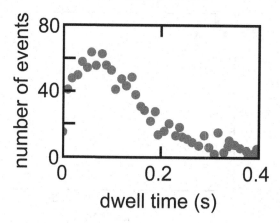

Figure 24.6 Experimental data for the dwell time histogram for the myosin V motor protein. [Redrawn from Purcell *et al.*, *Proc. Natl. Acad. Sci. U.S.A.* **102** (2005) 13873.]

As a further quantity of interest, we consider the so-called randomness parameter,

$$r = \frac{D_0}{aA_0}, \tag{24.19}$$

as a function of the load. This dimensionless parameter compares the diffusion coefficient with the average velocity, or stochastic with deterministic effects. Figure 24.5 shows that the models (24.2) and (24.3) make almost undistinguishable predictions for this parameter.

Note that we have considered the motion of individual molecules to understand molecular motors. When viewed from a larger distance and on a correspondingly longer time scale, the motion of the motor protein molecules results from many cycles. As the sum of independent cycle random variables, the total displacement tends to become Gaussian. The parameters A_0 and D_0 introduced in (24.15) and (24.16) should be recognized as macroscopic transport properties. By comparison with (2.9) and (2.10), we actually find that the long-time behavior is governed by a diffusion equation with the constant drift and diffusion coefficients A_0 and D_0. We have thus managed to express these coefficients in terms of the step size and molecular transition rates. The master equation describing atomistic phenomena is thus related to the macroscopic diffusion equation of Chapter 2. It is the general goal of kinetic theory to express macroscopic material properties in terms of molecular properties, which is a further step in our development of the theory of transport phenomena.

Let us now compare our model predictions with experimental results. Experimental data for the dwell time distribution for a myosin motor protein

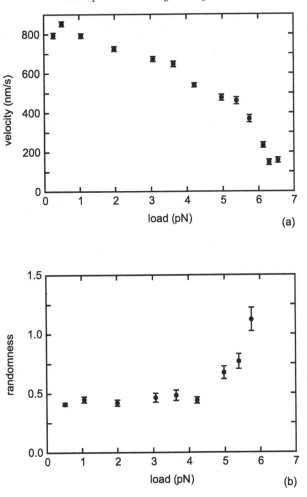

Figure 24.7 Experimental data for (a) the molecular motor velocity and (b) the randomness parameter for kinesin moving along microtubules at high ATP concentration. [Redrawn from Visscher *et al.*, *Nature* **400** (1999) 184.]

are shown in Figure 24.6. The data demonstrate nicely that at least a two-state model is required in order to reproduce the observed behavior (cf. Figure 24.3). Actually, the rate parameters $(11 \pm 1)\,\mathrm{s}^{-1}$ and $(24 \pm 3)\,\mathrm{s}^{-1}$ were extracted in the original paper.[3] This is a nice example of a direct experimental determination of two rate parameters for the two-state model.

We next discuss two ways to distinguish experimentally between the models for the rate parameters introduced in (24.2) and (24.3). Figure 24.7 shows

[3] Purcell *et al.*, *Proc. Natl. Acad. Sci. U.S.A.* **102** (2005) 13873.

experimental data for the molecular motor velocity and the randomness parameter for kinesin at high ATP concentration. The stall force is around 7.5 pN. Figure 24.7(a) suggests a significant decrease of the motor velocity at loads significantly smaller than the stall force. This situation corresponds to the continuous curve in Figure 24.4, that is, for the rate parameters in (24.2). The experimentally observed randomness shown in Figure 24.7(b) is nicely consistent with the predictions of Figure 24.5.

Our comparison with experimental data demonstrates convincingly that the master equation for the two-state model provides an excellent starting point for understanding molecular motors. For more complete data sets, it may turn out that more than two states are needed to fit experimental observations. However, the tools developed in this chapter still remain applicable. We have thus successfully developed a detailed kinetic theory in the context of molecular motors.

Exercise 24.1   Number of Cycles in a Given Time
Derive (24.12) and (24.13).

Exercise 24.2   Dwell Time Distribution
Verify the expression (24.18) for the dwell time distribution for two successive rate processes with rates Γ_1 and Γ_2.

Exercise 24.3   Dispersion Relation
Try the *Ansatz* $p(t, z, j) = C_j \exp[i(kza - \omega t)]$ for solving the two-state model. Nontrivial solutions $C_j \neq 0$ are obtained only for a special function $k(\omega)$ which is known as the dispersion relation. Expand the dispersion relation for small k, and verify that you obtain $\omega = A_0 k - (i/2)D_0 k^2$ [cf. (2.28) in Section 2.5].

Exercise 24.4   Energy Burned by Muscle
Estimate the maximum energy that can be burned by 1 kg of muscle in one hour.

Exercise 24.5   Time Taken to Lift a Weight
How long does it take to lift a weight of 1 kg by 1 m if 5 kg of muscle mass are involved. [Hint: make the realistic assumption that molecular motors work at close to 100 % efficiency.]

24.2 Active Transport: Irreversible Thermodynamics

Throughout this book, we have examined a wide range of phenomena involving fluxes of momentum, energy, and mass that are driven by forces resulting from gradients of velocity, temperature, and concentration. In Chapter 6, we learned that, to ensure nonnegative entropy production, the direction of the flux coincided with the direction of minus the gradient creating the force. For example, heat is transported from regions of high temperature to regions of low temperature, and species mass is transported from regions of high concentration to regions of low concentration. Is it possible for mass to be transferred from regions of low concentration to regions of high concentration? Fortunately, for all living things, the answer is yes, and it occurs because of what is known as *active transport*.

Active transport is the transport of mass from regions of low concentration regions of to high concentration made possible by a coupling of diffusive transport and chemical reaction. This statement might appear to be at odds with two important concepts from Chapter 6: the requirement for nonnegative entropy production (violation of the second law of thermodynamics); and the coupling of force–flux pairs having different tensorial orders is not allowed (violation of the Curie principle). In this section, these issues will be addressed as we examine an example of active transport that occurs within biological cells.[4]

The key to resolving the apparent violation of the Curie principle is that active transport occurs at *cell membranes*, which can be treated as interfaces. Cells are separated from the extracellular matrix by membranes, and also organelles within cells (such as mitochondria and vacuoles) are separated from the surrounding cytosol by membranes. These membranes, which have a thickness of roughly 10 nm, are comprised of phospholipids having two hydrophobic tails connected to a hydrophilic head that self-assemble to form bilayers. Phospholipid bilayers, which have imbedded proteins within them, regulate the transfer of various components such as water, protons, and cations (see Figure 24.8).

We begin by considering the rate of entropy production at an interface where mass transfer and chemical reaction occur. Neglecting mechanical and thermal effects ($T^{\mathrm{I}} = T^{\mathrm{s}} = T^{\mathrm{II}} = T$), we can write (15.23) as

$$T\sigma^{\mathrm{s}} = \sum_{\alpha=1}^{k} \boldsymbol{n} \cdot [(\boldsymbol{v}^{\mathrm{I}} - \boldsymbol{v}^{\mathrm{s}})\rho_{\alpha}^{\mathrm{I}} - \boldsymbol{m} + \boldsymbol{j}_{\alpha}^{\mathrm{I}}] \left[(\hat{\mu}_{\alpha}^{\mathrm{s}} - \hat{\mu}^{\mathrm{s}}) - (\hat{\mu}_{\alpha}^{\mathrm{I}} - \hat{\mu}^{\mathrm{I}})\right] - \tilde{\Gamma}^{\mathrm{s}}\mathcal{A}^{\mathrm{s}}$$

$$- \sum_{\alpha=1}^{k} \boldsymbol{n} \cdot [(\boldsymbol{v}^{\mathrm{II}} - \boldsymbol{v}^{\mathrm{s}})\rho_{\alpha}^{\mathrm{II}} - \boldsymbol{m} + \boldsymbol{j}_{\alpha}^{\mathrm{II}}] \left[(\hat{\mu}_{\alpha}^{\mathrm{s}} - \hat{\mu}^{\mathrm{s}}) - (\hat{\mu}_{\alpha}^{\mathrm{II}} - \hat{\mu}^{\mathrm{II}})\right], \quad (24.20)$$

---

[4] Katchalsky & Curran, *Nonequilibrium Thermodynamics in Biophysics* (Cambridge, 1965); Caplan & Essig, *Bioenergetics and Linear Nonequilibrium Thermodynamics* (Cambridge, 1983); Kjelstrup *et al.*, *J. Theor. Biol.* **234** (2005) 7.

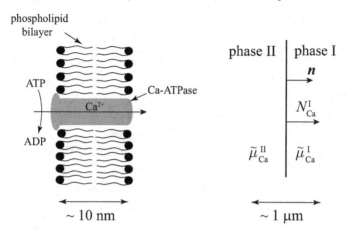

Figure 24.8 Schematic diagram of phospholipid bilayer membrane with Ca-ATPase protein for active transport of Ca^{2+} (left); idealization of membrane as an interface with molar flux N_{Ca}^I from interface into phase I and $\tilde{\mu}_{Ca}^I > \tilde{\mu}_{Ca}^{II}$ (right).

where $\mathcal{A}^s = \sum_{\alpha=1}^{k} \tilde{\nu}_\alpha^s \tilde{\mu}_\alpha^s$ is the affinity for the surface reaction. The expression for entropy production in (24.20) involves only scalar force–flux pairs, so that coupling between chemical reaction and species mass transfer is allowed.

In our analysis of active transport, we consider steady-state situations and assume the interface is uniform. Hence, the excess species mass densities are independent of position and time, and we can use the jump balances for species mass in (14.25). This allows us to rewrite (24.20) as

$$T\sigma^s = \sum_{\alpha=1}^{k} \boldsymbol{n} \cdot [(\boldsymbol{v}^I - \boldsymbol{v}^s)\rho_\alpha^I - \boldsymbol{m} + \boldsymbol{j}_\alpha^I] \left[(\hat{\mu}_\alpha^{II} - \hat{\mu}^{II}) - (\hat{\mu}_\alpha^I - \hat{\mu}^I) \right] - \tilde{\Gamma}^s \sum_{\alpha=1}^{k} \tilde{\nu}_\alpha^s \tilde{\mu}_\alpha^{II},$$
(24.21)

in which the surface chemical potentials have disappeared and species mass transfer occurs directly between the bulk phases. For simplicity, we assume that the interface is equilibrated with the bulk solution on the incoming side, $\tilde{\mu}_\alpha^{II} = \tilde{\mu}_\alpha^s$ (see Figure 24.8), so that the reaction term in (24.21) can be rewritten as $-\tilde{\Gamma}^s \mathcal{A}^s$. To ensure nonnegative entropy production for the two force–flux pairs in (24.21), we write the linear force–flux relations

$$\tilde{\Gamma}^s = -L_{\mathcal{A}\mathcal{A}} \mathcal{A}^s - \sum_{\beta=1}^{k} L_{\mathcal{A}\beta} (\hat{\mu}_\beta^I - \hat{\mu}_\beta^{II}),$$
(24.22)

$$\boldsymbol{n} \cdot [(\boldsymbol{v}^I - \boldsymbol{v}^s)\rho_\alpha^I - \boldsymbol{m} + \boldsymbol{j}_\alpha^I] = -L_{\alpha\mathcal{A}} \mathcal{A}^s - \sum_{\beta=1}^{k} L_{\alpha\beta} (\hat{\mu}_\beta^I - \hat{\mu}_\beta^{II}),$$
(24.23)

where $L_{\alpha\beta}$, with $\alpha, \beta = 1, 2, \ldots, k$, define a matrix of phenomenological coefficients having rank $k - 1$ because it satisfies the constraints $\sum_{\alpha=1}^{k} L_{\alpha\beta} = \sum_{\alpha=1}^{k} L_{\beta\alpha} = 0$ (see Section 15.3). This matrix is augmented by the column vector $L_{\alpha A}$, with $\alpha = 1, 2, \ldots, k$ and the corresponding row vector $L_{A\alpha}$, having the property $\sum_{\alpha=1}^{k} L_{\alpha A} = \sum_{\alpha=1}^{k} L_{A\alpha} = 0$, as well as the diagonal element L_{AA}. The degeneracy of the augmented matrix causes the flat averages of the chemical potentials to drop out.

Typically, active transport occurs in systems that are dilute, multicomponent, aqueous mixtures. We assume that species 1 is the only solute that is transported actively, and we take the solvent as species k. More precisely, we assume that $L_{11} = -L_{1k} = -L_{k1} = L_{kk}$, $L_{1A} = -L_{kA} = -L_{Ak} = L_{A1}$, and L_{AA} are the only non-vanishing phenomenological coefficients. Note that the solvent needs to be taken into account in order to satisfy the constraints on the matrix of phenomenological coefficients. The force–flux relations (24.22) and (24.23) then take the much simpler form

$$\tilde{\Gamma}^{\mathrm{s}} = -L_{AA}\mathcal{A}^{\mathrm{s}} - L_{A1}\,\Delta\hat{\mu}, \tag{24.24}$$

$$\boldsymbol{n} \cdot [(\boldsymbol{v}^{\mathrm{I}} - \boldsymbol{v}^{\mathrm{s}})\rho_1^{\mathrm{I}} - \boldsymbol{m} + \boldsymbol{j}_1^{\mathrm{I}}] = -L_{1A}\mathcal{A}^{\mathrm{s}} - L_{11}\,\Delta\hat{\mu}, \tag{24.25}$$

with $\Delta\hat{\mu} = \hat{\mu}_1^{\mathrm{I}} - \hat{\mu}_1^{\mathrm{II}} - (\hat{\mu}_k^{\mathrm{I}} - \hat{\mu}_k^{\mathrm{II}})$. If the solvent is equilibrated, this definition reduces to $\Delta\hat{\mu} = \hat{\mu}_1^{\mathrm{I}} - \hat{\mu}_1^{\mathrm{II}}$. However, it seems natural to assume that the chemical potential of the solute (species 1) is affected by the nature of the solvent (species k). We further assume that the interface is at rest and that only the diffusive flux $\boldsymbol{n} \cdot \boldsymbol{j}_1^{\mathrm{I}}$ matters: $\boldsymbol{v}^{\mathrm{I}} = \boldsymbol{v}^{\mathrm{s}} = \boldsymbol{m} = \boldsymbol{0}$. The diffusive flux of species 1 is, of course, properly compensated for by a diffusive solvent flux.

Let us consider the specific case of the active transport of Ca^{2+} across the membrane of the sarcoplasmic reticulum, an organelle found in muscle cells. After muscle contraction, Ca^{2+} is transferred from the cytosol to the lumen, or from phase II to phase I (see Figure 24.8), by the hydrolysis of adenosine triphosphate (ATP) to adenosine diphosphate (ADP). This reaction is $ATP + H_2O \rightleftarrows ADP + P$, where P is a phosphate, and has a standard Gibbs free energy of $\Delta\tilde{g}^0 \approx -30$ kJ/mol. The process, which is sometimes referred to as a *calcium ion pump*, occurs when two Ca^{2+} ions from cytosol bind to the head of the protein Ca-ATPase, which triggers the hydrolysis of ATP, causing two Ca^{2+} ions to be transferred to the lumen. Electroneutrality is maintained by the counter-transfer of two protons, H^+, for each Ca^{2+} ion transferred.

For the dilute solution limit, we rewrite the force–flux relations (24.24) and (24.25) in a form based entirely on molar quantities,

$$\tilde{\Gamma}^{\mathrm{s}} = -L_{\mathcal{A}\mathcal{A}}\mathcal{A}^{\mathrm{s}} - \frac{L_{\mathcal{A}\mathrm{Ca}}}{\tilde{M}_{\mathrm{Ca}}}\Delta\tilde{\mu}_{\mathrm{Ca}}, \tag{24.26}$$

$$N^{\mathrm{I}}_{\mathrm{Ca}} = -\frac{L_{\mathrm{Ca}\mathcal{A}}}{\tilde{M}_{\mathrm{Ca}}}\mathcal{A}^{\mathrm{s}} - \frac{L_{\mathrm{Ca}\mathrm{Ca}}}{\tilde{M}^2_{\mathrm{Ca}}}\Delta\tilde{\mu}_{\mathrm{Ca}}, \tag{24.27}$$

with the chemical potential difference $\Delta\tilde{\mu}_{\mathrm{Ca}} = \tilde{\mu}^{\mathrm{I}}_{\mathrm{Ca}} - \tilde{\mu}^{\mathrm{II}}_{\mathrm{Ca}}$ and the molar flux $N^{\mathrm{I}}_{\mathrm{Ca}} = \boldsymbol{n} \cdot \boldsymbol{j}^{\mathrm{I}}_1/\tilde{M}_{\mathrm{Ca}}$. Active transport implies that $N^{\mathrm{I}}_{\mathrm{Ca}}$ and $\Delta\tilde{\mu}_{\mathrm{Ca}}$ have the same sign. The first term on the right-hand side of (24.27) hence has to revert the sign of the second term. By construction, this is consistent with a nonnegative rate of entropy production.

It is of interest to examine the thermodynamic efficiency of active transport processes. Here, we define the efficiency η as the ratio of the energy used to transfer Ca^{2+} across the membrane to the energy released by ATP hydrolysis,

$$\eta = -\frac{N^{\mathrm{I}}_{\mathrm{Ca}}\Delta\tilde{\mu}_{\mathrm{Ca}}}{\tilde{\Gamma}^{\mathrm{s}}\mathcal{A}^{\mathrm{s}}}. \tag{24.28}$$

Using (24.26) and (24.27), we can write (24.28) as

$$\eta = \frac{(q - Z)Z}{1 - qZ}, \tag{24.29}$$

where

$$q = \frac{L_{\mathcal{A}\mathrm{Ca}}}{\sqrt{L_{\mathcal{A}\mathcal{A}}L_{\mathrm{Ca}\mathrm{Ca}}}}, \qquad Z = -\sqrt{\frac{L_{\mathrm{Ca}\mathrm{Ca}}}{L_{\mathcal{A}\mathcal{A}}\tilde{M}^2_{\mathrm{Ca}}}}\,\frac{\Delta\tilde{\mu}_{\mathrm{Ca}}}{\mathcal{A}^{\mathrm{s}}}. \tag{24.30}$$

To ensure the matrix of phenomenological coefficients is positive semidefinite, the coupling coefficient is bound by $q \leq 1$; q characterizes the strength of the cross effect, and Z^2 is the ratio of the entropy decrease from the ion transport and the entropy increase from the chemical reaction without consideration of cross effects. According to (24.27), positive $N^{\mathrm{I}}_{\mathrm{Ca}}$ requires $Z < q$. For $q = 1$, the determinant of the matrix of phenomenological coefficients vanishes. This implies that there is only a single combined dissipative process at work rather than the two coupled processes of chemical reaction and ion transport. We hence say that $q = 1$ corresponds to a tight pump, whereas $q < 1$ describes slipping pumps.

Figure 24.9 shows the dependence of a Ca^{2+} ion pump efficiency as a function of Z for different values of q. As expected, the efficiency increases with increasing values of the coupling coefficient, but only if Z is not too large. Actual values of Ca^{2+} ion pump efficiency are in the range 5%–10%.[5]

[5] Lervik et al., Eur. Biophys. J. **41** (2012) 437; Lervik et al., Biophys. J. **103** (2012) 1218.

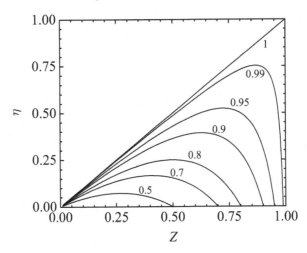

Figure 24.9 Thermodynamic efficiency η of Ca^{2+} ion pump versus Z for different values of the coupling coefficient q given by (24.29) and (24.30).

Exercise 24.6   Maximum Efficiency of Ca^{2+} *Ion Pumps*

Derive an expression for $Z_{\max}$, the value of Z at which the efficiency η is maximized (see Figure 24.9). Assume ideal solution behavior and derive the following expression for the ratio of Ca^{2+} ion mole fractions, and verify $x^{I}_{Ca}/x^{II}_{Ca} > 1$,

$$\ln\left(\frac{x^{I}_{Ca}}{x^{II}_{Ca}}\right) = -\frac{\mathcal{A}^{s}}{\tilde{R}T}\sqrt{\frac{L_{\mathcal{AA}}\tilde{M}^{2}_{Ca}}{L_{CaCa}}}\,\frac{1-\sqrt{1-q^{2}}}{q}.$$

Exercise 24.7   Michaelis–Menten Model for Enzymatic Reactions

An important class[6] of biochemical reactions are modeled as $E+S \rightleftarrows ES \rightarrow E+P$, where enzyme E binds to substrate S to form the complex ES, which then dissociates to form E and the product P. Since we assume the reactions are coupled only by stoichiometry, the reaction fluxes in (6.60) can be written as $\tilde{\Gamma}_{i} = -L_{i}(e^{x_{i}} - e^{y_{i}})$, where $\tilde{R}T\mathcal{A}_{i} = x_{i} - y_{i}$. If ideal solution behavior is assumed, then the reaction fluxes have the form of a mass action law with rate constants k_{1f}, k_{1r}, k_{2f}. Use the results from Exercise 5.6, assuming both convective and diffusive mass transport can be neglected, to derive evolution equations for the species mole fractions. Further assume x_{ES} is at a quasi-steady state to obtain

$$c\frac{dx_{P}}{dt} = \frac{k_{2f}(x_{E} + x_{ES})x_{S}}{K_{M} + x_{S}},$$

where $K_{M} = (k_{1r} + k_{2f})/k_{1f}$ is the Michaelis–Menten dissociation constant.

---

[6] Michaelis & Menten, *Biochem. Z.* **49** (1913) 333.

24.3 Active Transport: Kinetic Theory

As a next step, we would like to describe the calcium ion pump on a more detailed level. The purpose of a more microscopic study is two-fold: we would like (i) to provide a natural way to extend the description of the energy-providing chemical reaction from the linear to the nonlinear regime; (ii) to give more microscopic expressions for the parameters of the macroscopic description developed in the previous section.

For developing a more detailed description, we adopt the ideas presented in a review article by Rubi *et al.*[7] and focus on the reacting state of the enzyme molecules. The state of these molecules is characterized by a reaction coordinate γ. For $\gamma = 0$, ATP and water as reactants come into contact with an enzyme molecule, and the chemical reaction begins. For $\gamma = 1$, the products ADP and P have been formed and are released from the enzyme molecule. Intermediate values of γ describe the continuous transition of the reacting enzyme complex during the reaction. Our description hence rests on the fundamental variable $c_r^s(\gamma)$, which is the molar surface density of reacting enzyme molecules in the state γ. The total molar density of reacting enzyme complexes is given by

$$c_r^{s,\text{tot}} = \int_0^1 c_r^s(\gamma)d\gamma. \tag{24.31}$$

The normalized molar density $p(\gamma) = c_r^s(\gamma)/c_r^{s,\text{tot}}$ is the probability for finding a reacting enzyme molecule in the state γ.

We next consider the Ca^{2+} ions that are passing through the membrane by being bound to the enzyme complex. Again, we spilt the molar concentration into contributions associated with enzymes in different states,

$$c_{Ca}^{s,\text{tot}} = \int_0^1 c_{Ca}^s(\gamma)d\gamma. \tag{24.32}$$

The Ca^{2+} ions are not fully involved in the chemical reactions. This means that there is no stoichiometric coefficient telling us how many Ca^{2+} ions are moved through the membrane per ATP molecule hydrolyzed. Rather there is a stochastic rate for Ca^{2+} ions to escape from the enzyme and to pass through the membrane depending on the state of the enzyme molecule. This feature leads to slippage of the calcium ion pump. Slippage is expected to become relevant for large thermodynamic driving forces.

Our further development follows the approach to transport phenomena used throughout this book. We first formulate balance equations for the

---

[7] Rubi *et al.*, *J. Non-Equilib. Thermodyn.* **32** (2007) 351.

molar densities $c_r^s(\gamma)$ and $c_{Ca}^s(\gamma)$. We then study the redundant entropy balance in order to identify the entropy production and the natural constitutive assumptions for the unknown fluxes occurring in the balance equations.

The balance equations are written in the form

$$\frac{dc_r^s(\gamma)}{dt} = -\frac{\partial}{\partial\gamma}\tilde{\Gamma}^s(\gamma) \tag{24.33}$$

and

$$\frac{dc_{Ca}^s(\gamma)}{dt} = N_{Ca}^{II}(\gamma) - N_{Ca}^I(\gamma), \tag{24.34}$$

where $\tilde{\Gamma}^s(\gamma)$ is a local flux along the reaction coordinate and $N_{Ca}^{I,II}(\gamma)$ are the contributions to the molar fluxes of Ca^{2+} ions from both sides of the membrane into enzymes in the state γ. By integrating over γ, we obtain the total bulk molar fluxes of Ca^{2+} ions. If (24.33) is divided by $c_r^{s,tot}$, it corresponds to the one-dimensional diffusion equation of Section 2.2 with the probability flux $J(\gamma) = \tilde{\Gamma}^s(\gamma)/c_r^{s,tot}$. Equation (24.34) is a simplified version of the balance equations for excess species densities considered in Section 14.3.

The redundant entropy balance equation is obtained from Gibbs' fundamental equation. Under isothermal conditions, we have

$$T\,ds^s(\gamma) = -\tilde{G}(\gamma)dc_r^s(\gamma) - \tilde{\mu}_{Ca}^s(\gamma)dc_{Ca}^s(\gamma), \tag{24.35}$$

where we have resolved also the contributions to the entropy density according to the state of the enzyme. From (24.35), we recognize $\tilde{G}(\gamma) = (-T)\partial s^s(\gamma)/\partial c_r^s(\gamma)$ as a chemical-potential-like quantity. Writing such an equation is based on the assumption that a local equilibrium state is reached at any stage of the chemical reaction. By inserting the balance equations (24.33) and (24.34), we can write the entropy balance as

$$\frac{ds^s(\gamma)}{dt} = \frac{1}{T}\frac{\partial}{\partial\gamma}[\tilde{\Gamma}^s(\gamma)\tilde{G}(\gamma)] + \frac{\tilde{\mu}_{Ca}^I}{T}N_{Ca}^I(\gamma) - \frac{\tilde{\mu}_{Ca}^{II}}{T}N_{Ca}^{II}(\gamma) + \sigma^s(\gamma), \tag{24.36}$$

with the production term

$$\sigma^s(\gamma) = -\tilde{\Gamma}^s(\gamma)\frac{1}{T}\frac{\partial\tilde{G}(\gamma)}{\partial\gamma} - \frac{\tilde{\mu}_{Ca}^I - \tilde{\mu}_{Ca}^s(\gamma)}{T}N_{Ca}^I(\gamma) + \frac{\tilde{\mu}_{Ca}^{II} - \tilde{\mu}_{Ca}^s(\gamma)}{T}N_{Ca}^{II}(\gamma). \tag{24.37}$$

As always, the splitting into flux and source terms is not unique and hence requires physical insight. The first term on the right-hand side of (24.36) represents a flux of entropy along the reaction coordinate as the reaction proceeds. The remaining terms consider the exchange of Ca^{2+} ions between the bulk phases and the enzymes in the interface in the state γ and the exchange of entropy associated with that. Note that the fluxes $N_{Ca}^{I,II}$ are functions of the reaction coordinate γ, whereas the $\tilde{\mu}_{Ca}^{I,II}$ are not, so that slippage is possible. We can simplify the entropy production rate (24.37) by assuming that the Ca^{2+} ions bound to enzymes are always equilibrated with those in the bulk solution on the incoming side, $\tilde{\mu}_{Ca}^{s}(\gamma) = \tilde{\mu}_{Ca}^{II}$ (see Figure 24.8 and the corresponding assumption leading to (24.21)),

$$\sigma^{s}(\gamma) = -\tilde{\Gamma}^{s}(\gamma)\frac{1}{T}\frac{\partial \tilde{G}(\gamma)}{\partial \gamma} - N_{Ca}^{I}(\gamma)\frac{\tilde{\mu}_{Ca}^{I} - \tilde{\mu}_{Ca}^{II}}{T}. \qquad (24.38)$$

The entropy production rate (24.38) is our starting point for the formulation of constitutive equations for the fluxes. The natural linear local force–flux relations are given by

$$\tilde{\Gamma}^{s}(\gamma) = -l_{\mathcal{A}\mathcal{A}}(\gamma)\frac{1}{T}\frac{\partial \tilde{G}(\gamma)}{\partial \gamma} - l_{\mathcal{A}Ca}(\gamma)\frac{\tilde{\mu}_{Ca}^{I} - \tilde{\mu}_{Ca}^{II}}{T}, \qquad (24.39)$$

$$N_{Ca}^{I}(\gamma) = -l_{Ca\mathcal{A}}(\gamma)\frac{1}{T}\frac{\partial \tilde{G}(\gamma)}{\partial \gamma} - l_{CaCa}(\gamma)\frac{\tilde{\mu}_{Ca}^{I} - \tilde{\mu}_{Ca}^{II}}{T}. \qquad (24.40)$$

As usual, the matrix of phenomenological coefficients is symmetric, $l_{\mathcal{A}Ca}(\gamma) = l_{Ca\mathcal{A}}(\gamma)$, and positive semidefinite. The Gibbs free energy introduced in (24.35) consists of two contributions,

$$\tilde{G}(\gamma) = \Phi(\gamma) + \tilde{R}T \ln p(\gamma). \qquad (24.41)$$

The contribution $\Phi(\gamma)$ is associated with the hydrolysis of ATP; we refer to it as the activation energy. Its characteristic behavior is sketched in Figure 24.10. In going from $\gamma = 0$ to $\gamma = 1$, the Gibbs free energy $\Phi_0 - \Phi_1$ associated with the hydrolysis of ATP is released, but a barrier needs to be overcome. The contribution $\tilde{R}T \ln p(\gamma)$ in (24.41) clearly is of entropic origin. For the further interpretation of $\tilde{G}(\gamma)$, we would like to argue that, for a general chemical reaction,

$$\tilde{G}(1) = \sum_{\alpha=q+1}^{k} \tilde{\nu}_{\alpha} \left[\tilde{\mu}_{\alpha}^{0}(T,p) + \tilde{R}T \ln x_{\alpha} \right], \qquad (24.42)$$

and

$$\tilde{G}(0) = \sum_{\alpha=1}^{q} |\tilde{\nu}_{\alpha}| \left[\tilde{\mu}_{\alpha}^{0}(T,p) + \tilde{R}T \ln x_{\alpha} \right], \qquad (24.43)$$

where we have used the chemical potential for ideal mixtures given in (4.60). To understand how the entropic contribution in (24.41) is related to the entropic contributions in (24.42), one should realize that $p(1)$ stands for the simultaneous release of all products according to their stoichiometric

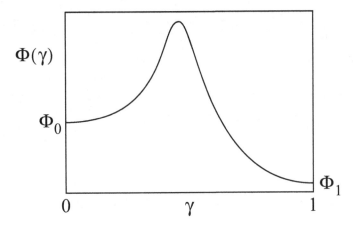

Figure 24.10 Qualitative behavior of the activation energy $\Phi(\gamma)$.

coefficients. Similarly, $p(0)$ requires the simultaneous occurrence of all reactants according to their stoichiometric coefficients. The assumption of an
ideal mixture is crucial to obtain products of $x_\alpha^{\tilde{\nu}_\alpha}$ and to find the intuitive
connection to $c_r^s(\gamma)$ at $\gamma = 0$ and 1. The chemical potentials $\tilde{\mu}_\alpha^0(T,p)$ of
the pure species α and possible normalization factors are included in the
definition of Φ_0 and Φ_1.

For a better understanding of the force–flux relations (24.39) and (24.40)
and of the two contributions in (24.41), we first consider these equations
in the absence of cross effects, that is, for $l_{A\text{Ca}}(\gamma) = l_{\text{Ca}A}(\gamma) = 0$. Equation (24.40) then is a discrete molar version of the constitutive equation
(6.11), where the assumption of an ideal mixture allows us to relate the phenomenological coefficient to a binary diffusion coefficient (see Exercise 6.3).
Equations (24.33) and (24.39) lead to

$$\frac{\partial p(\gamma)}{\partial t} = \frac{\partial}{\partial \gamma}\left\{ p(\gamma)\frac{\tilde{R}\,l_{AA}(\gamma)}{c_r^s(\gamma)}\left[\frac{1}{\tilde{R}T}\frac{\partial \Phi(\gamma)}{\partial \gamma} + \frac{\partial \ln p(\gamma)}{\partial \gamma}\right] \right\}. \qquad (24.44)$$

A comparison with (2.24) reveals that (24.44) describes diffusion in the
dimensionless activation energy landscape, where a constant diffusion coefficient would suggest the choice $l_{AA}(\gamma) \propto c_r^s(\gamma)$, which corresponds to
constant

$$D = \frac{\tilde{R}\,l_{AA}(\gamma)}{c_r^s(\gamma)}. \qquad (24.45)$$

Constant D implies uniform diffusion in γ; this simply suggests that the
otherwise quite arbitrary reaction coordinate has been introduced properly.

The entropic contribution $\tilde{R}T \ln p(\gamma)$ leads to diffusion in the activation energy landscape $\Phi(\gamma)$ (rather than in the full Gibbs free energy landscape $\tilde{G}(\gamma)$) and to the fluctuation–dissipation relation discussed in the remarks at the end of Section 2.4. It has been shown in a classic paper by Kramers[8] that diffusion over an activation-energy barrier is the origin of the chemical reaction rate, as we elaborate in the following.

We now return to the general force–flux relations (24.39) and (24.40) including cross effects. We assume that also $l_{CaA}(\gamma)$ is proportional to $c_r^s(\gamma)$ so that

$$n = \frac{l_{CaA}(\gamma)}{l_{AA}(\gamma)} \tag{24.46}$$

is a constant, which can be interpreted as the number of Ca^{2+} ions transported for every ATP molecule hydrolyzed. Without slippage, this number would be $n = 2$ because the ATPase has two binding sites for Ca^{2+} ions and both ions are transported when an ATP molecule is hydrolyzed. We further introduce

$$n' = \frac{l_{CaCa}(\gamma)}{l_{ACa}(\gamma)}, \tag{24.47}$$

such that, for $n = n'$, the determinant of the matrix of phenomenological coefficients vanishes and there is no slippage (see Exercise 24.9).

If the potential barrier is high, the reaction rate is small and the system will reach a quasi-stationary state, that is, $\tilde{\Gamma}^s(\gamma) = \tilde{\Gamma}^s$ becomes independent of γ. Following Rubi *et al.*[9], we then assume also $N_{Ca}^I(\gamma) = N_{Ca}^I$. In the absence of slippage, it is a plausible assumption; in the presence of slippage, this definitely is a useful approximation. Motivated by the identity (24.53) derived in Exercise 24.8, we multiply the force–flux relations (24.39) and (24.40) by $\exp[\Phi(\gamma)/(\tilde{R}T)]$ and integrate over γ from 0 to 1 to obtain

$$\tilde{\Gamma}^s = -\frac{D}{\mathcal{Z}} c_r^{s,\text{tot}} \left\{ \exp\left[\frac{\tilde{G}(1)}{\tilde{R}T}\right] - \exp\left[\frac{\tilde{G}(0)}{\tilde{R}T}\right] \right\} - \frac{nD\mathcal{N}}{\mathcal{Z}} \frac{\tilde{\mu}_{Ca}^I - \tilde{\mu}_{Ca}^{II}}{\tilde{R}T} \tag{24.48}$$

and

$$N_{Ca}^I = -\frac{nD}{\mathcal{Z}} c_r^{s,\text{tot}} \left\{ \exp\left[\frac{\tilde{G}(1)}{\tilde{R}T}\right] - \exp\left[\frac{\tilde{G}(0)}{\tilde{R}T}\right] \right\} - \frac{nn'D\mathcal{N}}{\mathcal{Z}} \frac{\tilde{\mu}_{Ca}^I - \tilde{\mu}_{Ca}^{II}}{\tilde{R}T}, \tag{24.49}$$

where $\tilde{G}(1)$ and $\tilde{G}(0)$ have been specified in (24.42) and (24.43), and where

$$\mathcal{Z} = \int_0^1 \exp\left[\frac{\Phi(\gamma)}{\tilde{R}T}\right] d\gamma \tag{24.50}$$

---

[8] Kramers, *Physica* **7** (1940) 284.
[9] Rubi *et al.*, *J. Non-Equilib. Thermodyn.* **32** (2007) 351.

and

$$\mathcal{N} = \int_0^1 \exp\left[\frac{\Phi(\gamma)}{\tilde{R}T}\right] c_{\mathrm{r}}^{\mathrm{s}}(\gamma)\,d\gamma. \tag{24.51}$$

The diffusion coefficient D and the integrals $\mathcal{Z}$ and $\mathcal{N}$ in the force–flux relations (24.48) and (24.49) provide a nice example of how phenomenological parameters can be obtained from a more microscopic description. The integral $\mathcal{Z}$ can become very large, as the peak in Figure 24.10 is amplified by the exponential. This integral contains the dominant effect of the activation energy on the reaction rate. In particular, it predicts the temperature dependence of the reaction rate to be governed by Arrhenius' law. In the integral $\mathcal{N}$, the exponentially amplified peak is compensated for by the behavior of $c_{\mathrm{r}}^{\mathrm{s}}(\gamma)$, which is exponentially suppressed in the peak region. At equilibrium, we know the exact form of $c_{\mathrm{r}}^{\mathrm{s}}(\gamma)$, and we obtain $\mathcal{N}_{\mathrm{eq}} = c_{\mathrm{r}}^{\mathrm{s,tot}} / \int_0^1 \exp[-\Phi(\gamma)/(\tilde{R}T)]d\gamma$ (see Exercise 24.10).

The first term of (24.48) contains the nonlinear description of a chemical reaction in the form of a mass-action law as we found in Section 6.6. The same exponential form also governs the cross effect that allows the chemical reaction to produce an ion flux against the chemical potential difference. The linearized version of the thermodynamic force is considered in Exercise 24.11; as a result, one obtains an explicit and more microscopic expression for the matrix of phenomenological coefficients in (24.26) and (24.27),

$$\begin{pmatrix} L_{\mathcal{A}\mathcal{A}} & L_{\mathcal{A}\mathrm{Ca}}/\tilde{M}_{\mathrm{Ca}} \\ L_{\mathrm{Ca}\mathcal{A}}/\tilde{M}_{\mathrm{Ca}} & L_{\mathrm{Ca}\mathrm{Ca}}/\tilde{M}_{\mathrm{Ca}}^2 \end{pmatrix} = \frac{1}{\tilde{R}T}\frac{D\mathcal{N}_{\mathrm{eq}}}{\mathcal{Z}}\begin{pmatrix} 1 & n \\ n & nn' \end{pmatrix}. \tag{24.52}$$

Exercise 24.8   An Integrable Combination
Derive the identity

$$\exp\left[\frac{\Phi(\gamma)}{\tilde{R}T}\right] c_{\mathrm{r}}^{\mathrm{s}}(\gamma)\frac{1}{\tilde{R}T}\frac{\partial \tilde{G}(\gamma)}{\partial \gamma} = \frac{\partial}{\partial \gamma}\left\{\exp\left[\frac{\Phi(\gamma)}{\tilde{R}T}\right] c_{\mathrm{r}}^{\mathrm{s}}(\gamma)\right\} = c_{\mathrm{r}}^{\mathrm{s,tot}}\frac{\partial}{\partial \gamma}\exp\left[\frac{\tilde{G}(\gamma)}{\tilde{R}T}\right].$$
$$\tag{24.53}$$

Exercise 24.9   Condition for the Absence of Slippage
Show that, in the absence of slippage, we have $n = n'\ (= 2)$.

Exercise 24.10   An Equilibrium Integral
Derive the result

$$\mathcal{N}_{\mathrm{eq}} = c_{\mathrm{r}}^{\mathrm{s,tot}}\left\{\int_0^1 \exp\left[-\frac{\Phi(\gamma)}{\tilde{R}T}\right] d\gamma\right\}^{-1}. \tag{24.54}$$

Exercise 24.11 Linearization of the Thermodynamic Force

Linearize the driving force for the chemical reaction in (24.48) and (24.49), and derive the more microscopic expression (24.52) for the matrix of phenomenological coefficients in (24.26) and (24.27).

- Various kinds of molecular motors are of crucial importance for living organisms; for example, translational motors are responsible for the transport of cargo in nerve cells and for muscle contraction.

- A two-state model accounting for one-dimensional displacements and their coupling to an energy-providing chemical reaction can be formulated in terms of a master equation for a stochastic jump process, thus providing a simple kinetic theory of translational molecular motors.

- The two-state model predicts the dwell-time distribution, the drift velocity as a function of the load, and the randomness parameter comparing stochastic with deterministic effects reasonably well.

- In the presence of interfaces, cross effects between chemical reactions and mass transport can and do exist; our phenomenological equations for interfaces show how such a coupling can lead to activated transport against concentration gradients enabled by an energy-providing reaction, for example, in ion pumps.

- Kinetic-theory modeling of chemical reactions by diffusion in activation energy landscapes leads to a nonlinear description of activated transport; high activation energies lead to small reaction rates, where the temperature dependence of reaction rates is governed by Arrhenius' law.

Microbead Rheology

Our first exposure to rheology, the study of the flow behavior of materials, came in Section 7.2 when we saw how the viscosity η of a Newtonian fluid could be measured. In Chapter 12, we saw that viscoelastic fluids are characterized by a frequency-dependent function called the dynamic viscosity $\eta^*(\omega)$, or, equivalently, the dynamic modulus $G^*(\omega)$. Traditional methods for measuring these material functions use rheometers based on shear flows in geometries such as coaxial cylinders as described in Section 7.2 or parallel disks (see Exercise 7.14). A common method for obtaining dynamic rheological properties is to subject the sample to a small-amplitude oscillatory shear over a range of frequencies ω; from the ratio of the stress to strain amplitudes and the phase lag between the two signals, the dynamic modulus $G^*(\omega)$ can be determined (see Exercise 12.1). These "bulk" rheology measurements are made using samples with dimensions ≈ 1–10 mm on time scales in the range $\approx 10^{-2}$–10^3 s. Most commercially available devices can be used to study materials with modulus values in the range 1 kPa–10 MPa.

Microrheology is a relatively new experimental technique that allows the study of complex fluids on time and length scales that are not accessible using traditional bulk rheology methods. As the name implies, microrheology is a class of experimental methods in which the rheological behavior of complex fluids is studied on micron length scales. This is desirable for several reasons. One reason is because the rheological behavior of some complex fluids (see Chapter 12) is the result of structures having this length scale. A second advantage is that it is possible to access smaller time scales, or higher frequencies, because inertial effects are reduced with smaller dimensions. A third benefit of microrheology is that it allows the study of fluids which are available only in small quantities.

There are several ways one could conceive of performing a microrheology experiment. One approach would be to simply scale down the dimensions of conventional geometries like coaxial cylinders or parallel disks. This, however, would be a significant challenge for even the best machinists. A more elegant approach is to insert micron-sized spheres, which are readily available, into the sample and observe their motion. This explains the meaning of the term microbead rheology. The motion of a bead deforms the fluid and, if the forces on the bead are known, it is possible to study the rheological behavior of the fluid. Microbead rheology techniques fall into two broad categories. In active microbead rheology, the motion of the bead is controlled by an electromagnetic field, which is typically created using a laser and is known as an optical trap. The second category is known as passive microbead rheology, where bead motion results from thermal fluctuations, or Brownian forces, in the fluid. In other words, passive microbead rheology is essentially just watching a bead undergo Brownian motion. The technique was pioneered by Mason and Weitz,[1] and a number of reviews of microrheology and its applications are available.[2]

Throughout this book, we have made connections between fluctuations and diffusive transport, and in this chapter we shall see how passive microbead rheology is a beautiful example of this. In fact, the basic principle of this technique can be understood by simply combining a few results described in previous chapters. In Chapter 2, we examined the properties of the diffusion equation governing the probability density for the position of a Brownian particle (a trajectory for a Brownian particle is shown in Figure 2.1). Knowing the probability density, we saw in (2.7) that the average of the square of the particle displacement $\langle \Delta x^2 \rangle$, in the absence of drift, increased linearly in time, $\langle \Delta x^2 \rangle \propto Dt$, where D is the diffusion coefficient of the particle. In Exercise 6.6, we saw that the diffusion coefficient is inversely proportional to the friction coefficient on the particle, $D = k_B T / \zeta$. Finally, in Section 17.3, we derived Stokes' law (17.33) for the force exerted on a sphere moving relative to a fluid and found the friction coefficient to be given by $\zeta = 6\pi\eta R$. Combining these results means that the viscosity of a fluid can be obtained from measurements of the motion of a spherical particle dispersed within it.

Measuring the viscosity of a Newtonian fluid using microbead rheology is not very practical. Besides being limited to Newtonian fluids, the simple example just given is based on several assumptions that should be examined. In this chapter, we carry out an analysis of passive microbead rheology that

[1] Mason & Weitz, *Phys. Rev. Lett.* **74** (1995) 1250.

[2] Waigh, *Rep. Prog. Phys.* **68** (2005) 685; Kimura, *J. Phys. Soc. Jpn.* **78** (2009) 041005; Squires & Mason, *Annu. Rev. Fluid Mech.* **42** (2010) 413.

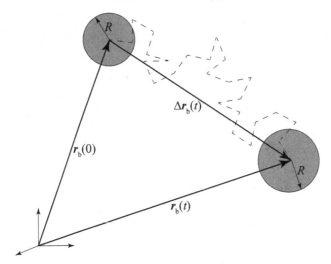

Figure 25.1 Schematic diagram of Brownian motion of spherical particle (bead) with radius R showing particle position $\boldsymbol{r}_\mathrm{b}$ and displacement $\Delta\boldsymbol{r}_\mathrm{b}$ vectors.

is influenced heavily by the work of Schieber and coworkers.[3] The importance of fluctuations in transport phenomena is further reinforced, and the result is a generalization of the Stokes–Einstein relation.

25.1 The Einstein Part

From the preceding discussion, it is clear that we require an equation that governs the motion of a particle in a fluid that experiences Brownian forces. Motion of the particle deforms the fluid, and the resulting stresses in the fluid will be transmitted to the particle. Hence, the particle experiences forces on very different scales: Brownian forces due to molecular interactions, and drag forces from the fluid continuum. An equation that includes these two different forces is called a *Langevin equation*.

Let $\boldsymbol{r}_\mathrm{b}(t)$ be the position of a particle at time t (see Figure 25.1). The stochastic motion of a particle in a viscoelastic medium is governed by the generalized Langevin equation

$$m\frac{d^2}{dt^2}\boldsymbol{r}_\mathrm{b}(t) = -\int_{-\infty}^{t}\zeta(t-t')\frac{d}{dt'}\boldsymbol{r}_\mathrm{b}(t')dt' - H\boldsymbol{r}_\mathrm{b}(t) + \boldsymbol{f}_\mathrm{B}(t), \qquad (25.1)$$

where m is an appropriate mass of the particle, $\zeta(t)$ is a memory function that accounts for the frictional interaction between the particle and the

[3] Córdoba *et al.*, *J. Rheol.* **56** (2012) 185; Indei *et al.*, *Phys. Rev. E* **85** (2012) 021504.

medium, $-H\boldsymbol{r}_b(t)$ is a linear spring force modeling an optical trap located at the origin, and $\boldsymbol{f}_B(t)$ is the Brownian force exerted by the medium on the particle. It can be proven using projection operators[4] that the Brownian force, by definition, has zero first moments,

$$\langle \boldsymbol{f}_B(t) \rangle_{eq} = \boldsymbol{0}, \tag{25.2}$$

and second moments given by

$$\langle \boldsymbol{f}_B(t)\boldsymbol{f}_B(t') \rangle_{eq} = k_B T \zeta(t - t')\boldsymbol{\delta}, \tag{25.3}$$

where the angular brackets $\langle . \rangle_{eq}$ indicate taking an ensemble average at equilibrium. The result in (25.3) giving the relationship between the Brownian forces and the frictional memory function is a form of the *fluctuation–dissipation theorem*. The memory function in (25.1) takes into account past events, or is subject to causality (cf. footnote on p. 219), whereas (25.3) suggests the symmetric continuation $\zeta(-t) = \zeta(t)$. The generalized Langevin equation in (25.1) assumes that the medium is isotropic and that there are no external forces on the particle except for the optical trap.

The term Langevin equation is synonymous with stochastic differential equation; such equations have been discussed in detail in Chapter 3 (see also the bead evolution equations (22.3) and (22.4) in Chapter 22). The presence of a memory integral in (25.1) gives it the designation as a generalized Langevin equation. If the interaction between the particle and the fluid is instantaneous, that is, the memory function has the form $\zeta(t) = \zeta_0 \delta(t)$, then (25.1) is a Langevin equation (see Exercise 25.1). Note also the similarity between the properties of the Brownian force in (25.2) and (25.3) and the random variables (3.2) appearing in the stochastic difference equations.

In passive microbead rheology, the measured quantity is the mean-square displacement of a particle $\langle [\Delta \boldsymbol{r}_b(t)]^2 \rangle_{eq}$, or a related quantity. The particle displacement is given by $\Delta \boldsymbol{r}_b(t) = \boldsymbol{r}_b(t) - \boldsymbol{r}_b(0)$, so we can write

$$\langle [\Delta \boldsymbol{r}_b(t)]^2 \rangle_{eq} = \langle \Delta \boldsymbol{r}_b(t) \cdot \Delta \boldsymbol{r}_b(t) \rangle_{eq} = 2 \left[\langle \boldsymbol{r}_b(0)^2 \rangle_{eq} - \langle \boldsymbol{r}_b(t) \cdot \boldsymbol{r}_b(0) \rangle_{eq} \right], \tag{25.4}$$

where have used the fact that the process is stationary so that $\langle \boldsymbol{r}_b(t)^2 \rangle_{eq} = \langle \boldsymbol{r}_b(0)^2 \rangle_{eq}$. The presence of an optical trap is crucial for obtaining a stationary process because otherwise the particle would diffuse freely. To evaluate (25.4), we must solve (25.1), which is most easily done in the frequency

---

[4] See, for example, Grabert, *Projection Operators* (Springer, 1982) or Chapter 6 of Öttinger, *Beyond Equilibrium Thermodynamics* (Wiley, 2005).

domain. Taking the two-sided Fourier transform[5] of (25.1), we obtain

$$r_b[\omega] = \frac{f_B[\omega]}{H + i\omega\left(\bar{\zeta}[\omega] + mi\omega\right)}, \tag{25.5}$$

where the denominator has the one-sided Fourier transform of the memory function $\bar{\zeta}[\omega]$, which arises from the convolution theorem (see Exercise 25.2).

According to (25.4), we need to know the correlation of particle positions at two different times. Since we are now working in the frequency domain, we use (25.5) to obtain the correlation at two different frequencies, which gives

$$\langle r_b[\omega] r_b[\omega']\rangle_{eq} = \frac{\langle f_B[\omega] f_B[\omega']\rangle_{eq}}{\left[H + i\omega\left(\bar{\zeta}[\omega] + mi\omega\right)\right]\left[H + i\omega'\left(\bar{\zeta}[\omega'] + mi\omega'\right)\right]}. \tag{25.6}$$

To evaluate the numerator, we must transform the fluctuation–dissipation theorem to the frequency domain. Multiplying each side of (25.3) by $e^{-i\omega t}e^{-i\omega' t'}$ and integrating over t and t' from $-\infty$ to ∞, we obtain

$$\langle f_B[\omega] f_B[\omega']\rangle_{eq} = k_B T \boldsymbol{\delta} \int_{-\infty}^{+\infty}\int_{-\infty}^{+\infty} \zeta(t - t')e^{-i\omega t - i\omega' t'}\, dt\, dt'$$

$$= 2\pi k_B T \boldsymbol{\delta}\, \zeta[\omega]\, \delta(\omega + \omega')$$

$$= 4\pi k_B T \boldsymbol{\delta}\, \mathcal{R}\left\{\bar{\zeta}[\omega]\right\}\, \delta(\omega + \omega'), \tag{25.7}$$

where the second line has been obtained by changing the variables of integration from t and t' to $s = t - t'$ and $u = (t + t')/2$, and by noting that the inverse Fourier transform of the Dirac delta function is simply 2π. The last line in (25.7) follows since $\zeta(-t) = \zeta(t)$, and $\mathcal{R}$ indicates taking the real part of a complex quantity (see Exercise 25.3). Substituting the last line of (25.7) into (25.6) gives, after some manipulation (see Exercise 25.4),

$$\langle r_b[\omega] r_b[\omega']\rangle_{eq} = \mathcal{R}\left\{\frac{4\pi k_B T \delta(\omega + \omega')\boldsymbol{\delta}}{\omega\left[\omega\bar{\zeta}[\omega] + i(m\omega^2 - H)\right]}\right\}. \tag{25.8}$$

Taking the two-sided Fourier transform of (25.4), we obtain[6]

$$\langle \Delta r_b^2[\omega]\rangle_{eq} = 4\pi\,\langle r_b(0)^2\rangle_{eq}\,\delta(\omega) - 2\,\langle r_b[\omega]\cdot r_b(0)\rangle_{eq}. \tag{25.9}$$

[5] The one- and two-sided Fourier transforms of the function $f(t)$ are, respectively, defined as

$$\bar{f}[\omega] = \bar{\mathcal{F}}[f(t)] = \int_0^\infty f(t)e^{-i\omega t}\, dt, \qquad f[\omega] = \mathcal{F}[f(t)] = \int_{-\infty}^\infty f(t)e^{-i\omega t}\, dt.$$

The transformed function is indicated by its dependence on ω in square brackets. Using integration by parts, the Fourier transform of the derivative of f gives $\mathcal{F}[df/dt] = i\omega f[\omega]$. Note that in this chapter we use the notation $f[\omega]$ to denote the (two-sided) Fourier transform of the function f, whereas in other parts of the book (cf. footnote on p. 17) we use $\tilde{f}$.

[6] Here, we use $\langle \Delta r_b^2[\omega]\rangle_{eq} = \mathcal{F}\{\langle(\Delta r_b(t))^2\rangle_{eq}\}$ and $\overline{\langle \Delta r_b^2[\omega]\rangle_{eq}} = \bar{\mathcal{F}}\{\langle(\Delta r_b(t))^2\rangle_{eq}\}$.

To evaluate the second term in (25.9), we write $r_b(0) = \int r_b[\omega']d\omega'/(2\pi)$ in terms of its Fourier transform. We can then integrate (25.8) over ω' from $-\infty$ to ∞ and perform a contraction to rewrite (25.9) as

$$\left\langle \Delta r_b^2[\omega] \right\rangle_{eq} = 4\pi \left\langle r_b(0)^2 \right\rangle_{eq} \delta(\omega) - \mathcal{R} \left\{ \frac{4dk_BT}{\omega\left[\omega\bar{\zeta}[\omega] + i(m\omega^2 - H)\right]} \right\}, \quad (25.10)$$

where d is the number of dimensions in which the particle is tracked. Since the mean-square displacement is symmetric in time, we can rewrite (25.10) as follows,[7]

$$\left\langle \overline{\Delta r_b^2[\omega]} \right\rangle_{eq} = -\frac{2dk_BT}{\omega\left[\omega\bar{\zeta}[\omega] + i(m\omega^2 - H)\right]}. \quad (25.11)$$

This result may be regarded as a generalization of the Nernst–Einstein equation (6.31). To go further, we need an expression for the memory function that contains information about the rheological behavior of the medium.

Exercise 25.1   Particle Motion in Newtonian Fluid
If the memory function has the form $\zeta(t) = \zeta_0\delta(t)$, where ζ_0 is a constant, then the fluid is Newtonian and (25.1) simplifies to a Langevin equation. Use (25.11) to find $\left\langle [\Delta r_b(t)]^2 \right\rangle_{eq}$ and identify the criterion when the particle motion is sub-diffusive, or ballistic.

Exercise 25.2   Convolution Theorem
Determine the two-sided Fourier transform of

$$h(t) = \int_{-\infty}^{t} f(t - t')g(t')dt'.$$

Exercise 25.3   One- and Two-Sided Fourier Transforms
Derive the identities $\bar{\zeta}[-\omega] = \bar{\zeta}[\omega]^*$, where the asterisk denotes complex conjugation, and $\zeta[\omega] = 2\mathcal{R}\left\{\bar{\zeta}[\omega]\right\}$.

Exercise 25.4   Frequency-Dependent Correlations
Derive the expression in (25.8).

---

[7] In Exercise 25.3, we show how to obtain the two-sided Fourier transform from the one-sided Fourier transform by taking the real part. For going in the opposite direction, one needs to find the proper imaginary part, which can be obtained from the real part by means of the Kramers–Kronig relations (see p. 219). In writing (25.11), we follow the procedure described on p. 4 of Indei *et al.*, *Phys. Rev. E* **85** (2012) 021504, who moreover argue that the contribution proportional to $\delta(\omega)$ becomes irrelevant for the one-sided Fourier transform.

25.2 The Stokes Part

As noted earlier in this chapter, we already have experience with flow around a spherical particle. In Section 17.3, we found the steady-state solution for creeping flow of a Newtonian fluid around a sphere. For a sphere moving with constant velocity $v_b = dr_b/dt$, the general form of Stokes' law (17.33) can be written as $\mathcal{F}_s = -\zeta_0 v_b$, where $\zeta_0 = 6\pi R\eta$ (see Exercise 17.6).

We now determine the memory function $\zeta(t)$ appearing in (25.1) for the drag force exerted by the medium on the particle. This means we must solve for the velocity and stress fields for the time-dependent flow of a viscoelastic fluid around a sphere. Motion of the sphere is due only to thermal fluctuations, so we assume creeping flow (see Section 17.3) and can therefore neglect terms that are nonlinear in the velocity. We further assume incompressible flow and neglect gravity. To find the drag force on the particle, we must find the solution to (5.33), which in the frequency domain takes the form

$$i\omega\rho v[\omega] = -\nabla \cdot \tau[\omega] - \nabla p^{\mathrm{L}}[\omega], \tag{25.12}$$

subject to the incompressibility constraint in (5.36), which we can write as $\nabla \cdot v[\omega] = 0$. The no-slip boundary condition on the particle can be expressed as $v(R) = v_b$.

Microbead rheology is primarily used to measure the properties of viscoelastic fluids. For reasons that will soon become clear, it is useful to consider first the Newtonian fluid case. For a constant density Newtonian fluid, we have from (6.5)

$$\tau[\omega] = -\eta \left[\nabla v[\omega] + (\nabla v[\omega])^{\mathrm{T}} \right]. \tag{25.13}$$

Albeit more complicated, solutions for time-dependent creeping flow around a sphere have been known for more than a century. An elegant solution of the problem and its historical background are given by Schieber *et al.*[8] Here, we give a brief outline of the solution, which involves writing the velocity in terms of the stream function defined by (17.19). The functional form of the time-dependent stream function given in (17.27) leads to a fourth-order equation for $f(r, t)$, which is solved in the frequency domain. The velocity field decays spatially and temporally in shear waves with wave number squared $k^2 = -i\omega\rho/\eta$. One then obtains the force exerted by the fluid on the sphere moving with time-dependent velocity $v_b(t)$, which in the frequency domain is given by

$$\mathcal{F}_s[\omega] = -\left(6\pi R\eta + 6\pi R^2 \sqrt{\rho i\omega\eta} + \frac{2}{3}\pi R^3 \rho i\omega \right) v_b[\omega]. \tag{25.14}$$

---

[8] Schieber *et al.*, *J. Non-Newtonian Fluid Mech.* **200** (2013) 3.

The first term in (25.14) is the Stokes law drag, and the second is the viscous force due to transient shear waves, sometimes called the Basset force. The third term in (25.14) is acceleration of the fluid near the sphere, which can be added to the mass of the sphere, $m_{\text{eff}} = \frac{4}{3}\pi R^3(\rho_{\text{b}} + \rho/2)$, where ρ_{b} is the density of the particle. Hence, it is natural to include the first two terms in (25.14) in the friction coefficient

$$\bar{\zeta}[\omega] = 6\pi R\eta + 6\pi R^2\sqrt{\rho i\omega\eta}. \tag{25.15}$$

Now we are ready to consider flow around a sphere imbedded in a viscoelastic fluid. As mentioned earlier, because particle motion is driven by Brownian forces, the fluid experiences relatively small deformations. Hence, it is appropriate to use the rheological constitutive equation for linear viscoelasticity. Taking the Fourier transform of (12.13), we obtain

$$\boldsymbol{\tau}[\omega] = -\eta^*(\omega)\left[\boldsymbol{\nabla v}[\omega] + (\boldsymbol{\nabla v}[\omega])^{\mathrm{T}}\right], \tag{25.16}$$

where we have used (12.9) and the convolution theorem. Clearly, (25.16) has the same form as (25.13), and both express the stress as a linear function of velocity. Since (25.12) is also linear in velocity and stress, its solution for given boundary conditions in the frequency domain with either (25.13) or (25.16) is the same, if η is replaced by $\eta^*(\omega)$. This is known as the *correspondence theorem*,[9] which means the spatial dependence of creeping flows of Newtonian and linear viscoelastic fluids are the same. The difference in rheological behavior between purely viscous and viscoelastic fluids appears when the solution is transformed back to the time domain. Hence, the memory function for time-dependent flow around a sphere in a viscoelastic fluid is obtained from (25.15) with η replaced by $\eta^*(\omega)$,

$$\bar{\zeta}[\omega] = 6\pi R\eta^*(\omega) + 6\pi R^2\sqrt{\rho i\omega\eta^*(\omega)} = \frac{6\pi R}{i\omega}\left[G^*(\omega) + Ri\omega\sqrt{\rho G^*(\omega)}\right], \tag{25.17}$$

where the second equality has used the relation $G^*(\omega) = i\omega\eta^*(\omega)$.

Exercise 25.5   Time-Dependent Creeping Flow around a Sphere
The expression in (25.14) can be obtained by solving the problem of time-dependent creeping flow around a sphere with velocity $v_{\text{b}} = v_{\text{b}}(t)\boldsymbol{\delta}_3$. As discussed in Exercise 17.6, it is convenient to solve the governing equations in a moving, or non-inertial, coordinate system. Since the form of the constraint on velocity is unchanged, the expressions for velocity in (17.19) can still be used. Show that the time-dependent

[9] Zwanzig & Bixon, *Phys. Rev. A* **2** (1970) 2005; Xu *et al.*, *J. Non-Newtonian Fluid Mech.* **145** (2007) 150.

stream function is governed by

$$E^2 \frac{\partial \psi}{\partial t} = \frac{\eta}{\rho} E^4 \psi,$$

where E^2 is the operator appearing in (17.20) and defined in Exercise 7.6. Hint: write (11.3) for creeping flow in terms of a modified pressure (dropping primes),

$$\mathcal{P} = p^{\mathrm{L}} + \rho x_3 \frac{dv_{\mathrm{b}}}{dt},$$

and use the results in Exercise 17.5.

25.3 Experimental Results

We have generalized the relationship between the mean-square displacement and the memory function, and we found an expression for the memory function for the time-dependent motion of a particle in a viscoelastic fluid. Combining these results, which means substitution of (25.17) into (25.11), in the absence of a trap gives

$$\left\langle \overline{\Delta r_{\mathrm{b}}^2[\omega]} \right\rangle_{\mathrm{eq}} = \frac{k_{\mathrm{B}}T}{\pi R i \omega \left[G^*(\omega) + R i \omega \sqrt{\rho G^*(\omega)} \right] - \frac{1}{6} m_{\mathrm{eff}} i \omega^3}, \qquad (25.18)$$

which is known as the generalized Stokes–Einstein relation.[10] In this section, we set $d = 3$, assuming particle motion is tracked in three dimensions.

Before we apply (25.18) to analyze microbead rheology experiments, we look at a special case. In the study of Mason and Weitz,[11] the following version of the generalized Stokes–Einstein relation was proposed,

$$\left\langle \overline{\Delta r_{\mathrm{b}}^2[\omega]} \right\rangle_{\mathrm{eq}} = \frac{k_{\mathrm{B}}T}{\pi R i \omega G^*(\omega)}, \qquad (25.19)$$

and it has been used frequently in subsequent research to interpret passive microbead rheology experiments. Comparison of (25.18) and (25.19) suggests the latter should be valid for frequencies at which the Basset force can be neglected, $\omega \ll \sqrt{G^*/(\rho R^2)}$, and when inertia can be neglected, $\omega \ll \sqrt{6\pi G^*/[(\rho_{\mathrm{b}} + \rho/2) R^2]}$. These criteria indicate the upper frequency is decreased for softer materials (lower G^*) and for larger beads (R). For an experiment on a material with $G^* \approx 10^2$ Pa and $\rho \approx 10^3$ kg/m^3, and a bead with $R \approx 10^{-6}$ m and $\rho_{\mathrm{b}} \approx \rho$, the upper limit on frequency is $\omega \approx 10^5$ s^{-1}.

It has been noted that there are more subtle issues behind (25.19) than simply neglecting inertia.[12] For example, consider a single-mode Maxwell

[10] Indei *et al.*, *Phys. Rev. E* **85** (2012) 021504.
[11] Mason & Weitz, *Phys. Rev. Lett.* **74** (1995) 1250.
[12] Córdoba *et al.*, *J. Rheol.* **56** (2012) 185.

fluid with complex modulus $G^*(\omega) = g\lambda i\omega/(1+\lambda i\omega)$. Substitution in (25.19) and transformation back to the time domain gives

$$\langle [\Delta r_{\mathrm{b}}(t)]^2 \rangle_{\mathrm{eq}} = \frac{dk_{\mathrm{B}}T}{3\pi Rg} \left(1 + \frac{t}{\lambda} \right), \tag{25.20}$$

which, as a consequence of neglecting inertia, does not give the correct initial condition $\langle [\Delta r_{\mathrm{b}}(0)]^2 \rangle_{\mathrm{eq}} = 0$. Also, if the second term (Basset force) in (25.17) is neglected and this is substituted into (25.11), then for a single-mode Maxwell fluid the correct initial condition is recovered; however, oscillatory behavior is predicted (see Exercise 25.6). This behavior, which is not observed in passive microbead rheology, is due to a resonance between the bead motion and the elastic response of the medium.

A variety of experimental methods are used to observe bead motion in passive microbead rheology. Each of these methods has its relative strengths and weaknesses that depend to some extent on the material and frequency range of interest. These include particle imaging microscopy, laser tracking, dynamic light scattering, and diffuse wave spectroscopy.

As an example, we present results from a study where both passive microbead rheology and mechanical rheometry were used to measure the complex modulus over a wide range of frequencies.[13] The experiments were performed on a wormlike micellar solution, which is an aqueous solution of surfactant molecules that physically associate to form chainlike structures, that displays viscoelastic behavior. Figure 25.2 shows the mean-square displacement data obtained using diffuse wave spectroscopy. In this figure, diffusive bead motion is observed at small and large times joined by a relatively flat region at intermediate times.

Typically, mean-square displacement data are converted to the frequency domain either numerically or by curve fitting. After transformation, the power spectrum is fit to the generalized Stokes–Einstein relation, usually the approximate version given in (25.19). For the mean-square displacement data in Figure 25.2, the issues discussed above with this expression mentioned above do not play a major role. The resulting complex modulus obtained from the data in Figure 25.2 and (25.19) are shown in Figure 25.3. The agreement with $G^*(\omega)$ obtained from mechanical rheometry is impressive, with the exception of at low frequencies. Also shown in Figure 25.3 are $G^*(\omega)$ obtained from the data in Figure 25.2 and the rigorous version of the generalized Stokes–Einstein relation in (25.18). Small differences are observed only at high frequencies; they are due to the inclusion of

---

[13] Willenbacher *et al.*, *Phys. Rev. Lett.* **99** (2007) 068302.

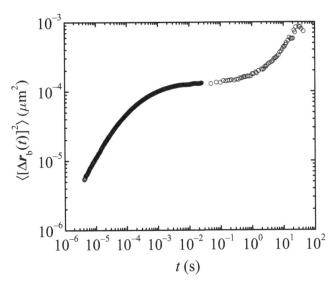

Figure 25.2 Mean-square displacement for a microbead rheology experiment performed on a wormlike micellar solution measured using diffuse wave spectroscopy [see Willenbacher *et al.*, *Phys. Rev. Lett.* **99** (2007) 068302].

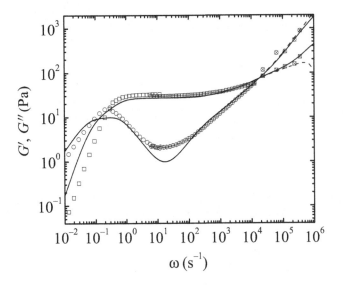

Figure 25.3 Complex modulus for a wormlike micellar solution from the mean-square displacement data in Figure 25.2 and from mechanical rheology: G' (squares); G'' (circles) [see Willenbacher *et al.*, *Phys. Rev. Lett.* **99** (2007) 068302]. The solid curves are obtained using (25.18) and the dashed curves using (25.19) (curves provided by Tsutomu Indei at IIT, Chicago).

inertia, which appears to improve the agreement with data from mechanical rheometry.

The experimental results for a wormlike micellar solution demonstrate that microbead rheology is a remarkably successful technique. Passive microbead rheology can be further developed into a two-point technique, where two micron-sized beads are observed and their relative motion is used to estimate the dynamic viscosity.[14] The two-point technique allows sampling of larger length scales and can hence be applied in materials with a coarser microstructure.

Exercise 25.6   Corrected Mean-Square Displacement Expression
Show, by neglecting the second term (Basset force) in (25.17) and substitution of this into (25.11), that for a single-mode Maxwell fluid the mean-square displacement is, for $m \ll 6\pi Rg\lambda^2$, given by

$$\langle [\Delta r_{\mathrm{b}}(t)]^2 \rangle_{\mathrm{eq}} \approx \frac{k_{\mathrm{B}}T}{\pi Rg} \left\{ 1 + \frac{t}{\lambda} - \exp\left(\frac{-t}{2\lambda}\right) \left[\cos(\omega_m t) + \frac{3\tau_m}{2\lambda} \sin(\omega_m t) \right] \right\},$$

where $\tau_m = \sqrt{m/(6\pi Rg)}$, and $\omega_m = \sqrt{4\lambda^2 - \tau_m^2}/(2\lambda\tau_m) \approx 1/\tau_m$.

- The diffusive motion of a particle with inertia and frictional interaction with memory can be discussed in terms of a generalized Langevin equation; the resulting one-sided Fourier transform of the mean-square displacement suggests a generalization of the Nernst-Einstein equation in which the friction coefficient is replaced by a frequency- and mass-dependent quantity.
- Results for steady creeping flow of an incompressible Newtonian fluid around a sphere are first generalized to time-dependent flows and then to a linear viscoelastic fluid; as a result, the frequency-dependent friction is related to the dynamic viscosity of the viscoelastic fluid through a generalized Stokes law.
- In passive microbead rheology, observation of the mean-square displacement of a Brownian particle is used to measure the complex modulus, or dynamic viscosity, of a linear viscoelastic fluid.

[14] Córdoba *et al.*, *Phys. Fluids* **24** (2012) 073103.

26

Dynamic Light Scattering

Chapter 7 illustrated how the transport properties of materials can be measured using experiments that could be interpreted using simplified forms of the evolution equations for velocity, temperature, or concentration. In each experiment, an apparatus was used to impose a single driving force that would induce a corresponding flux in the sample. We assumed that the materials could be described by linear force–flux relations and that the phenomenological coefficients were constant. The validity of these assumptions depends on the type of material being studied and on the magnitude of the imposed driving force. We imposed no restrictions on how large the applied driving force was, or, in other words, how far the system was from equilibrium. In this chapter, we will see that it is possible to measure the transport properties of materials that are *at* equilibrium. The fact that this is possible is actually quite profound. Put another way, we will show that it is possible to obtain material properties that govern behavior away from equilibrium by performing measurements on a material in a state of equilibrium.

The reason why we are able to do this is because, as we discussed in Section 20.4, systems at equilibrium undergo small fluctuations as they sample the different microstates that are possible for a given equilibrium state. As the position and velocity of all the particles in a system change, the energy, number of particles, and entropy of some small material element all fluctuate around some average value. It is natural to assume that the dynamics of a system as it relaxes to an equilibrium state from a state slightly removed from equilibrium are the same as those that govern the fluctuations of a system at equilibrium.

There is a second, and equally important, difference between this chapter and Chapter 7. In Chapter 7, we made assumptions such that different transport processes were decoupled. For example, to measure viscosity we assumed that the flow was isothermal so that we needed only to consider the

Figure 26.1 Lord (John William Strutt) Rayleigh (1842–1919).

momentum balance. Similarly, we considered the energy balance to develop a model for thermal conductivity measurements, which was possible because we assumed the sample was at rest. In the present chapter, we do not make assumptions to decouple different transport processes. In fact, *dynamic light scattering* as a method for measuring transport coefficients relies on the existence of coupling.

Scattering is the interaction of waves with matter. These interactions can result in changes in the direction, the amplitude, and/or the wavelength of the propagating waves. The ability to produce collimated and nearly monochromatic electromagnetic waves makes it possible to study the structure and properties of matter. The most commonly used portions of the electromagnetic spectrum are X-rays (0.05–0.2 nm), which probe length scales similar to those accessible by neutron scattering (0.1–1.0 nm), and visible light (400–800 nm). The latter is usually referred to as light scattering, or Rayleigh scattering, and the typical length scale probed by such experiments is on the order of several hundred nanometers. The technique of dynamic light scattering usually refers to an experiment in which the intensity of the scattered light fluctuates in time as a result of the motion of particles with which the electromagnetic waves interact.

In the first two sections of this chapter, we briefly discuss fluctuations, correlation functions, and the basic principles of light scattering. The third section describes in some detail how the equations of transport phenomena can be formulated as an elegant set of linear equations that govern hydrodynamic fluctuations and how the solution of these equations can be used to interpret dynamic light scattering experiments. We refer readers interested in more extensive discussions of dynamic light scattering to the classic texts by Boon and Yip[1] and by Berne and Pecora.[2]

26.1 Fluctuations and Correlation Functions

As we have discussed (see Section 20.4), the properties of a system at equilibrium fluctuate around some average value. These fluctuations are the result of changes in particle positions and momenta and therefore occur on a time scale that is comparable to the time between particle collisions. Let $A(t)$ be some property of a system at equilibrium, the mass density for example, that depends on the positions of the particles. The motion of each particle is governed by Newton's laws of motion and is therefore deterministic. However, because a system is comprised of a large number of particles with unknown initial conditions, the collective motion of the particles appears to be stochastic, or random. The trajectory of a Brownian particle (see Figure 2.1) is an example of stochastic motion caused by the interactions between a large number of particles.

The dependence of property $A(t)$ on the positions and momenta of a large number of particles makes the time-dependent behavior of $A(t)$ appear to be stochastic as shown schematically in Figure 26.2. The average value of $A(t)$ about which it fluctuates is given by

$$\langle A \rangle = \lim_{T \to \infty} \frac{1}{T} \int_0^T A(t)dt, \qquad (26.1)$$

where the period T over which the average is taken must be large compared with the time scale of the fluctuations. In writing (26.1), we have assumed $A(t)$ is a stationary property, which means its average value does not depend on the absolute time at which the average is computed.

The average $\langle A \rangle$ contains information about the equilibrium system, but more important for transport phenomena is the information contained in the correlations of the property at different times, say $A(t)$ and $A(t + \tau)$. For small τ, the difference between $A(t)$ and $A(t + \tau)$ is likely to be small, while for large τ the difference is likely to be large. In other words, the correlation of a property with itself becomes weaker as the difference in the times at

---

[1] Boon & Yip, *Molecular Hydrodynamics* (Dover, 1991).
[2] Berne & Pecora, *Dynamic Light Scattering* (Dover, 2000).

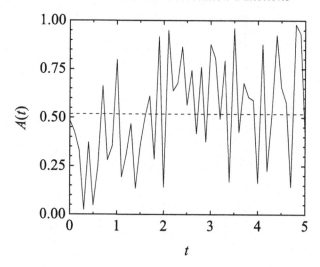

Figure 26.2 Plot òf the time dependence of a property $A(t)$ fluctuating about its average value $\langle A \rangle$ indicated by the dashed line.

which the property is observed increases. A measure of this correlation is

$$\langle A(0)A(\tau)\rangle = \lim_{T\to\infty} \frac{1}{T} \int_0^T A(t)A(t+\tau)dt, \qquad (26.2)$$

which is called the *autocorrelation function* of A. From the discussion above, the property $A(t)$ is strongly correlated with itself for small τ, so the largest value of the autocorrelation function is given by $\langle A(0)A(0)\rangle = \langle A^2\rangle$. For large τ, the correlation of $A(t)$ with itself is lost so that

$$\lim_{\tau\to\infty} \langle A(0)A(\tau)\rangle = \langle A\rangle^2. \qquad (26.3)$$

Figure 26.3 shows a representative dynamic behavior of the autocorrelation function $\langle A(0)A(\tau)\rangle$, which decays from its initial value $\langle A^2\rangle$ to the average value $\langle A\rangle^2$ (which we have assumed to be zero). If $\langle A\rangle \neq 0$, it is sometimes more useful to express a property $A(t)$ as

$$A(t) = \langle A\rangle + \delta A(t), \qquad (26.4)$$

where $\delta A(t)$ is the fluctuation of $A(t)$ from its average value. From (26.1) and (26.4), it is clear that $\langle \delta A(t)\rangle = 0$, and from (26.2) and (26.4), we have

$$\langle \delta A(0)\delta A(\tau)\rangle = \langle A(0)A(\tau)\rangle - \langle A\rangle^2, \qquad (26.5)$$

which is the autocorrelation function of the fluctuation of $A(t)$.

The autocorrelation function measured in a typical light scattering experiment is more complicated than the monotonic decay shown in

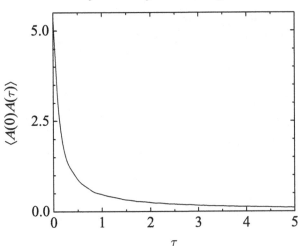

Figure 26.3 Schematic plot of the time dependence of the autocorrelation function $\langle \delta A(0) \delta A(\tau) \rangle$ obtained from the time dependence of $A(t)$ in Figure 26.2.

Figure 26.3. A commonly used quantity for the analysis of complex, time-dependent functions is known as the *spectral density*, or power spectrum. The spectral density of the autocorrelation function of $\delta A(t)$ is given by

$$I_A(\omega) = \frac{1}{2\pi} \int_{-\infty}^{\infty} \langle \delta A(0) \delta A(t) \rangle e^{-i\omega t} \, dt. \tag{26.6}$$

Hence, the spectral density is the time-Fourier transform[3] of the autocorrelation function of a time-dependent function – in this case the fluctuation $\delta A(t)$.

The average and autocorrelation function just introduced are time averages of the property A, which depends on the time-dependent positions and momenta of all the particles. The time evolution of this large set of variables, known as the *phase-space trajectory*, is denoted by $\boldsymbol{\Gamma}_t$, and hence we can write $A(t) = A(\boldsymbol{\Gamma}_t)$. Quantities of interest, the mass density for example, represent the averages taken over a large number of particles. The latter, which are known as ensemble-average properties, can be obtained if the positions and momenta of all the particles are known. As noted above, the motion of the particles is deterministic and is governed by Newton's law of motion. For systems at equilibrium, a solution of these equations can be found which yields the equilibrium distribution function of particle positions

---

[3] Note that this definition for the Fourier transform, which is commonly used in the analysis of dynamic light scattering, differs from the one used in the previous chapter (cf. footnote on p. 442) by a factor of $1/(2\pi)$.

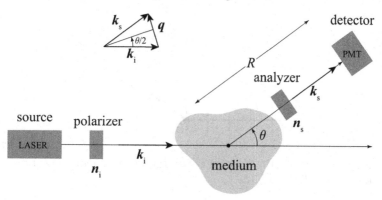

Figure 26.4 Schematic diagram of dynamic light scattering (DLS) setup.

and momenta $p_{eq}(\mathbf{\Gamma}_t)$. The ensemble-average autocorrelation function of A can be expressed as

$$\langle A(0)A(t)\rangle = \int p_{eq}(\mathbf{\Gamma}_0)A(\mathbf{\Gamma}_0)A(\mathbf{\Gamma}_t)d\mathbf{\Gamma}_0. \tag{26.7}$$

Systems for which the time-average (26.2) and ensemble-average (26.7) auto-correlation function are the same are called ergodic. As discussed in Section 20.1, an ergodic system is one that samples all available microstates with equal probability.

26.2 Light Scattering Basics

The basic configuration of a light scattering experiment is shown schematically in Figure 26.4. The three main elements of a typical setup are a radiation (light) source, the medium, and a detector. Typically, the light source is a laser that emits a coherent and nearly monochromatic beam of light with a wavelength in the visible part of the electromagnetic spectrum. A device known as a photomultiplier tube (PMT), which converts light (photons) to an electrical current by the photoelectric effect, is commonly used as a detector.

While our interest is in light scattering, it will be convenient to develop the results for general scattering phenomena. The incident electromagnetic waves from the source are assumed to be planar and have intensity I_i and wave vector $\mathbf{k}_i$ with magnitude $k_i = 2\pi/\lambda_i$. If the scattering particle is small compared with the wavelength λ_i, the scattered waves are spherical. The detector at distance R from the sample sees electromagnetic waves with intensity I_s and wave vector $\mathbf{k}_s$. The change in the wave vector is known as

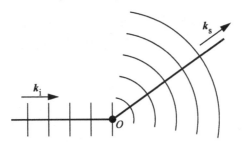

Figure 26.5 Schematic diagram of scattering from a single stationary particle located at the origin.

the scattering vector,

$$q = k_s - k_i, \tag{26.8}$$

and for quasi-elastic scattering $k_s \approx k_i$ the magnitude of the scattering vector is given by $q = (4\pi/\lambda_i)\sin(\theta/2)$, where θ is known as the scattering angle. This expression, which comes from elementary geometric arguments (see Figure 26.4), is known as the Bragg condition. The quantity measured in a typical scattering experiment is related to the differential cross section, which can be written as

$$\frac{d\sigma}{d\Omega} = \frac{I_s R^2}{I_i}. \tag{26.9}$$

For the purpose of this section, it is sufficient to realize that the differential cross section $d\sigma/d\Omega$ expresses the idea that the incident beam has to hit an area element $d\sigma$ to be scattered into a solid angle $d\Omega$. More details and some explicit calculations can be found elsewhere.[4]

We first consider the scattering process from a single, stationary particle located at the origin as shown schematically in Figure 26.5. The magnitude of the planar incident electromagnetic wave can be written as

$$\mathcal{A}_i(r) = \mathcal{A}_0 e^{ik_i \cdot r}, \tag{26.10}$$

where $\mathcal{A}_0$ is the field amplitude; the normalization of the amplitude is chosen such that the incident intensity is $I_i = \mathcal{A}_i^* \mathcal{A}_i = \mathcal{A}_0^2$. The magnitude of the spherical scattered electromagnetic wave at the detector is

$$\mathcal{A}_s(R) = \mathcal{A}_i(0)b\frac{e^{ik_s R}}{R} = \mathcal{A}_0 b\frac{e^{ik_s R}}{R}, \tag{26.11}$$

where b is the *scattering length*, which depends on the details of the scattering experiment. The intensity of the scattered radiation is $I_s = b^2 \mathcal{A}_0^2/R^2$

---

[4] See Section 7.1 of Öttinger, *Beyond Equilibrium Thermodynamics* (Wiley, 2005).

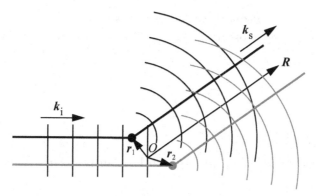

Figure 26.6 Schematic diagram of scattering from two stationary particles located near the origin.

so that from (26.9) we have $d\sigma/d\Omega = b^2$. For the scattering at a particle small compared with the wavelength λ_i, the scattering length b thus contains the complete information about the scattering process. Our development in terms of b is very general – only the particular form of b depends on the particular type of scattering. For light scattering, b can be expressed in terms of the polarizability of atoms which then act as Hertz dipoles, driven by the incident electromagnetic wave and emitting the scattered electromagnetic radiation.[5] For a detailed calculation, one solves the Maxwell equations in the far-field approximation (see, for example, Section 13.4 of the textbook by Honerkamp and Römer[6]). The result can then be further related to the refractive index. Scattering length, polarizability, and refractive index all contain equivalent information about the interaction of electromagnetic waves with matter. For neutron scattering, the scattering length b has to be obtained from the strong interactions of neutrons within the nuclei in the sample.

Next, we generalize to a scattering process involving two stationary particles located at positions r_1 and r_2 as shown schematically in Figure 26.6. The magnitude of the scattered electromagnetic wave at the detector is the superposition of the scattered waves for each particle,

$$\mathcal{A}_s(\boldsymbol{R}) = \mathcal{A}_0 b_1 \frac{e^{i\boldsymbol{k}_i \cdot \boldsymbol{r}_1 + i k_s |\boldsymbol{R} - \boldsymbol{r}_1|}}{|\boldsymbol{R} - \boldsymbol{r}_1|} + \mathcal{A}_0 b_2 \frac{e^{i\boldsymbol{k}_i \cdot \boldsymbol{r}_2 + i k_s |\boldsymbol{R} - \boldsymbol{r}_2|}}{|\boldsymbol{R} - \boldsymbol{r}_2|}. \tag{26.12}$$

---

[5] For light scattering, $b^2 \propto \lambda_i^{-4}$ so that smaller wavelengths of the visible spectrum are scattered more strongly than longer wavelengths, which partially explains why the sky is blue.

[6] Honerkamp & Römer, *Theoretical Physics* (Springer, 1993).

Exercise 26.1   *Scattered Wave at Large Distances*
Justify the following approximation for the scattered-wave contributions in (26.12)
for $R \gg |r_j|$:

$$\frac{e^{ik_s|R-r_j|}}{|R-r_j|} \approx \frac{e^{ik_s R}}{R} e^{-ik_s \cdot r_j}, \tag{26.13}$$

where $k_s = k_s R/R$.

By means of (26.13), Equation (26.12) can be rewritten as

$$A_s(R) = \frac{A_0 e^{ik_s R}}{R} \left(b_1 e^{iq \cdot r_1} + b_2 e^{iq \cdot r_2} \right). \tag{26.14}$$

If we consider N particles with identical scattering amplitude $b_j = b$, we
further obtain

$$A_s(R) = \frac{bA_0 e^{ik_s R}}{R} \sum_{j=1}^{N} e^{iq \cdot r_j}. \tag{26.15}$$

Using (26.15) in (26.9), we obtain the following expression for the differential
scattering cross section for N fixed particles:

$$\frac{d\sigma}{d\Omega} = b^2 \sum_{j=1}^{N} e^{iq \cdot r_j} \sum_{k=1}^{N} e^{-iq \cdot r_k} = b^2 \sum_{j,k=1}^{N} e^{iq \cdot (r_j - r_k)}. \tag{26.16}$$

So far, we have developed a static description of the scattering process.
For quasi-elastic scattering with $k_s \approx k_i$ (instead of elastic scattering with
$k_s = k_i$), however, the observed intensity can fluctuate in time. This time
dependence results from the motion of the scattering particles. As the particle velocities are small compared with the speed of light, we can obtain
the relevant information for coherent dynamic light scattering by simply replacing r_j and r_k in (26.16) by $r_j(0)$ and $r_k(t)$, respectively. For the second
sum with $r_k \rightarrow r_k(t)$, we then write

$$\sum_{k=1}^{N} e^{-iq \cdot r_k(t)} = \int_V \sum_{k=1}^{N} \delta(r - r_k(t)) e^{-iq \cdot r} \, d^3r$$

$$= \frac{1}{m} \int_V \rho(r,t) e^{-iq \cdot r} \, d^3r = \frac{1}{m} \tilde{\rho}(q,t), \tag{26.17}$$

where m is the mass of a particle and $\tilde{\rho}(q,t)$ is the spatial Fourier transform
of $\rho(r,t)$ (cf. footnote on p. 17). The scattering volume V is determined by
the intersection of the incident beam and the sample. Using this result for

the other sum in (26.16) with $r_j \to r_j(0)$, we can write (26.16) as

$$b^2 \sum_{j,k=1}^{N} e^{i\mathbf{q}\cdot[\mathbf{r}_j(0)-\mathbf{r}_k(t)]} = \frac{b^2}{m^2}\tilde{\rho}^*(\mathbf{q},0)\tilde{\rho}(\mathbf{q},t). \qquad (26.18)$$

By generalizing the latter result, which holds for a given initial particle configuration, to an equilibrium ensemble of particle configurations, we can write (26.18) as an autocorrelation function (see Section 26.1). We further write $\rho(\mathbf{r},t) = \rho_0 + \delta\rho(\mathbf{r},t)$, where ρ_0 is the average density and $\delta\rho(\mathbf{r},t)$ is the density fluctuation, and rewrite (26.18) as

$$b^2 \sum_{j,k=1}^{N} \langle e^{i\mathbf{q}\cdot[\mathbf{r}_j(0)-\mathbf{r}_k(t)]} \rangle = \frac{b^2}{m^2}\langle \widetilde{\delta\rho}^*(\mathbf{q},0)\widetilde{\delta\rho}(\mathbf{q},t) \rangle. \qquad (26.19)$$

A constant density cannot contribute to the scattering because its Fourier transform involves a factor of $\delta(\mathbf{q})$ and the wave vector $\mathbf{q} = \mathbf{0}$ corresponds to the unmodified incident beam without any scattering.

In practice, one measures the autocorrelation of the intensity $I(\mathbf{q},t)$ (see Figure 26.3 with $A \to I$). Assuming a Gaussian distribution for the intensity profiles, we can write

$$\frac{\langle I(\mathbf{q},0)I(\mathbf{q},t) \rangle}{\langle I(\mathbf{q},0)^2 \rangle} = 1 + \frac{|\langle \widetilde{\delta\rho}^*(\mathbf{q},0)\widetilde{\delta\rho}(\mathbf{q},t) \rangle|^2}{\langle |\widetilde{\delta\rho}(\mathbf{q},0)|^2 \rangle^2}, \qquad (26.20)$$

which is known as the Siegert relation.[7] In the following section, we obtain an expression for $\langle \widetilde{\delta\rho}^*(\mathbf{q},0)\widetilde{\delta\rho}(\mathbf{q},t) \rangle$ from a solution of the hydrodynamic equations.

It is common to characterize the measured time-dependent intensity in terms of the power spectral density of $\langle \widetilde{\delta\rho}^*(\mathbf{q},0)\widetilde{\delta\rho}(\mathbf{q},t) \rangle$,

$$S(\mathbf{q},\omega) = \frac{1}{2\pi} \int_{-\infty}^{\infty} \langle \widetilde{\delta\rho}^*(\mathbf{q},0)\widetilde{\delta\rho}(\mathbf{q},t) \rangle e^{-i\omega t}\, dt, \qquad (26.21)$$

which is known as the *dynamic structure factor*. Integrating $S(\mathbf{q},\omega)$ over ω gives

$$S(\mathbf{q}) = \int_{-\infty}^{\infty} S(\mathbf{q},\omega)d\omega = \langle |\widetilde{\delta\rho}(\mathbf{q},0)|^2 \rangle, \qquad (26.22)$$

which is known as the static structure factor.

---

[7] The relationship between fourth and second moments is discussed in Section 4.3 (and the appendices) of Berne & Pecora, *Dynamic Light Scattering* (Dover, 2000).

26.3 Hydrodynamic Fluctuations

Our goal is to develop an expression for the dynamic structure factor $S(\boldsymbol{q}, \omega)$ that, as shown in (26.21), requires that we have an evolution equation for the density fluctuation $\delta\rho(\boldsymbol{r}, t)$. For a one-component fluid, the equation of state is given by $\rho = \rho(T, p)$; using standard thermodynamic relations that can be found in Appendix A, we can write

$$\delta\rho = \frac{\gamma}{c_{\hat{s}}^2}\,\delta p - \rho\alpha_p\,\delta T, \qquad (26.23)$$

where α_p is the isobaric coefficient of thermal expansion, $c_{\hat{s}}$ is the isentropic speed of sound, and $\gamma = \hat{c}_p/\hat{c}_{\hat{v}}$ is the ratio of specific heats at constant pressure and volume. Hence, density fluctuations occur as a result of both mechanical and thermal fluctuations. Consequently, the evolution equation for density is coupled to evolution equations for momentum and energy.

The choice of variables used to formulate the set of evolution equations is a matter of preference as long as the fluctuating variables chosen are independent (see Section 20.4). Here, we choose density, velocity, and temperature since in previous chapters we have already gained some experience with evolution equations for these variables. Since we are interested in describing small fluctuations around an equilibrium state, we use linear expansions (see Exercise 5.15) and write

$$\rho = \rho_0 + \delta\rho, \qquad \boldsymbol{v} = \boldsymbol{0} + \boldsymbol{v}, \qquad T = T_0 + \delta T, \qquad (26.24)$$

where ρ_0, $\boldsymbol{0}$, and T_0 are equilibrium values of density, velocity, and temperature, respectively, and $\delta\rho$, $\boldsymbol{v}$, and δT are the fluctuations in these fields. Substitution of (26.24) into (7.1), (7.7), and (7.3) gives, retaining only first-order terms in the fluctuating variables,

$$\frac{\partial}{\partial t}\delta\rho = -\rho_0\,\boldsymbol{\nabla}\cdot\boldsymbol{v}, \qquad (26.25)$$

$$\frac{\partial}{\partial t}\boldsymbol{v} = -\nu\,\boldsymbol{\nabla}\times(\boldsymbol{\nabla}\times\boldsymbol{v}) + \nu_1\,\boldsymbol{\nabla}(\boldsymbol{\nabla}\cdot\boldsymbol{v}) - \frac{c_{\hat{s}}^2}{\gamma}\left(\alpha_p\,\boldsymbol{\nabla}\delta T + \frac{1}{\rho_0}\,\boldsymbol{\nabla}\delta\rho\right), \quad (26.26)$$

$$\frac{\partial}{\partial t}\delta T = \gamma\chi\,\nabla^2\delta T - \frac{\gamma - 1}{\alpha_p}\,\boldsymbol{\nabla}\cdot\boldsymbol{v}, \qquad (26.27)$$

where we have used (26.23) to eliminate pressure. Note that we have assumed the gravitational force to be negligible and that terms from $\boldsymbol{\tau} : \boldsymbol{\nabla}\boldsymbol{v}$ do not appear since they are of second order in the fluctuating variables. The linearized evolution equations in (26.25)–(26.27) contain the longitudinal kinematic viscosity $\nu_1 = (\frac{4}{3}\eta + \eta_d)/\rho_0$ and the thermal diffusivity $\chi = \lambda/(\rho_0 \hat{c}_p)$, which we assume are constants.

Exercise 26.2   Linearized Hydrodynamic Equations
Starting with (7.1), (7.7), and (7.3), derive the linearized equations (26.25), (26.26), and (26.27). In the derivation, make use of the identity given in (5.74).

The coupled evolution equations in (26.25)–(26.27) govern two scalar quantities and one vector quantity. Notice, however, that the coupling of the scalars to the velocity vector occurs only through the scalar divergence term, $\nabla \cdot \boldsymbol{v}$. This suggests that the vector equation (26.26) can be split into two separate equations. The first is obtained by taking the curl ($\nabla \times$) of (26.26), which, using (5.74), can be written as

$$\frac{\partial}{\partial t}\boldsymbol{w} = \nu \nabla^2 \boldsymbol{w}, \tag{26.28}$$

where $\boldsymbol{w} = \nabla \times \boldsymbol{v}$ is the vorticity vector (see Exercise 7.6). Vorticity is a measure of the rotation of fluid elements, and from (26.28) we see that vorticity fluctuations are governed by a diffusion equation. More importantly, vorticity fluctuations $\boldsymbol{w}$ are completely decoupled from the evolution equations for density and temperature fluctuations. The second equation must be a scalar equation and is obtained by taking the divergence ($\nabla \cdot$) of (26.26). This allows (26.25)–(26.27) to be written as

$$\frac{\partial}{\partial t}\delta\rho = -\rho_0 \psi, \tag{26.29}$$

$$\frac{\partial}{\partial t}\psi = \nu_1 \nabla^2 \psi - \frac{c_{\hat{s}}^2}{\gamma}\left(\alpha_p \nabla^2 \delta T + \frac{1}{\rho_0}\nabla^2 \delta\rho\right), \tag{26.30}$$

$$\frac{\partial}{\partial t}\delta T = \gamma\chi \nabla^2 \delta T - \frac{\gamma - 1}{\alpha_p}\psi, \tag{26.31}$$

where $\psi = \nabla \cdot \boldsymbol{v}$.

In (26.29)–(26.31), we now have a system of coupled linear evolution equations for the scalar fluctuations $\delta\rho$, ψ, and δT. Note that density fluctuations are driven by a drift term only, velocity fluctuations are driven by diffusion, and temperature fluctuations by both diffusion and drift terms.

Exercise 26.3   Speed of Sound Revisited
Consider the isothermal ($\delta T = 0$) case of (26.29)–(26.31) to obtain

$$\frac{\partial^2}{\partial t^2}\delta\rho = c_T^2 \nabla^2 \delta\rho + \nu_1 \frac{\partial}{\partial t}\nabla^2 \delta\rho, \tag{26.32}$$

where $c_T^2 = c_s^2/\gamma$ is the isothermal speed of sound squared. Give the criterion for the wave equation (26.32) to be consistent with (5.55) from Exercise 5.15, and discuss its significance for common fluids like air and water.

Exercise 26.4   Helmholtz Decomposition
The procedure we have just used to decompose (26.26) is closely related to the general procedure of expressing a vector as the sum of divergence-free and curl-free vectors, which is known as Helmholtz' (decomposition) theorem. According to this theorem, any continuous vector $\boldsymbol{v}$ field with continuous first derivatives can be written as

$$\boldsymbol{v} = -\boldsymbol{\nabla}\xi + \boldsymbol{\nabla} \times \boldsymbol{a}, \qquad (26.33)$$

where ξ is a scalar potential field and $\boldsymbol{a}$ is a vector potential field. Show that the first term in (26.33), or the longitudinal part of the velocity, is curl-free, and that the second term, or the transverse part, is divergence-free. Also, find relationships between the longitudinal part of the velocity and ψ and between the transverse part of the velocity and $\boldsymbol{w}$.

We now begin the task of solving the simplified set of hydrodynamic equations and start with the evolution equation for vorticity fluctuations (26.28). As we learned in Chapter 2, a convenient way to solve linear partial differential equations with constant coefficients is by using the Fourier transform. The spatial Fourier transform of a function has already been defined in (26.17). The Fourier transform of the Laplacian of $f(\boldsymbol{r}, t)$, $\nabla^2 f = \boldsymbol{\nabla} \cdot \boldsymbol{\nabla} f$, gives

$$\int_V e^{-i\boldsymbol{q}\cdot\boldsymbol{r}} \, \nabla^2 f(\boldsymbol{r}, t) d^3 r = -q^2 \tilde{f}(\boldsymbol{q}, t), \qquad (26.34)$$

which is obtained from an integration by parts over an infinite domain (without boundary contributions). The Fourier transform of (26.28), using (26.34), gives

$$\frac{\partial}{\partial t}\tilde{w}(\boldsymbol{q}, t) = -q^2 \nu \tilde{w}(\boldsymbol{q}, t), \qquad (26.35)$$

which has the solution

$$\tilde{w}(\boldsymbol{q}, t) = \tilde{w}(\boldsymbol{q}, 0)e^{-q^2\nu t}, \qquad (26.36)$$

where $\tilde{w}(\boldsymbol{q}, 0)$ is the Fourier transform of the initial value of the fluctuation. Hence, the decoupled vorticity fluctuations decay exponentially in time with rate $q^2\nu$. Since vorticity fluctuations are not coupled to density fluctuations, we no longer consider them.

To solve the system of coupled evolution equations (26.29)–(26.31) we use the Fourier transform and (26.34), which gives

$$\frac{\partial}{\partial t}\begin{pmatrix} \widetilde{\delta\rho}(q,t) \\ \widetilde{\psi}(q,t) \\ \widetilde{\delta T}(q,t) \end{pmatrix} = \begin{pmatrix} 0 & -\rho_0 & 0 \\ c_{\hat{s}}^2 q^2/(\gamma\rho_0) & -\nu_l q^2 & \alpha_p c_{\hat{s}}^2 q^2/\gamma \\ 0 & -(\gamma-1)/\alpha_p & -\gamma\chi q^2 \end{pmatrix} \cdot \begin{pmatrix} \widetilde{\delta\rho}(q,t) \\ \widetilde{\psi}(q,t) \\ \widetilde{\delta T}(q,t) \end{pmatrix}. \tag{26.37}$$

We now have a coupled system of ordinary differential equations (26.37), which can be expressed in compact form as $\dot{x} = A \cdot x$, where the solution vector has the components $x(q,t) = [\widetilde{\delta\rho}(q,t), \widetilde{\psi}(q,t), \widetilde{\delta T}(q,t)]^{\mathrm{T}}$ and A is the non-symmetric 3×3 matrix in (26.37). The solution of (26.37) can be expressed as

$$x = \sum_{i=1}^{3} c_i e_i e^{\alpha_i t}, \tag{26.38}$$

where e_i ($i = 1, 2, 3$) are the (right) eigenvectors associated with eigenvalues α_i. In the discussion of time-dependent equilibrium fluctuations, the weight factors c_i are independent random numbers determined by initial conditions.

The eigenvalues are found by solving $\det(A - \alpha\delta) = 0$, and the eigenvector for each eigenvalue can be verified by $A \cdot e_i = \alpha_i e_i$. This procedure, the details of which are left as an exercise, results in

$$e_1 = \begin{pmatrix} \rho_0 \\ iqc_{\hat{s}}f_1(iq) \\ (\gamma-1)/\alpha_p - (1/\hat{c}_{\hat{v}})f_2(iq) \end{pmatrix}, \qquad \alpha_1 = -iqc_{\hat{s}}f_1(iq), \tag{26.39}$$

$$e_2 = \begin{pmatrix} \rho_0 \\ -iqc_{\hat{s}}f_1(-iq) \\ (\gamma-1)/\alpha_p - (1/\hat{c}_{\hat{v}})f_2(iq) \end{pmatrix}, \qquad \alpha_2 = iqc_{\hat{s}}f_1(-iq), \tag{26.40}$$

and

$$e_3 = \begin{pmatrix} \rho_0 \\ \chi q^2/[f_1(iq)f_1(-iq)] \\ -(1/\alpha_p)(\gamma-1)/[\gamma f_1(iq)f_1(-iq) - 1] \end{pmatrix}, \qquad \alpha_3 = -\frac{\chi q^2}{f_1(iq)f_1(-iq)}, \tag{26.41}$$

where the functions f_1 and f_2 are determined by two nonlinear equations,

$$[f_1(iq)]^2 = 1 - \frac{\alpha_p}{\gamma\hat{c}_{\hat{v}}}f_2(iq) - \frac{\nu_l}{c_{\hat{s}}}iqf_1(iq), \tag{26.42}$$

$$f_2(iq) = \frac{(\gamma-1)\hat{c}_{\hat{v}}}{\alpha_p} \frac{(\gamma\chi/c_{\hat{s}})iq}{f_1(iq) + (\gamma\chi/c_{\hat{s}})iq}. \tag{26.43}$$

While it is possible to solve the cubic equation implied by (26.42) and (26.43) and obtain an exact expression for $f_1(iq)$, the closed-form solution would be quite complicated. For a typical dynamic light scattering experiment, however, $c_{\hat{s}} \approx 10^5$ cm/s, $q \approx 10^5$ cm^{-1}, and $\gamma\chi \approx \nu_l \approx 10^{-3}$ cm^2/s. This means $\gamma\chi \ll c_{\hat{s}}/q$ and $\nu_l \ll c_{\hat{s}}/q$, which corresponds to the small-q, or large-length-scale, limit (from a molecular perspective). Hence, it is sufficient to keep only the lowest-order terms of an expansion of (26.42) and (26.43) in terms of iq, which gives

$$f_1(iq) = 1 - \frac{1}{2}\frac{\Gamma}{c_{\hat{s}}}iq + O(q^2), \qquad f_2(iq) = \frac{\chi}{c_{\hat{s}}}iq + O(q^2), \qquad (26.44)$$

where $\Gamma = \frac{1}{2}[(\gamma-1)\chi+\nu_l]$ is the attenuation coefficient. Substituting (26.44) into (26.39)–(26.41), we then arrive at the approximate eigenvalues

$$\alpha_{1,2} = \mp ic_{\hat{s}}q - \Gamma q^2, \qquad \alpha_3 = -\chi q^2, \qquad (26.45)$$

and the approximate eigenvectors

$$\boldsymbol{e}_1 = \begin{pmatrix} \rho_0 \\ ic_{\hat{s}}q \\ (\gamma-1)/\alpha_p \end{pmatrix}, \qquad \boldsymbol{e}_2 = \begin{pmatrix} \rho_0 \\ -ic_{\hat{s}}q \\ (\gamma-1)/\alpha_p \end{pmatrix}, \qquad \boldsymbol{e}_3 = \begin{pmatrix} \rho_0 \\ 0 \\ -(1/\alpha_p) \end{pmatrix}. \quad (26.46)$$

Note that the eigenvalues in (26.45) are of second order in q and that the second entry of $\boldsymbol{e}_1$ and $\boldsymbol{e}_2$ in (26.46) is of a higher order in q than the other rows, which is due to the fact that $\widetilde{\psi}$ is obtained from a divergence.

The modes (26.46) evolve and fluctuate independently, and hence they separately contribute to the fluctuations. To find the weights of the different contributions, we examine the properties of fluctuations at equilibrium. As discussed in Section 20.4, fluctuations at equilibrium are related to second moments, and fluctuations in density and temperature are independent. We choose the weights of modes 1 and 2 to be equal because they describe sound waves differing only in the direction of their propagation. The relative weight of mode 3 is chosen such that the fluctuations $\widetilde{\delta\rho}$ and $\widetilde{\delta T}$ are independent. Second moments can be expressed as a dyadic $\langle \boldsymbol{x}^*(\boldsymbol{q},0)\boldsymbol{x}(\boldsymbol{q},0)\rangle$, which, since $\langle c_i c_j\rangle = 0$ for $i \neq j$, can be expressed as

$$\frac{1}{2}(\boldsymbol{e}_1^*\boldsymbol{e}_1 + \boldsymbol{e}_2^*\boldsymbol{e}_2) + (\gamma-1)\,\boldsymbol{e}_3^*\boldsymbol{e}_3 = \begin{pmatrix} \gamma\rho_0^2 & 0 & 0 \\ 0 & c_{\hat{s}}^2 q^2 & 0 \\ 0 & 0 & \gamma(\gamma-1)/\alpha_p^2 \end{pmatrix}. \quad (26.47)$$

Multiplying (26.47) by $\alpha_p^2 k_B T_0^2 / [\gamma(\gamma-1)\hat{c}_{\hat{v}}\rho_0 V]$, we obtain the equilibrium fluctuations

$$\langle |\widetilde{\delta\rho}(\boldsymbol{q}, 0)|^2 \rangle = \frac{\alpha_p^2 \rho_0 k_B T_0^2}{(\gamma-1)\hat{c}_{\hat{v}}V} = \frac{\kappa_T \rho_0^2 k_B T_0}{V}, \tag{26.48}$$

$$\langle |\widetilde{\delta T}(\boldsymbol{q}, 0)|^2 \rangle = \frac{k_B T_0^2}{\rho_0 \hat{c}_{\hat{v}} V}. \tag{26.49}$$

The second equation in (26.48), which contains the isothermal compressibility κ_T, is obtained using standard thermodynamic relations (see Appendix A). Equations (26.48) and (26.49) describe the equilibrium fluctuations in a volume V, which were discussed in Section 20.4. The expression in (26.48) gives the static structure factor $S(\boldsymbol{q})$ defined in (26.22). For local equilibrium, the fluctuations are δ correlated in space. The fluctuations of the Fourier modes are hence independent of $\boldsymbol{q}$, which is why the expressions in (26.48) and (26.49) are the same as those found in Section 20.4.

Evolving the eigenmodes in time according to (26.38), we obtain the time-dependent fluctuation correlations

$$\langle \boldsymbol{x}^*(\boldsymbol{q}, 0)\boldsymbol{x}(\boldsymbol{q}, t) \rangle = \sum_{i=1}^{3} \sum_{i=j}^{3} \langle c_i c_j \rangle e_i^* e_j e^{\alpha_j t}, \tag{26.50}$$

where $\langle c_i c_j \rangle = c_i' \delta_{ij}$. We are interested in the autocorrelation function for the density fluctuation, which from (26.50) is given by

$$\langle \widetilde{\delta\rho}^*(\boldsymbol{q}, 0)\widetilde{\delta\rho}(\boldsymbol{q}, t) \rangle = \frac{\kappa_T \rho_0^2 k_B T_0}{V} \left[\frac{\gamma-1}{\gamma} e^{-\chi q^2 t} + \frac{1}{\gamma} \cos(c_{\hat{s}} q t) e^{-\Gamma q^2 t} \right], \tag{26.51}$$

where the weight factors c_i' have been chosen so that the prefactor produces the equilibrium fluctuations given in (26.48). Note that the density fluctuations evolve deterministically in time and that all the randomness comes only from the initial condition. More importantly, the decay of fluctuations towards equilibrium is governed by the macroscopic balance equations. This is known as the *Onsager regression hypothesis*. The behavior of the autocorrelation function for the density fluctuation predicted by (26.51) is shown in Figure 26.7. Clearly, this is a more complicated behavior than the monotonic decay shown in Figure 26.3. The oscillations are the result of sound wave propagation and are damped by thermal and mechanical dissipation, and the weak, long-time decay is solely due to thermal fluctuations.

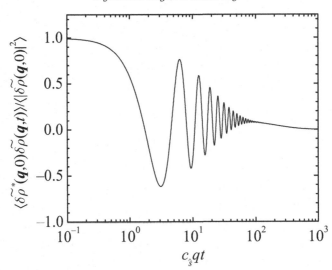

Figure 26.7 Normalized autocorrelation function of density fluctuation $\langle \widetilde{\delta\rho}^*(\boldsymbol{q},0)\widetilde{\delta\rho}(\boldsymbol{q},t)\rangle$ predicted by (26.51) with $\chi q/c_{\hat{s}} = 5 \times 10^{-3}$, $\Gamma q/c_{\hat{s}} = 5 \times 10^{-2}$, and $\gamma = 1.15$.

Substitution of (26.51) into the expression for the dynamic structure factor $S(\boldsymbol{q},\omega)$ defined in (26.21) gives

$$\frac{S(\boldsymbol{q},\omega)}{S(\boldsymbol{q})} = \frac{\gamma-1}{\gamma}\frac{2\chi q^2}{\omega^2 + (\chi q^2)^2} + \frac{1}{\gamma}\left[\frac{\Gamma q^2}{(\omega + c_{\hat{s}}q)^2 + (\Gamma q^2)^2} + \frac{\Gamma q^2}{(\omega - c_{\hat{s}}q)^2 + (\Gamma q^2)^2}\right].$$
$$(26.52)$$

The dynamic structure factor, or spectral density of the autocorrelation function of the density fluctuations, given in (26.52) is plotted in Figure 26.8. From (26.52), we see the spectrum is the sum of three Lorentzians that produce the three peaks in Figure 26.8. The unshifted peak is known as the Rayleigh line and has a half-width at half-maximum equal to $\Delta\omega_R = \chi q^2$. The Rayleigh peak is due to entropy (thermal) fluctuations at constant pressure. The remaining two peaks are the Brillouin doublet that are symmetrically shifted by $\omega_B = \pm c_{\hat{s}}q$ and have half-width at half-maximum $\Delta\omega_B = \Gamma q^2$. The Brillouin doublet is due to pressure (mechanical) fluctuations, or sound waves, at constant entropy. The final piece of information in the spectrum is obtained from the relative areas of the peaks. Integrations over frequency of the first and of the remaining two terms of (26.52) give I_R and $2I_B$, respectively. The ratio of these is known as the Landau–Placzek ratio, $I_R/2I_B = \gamma - 1$. The superposition of thermal and mechanical fluctuations shown in Figure 26.8 is known as the Rayleigh–Brillouin spectrum.

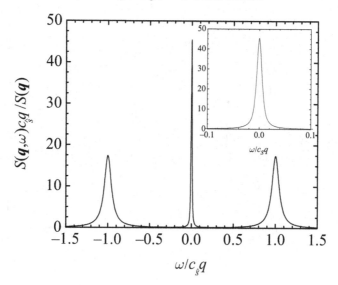

Figure 26.8 Normalized dynamic structure factor $S(\boldsymbol{q}, \omega)/S(\boldsymbol{q})$ (scaled by $c_{\hat{s}}q$) predicted by (26.52) with $\chi q/c_{\hat{s}} = 5 \times 10^{-3}$, $\Gamma q/c_{\hat{s}} = 5 \times 10^{-2}$, and $\gamma = 1.15$. Inset shows a magnified view of the Rayleigh peak.

We have just demonstrated that it is possible to obtain both thermodynamic equilibrium properties ($\gamma, c_{\hat{s}}$) and transport coefficients (χ, ν_l) from a dynamic light scattering experiment. This is quite remarkable since in effect all we did is simply "look" at the sample. Thermodynamic properties such as ρ_0 and $\hat{c}_p$ are relatively easy to obtain, so, when they are combined with χ and γ from a dynamic light scattering experiment, the thermal conductivity λ can be determined (see Section 7.3). If the shear viscosity η is also known (see Section 7.2), then from ν_l it is possible to determine the bulk viscosity $\eta_{\rm d}$, which is an otherwise difficult property to measure. Our analysis is based on the assumption that the system can be described by linearized forms of the balance equations, which seems reasonable because these describe the relaxation of small fluctuations. For convenience, we assumed $\gamma\chi \ll c_{\hat{s}}/q$ and $\nu_l \ll c_{\hat{s}}/q$, which implies the widths of the peaks are small compared with the shifts in the Brillouin peaks.

As an example, in Figure 26.9 we show the Rayleigh–Brillouin spectrum for water at 293 K.[8] From the Brillouin doublet, with shift $\omega_{\rm B}$ and half-width at half-maximum $\Delta\omega_{\rm B}$, it is possible to determine the isentropic speed of sound $c_{\hat{s}}$ and bulk viscosity $\eta_{\rm d}$ (see Exercise 26.7).

---

[8] Xu *et al.*, *Appl. Optics* **42** (2003) 6704.

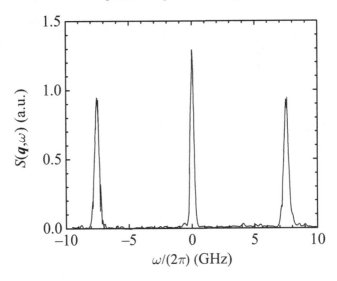

Figure 26.9 Rayleigh–Brillouin spectrum for water at 293 K with $q = 3.0 \times 10^7$ m^{-1}. The Brillouin doublet shift is $\omega_B/(2\pi) \approx 7.4 \times 10^9$ Hz and the half-width at half maximum $\Delta\omega_B/(2\pi) \approx 0.31 \times 10^9$ Hz. [Adapted from Xu *et al.*, *Appl. Optics*, **42** (2003) 6704.]

Dynamic light scattering is commonly used to study dilute polymer solutions and colloidal dispersions. To analyze such systems, the linearized hydrodynamic equations in (26.29)–(26.31) must be generalized to two-component mixtures, which includes an evolution equation for the fluctuation of the solute mass density $\delta\rho_1$ (see Exercise 26.8). In most cases, since for these systems $\nu/D_{12} \gg \nu/\chi \gg 1$ (see Table 7.1), it is possible to decouple concentration fluctuations $\delta\rho_1$ from $\delta\rho$, ψ, and δT. This leads to a diffusion equation for $\delta\rho_1$ having the solution

$$\widetilde{\delta\rho_1}(\boldsymbol{q}, t) = \widetilde{\delta\rho_1}(\boldsymbol{q}, 0)e^{-q^2 D_{12}t}, \qquad (26.53)$$

which can be used to obtain the autocorrelation function for the solute density fluctuation. Once found, the diffusivity D_{12} can be used with the Stokes–Einstein relation (17.46) to determine the (effective) size of the polymer molecule or colloidal particle in the solution.

Exercise 26.5   Solution of the Hydrodynamic Equations
Verify the expressions for the eigenvectors and eigenvalues in (26.39)–(26.43).

Exercise 26.6   Expression for the Dynamic Structure Factor
Derive the expression in (26.51) and the expression in (26.52) for the dynamic structure factor $S(\boldsymbol{q}, \omega)$.

Exercise 26.7   Speed of Sound and Bulk Viscosity for Water
From the Rayleigh–Brillouin spectrum data in Figure 26.9, determine the isentropic speed of sound $c_{\hat{s}}$ and bulk viscosity η_d for water.

Exercise 26.8   Linearized Hydrodynamic Equations for Two-Component Systems
Generalize the linearized hydrodynamic equations in (26.29)–(26.31) for a two-component mixture to include an evolution equation for $\delta\rho_1$, where $\rho_1 = \rho_{10} + \delta\rho_1$. Assume dilute, ideal mixture behavior and that cross effects (see Section 6.3) can be neglected. Identify the conditions for which the evolution equation for $\delta\rho_1$ simplifies to a diffusion equation, which has the solution given in (26.53).

- For systems at equilibrium, fluctuations in properties such as mass density due to molecular motion are described in terms of the autocorrelation function defined in terms of a time average or equivalently for ergodic systems as an ensemble average.
- The interaction of electromagnetic waves with matter, or scattering, leads to changes in the propagation direction, amplitude, and wavelength of the waves, which can be measured and used to investigate both the static and the dynamic structure of the matter.
- Dynamic light scattering is an experimental technique that allows the investigation of density fluctuations, expressed in terms of an autocorrelation function, from an analysis of time-dependent measurements of the intensity of scattered light.
- Linearized forms of balance equations for mass, momentum, and energy can be expressed as a coupled system of evolution equations for density, velocity, and temperature fluctuations for systems at equilibrium.
- Decoupled solutions of the evolution equations can be expressed as the autocorrelation function for density fluctuations, or as the dynamic structure factor, which can be matched to dynamic light scattering experiments and used to obtain equilibrium thermodynamic properties and nonequilibrium transport coefficients.

Appendix A

Thermodynamic Relations

According to Chapter 4, thermodynamics is all about the relationships between a few macroscopic variables. For a one-component system, we are dealing with s, ρ, u, T, μ, p and combinations of these variables, such as $\hat{s} = s/\rho$ or $f = u - Ts$. Only two of these variables are independent, and all the other ones can be obtained from the proper thermodynamic potential for the chosen independent variables, such as $s(\rho, u)$, $u(\rho, s)$, or $f(\rho, T)$. Thermodynamic material properties are defined in terms of second-order derivatives of thermodynamic potentials (see Section 4.4). There are four second-order derivatives, where the two mixed ones coincide (the Maxwell relation), so that there are only three independent thermodynamic material properties. As a much larger variety of thermodynamic material properties is used in practice, the question of how to obtain and relate all these properties in an efficient and straightforward way arises. This question is addressed in the present appendix.

The equations of this appendix for one-component systems are particularly useful in the discussion of dynamic light scattering. For example, we can use them to relate equivalent ways of constructing linearized hydrodynamic equations in Section 26.3 or to identify independently fluctuating variables in (26.47). Our concluding remarks on two-component systems clarify the role of differences between the partial specific quantities introduced in Section 4.5 as thermodynamic material properties and are particularly useful in the discussion of binary diffusion in Section 6.3.

Thermodynamic Material Properties

A number of frequently used thermodynamic material properties for single-component systems are compiled in Table A.1. Further derivatives can be

Table A.1 *Definition of thermodynamic material properties*

Symbol	Definition	Name
α_p	$-\dfrac{1}{\rho}\dfrac{\partial\rho(p,T)}{\partial T}$	isobaric thermal expansion coefficient
$\alpha_{\hat{s}}$	$\dfrac{1}{\rho}\dfrac{\partial\rho(\hat{s},T)}{\partial T}$	adiabatic thermal expansivity
$\hat{c}_p$	$T\dfrac{\partial\hat{s}(p,T)}{\partial T}$	specific heat at constant pressure
$\hat{c}_{\hat{v}}$	$\dfrac{T}{\rho}\dfrac{\partial s(\rho,T)}{\partial T}=T\dfrac{\partial\hat{s}(\rho,T)}{\partial T}$	specific heat at constant specific volume
γ	$\dfrac{\hat{c}_p}{\hat{c}_{\hat{v}}}$	heat capacity ratio (adiabatic index)
κ_T	$\dfrac{1}{\rho}\dfrac{\partial\rho(p,T)}{\partial p}$	isothermal compressibility
$\kappa_{\hat{s}}$	$\dfrac{1}{\rho}\dfrac{\partial\rho(\hat{s},p)}{\partial p}$	isentropic compressibility
$c_{\hat{s}}^2$	$\dfrac{\partial p(\rho,\hat{s})}{\partial\rho}=\dfrac{\partial p(\rho,s)}{\partial\rho}+\dfrac{s}{\rho\alpha_{\hat{s}}}$	(adiabatic) speed of sound squared
c_T^2	$\dfrac{\partial p(\rho,T)}{\partial\rho}$	isothermal speed of sound squared

expressed in terms of the properties defined in in Table A.1, for example,

$$\frac{\partial p(\hat{s},T)}{\partial T}=\rho\alpha_{\hat{s}}c_{\hat{s}}^2,\qquad\frac{\partial p(\rho,T)}{\partial T}=\rho\alpha_p c_T^2.\qquad\text{(A.1)}$$

The first identity is obtained by differentiating $p(\rho(\hat{s},T)),\hat{s})$ with respect to T by means of the chain rule; the second identity is a special case of the general differentiation rule

$$\frac{\partial y(x,z)}{\partial x}=-\frac{\partial y(x,z)}{\partial z}\frac{\partial z(x,y)}{\partial x}.\qquad\text{(A.2)}$$

In order to give the reader an intuitive feeling for the various properties and to check the relationships to be derived, we consider the concrete examples of the ideal gas and a more general hard-sphere system. This system is defined by (see Exercise 4.4)

$$s(\rho,u)=\rho\frac{k_{\text{B}}}{m}\ln\left[R_0(\rho)u^{3/2}\rho^{-5/2}\right],\qquad\text{(A.3)}$$

with a given function $R_0(\rho)$ and the derived functions

$$R_1=-\frac{\rho^2}{R_0}\frac{\partial}{\partial\rho}\left(\frac{R_0}{\rho}\right)=1-\frac{\rho}{R_0}\frac{\partial R_0}{\partial\rho},\quad R_2=\frac{\partial}{\partial\rho}(\rho R_1)=1-\frac{\partial}{\partial\rho}\frac{\rho^2}{R_0}\frac{\partial R_0}{\partial\rho}.$$
$$\text{(A.4)}$$

Table A.2 *Thermodynamic material properties for ideal gases and hard-sphere systems*

Symbol	Ideal Gas	Hard Spheres
α_p	$\dfrac{1}{T}$	$\dfrac{R_1}{TR_2}$
$\alpha_{\hat{s}}$	$\dfrac{3}{2T}$	$\dfrac{3}{2TR_1}$
$\hat{c}_p$	$\dfrac{5}{2}\dfrac{k_{\mathrm{B}}}{m}$	$\left(\dfrac{3}{2}+\dfrac{R_1^2}{R_2}\right)\dfrac{k_{\mathrm{B}}}{m}$
$\hat{c}_{\hat{v}}$	$\dfrac{3}{2}\dfrac{k_{\mathrm{B}}}{m}$	$\dfrac{3}{2}\dfrac{k_{\mathrm{B}}}{m}$
γ	$\dfrac{5}{3}$	$1+\dfrac{2}{3}\dfrac{R_1^2}{R_2}$
κ_T	$\dfrac{1}{p}$	$\dfrac{R_1}{pR_2}$
$\kappa_{\hat{s}}$	$\dfrac{3}{5p}$	$\left[p\left(\dfrac{2}{3}R_1+\dfrac{R_2}{R_1}\right)\right]^{-1}$
$c_{\hat{s}}^2$	$\dfrac{5}{3}\dfrac{k_{\mathrm{B}}T}{m}$	$\dfrac{k_{\mathrm{B}}T}{m}\left(\dfrac{2}{3}R_1^2+R_2\right)$
c_T^2	$\dfrac{k_{\mathrm{B}}T}{m}$	$\dfrac{k_{\mathrm{B}}T}{m}R_2$

The thermodynamic potential (A.3) leads to the following equations of state,

$$u = \frac{3}{2}\frac{\rho}{m}k_{\mathrm{B}}T, \qquad p = \frac{\rho}{m}k_{\mathrm{B}}T\,R_1(\rho). \tag{A.5}$$

For an ideal gas, we have $R_0 = R_1 = R_2 = 1$. The function $R_1(\rho)$ expresses the pressure increase associated with hard spheres of finite size. The specific results for the thermodynamic material properties defined in Table A.1 for the ideal gas and more general hard spheres are compiled in Table A.2.

Systematic Relations

Based on $u(\rho, s)$

Three basic material properties ($\hat{c}_{\hat{v}}$, $\alpha_{\hat{s}}$, $c_{\hat{s}}^2$) can be successively obtained from the second derivatives

$$\frac{\partial^2 u(\rho, s)}{\partial s^2} = \frac{T}{\rho\hat{c}_{\hat{v}}}, \tag{A.6}$$

$$\frac{\partial^2 u(\rho, s)}{\partial\rho\,\partial s} = \frac{1}{\rho}\left(\frac{1}{\alpha_{\hat{s}}} - \frac{sT}{\rho\hat{c}_{\hat{v}}}\right), \tag{A.7}$$

$$\frac{\partial^2 u(\rho, s)}{\partial \rho^2} = \frac{1}{\rho}\left(c_{\hat{s}}^2 - \frac{2s}{\rho\alpha_{\hat{s}}} + \frac{s^2 T}{\rho^2 \hat{c}_{\hat{v}}}\right). \tag{A.8}$$

Based on these expressions, a number of useful identities can be developed for the partial derivatives of p and T,

$$p(\rho, s) = \rho\frac{\partial u(\rho, s)}{\partial \rho} + s\frac{\partial u(\rho, s)}{\partial s} - u(\rho, s), \tag{A.9}$$

$$\frac{\partial p(\rho, s)}{\partial s} = \rho\frac{\partial^2 u(\rho, s)}{\partial \rho \partial s} + s\frac{\partial^2 u(\rho, s)}{\partial s^2} = \frac{1}{\alpha_{\hat{s}}}, \tag{A.10}$$

$$\frac{\partial p(\rho, s)}{\partial \rho} = \rho\frac{\partial^2 u(\rho, s)}{\partial \rho^2} + s\frac{\partial^2 u(\rho, s)}{\partial \rho \partial s} = c_{\hat{s}}^2 - \frac{s}{\rho\alpha_{\hat{s}}}, \tag{A.11}$$

$$\frac{\partial T(\rho, s)}{\partial s} = \frac{\partial^2 u(\rho, s)}{\partial s^2} = \frac{T}{\rho\hat{c}_{\hat{v}}}, \tag{A.12}$$

$$\frac{\partial T(\rho, s)}{\partial \rho} = \frac{\partial^2 u(\rho, s)}{\partial \rho \partial s} = \frac{1}{\rho}\left(\frac{1}{\alpha_{\hat{s}}} - \frac{sT}{\rho\hat{c}_{\hat{v}}}\right). \tag{A.13}$$

We further observe a nice identity for the speed of sound,

$$\rho^2\frac{\partial^2 u(\rho, s)}{\partial \rho^2} + 2\rho s\frac{\partial^2 u(\rho, s)}{\partial \rho \partial s} + s^2\frac{\partial^2 u(\rho, s)}{\partial s^2} = \rho c_{\hat{s}}^2. \tag{A.14}$$

Based on $s(\rho, u)$

Three basic material properties $(\hat{c}_{\hat{v}}, \alpha_{\hat{s}}, c_{\hat{s}}^2)$ can be successively obtained from the second derivatives

$$\frac{\partial^2 s(\rho, u)}{\partial u^2} = -\frac{1}{\rho T^2 \hat{c}_{\hat{v}}}, \tag{A.15}$$

$$\frac{\partial^2 s(\rho, u)}{\partial \rho \partial u} = -\frac{1}{\rho T^2}\left(\frac{1}{\alpha_{\hat{s}}} - \frac{\hat{h}}{\hat{c}_{\hat{v}}}\right), \tag{A.16}$$

$$\frac{\partial^2 s(\rho, u)}{\partial \rho^2} = -\frac{1}{\rho T^2}\left(T c_{\hat{s}}^2 - \frac{2\hat{h}}{\alpha_{\hat{s}}} + \frac{\hat{h}^2}{\hat{c}_{\hat{v}}}\right), \tag{A.17}$$

where $\hat{h} = (u + p)/\rho$ is the specific enthalpy. Based on these expressions, a number of useful identities can be developed for the partial derivatives of p and T,

$$\frac{\partial s(\rho, u)}{\partial u}[u + p(\rho, u)] = s(\rho, u) - \rho\frac{\partial s(\rho, u)}{\partial \rho}, \tag{A.18}$$

$$\frac{\partial p(\rho, u)}{\partial u} = -T\rho \left[\frac{\partial^2 s(\rho, u)}{\partial \rho \, \partial u} + \hat{h} \frac{\partial^2 s(\rho, u)}{\partial u^2} \right] = \frac{1}{T\alpha_{\hat{s}}}, \tag{A.19}$$

$$\frac{\partial p(\rho, u)}{\partial \rho} = -T\rho \left[\frac{\partial^2 s(\rho, u)}{\partial \rho^2} + \hat{h} \frac{\partial^2 s(\rho, u)}{\partial \rho \, \partial u} \right] = c_{\hat{s}}^2 - \frac{\hat{h}}{T\alpha_{\hat{s}}}, \tag{A.20}$$

$$\frac{\partial T(\rho, u)}{\partial u} = -T^2 \frac{\partial^2 s(\rho, u)}{\partial u^2} = \frac{1}{\rho \hat{c}_{\hat{v}}}, \tag{A.21}$$

$$\frac{\partial T(\rho, u)}{\partial \rho} = -T^2 \frac{\partial^2 s(\rho, u)}{\partial \rho \, \partial u} = \frac{1}{\rho} \left(\frac{1}{\alpha_{\hat{s}}} - \frac{\hat{h}}{\hat{c}_{\hat{v}}} \right). \tag{A.22}$$

We further observe another nice identity for the speed of sound,

$$-T\rho \left[\frac{\partial^2 s(\rho, u)}{\partial \rho^2} + 2\hat{h} \frac{\partial^2 s(\rho, u)}{\partial \rho \, \partial u} + \hat{h}^2 \frac{\partial^2 s(\rho, u)}{\partial u^2} \right] = c_{\hat{s}}^2. \tag{A.23}$$

Based on $f(\rho, T)$

Three basic material properties ($\hat{c}_{\hat{v}}$, $\alpha_{\hat{s}}$, c_T^2) can be successively obtained from the second derivatives of the Helmholtz free energy density $f(\rho, T)$,

$$\frac{\partial^2 f(\rho, T)}{\partial T^2} = -\frac{\rho \hat{c}_{\hat{v}}}{T}, \qquad \frac{\partial^2 f(\rho, T)}{\partial \rho \, \partial T} = \frac{\hat{c}_{\hat{v}}}{T\alpha_{\hat{s}}} - \frac{s}{\rho}, \qquad \frac{\partial^2 f(\rho, T)}{\partial \rho^2} = \frac{c_T^2}{\rho}. \tag{A.24}$$

We postpone the discussion of the specific Gibbs free energy density $\hat{g}(T, p)$ as a thermodynamic potential to the final remarks on two-component systems.

Summary of Basic Relationships

The following six relationships allow us to express all nine thermodynamic material properties introduced in Table A.1 in terms of three basic ones:

$$\rho \kappa_{\hat{s}} = \frac{1}{c_{\hat{s}}^2}, \tag{A.25}$$

$$\gamma = \frac{\hat{c}_p}{\hat{c}_{\hat{v}}}, \tag{A.26}$$

$$\gamma = \frac{c_{\hat{s}}^2}{c_T^2}, \tag{A.27}$$

$$\gamma = \frac{\kappa_T}{\kappa_{\hat{s}}}, \tag{A.28}$$

$$\gamma = 1 + \frac{\alpha_p}{\alpha_{\hat{s}}}, \tag{A.29}$$

$$\gamma = 1 + \frac{\hat{c}_p}{T\alpha_{\hat{s}}^2 c_{\hat{s}}^2}. \tag{A.30}$$

Equations (A.26)–(A.30) give five different expressions for γ. Equations (A.29) and (A.30) further imply

$$\hat{c}_p = T\alpha_p \alpha_{\hat{s}} c_{\hat{s}}^2, \tag{A.31}$$

and, after dividing by γ,

$$\hat{c}_{\hat{v}} = T\alpha_p \alpha_{\hat{s}} c_T^2. \tag{A.32}$$

As an illustration of how the results collected in this appendix can be verified, we look at the Maxwell relation

$$\frac{\partial^2 f(\rho, T)}{\partial \rho \partial T} = -\frac{\partial s(\rho, T)}{\partial \rho} = -\rho \frac{\partial \hat{s}(\rho, T)}{\partial \rho} - \hat{s} = \rho \frac{\partial \hat{s}(\rho, T)/\partial T}{\partial \rho(\hat{s}, T)/\partial T} - \hat{s} = \frac{\hat{c}_{\hat{v}}}{T\alpha_{\hat{s}}} - \frac{s}{\rho}$$

$$= \frac{\partial \hat{\mu}(\rho, T)}{\partial T} = \frac{1}{\rho} \frac{\partial [u(\rho, T) + p(\rho, T) - Ts(\rho, T)]}{\partial T} = \frac{1}{\rho} \frac{\partial p(\rho, T)}{\partial T} - \frac{s}{\rho}$$

$$= -\frac{1}{\rho} \frac{\partial p(\rho, T)}{\partial \rho} \frac{\partial \rho(p, T)}{\partial T} - \frac{s}{\rho} = \alpha_p c_T^2 - \frac{s}{\rho}. \tag{A.33}$$

The first line leads to the mixed derivative given in (A.24). By comparing this result with the very last expression in (A.33), we obtain the relationship (A.32), which is hence recognized as a Maxwell relation. In a slightly simpler way, (A.31) can be recognized as the Maxwell relation associated with the mixed derivatives of $\hat{g}(T, p)$.

Remarks on Two-Component Systems

For two-component systems, thermodynamic potentials depend on three independent variables and there exist six different second-order derivatives, that is, six independent thermodynamic material properties. As a first step, it is natural to generalize the thermodynamic material properties defined in Table A.1 by keeping the composition fixed, as we have done in the definitions of $\hat{c}_{\hat{v}}$, α_p, and κ_T given after (4.53).

For any equation of state $\hat{a}(T, p, w_1)$ for a specific quantity $\hat{a}$, we have introduced the derivative with respect to the mass fraction w_1 in (4.52) in terms of the partial specific quantities

$$\hat{a}_1 = \hat{a} + (1 - w_1)\frac{\partial \hat{a}(T, p, w_1)}{\partial w_1}, \qquad \hat{a}_2 = \hat{a} - w_1\frac{\partial \hat{a}(T, p, w_1)}{\partial w_1}, \tag{A.34}$$

so that $\hat{a}_1 - \hat{a}_2$ is the derivative with respect to w_1 at constant T and p. The partial specific properties associated with enthalpy, volume, and entropy have been found to be of special interest and could be used to introduce independent thermodynamic material properties associated with composition effects for two-component systems (see Exercises 4.5 and 4.6).

As there now are too many ways of choosing three independent variables, we here consider only the particularly convenient thermodynamic potential $\hat{g}(T, p, w_1)$. Its second-order derivatives lead us to the following six independent thermodynamic material properties:

$$\frac{\partial^2 \hat{g}(T, p, w_1)}{\partial T^2} = -\frac{\hat{c}_p}{T}, \tag{A.35}$$

$$\frac{\partial^2 \hat{g}(T, p, w_1)}{\partial T \, \partial p} = \frac{\alpha_p}{\rho}, \tag{A.36}$$

$$\frac{\partial^2 \hat{g}(T, p, w_1)}{\partial p^2} = -\frac{\kappa_T}{\rho}, \tag{A.37}$$

$$\frac{\partial^2 \hat{g}(T, p, w_1)}{\partial T \, \partial w_1} = -(\hat{s}_1 - \hat{s}_2), \tag{A.38}$$

$$\frac{\partial^2 \hat{g}(T, p, w_1)}{\partial p \, \partial w_1} = \hat{v}_1 - \hat{v}_2, \tag{A.39}$$

$$\frac{\partial^2 \hat{g}(T, p, w_1)}{\partial w_1^2} = \frac{1}{1 - w_1} \frac{\partial \hat{\mu}_1(T, p, w_1)}{\partial w_1} = \Xi, \tag{A.40}$$

where, in addition to the differences of partial specific entropies and volumes, we have introduced the miscibility parameter Ξ as a new thermodynamic material property for two-component systems. It plays an important role in the discussion of binary diffusion and thermoelectric effects (see Sections 6.3 and 6.4).

For k-component systems, the number of independent thermodynamic material properties is $(k+1)(k+2)/2$. For each pair of components, there is an associated miscibility parameter. A more detailed explicit characterization of the thermodynamics of k-component systems is the goal of the CALPHAD (CALculation of PHAse Diagrams) project,[1] within which a huge variety of experimental data is condensed into consistent thermodynamic potentials.

---

[1] See the official website www.calphad.org.

Appendix B

Differential Operations in Coordinate Form

Rectangular Coordinates (x_1, x_2, x_3)

$$\boldsymbol{\nabla} \cdot \boldsymbol{a} = \frac{\partial a_1}{\partial x_1} + \frac{\partial a_2}{\partial x_2} + \frac{\partial a_3}{\partial x_3} \tag{B.1}$$

$$\nabla^2 a = \frac{\partial^2 a}{\partial x_1^2} + \frac{\partial^2 a}{\partial x_2^2} + \frac{\partial^2 a}{\partial x_3^2} \tag{B.2}$$

$$[(\boldsymbol{\nabla} a)_i] = \begin{bmatrix} \dfrac{\partial a}{\partial x_1} \\[2ex] \dfrac{\partial a}{\partial x_2} \\[2ex] \dfrac{\partial a}{\partial x_3} \end{bmatrix} \tag{B.3}$$

$$[(\boldsymbol{\nabla} \times \boldsymbol{a})_i] = \begin{bmatrix} \dfrac{\partial a_3}{\partial x_2} - \dfrac{\partial a_2}{\partial x_3} \\[2ex] \dfrac{\partial a_1}{\partial x_3} - \dfrac{\partial a_3}{\partial x_1} \\[2ex] \dfrac{\partial a_2}{\partial x_1} - \dfrac{\partial a_1}{\partial x_2} \end{bmatrix} \tag{B.4}$$

$$[(\nabla^2 \boldsymbol{a})_i] = \begin{bmatrix} \dfrac{\partial^2 a_1}{\partial x_1^2} + \dfrac{\partial^2 a_1}{\partial x_2^2} + \dfrac{\partial^2 a_1}{\partial x_3^2} \\[2ex] \dfrac{\partial^2 a_2}{\partial x_1^2} + \dfrac{\partial^2 a_2}{\partial x_2^2} + \dfrac{\partial^2 a_2}{\partial x_3^2} \\[2ex] \dfrac{\partial^2 a_3}{\partial x_1^2} + \dfrac{\partial^2 a_3}{\partial x_2^2} + \dfrac{\partial^2 a_3}{\partial x_3^2} \end{bmatrix} \tag{B.5}$$

$$[(\boldsymbol{\nabla} \cdot \boldsymbol{\alpha})_i] = \begin{bmatrix} \dfrac{\partial \alpha_{11}}{\partial x_1} + \dfrac{\partial \alpha_{21}}{\partial x_2} + \dfrac{\partial \alpha_{31}}{\partial x_3} \\[2mm] \dfrac{\partial \alpha_{12}}{\partial x_1} + \dfrac{\partial \alpha_{22}}{\partial x_2} + \dfrac{\partial \alpha_{32}}{\partial x_3} \\[2mm] \dfrac{\partial \alpha_{13}}{\partial x_1} + \dfrac{\partial \alpha_{23}}{\partial x_2} + \dfrac{\partial \alpha_{33}}{\partial x_3} \end{bmatrix} \tag{B.6}$$

$$[(\boldsymbol{\nabla} a)_{ij}] = \begin{bmatrix} \dfrac{\partial a_1}{\partial x_1} & \dfrac{\partial a_2}{\partial x_1} & \dfrac{\partial a_3}{\partial x_1} \\[2mm] \dfrac{\partial a_1}{\partial x_2} & \dfrac{\partial a_2}{\partial x_2} & \dfrac{\partial a_3}{\partial x_2} \\[2mm] \dfrac{\partial a_1}{\partial x_3} & \dfrac{\partial a_2}{\partial x_3} & \dfrac{\partial a_3}{\partial x_3} \end{bmatrix} \tag{B.7}$$

Cylindrical Coordinates (r, θ, z)

$$\boldsymbol{\nabla} \cdot \boldsymbol{a} = \frac{1}{r}\frac{\partial}{\partial r}(r a_r) + \frac{1}{r}\frac{\partial a_\theta}{\partial \theta} + \frac{\partial a_z}{\partial z} \tag{B.8}$$

$$\nabla^2 a = \frac{1}{r}\frac{\partial}{\partial r}\left(r \frac{\partial a}{\partial r}\right) + \frac{1}{r^2}\frac{\partial^2 a}{\partial \theta^2} + \frac{\partial^2 a}{\partial z^2} \tag{B.9}$$

$$[(\boldsymbol{\nabla} a)_i] = \begin{bmatrix} \dfrac{\partial a}{\partial r} \\[2mm] \dfrac{1}{r}\dfrac{\partial a}{\partial \theta} \\[2mm] \dfrac{\partial a}{\partial z} \end{bmatrix} \tag{B.10}$$

$$[(\boldsymbol{\nabla} \times \boldsymbol{a})_i] = \begin{bmatrix} \dfrac{1}{r}\dfrac{\partial a_z}{\partial \theta} - \dfrac{\partial a_\theta}{\partial z} \\[2mm] \dfrac{\partial a_r}{\partial z} - \dfrac{\partial a_z}{\partial r} \\[2mm] \dfrac{1}{r}\dfrac{\partial}{\partial r}(r a_\theta) - \dfrac{1}{r}\dfrac{\partial a_r}{\partial \theta} \end{bmatrix} \tag{B.11}$$

$$[(\nabla^2 \boldsymbol{a})_i] = \begin{bmatrix} \dfrac{\partial}{\partial r}\left(\dfrac{1}{r}\dfrac{\partial}{\partial r}(r a_r)\right) + \dfrac{1}{r^2}\dfrac{\partial^2 a_r}{\partial \theta^2} - \dfrac{2}{r^2}\dfrac{\partial a_\theta}{\partial \theta} + \dfrac{\partial^2 a_r}{\partial z^2} \\[2mm] \dfrac{\partial}{\partial r}\left(\dfrac{1}{r}\dfrac{\partial}{\partial r}(r a_\theta)\right) + \dfrac{1}{r^2}\dfrac{\partial^2 a_\theta}{\partial \theta^2} + \dfrac{2}{r^2}\dfrac{\partial a_r}{\partial \theta} + \dfrac{\partial^2 a_\theta}{\partial z^2} \\[2mm] \dfrac{1}{r}\dfrac{\partial}{\partial r}\left(r \dfrac{\partial a_z}{\partial r}\right) + \dfrac{1}{r^2}\dfrac{\partial^2 a_z}{\partial \theta^2} + \dfrac{\partial^2 a_z}{\partial z^2} \end{bmatrix} \tag{B.12}$$

$$[(\boldsymbol{\nabla} \cdot \boldsymbol{\alpha})_i] = \begin{bmatrix} \frac{1}{r}\frac{\partial}{\partial r}(r\alpha_{rr}) + \frac{1}{r}\frac{\partial \alpha_{\theta r}}{\partial \theta} - \frac{\alpha_{\theta\theta}}{r} + \frac{\partial \alpha_{zr}}{\partial z} \\[2mm] \frac{1}{r^2}\frac{\partial}{\partial r}(r^2\alpha_{r\theta}) + \frac{1}{r}\frac{\partial \alpha_{\theta\theta}}{\partial \theta} + \frac{\alpha_{\theta r} - \alpha_{r\theta}}{r} + \frac{\partial \alpha_{z\theta}}{\partial z} \\[2mm] \frac{1}{r}\frac{\partial}{\partial r}(r\alpha_{rz}) + \frac{1}{r}\frac{\partial \alpha_{\theta z}}{\partial \theta} + \frac{\alpha_{zz}}{\partial z} \end{bmatrix} \quad \text{(B.13)}$$

$$[(\boldsymbol{\nabla} a)_{ij}] = \begin{bmatrix} \frac{\partial a_r}{\partial r} & \frac{\partial a_\theta}{\partial r} & \frac{\partial a_z}{\partial r} \\[2mm] \frac{1}{r}\frac{\partial a_r}{\partial \theta} - \frac{a_\theta}{r} & \frac{1}{r}\frac{\partial a_\theta}{\partial \theta} + \frac{a_r}{r} & \frac{1}{r}\frac{\partial a_z}{\partial \theta} \\[2mm] \frac{\partial a_r}{\partial z} & \frac{\partial a_\theta}{\partial z} & \frac{\partial a_z}{\partial z} \end{bmatrix} \quad \text{(B.14)}$$

Spherical Coordinates (r, θ, ϕ)

$$\boldsymbol{\nabla} \cdot \boldsymbol{a} = \frac{1}{r^2}\frac{\partial}{\partial r}(r^2 a_r) + \frac{1}{r\sin\theta}\frac{\partial}{\partial \theta}(a_\theta \sin\theta) + \frac{1}{r\sin\theta}\frac{\partial a_\phi}{\partial \phi} \quad \text{(B.15)}$$

$$\nabla^2 a = \left[\frac{1}{r^2}\frac{\partial}{\partial r}\left(r^2\frac{\partial a}{\partial r}\right) + \frac{1}{r^2\sin\theta}\frac{\partial}{\partial \theta}\left(\sin\theta\frac{\partial a}{\partial \theta}\right) + \frac{1}{r^2\sin^2\theta}\frac{\partial^2 a}{\partial \phi^2}\right] \quad \text{(B.16)}$$

$$[(\boldsymbol{\nabla} a)_i] = \begin{bmatrix} \frac{\partial a}{\partial r} \\[2mm] \frac{1}{r}\frac{\partial a}{\partial \theta} \\[2mm] \frac{1}{r\sin\theta}\frac{\partial a}{\partial \phi} \end{bmatrix} \quad \text{(B.17)}$$

$$[(\boldsymbol{\nabla} \times \boldsymbol{a})_i] = \begin{bmatrix} \frac{1}{r\sin\theta}\frac{\partial}{\partial \theta}(a_\phi \sin\theta) - \frac{1}{r\sin\theta}\frac{\partial a_\theta}{\partial \phi} \\[2mm] \frac{1}{r\sin\theta}\frac{\partial a_r}{\partial \phi} - \frac{1}{r}\frac{\partial}{\partial r}(ra_\phi) \\[2mm] \frac{1}{r}\frac{\partial}{\partial r}(ra_\theta) - \frac{1}{r}\frac{\partial a_r}{\partial \theta} \end{bmatrix} \quad \text{(B.18)}$$

$$[(\nabla^2 \boldsymbol{a})_i] = \begin{bmatrix} \nabla^2 a_r - \frac{2}{r^2}a_r - \frac{2}{r^2}\frac{\partial a_\theta}{\partial \theta} - \frac{2\cot\theta}{r^2}a_\theta - \frac{2}{r^2\sin\theta}\frac{\partial a_\phi}{\partial \phi} \\[2mm] \nabla^2 a_\theta + \frac{2}{r^2}\frac{\partial a_r}{\partial \theta} - \frac{a_\theta}{r^2\sin^2 0} - \frac{2\cos\theta}{r^2\sin^2\theta}\frac{\partial a_\phi}{\partial \phi} \\[2mm] \nabla^2 a_\phi - \frac{a_\phi}{r^2\sin^2\theta} + \frac{2}{r^2\sin\theta}\frac{\partial a_r}{\partial \phi} + \frac{2\cos\theta}{r^2\sin^2\theta}\frac{\partial a_\theta}{\partial \phi} \end{bmatrix} \quad \text{(B.19)}$$

$[(\boldsymbol{\nabla} \cdot \boldsymbol{\alpha})_i] =$

$$\begin{bmatrix} \frac{1}{r^2}\frac{\partial}{\partial r}(r^2\alpha_{rr}) + \frac{1}{r\sin\theta}\frac{\partial}{\partial\theta}(\alpha_{\theta r}\sin\theta) + \frac{1}{r\sin\theta}\frac{\partial\alpha_{\phi r}}{\partial\phi} - \frac{\alpha_{\theta\theta}+\alpha_{\phi\phi}}{r} \\ \frac{1}{r^3}\frac{\partial}{\partial r}(r^3\alpha_{r\theta}) + \frac{1}{r\sin\theta}\frac{\partial}{\partial\theta}(\alpha_{\theta\theta}\sin\theta) + \frac{1}{r\sin\theta}\frac{\partial\alpha_{\phi\theta}}{\partial\phi} + \frac{\alpha_{\theta r}-\alpha_{r\theta}-\alpha_{\phi\phi}\cot\theta}{r} \\ \frac{1}{r^3}\frac{\partial}{\partial r}(r^3\alpha_{r\phi}) + \frac{1}{r\sin\theta}\frac{\partial}{\partial\theta}(\alpha_{\theta\phi}\sin\theta) + \frac{1}{r\sin\theta}\frac{\partial\alpha_{\phi\phi}}{\partial\phi} + \frac{\alpha_{\phi r}-\alpha_{r\phi}+\alpha_{\phi\theta}\cot\theta}{r} \end{bmatrix}$$

$$\text{(B.20)}$$

$[(\boldsymbol{\nabla}\boldsymbol{a})_{ij}] = \begin{bmatrix} \frac{\partial a_r}{\partial r} & \frac{\partial a_\theta}{\partial r} & \frac{\partial a_\phi}{\partial r} \\ \frac{1}{r}\frac{\partial a_r}{\partial\theta} - \frac{a_\theta}{r} & \frac{1}{r}\frac{\partial a_\theta}{\partial\theta} + \frac{a_r}{r} & \frac{1}{r}\frac{\partial a_\phi}{\partial\theta} \\ \frac{1}{r\sin\theta}\frac{\partial a_r}{\partial\phi} - \frac{a_\phi}{r} & \frac{1}{r\sin\theta}\frac{\partial a_\theta}{\partial\phi} - \frac{\cot\theta}{r}a_\phi & \frac{1}{r\sin\theta}\frac{\partial a_\phi}{\partial\phi} + \frac{a_r}{r} + \frac{\cot\theta}{r}a_\theta \end{bmatrix}$$

$$\text{(B.21)}$$

References

Agassant, J. F., Avenas, P., Carreau, P. J., and Sergent, J. P. 1991. *Polymer Processing, Principles and Modeling.* Munich: Hanser.

Albano, A. M., and Bedeaux, D. 1987. Non-Equilibrium Electro-Thermodynamics of Polarizable Multicomponent Fluids with an Interface. *Physica A*, **147**, 407–435.

Anderson, T. B., and Jackson, R. 1967. A Fluid Mechanical Description of Fluidized Beds. *AIChE J.*, **6**, 527–539.

Archibald, W. J. 1938a. The Process of Diffusion in a Centrifugal Field of Force. *Phys. Rev.*, **53**, 746–752.

Archibald, W. J. 1938b. The Process of Diffusion in a Centrifugal Field of Force. II. *Phys. Rev.*, **54**, 371–374.

Aris, R. 1956. On the Dispersion of a Solute in a Fluid Flowing through a Tube. *Proc. Roy. Soc. A*, **235**, 67–77.

Aris, R. 1957. On Shape Factors for Irregular Particles I: The Steady State Problem. Diffusion and Reaction. *Chem. Eng. Sci.*, **6**, 262–268.

Atkinson, B., Kemblowski, Z., and Smith, J. M. 1967. Measurements of Velocity Profile in Developing Liquid Flows. *AIChE J.*, **13**, 17–20.

Atkinson, B., Brocklebank, M. P., Card, C. C. H., and Smith, J. M. 1969. Low Reynolds Number Developing Flows. *AIChE J.*, **15**, 548–553.

Baier, R., Romatschke, P., and Wiedemann, U. A. 2006. Dissipative Hydrodynamics and Heavy-Ion Collisions. *Phys. Rev. C*, **73**, 064903.

Balasubramanian, V., Bush, K., Smoukov, S., Venerus, D. C., and Schieber, J. D. 2005. Measurements of Flow-Induced Anisotropic Thermal Conduction in a Polyisobutylene Melt Following Step Shear Flow. *Macromolecules*, **38**, 6210–6215.

Barlow, E. J., and Langlois, W. E. 1962. Diffusion of Gas from a Liquid into an Expanding Bubble. *IBM J.*, **6**, 329–337.

Barnes, H. A., Hutton, J. F., and Walters, K. 1989. *An Introduction to Rheology.* Amsterdam: Elsevier.

Bataille, J., Edelen, D. G. B., and Kestin, J. 1978. Nonequilibrium Thermodynamics of the Nonlinear Equations of Chemical Kinetics. *J. Non-Equilib. Thermodyn.*, **3**, 153–168.

Batchelor, G. K. 1967. *An Introduction to Fluid Dynamics.* Cambridge: Cambridge University Press.

Bear, J., and Bachmat, Y. 1990. *Introduction to Modeling of Transport Phenomena in Porous Media*. Dordrecht: Kluwer Academic Publishers.

Bedeaux, D. 1986. Nonequilibrium Thermodynamics and Statistical Physics of Surfaces. *Adv. Chem. Phys.*, **64**, 47–109.

Bedeaux, D., Albano, A. M., and Mazur, P. 1976. Boundary Conditions and Non-Equilibrium Thermodynamics. *Physica A*, **82**, 438–462.

Beris, A. N., and Edwards, B. J. 1994. *The Thermodynamics of Flowing Systems*. New York: Oxford University Press.

Berne, B. J., and Pecora, R. 2000. *Dynamic Light Scattering with Applications to Chemistry, Biology, and Physics*. New York: Dover Publications Inc.

Biller, P., and Petruccione, F. 1990. Continuous Time Simulation of Transient Polymer Network Models. *J. Chem. Phys.*, **92**, 6322–6326.

Bingham, E. C. 1922. *Fluidity and Plasticity*. New York: McGraw-Hill.

Bird, R. B., and Curtiss, C. F. 1959. Tangential Newtonian Flow in Annuli – I: Unsteady State Velocity Profiles. *Chem. Eng. Sci.*, **11**, 108–113.

Bird, R. B., and Curtiss, C. F. 1984. Fascinating Polymeric Liquids. *Physics Today*, **37/1**, 36–43.

Bird, R. B., Stewart, W. E., and Lightfoot, E. N. 1960. *Transport Phenomena*. New York: Wiley.

Bird, R. B., Stewart, W. E., and Lightfoot, E. N. 2001. *Transport Phenomena*. Second edn. New York: Wiley.

Bird, R. B., Armstrong, R. C., and Hassager, O. 1987a. *Dynamics of Polymeric Liquids. Volume 1. Fluid Mechanics*. Second edn. New York: Wiley.

Bird, R. B., Curtiss, C. F., Armstrong, R. C., and Hassager, O. 1987b. *Dynamics of Polymeric Liquids. Volume 2. Kinetic Theory*. Second edn. New York: Wiley.

Blasius, H. 1913. Das Ähnlichkeitsgesetz bei Reibungsvorgängen in Flüssigkeiten. *Mitteilungen über Forschungsarbeiten auf dem Gebiete des Ingenieurwesens*, **131**, 1–40.

Boltzmann, L. 1896. *Vorlesungen über Gastheorie, I. Theil*. Leipzig: Barth.

Boon, J. P., and Yip, S. 1991. *Molecular Hydrodynamics*. New York: Dover.

Bornhorst, W. J., and Hatsopoulos, G. N. 1967. Bubble-Growth Calculation without Neglect of Interfacial Discontinuities. *J. Appl. Mech.*, **34**, 847–853.

Boussinesq, M. J. 1914. Sur le calcul de plus en plus approché des vitesses bien continues de régime uniforme par des polynomes, dans un tube prismatique à section carrée. *Comptes Rendus*, **158**, 1743–1749.

Brochard, F., and de Gennes, P. G. 1992. Shear-Dependent Slippage at a Polymer/Solid Interface. *Langmuir*, **8**, 3033–3037.

Brown, R. A. 1988. Theory of Transport Processes in Single-Crystal Growth from the Melt. *AIChE J.*, **34**, 881–910.

Cahill, D. G., Ford, W. K., Goodson, K. E., Mahan, G. D., Majumdar, A., Maris, H. J., Merlin, R., and Phillpot, S. R. 2003. Nanoscale Thermal Transport. *J. Appl. Phys.*, **93**, 793–818.

Cahill, D. G., Braun, P. V., Chen, G., Clarke, D. R., Fan, S., Goodson, K. E., Keblinski, P., King, W. P., Mahan, G. D., Majumdar, A., Maris, H. J., Phillpot, S. R., Pop, E., and Shi, L. 2014. Nanoscale Thermal Transport. II 2003–2012. *Appl. Phys. Rev.*, **1**, 011305.

Callen, H. B. 1985. *Thermodynamics and an Introduction to Thermostatistics*. Second edn. New York: Wiley.

Caplan, S. R., and Essig, A. 1983. *Bioenergetics and Linear Nonequilibrium Thermodynamics: The Steady State*. Cambridge: Harvard University Press.

Carreau, P. J. 1972. Rheological Equations from Molecular Network Theories. *Trans. Soc. Rheol.*, **16**, 99–127.

Carslaw, H. S., and Jaeger, J. C. 1959. *Conduction of Heat in Solids*. Second edn. New York: Oxford University Press.

Casimir, H. B. G. 1945. On Onsager's Principle of Microscopic Reversibility. *Rev. Mod. Phys.*, **17**, 343–350.

Chandrasekhar, S. 1943. Stochastic Problems in Physics and Astronomy. *Rev. Mod. Phys.*, **15**, 1–89.

Chang, H.-C. 1983. Effective Diffusion and Conduction in Two-Phase Media – A Unified Approach. *AIChE J.*, **29**, 846–853.

Córdoba, A., Schieber, J. D., and Indei, T. 2012a. The Effects of Hydrodynamic Interaction and Inertia in Determining the High-Frequency Dynamic Modulus of a Viscoelastic Fluid with Two-Point Passive Microrheology. *Phys. Fluids*, **24**, 073103.

Córdoba, A., Indei, T., and Schieber, J. D. 2012b. Elimination of Inertia from a Generalized Langevin Equation: Applications to Microbead Rheology Modeling and Data Analysis. *J. Rheol.*, **56**, 185–212.

Crank, J. 1975. *The Mathematics of Diffusion*. Second edn. New York: Oxford University Press.

Curie, P. 1894. Sur la symétrie dans les phénomènes physiques. Symétrie d'un champ électrique et d'un champ magnétique. *J. de Physique*, **3**, 393–415.

Curtiss, C. F., and Bird, R. B. 1996. Thermal Conductivity of Dilute Solutions of Chainlike Polymers. *J. Chem. Phys.*, **107**, 5254–5267.

Dahler, J. S., and Scriven, L. E. 1961. Angular Momentum of Continua. *Nature*, **192**, 36–37.

Damköhler, G. 1936. Einflüsse der Strömung, Diffusion und des Wärmeüberganges auf die Leistung von Reaktionsöfen. *Z. Elektrochemie*, **42**, 846–862.

Danckwerts, P. V. 1950. Absorption by Simultaneous Diffusion and Chemical Reaction. *Trans. Faraday Soc.*, **46**, 300–304.

Darcy, H. 1856. *Les fontaines publiques de la ville de Dijon*. Paris: Dalmont.

de Groot, S. R., and Mazur, P. 1984. *Non-Equilibrium Thermodynamics*. Second edn. New York: Dover.

de Pablo, J. J., and Schieber, J. D. 2014. *Molecular Engineering Thermodynamics*. Cambridge: Cambridge University Press.

de Sénarmont, H. H. 1848. Über die Wärmeleitung in kristallisierten Substanzen. *Pogg. Ann. Phys. Chem.*, **73**, 191–202.

de Sénarmont, H. H. 1849. Versuche über die Abänderungen, welche die mechanischen Agentien der Wärmeleitungsfähigkeit homogener Korper einprägen. *Pogg. Ann. Phys. Chem.*, **76**, 119–128.

Deal, B. E., and Grove, A. S. 1965. General Relationship for the Thermal Oxidation of Silicon. *J. Appl. Phys.*, **36**, 3770–3778.

Deen, W. M. 2012. *Analysis of Transport Phenomena*. Second edn. New York: Oxford University Press.

Denn, M. M. 1980. Continuous Drawing of Liquids to Form Fibers. *Annu. Rev. Fluid Mech.*, **12**, 365–387.

Denn, M. M. 2001. Extrusion Instabilities and Wall Slip. *Annu. Rev. Fluid Mech.*, **33**, 265–287.

Denn, M. M. 2008. *Polymer Melt Processing: Foundations in Fluid Mechanics and Heat Transfer*. New York: Cambridge University Press.

Denn, M. M., Petrie, C. J. S., and Avenas, P. 1975. Mechanics of Steady Spinning of a Viscoelastic Liquid. *AIChE J.*, **21**, 791–799.

Doi, M. 2013. *Soft Matter Physics.* Oxford: Oxford University Press.

Doi, M., and Edwards, S. F. 1986. *The Theory of Polymer Dynamics.* Oxford: Oxford University Press.

Duda, J. L., and Vrentas, J. S. 1969. Mathematical Analysis of Bubble Dissolution. *AIChE J.*, **15**, 351–356.

Duhamel, J. M. C. 1832. Sur les équations générales de la propagation de chaleur dans les corps solides dont la conductibilité n'est pas la même dans tous les sens. *J. de l'École Polytech.*, **13**, 356–399.

Dutcher, C. S., and Muller, S. J. 2009. Spatio-temporal Mode Dynamics and Higher Order Transitions in High Aspect Ratio Newtonian Taylor Couette Flows. *J. Fluid Mech.*, **641**, 85–113.

Edwards, D. A., Brenner, H., and Wasan, D. T. 1991. *Interfacial Transport Processes and Rheology.* Boston: Butterworth-Heinemann.

Einstein, A. 1906. Eine neue Bestimmung der Moleküldimensionen. *Ann. der Phys.*, **324**, 289–306.

Emslie, A. G., Bonner, F. T., and Peck, L. G. 1958. Flow of a Viscous Liquid on a Rotating Disk. *J. Appl. Phys.*, **29**, 858–862.

Epstein, P. S., and Plesset, M. S. 1950. On the Stability of Gas Bubbles in Liquid–Gas Systems. *J. Chem. Phys.*, **18**, 1505–1509.

Faxén, H. 1929. Über eine Differentialgleichung aus der physikalischen Chemie. *Ark. Math. Astr. Fys.*, **21B**, 1–6.

Ferry, J. D. 1980. *Viscoelastic Properties of Polymers.* Third edn. New York: Wiley.

Feynman, R. P., Leighton, R. B., and Sands, M. 1964. *The Feynman Lectures on Physics: Volume 1.* New York: Addison-Wesley.

Flack, W. W., Soong, D. S., Bell, A. T., and Hess, D. W. 1984. A Mathematical Model for Spin Coating of Polymer Resists. *J. Appl. Phys.*, **56**, 1199–1206.

Frisch, U. 1995. *Turbulence.* Cambridge: Cambridge University Press.

Fujita, H. 1975. *Foundations of Ultracentrifugal Analysis.* New York: Wiley.

Gardiner, C. W. 1990. *Handbook of Stochastic Methods for Physics, Chemistry and the Natural Sciences.* Second edn. Berlin: Springer.

Gibbs, J. W. 1902. *Elementary Principles in Statistical Mechanics.* New York: Charles Scribner's Sons.

Gibbs, J. W. 1993. *Thermodynamics. The Scientific Papers of J. Willard Gibbs, Vol. I.* Woodbridge: Ox Bow.

Giddings, J. C. 1966. A New Separation Concept Based on a Coupling of Concentration and Flow Nonuniformities. *Sep. Sci.*, **1**, 123–125.

Giddings, J. C. 1993. Field-Flow Fractionation: Analysis of Macromolecular, Colloidal, and Particulate Materials. *Science*, **260**, 1456–1465.

Giddings, J. C., Yang, F. J., and Myers, M. N. 1976. Field-Flow Fractionation: A Versatile New Separation Method. *Science*, **193**, 1244–1245.

Gill, W. N., and Sankarasubramanian, R. 1970. Exact Analysis of Unsteady Convective Diffusion. *Proc. Roy. Soc. A*, **316**, 341–350.

Goldstein, H. 1980. *Classical Mechanics.* Second edn. Reading: Addison-Wesley.

Goodridge, R. D., Tuck, C. J., and Hague, R. J. M. 2012. Laser Sintering of Polyamides and Other Polymers. *Prog. Mat. Sci.*, **57**, 229–267.

Gorban, A. N., Kazantzis, N., Kevrekidis, I. G., Öttinger, H. C., and Theodoropoulos, C. (eds.). 2006. *Model Reduction and Coarse-Graining Approaches for Multiscale Phenomena.* Berlin: Springer.

Grabert, H. 1982. *Projection Operator Techniques in Nonequilibrium Statistical Mechanics.* Berlin: Springer.

Grad, H. 1958. Principles of the Kinetic Theory of Gases. Pages 205–294 of: Flügge, S. (ed.), *Thermodynamics of Gases. Encyclopedia of Physics, Volume XII.* Berlin: Springer.

Grad, H. 1963. Asymptotic Theory of the Boltzmann Equation. *Phys. Fluids,* **6**, 147–181.

Graetz, L. 1883. Über die Wärmeleitungsfähigkeit von Flüssigkeiten. *Ann. Phys. Chem.,* **18**, 79–94.

Green, M. S., and Tobolsky, A. V. 1946. A New Approach to the Theory of Relaxing Polymeric Media. *J. Chem. Phys.,* **14**, 80–92.

Grmela, M., and Öttinger, H. C. 1997. Dynamics and Thermodynamics of Complex Fluids. I. Development of a General Formalism. *Phys. Rev. E,* **56**, 6620–6632.

Hagen, G. 1839. Über die Bewegung des Wassers in engen zylindrischen Röhren. *Pogg. Ann. Phys. Chem.,* **46**, 423–442.

Hagenbach, E. 1860. Über die Bestimmung der Zähigkeit einer Flüssigkeit durch den Ausfluß aus Röhren. *Pogg. Ann. Phys. Chem.,* **108**, 385–426.

Happel, J., and Brenner, H. 1973. *Low Reynolds Number Hydrodynamics.* Second edn. Leyden: Martinus Nijhoff.

Healy, J. J., de Groot, J. J., and Krestin, J. 1976. The Theory of the Transient Hot-Wire Method for Measuring Thermal Conductivity. *Physica C,* **82**, 392–408.

Hermans, J. J. 1943. Theoretische beschouwingen over de viskositeit en de stromings-dubbelebreking in oplossingen van macromoleculaire stoffen. *Physica,* **10**, 777–789.

Hill, T. L. 1994. *Thermodynamics of Small Systems.* New York: Dover.

Hinch, E. J. 1991. *Perturbation Methods.* Cambridge: Cambridge University Press.

Hirschfelder, J. O., Curtiss, C. F., and Bird, R. B. 1954. *Molecular Theory of Gases and Liquids.* New York: Wiley.

Honerkamp, J., and Römer, H. 1993. *Theoretical Physics: A Classical Approach.* Berlin: Springer.

Honerkamp, J., and Weese, J. 1993. A Nonlinear Regularization Method for the Calculation of Relaxation Spectra. *Rheol. Acta,* **32**, 65–73.

Howes, F. A., and Whitaker, S. 1985. The Spatial Averaging Theorem Revisited. *Chem. Eng. Sci.,* **40**, 1387–1392.

Indei, T., Schieber, J. D., Córdoba, A., and Pilyugina, E. 2012. Treating Inertia in Passive Microbead Rheology. *Phys. Rev. E,* **85**, 021504.

Jackson, J. D. 1975. *Classical Electrodynamics.* Second edn. New York: Wiley.

Jaeger, R. C. 2002. *Introduction to Microelectronic Fabrication.* Second edn. New Jersey: Prentice Hall.

Jongschaap, R. J. J., and Öttinger, H. C. 2001. Equilibrium Thermodynamics – Callen's Postulational Approach. *J. Non-Newtonian Fluid Mech.,* **96**, 5–17.

Karniadakis, G. E., Beskök, A., and Aluru, N. R. 2005. *Microflows and Nanoflows.* New York: Springer.

Katchalsky, A. K., and Curran, P. F. 1965. *Nonequilibrium Thermodynamics in Biophysics.* Cambridge: Harvard University Press.

Kesner, L. F., Caldwell, K. D., Meyers, M. N., and Giddings, J. C. 1976. Performance Characteristics of Electrical Field-Flow Fractionation in a Flexible Membrane Channel. *Anal. Chem.,* **48**, 1834–1839.

Kim, J., Löwe, T., and Hoppe, T. 2008. Protein Quality Control Gets Muscle into Shape. *Trends Cell Biol.*, **18**, 264–272.

Kimura, Y. 2009. Microrheology of Soft Matter. *J. Phys. Soc. Jpn.*, **78**, 041005.

Kirby, B. J. 2010. *Micro- and Nanoscale Fluid Mechanics: Transport in Microfluidic Devices*. New York: Cambridge University Press.

Kjelstrup, S., and Bedeaux, D. 2008. *Non-Equilibrium Thermodynamics of Heterogeneous Systems*. New Jersey: World Scientific.

Kjelstrup, S., Rubi, J. M., and Bedeaux, D. 2005. Active Transport: A Kinetic Description Based on Thermodynamic Grounds. *J. Theor. Biol.*, **234**, 7–12.

Kjelstrup, S., Bedeaux, D., Johannessen, E., and Gross, J. 2010. *Non-Equilibrium Thermodynamics for Engineers*. New Jersey: World Scientific.

Kloeden, P. E., and Platen, E. 1992. *Numerical Solution of Stochastic Differential Equations*. Berlin: Springer.

Knudsen, M. 1909. Die Gesetze der Molekularströmung und der inneren Reibungsströmung der Gase durch Röhren. *Ann. der Phys.*, **331**, 75–130.

Köhler, W., and Wiegand, S. 2002. *Thermal Nonequilibrium Phenomena in Fluid Mixtures*. Berlin: Springer.

Koopmans, R. J., den Doelder, J. C. F., and Paquet, A. 2000. Modeling Foam Growth in Thermoplastics. *Adv. Mater.*, **12**, 1873–1880.

Kramers, H. A. 1940. Brownian Motion in a Field of Force and the Diffusion Model of Chemical Reactions. *Physica*, **7**, 284–304.

Krishnamurthy, S., and Subramanian, R. S. 1977. Exact Analysis of Field-Flow Fractionation. *Sep. Sci.*, **12**, 347–379.

Kubo, R., Toda, M., and Hashitsume, N. 1991. *Nonequilibrium Statistical Mechanics*. Second Edn. Berlin: Springer.

Lamb, H. 1932. *Hydrodynamics*. Sixth edn. New York: Dover Publications.

Lamm, O. 1929. Die Differentialgleichung der Ultrazentrifugierung. *Ark. Math. Astr. Fys.*, **21B(2)**, 1–4.

Landau, L. D., and Lifshitz, E. M. 1987. *Fluid Mechanics*. Second edn. Oxford: Pergamon.

Larson, R. G. 1999. *The Structure and Rheology of Complex Fluids*. Oxford: Oxford University Press.

Laue, T. M., and Stafford III, W. F. 1999. Modern Applications of Analytical Ultracentrifugation. *Annu. Rev. Biophys. Biomol. Struct.*, **28**, 75–100.

Lervik, A., Bedeaux, D., and Kjelstrup, S. 2012a. Kinetic and Mesoscopic Nonequilibrium Description of the Ca^{2+} Pump. *Eur. Biophys. J.*, **41**, 437–448.

Lervik, A., Bedeaux, D., and Kjelstrup, S. 2012b. On the Thermodynamic Efficiency of Ca^{2+}–ATPase Molecular Motors. *Biophys. J.*, **103**, 1218–1226.

Lodge, A. S. 1956. A Network Theory of Flow Birefringence and Stress in Concentrated Polymer Solutions. *Trans. Faraday Soc.*, **52**, 120–130.

Macosko, C. W. 1994. *Rheology: Principles, Measurements and Applications*. New York: VCH Publishers.

Mahan, G., Sales, B., and Sharp, J. 1997. Thermoelectric Materials: New Approaches to an Old Problem. *Physics Today*, **50**, 42–47.

Mason, T. G., and Weitz, D. A. 1995. Optical Measurements of Frequency-Dependent Linear Viscoelastic Moduli of Complex Fluids. *Phys. Rev. Lett.*, **74**, 1250–1253.

Matovich, M. A., and Pearson, J. R. A. 1969. Spinning a Molten Threadline. Steady-State Isothermal Viscous Flows. *Ind. Eng. Chem. Fund.*, **8**, 512–520.

Maxwell, J. C. 1904. *A Treatise on Electricity and Magnetism*. Third edn. Oxford: Oxford University Press.

Maxwell, J. C. 2001. *Theory of Heat*. New York: Dover.

Mheta, V. R., and Archer, L. A. 1998. Perturbation Solution for the Interfacial Oxidation of Silicon. *J. Vac. Sci. Tech. B*, **16**, 2121–2124.

Michaelis, L., and Menten, M. L. 1913. Die Kinetik der Invertinwirkung. *Biochem. Z.*, **49**, 333–369.

Middleman, S. 1977. *Fundamentals of Polymer Processing*. New York: McGraw-Hill.

Mihaljan, J. M. 1962. A Rigorous Exposition of the Boussinesq Approximations Applicable to a Thin Layer of Fluid. *Astrophys. J.*, **136**, 1126–1133.

Miller, D. G. 1960. Thermodynamics of Irreversible Processes: The Experimental Verification of the Onsager Reciprocal Relations. *Chem. Rev.*, **60**, 15–37.

Moody, L. F. 1944. Friction Factors for Pipe Flow. *Trans. Am. Soc. Mech. Eng.*, **66**, 671–684.

Morrison, F. A. 2001. *Understanding Rheology*. Oxford: Oxford University Press.

Nan, C. W., Birringer, R, Clark, D. R., and Gleiter, H. 1997. Effective Thermal Conductivity of Particulate Composites with Interfacial Thermal Resistance. *J. Appl. Phys.*, **81**, 6692–6699.

Nayfeh, A. H. 1973. *Perturbation Methods*. New York: Wiley Interscience.

Nelson, E. 1967. *Dynamical Theories of Brownian Motion*. Princeton: Princeton University Press.

Newman, J., and Thomas-Alyea, K. E. 2004. *Electrochemical Systems*. Third edn. Hoboken: Wiley.

Nieto Simavilla, D., Schieber, J. D., and Venerus, D. C. 2012. Anisotropic Thermal Transport in a Cross-Linked Polyisoprene Rubber Subjected to Uniaxial Elongation. *J. Polym. Sci. Part B*, **50**, 1638–1644.

Nusselt, W. 1910. Die Abhängigkeit der Wärmeübergangszahl von der Rohrlänge. *Z. Ver. deutsch. Ing.*, **54**, 1154–1158.

Oldroyd, J. G. 1947. A Rational Formulation of the Equations of Plastic Flow for a Bingham Solid. *Proc. Cambridge Phil. Soc.*, **43**, 100–105.

Onsager, L. 1931a. Reciprocal Relations in Irreversible Processes. I. *Phys. Rev.*, **37**, 405–426.

Onsager, L. 1931b. Reciprocal Relations in Irreversible Processes. II. *Phys. Rev.*, **38**, 2265–2279.

Osswald, T. A., and Hernández-Ortiz, J. P. 2006. *Polymer Processing: Modeling and Simulation*. Cincinnati: Hanser Gardner.

Öttinger, H. C. 1996. *Stochastic Processes in Polymeric Fluids: Tools and Examples for Developing Simulation Algorithms*. Berlin: Springer.

Öttinger, H. C. 1998a. On the Structural Compatibility of a General Formalism for Nonequilibrium Dynamics with Special Relativity. *Physica A*, **259**, 24–42.

Öttinger, H. C. 1998b. Relativistic and Nonrelativistic Description of Fluids with Anisotropic Heat Conduction. *Physica A*, **254**, 433–450.

Öttinger, H. C. 2005. *Beyond Equilibrium Thermodynamics*. Hoboken: Wiley.

Öttinger, H. C. 2009. Constraints in Nonequilibrium Thermodynamics: General Framework and Application to Multicomponent Diffusion. *J. Chem. Phys.*, **130**, 114904, 1–9.

Öttinger, H. C., and Grmela, M. 1997. Dynamics and Thermodynamics of Complex Fluids. II. Illustrations of a General Formalism. *Phys. Rev. E*, **56**, 6633–6655.

Öttinger, H. C., and Petrillo, F. 1996. Kinetic Theory and Transport Phenomena for a Dumbbell Model under Nonisothermal Conditions. *J. Rheol.*, **40**, 857–874.

Öttinger, H. C., and Venerus, D. C. 2014. Thermodynamic Approach to Interfacial Transport Phenomena: Single-Component Systems. *AIChE J.*, **60**, 1424–1433.

Öttinger, H. C., Struchtrup, H., and Liu, M. 2009a. Inconsistency of a Dissipative Contribution to the Mass Flux in Hydrodynamics. *Phys. Rev. E*, **80**, 056303, 1–8.

Öttinger, H. C., Bedeaux, D., and Venerus, D. C. 2009b. Nonequilibrium Thermodynamics of Transport through Moving Interfaces with Application to Bubble Growth and Collapse. *Phys. Rev. E*, **80**, 021606, 1–16.

Peng, K.-W., Wang, L.-C., and Slattery, J. C. 1996. A New Theory for the Thermal Oxidation of Silicon. *J. Vac. Sci. Tech. B*, **14**, 3317–3320.

Phillips, R., Kondev, J., and Theriot, J. 2009. *Physical Biology of the Cell.* New York: Garland Science.

Pipkin, A. C. 1972. *Lectures on Viscoelastic Theory.* New York: Springer.

Poiseuille, J.-L. 1840. Recherches expérimentales sur le mouvement des liquides dans les tubes de très petits diamètres. *Comptes Rendus*, **11**, 961–967.

Pollard, W. G., and Present, R. D. 1948. On Gaseous Self-Diffusion in Long Capillary Tubes. *Phys. Rev.*, **73**, 762–774.

Prandtl, L. 1952. *Essentials of Fluid Mechanics.* New York: Hafner.

Prasad, V., Zhang, H., and Anselmo, A. P. 1997. Transport Phenomena in Czochralski Crystal Growth Processes. *Adv. Heat Trans.*, **30**, 313–435.

Prigogine, I., Lefever, R., Goldbeter, A., and Herschkowitz-Kaufman, M. 1969. Symmetry Breaking Instabilities in Biological Systems. *Nature*, **30**, 913–916.

Prosperetti, A. 1979. Boundary Conditions at a Liquid–Vapor Interface. *Meccanica*, **14**, 34–47.

Proussevitch, A. A., and Sahagian, D. L. 1996. Dynamics of Coupled Diffusive and Decompressive Bubble Growth in Magmatic Systems. *J. Geophys. Res.*, **101**, 17447–17455.

Purcell, T. J., Sweeney, H. L., and Spudich, J. A. 2005. A Force-Dependent State Controls the Coordination of Processive Myosin V. *Proc. Natl. Acad. Sci. U.S.A.*, **102**, 13873–13878.

Quintard, M., and Whitaker, S. 1993. Transport in Ordered and Disordered Porous Media: Volume-Averaged Equations, Closure Problems, and Comparison with Experiment. *Chem. Eng. Sci.*, **48**, 2537–2564.

Rayleigh, Lord. 1878. On the Instability of Jets. *Proc. Lon. Math. Soc.*, **10**, 4–13.

Rayleigh, Lord. 1917. On the Pressure Developed in a Liquid during the Collapse of a Spherical Cavity. *Phil. Mag.*, **34**, 94–98.

Reguera, D., Rubi, J. M., and Vilar, J. M. G. 1955. The Mesoscopic Dynamics of Thermodynamic Systems. *J. Phys. Chem. B*, **26**, 21502–21515.

Reichl, L. E. 1980. *A Modern Course in Statistical Physics.* Austin: University of Texas Press.

Reynolds, O. 1886. On the Theory of Lubrication and Its Application to Mr. Beauchamp Tower's Experiments, Including an Experimental Determination of the Viscosity of Olive Oil. *Phil. Trans. R. Soc. Lond. A*, **177**, 157–234.

Rouse, P. E. 1953. A Theory of the Linear Viscoelastic Properties of Dilute Solutions of Coiling Polymers. *J. Chem. Phys.*, **21**, 1272–1280.

Rubi, J. M. 2012. Mesoscopic Thermodynamics. *Physica Scripta*, **T151**, 014027.

Rubi, J. M., Nespreda, M., Kjelstrup, S., and Bedeaux, D. 2007. Energy Transduction in Biological Systems: A Mesoscopic Non-Equilibrium Thermodynamics Description. *J. Non-Equilib. Thermodyn.*, **32**, 351–378.

Ryan, D., Carbonell, R. G., and Whitaker, S. 1980. Effective Diffusivities for Catalyst Pellets under Reactive Conditions. *Chem. Eng. Sci.*, **35**, 10–16.

Sagis, L. M. C. 2011. Dynamic Properties of Interfaces in Soft Matter: Experiments and Theory. *Rev. Mod. Phys.*, **83**, 1367–1403.

Sarhangi Fard, A., Hulsen, M. A., and Anderson, P. D. 2012. Extended Finite Element Method for Viscous Flow inside Complex Three-Dimensional Geometries with Moving Internal Boundaries. *Int. J. Num. Meth. Fluids*, **70**, 775–792.

Savin, T., Glavatskiy, K. S., Kjelstrup, S., Öttinger, H. C., and Bedeaux, D. 2012. Local Equilibrium of the Gibbs Interface in Two-Phase Systems. *Europhys. Lett.*, **97**, 40002, 1–6.

Schieber, J. D., Venerus, D. C., Bush, K., Balasubramanian, V., and Smoukov, S. 2004. Measurement of Anisotropic Energy Transport in Flowing Polymers by Using a Holographic Technique. *Proc. Natl. Acad. Sci. U.S.A.*, **36**, 13142–13146.

Schieber, J. D., Córdoba, A., and Indei, T. 2013. The Analytic Solution of Stokes for Time-Dependent Creeping Flow around a Sphere: Application to Linear Viscoelasticity as an Ingredient for the Generalized Stokes–Einstein Relation and Microrheology Analysis. *J. Non-Newtonian Fluid Mech.*, **200**, 3–8.

Schimpf, M. E., Caldwell, K., and J. C. Giddings (eds.). 2000. *Field-Flow Fractionation Handbook*. New York: Wiley.

Schlichting, H. 1955. *Boundary Layer Theory*. New York: McGraw-Hill.

Schwarzl, F. R. 1990. *Polymermechanik: Struktur und mechanisches Verhalten von Polymeren*. Berlin: Springer.

Schweizer, M., Öttinger, H. C., and Savin, T. 2016. Nonequilibrium Thermodynamics of an Interface. *Phys. Rev. E*, **93**, 052803.

Schweizer, T., Hostettler, J., and Mettler, F. 2008. A Shear Rheometer for Measuring Shear Stress and Both Normal Stress Differences in Polymer Melts Simultaneously: The MTR 25. *Rheol. Acta*, **47**, 943–957.

Scott, G. D., and Kilgour, D. M. 1969. The Density of Random Close Packing of Spheres. *J. Phys. D: Appl. Phys.*, **2**, 863–866.

Scriven, L. E. 1959. On the Dynamics of Phase Growth. *Chem. Eng. Sci.*, **10**, 1–13.

Shah, R. K., and London, A. L. 1978. *Laminar Flow Forced Convection in Ducts*. New York: Academic Press.

Shakouri, A. 2011. Recent Developments in Semiconductor Thermoelectric Physics and Materials. *Annu. Rev. Mater. Sci.*, **41**, 399–431.

Sieder, E. N., and Tate, G. E. 1936. Heat Transfer and Pressure Drop of Liquids in Tubes. *Ind. Eng. Chem.*, **28**, 1429–1435.

Siegel, R., and Howell, J. R. 2002. *Thermal Radiation Heat Transfer*. Fourth edn. New York: Taylor and Francis.

Siegel, R., Sparrow, E. M., and Hallman, T. M. 1958. Steady Laminar Heat Transfer in a Circular Tube with Prescribed Wall Heat Flux. *Appl. Sci. Res.*, **7**, 386–392.

Slattery, J. C. 1967. Flow of Viscoelastic Fluids through Porous Media. *AIChE J.*, **13**, 1066–1071.

Slattery, J. C. 1999. *Advanced Transport Phenomena*. New York: Cambridge University Press.

Slattery, J. C., Sagis, L. M. C., and Oh, E. S. 2007. *Interfacial Transport Phenomena*. Second edn. New York: Springer.

Smith, N. O. 1965. The Difference between C_p and C_v for Liquids and Solids. *J. Chem. Ed.*, **42**, 654–655.

Spiegel, E. A., and Veronis, G. 1960. On the Boussinesq Approximation for a Compressible Fluid. *Astrophys. J.*, **131**, 442–447.

Squires, T. M., and Mason, T. G. 2010. Fluid Mechanics of Microrheology. *Annu. Rev. Fluid Mech.*, **42**, 413–438.

Srinivasan, R. S., Gerth, W. A., and Powell, M. P. 1999. Mathematical Models of Diffusion-Limited Gas Bubble Dynamics in Tissue. *J. Appl. Physiol.*, **86**, 732–741.

Straumann, H. 2005. On Einstein's Doctoral Thesis. arXiv:physics/0504201.

Struchtrup, H. 2005. *Macroscopic Transport Equations for Rarefied Gas Flows*. Berlin: Springer.

Sutera, S. P., and Skalak, R. 1993. The History of Poiseuille's Law. *Annu. Rev. Fluid Mech.*, **25**, 1–19.

Tadmor, Z., and Gogos, C. G. 2006. *Principles of Polymer Processing*. Second edn. New York: Wiley.

Tanner, R. 1970. A Theory of Die-Swell. *J. Polym. Sci. A*, **8**, 2067–2078.

Taylor, G. I. 1923. Flow Regimes in a Circular Couette System with Independently Rotating Cylinders. *Phil. Trans. R. Soc. Lond. A*, **223**, 289–343.

Taylor, G. I. 1953. Dispersion of Soluble Matter in Solvent Flowing Slowly through a Tube. *Proc. Roy. Soc. A*, **219**, 186–203.

Thiele, E. W. 1939. Relation between Catalytic Activity and Size of Particle. *Ind. Eng. Chem.*, **31**, 916–920.

Truesdell, C., and Noll, W. 1992. *The Non-Linear Field Theories of Mechanics*. Second edn. Berlin: Springer.

Tschoegl, N. W. 1989. *The Phenomenological Theory of Linear Viscoelastic Behavior*. Berlin: Springer.

Tucker, C. L. 1989. *Fundamentals of Computer Modeling for Polymer Processing*. New York: Hanser.

Turing, A. M. 1952. The Chemical Basis of Morphogenesis. *Phil. Trans. R. Soc. Lond. B*, **237**, 37–72.

Turner, B. N., Strong, R., and Gold, S. A. 2014. A Review of Melt Extrusion Additive Manufacturing Processes: I. Process Design and Modeling. *Rapid Proto. J.*, **20**, 192–204.

van den Brule, B. H. A. A. 1989. A Network Theory for the Thermal Conductivity of an Amorphous Polymeric Material. *Rheol. Acta*, **28**, 257–266.

van der Walt, C., Hulsen, M. A., Bogaerds, A. C. B., and Anderson, P. D. 2014. Transient Modeling of Fiber Spinning with Filament Pull-Out. *J. Non-Newtonian Fluid Mech.*, **208–209**, 72–87.

Van Dyke, M. 1964. *Perturbation Methods in Fluid Mechanics*. New York: Academic Press.

van Kampen, N. G. 1992. *Stochastic Processes in Physics and Chemistry*. Second edn. Amsterdam: North Holland.

Venerus, D. C. 2006. Laminar Capillary Flow of Compressible Viscous Fluids. *J. Fluid Mech.*, **555**, 59–80.

Venerus, D. C. 2015. Diffusion-Induced Bubble Growth in Yield-Stress Fluids. *J. Non-Newtonian Fluid Mech.*, **215**, 53–59.

Venerus, D. C., and Bugajsky, D. J. 2010. Compressible Laminar Flow in a Channel. *Phys. Fluids*, **22**, 046101.

Venerus, D. C., and Yala, N. 1997. Transport Analysis of Diffusion-Induced Bubble Growth and Collapse in Viscous Liquids. *AIChE J.*, **43**, 2948–2959.

Venerus, D. C., Yala, N., and Bernstein, B. 1998. Analysis of Diffusion-Induced Bubble Growth in Viscoelastic Liquids. *J. Non-Newtonian Fluid Mech.*, **75**, 55–75.

Venerus, D. C., Schieber, J. D., Iddir, H., Guzmán, J. D., and Broerman, A. W. 1999. Relaxation of Anisotropic Thermal Diffusivity in a Polymer Melt Following Step Shear Strain. *Phys. Rev. Lett.*, **82**, 366–369.

Visscher, K., Schnitzer, M. J., and Block, S. M. 1999. Single Kinesin Molecules Studied with a Molecular Force Clamp. *Nature*, **400**, 184–189.

Vrentas, J. S., and Vrentas, C. M. 2012. *Diffusion and Mass Transfer*. Boca Raton: CRC Press.

Vrentas, J. S., Vrentas, C. M., and Ling, H.-C. 1983. Equations for Predicting Growth or Dissolution Rates of Spherical Particles. *Chem. Eng. Sci.*, **38**, 1927–1934.

Waigh, T. A. 2005. Microrheology of Complex Fluids. *Rep. Prog. Phys.*, **68**, 685–742.

Waldmann, L. 1967. Non-Equilibrium Thermodynamics of Boundary Conditions. *Z. Naturforsch. A*, **22**, 1269–1286.

Weinberg, S. 1977. *The First Three Minutes: A Modern View Of The Origin Of The Universe*. New York: Basic Books.

Whitaker, S. 1967. Diffusion and Dispersion in Porous Media. *AIChE J.*, **13**, 420–427.

Whitaker, S. 1999. *The Method of Volume Averaging*. Norwell: Kluwer Academic Publishers.

Willenbacher, N., Oelschlaeger, C., Schopferer, M., Fischer, P., Cardinaux, F., and Scheffold, F. 2007. Broad Bandwidth Optical and Mechanical Rheometry of Wormlike Micelle Solutions. *Phys. Rev. Lett.*, **99**, 068302.

Winter, H. H. 1977. Viscous Dissipation in Shear Flows of Molten Polymers. *Adv. Heat Trans.*, **13**, 205–267.

Winter, H. H. 1997. Analysis of Dynamic Mechanical Data: Inversion into a Relaxation Time Spectrum and Consistency Check. *J. Non-Newtonian Fluid Mech.*, **68**, 225–239.

Xu, J., Gong, W., Dai, R., and Liu, D. 2003. Measurement of the Bulk Viscosity of Liquid by Brillouin Scattering. *Appl. Optics*, **42**, 6704–6709.

Xu, K., Forest, M. G., and Klapper, I. 2007. On the Correspondence between Creeping Flows of Viscous and Viscoelastic Fluids. *J. Non-Newtonian Fluid Mech.*, **145**, 150–172.

Zwanzig, R., and Bixon, M. 1970. Hydrodynamic Theory of the Velocity Correlation Function. *Phys. Rev. A*, **2**, 2005–2012.

Author Index

Subject Index